Arid Land Systems

Arid Land Systems: Sciences and Societies

Special Issue Editors

Troy Sternberg
Ariell Ahearn

MDPI • Basel • Beijing • Wuhan • Barcelona • Belgrade

Special Issue Editors
Troy Sternberg
Oxford University Centre for the Environment
UK

Ariell Ahearn
University of Oxford
UK

Editorial Office
MDPI
St. Alban-Anlage 66
4052 Basel, Switzerland

This is a reprint of articles from the Special Issue published online in the open access journal *Land* (ISSN 2073-445X) from 2017 to 2019 (available at: https://www.mdpi.com/journal/land/special_issues/deserts)

For citation purposes, cite each article independently as indicated on the article page online and as indicated below:

LastName, A.A.; LastName, B.B.; LastName, C.C. Article Title. *Journal Name* **Year**, *Article Number*, Page Range.

ISBN 978-3-03921-347-4 (Pbk)
ISBN 978-3-03921-348-1 (PDF)

Cover image courtesy of Jerome Mayaud.

Contents

About the Special Issue Editors

Klaus Zierer is an Associate Research Fellow of the ESRC-Funded Centre of Skills, Knowledge and Occupational Performance (SKOPE). He taught for five years in primary and secondary schools. In 2009 he was a Visiting Research Fellow at the Department of Education, University of Oxford. In 2011, he was Professor of Education at the Carl von Ossietzky University, Oldenburg, Germany. Since 2015 he has been Professor of School Education at the University of Augsburg

Preface to "Arid Land Systems"

Understanding deserts and drylands is essential, as arid landscapes cover >40% of the Earth and are home to two billion people. Today's problematic environment–human interaction needs contemporary knowledge to address dryland complexity. Physical dimensions in arid zones—land systems, climate and hazards, ecology—are linked with social processes that directly impact drylands, such as land management, livelihoods, and development. The challenges require integrated research that identifies systemic drivers across global arid regions. This book, based on a Special Issue in the journal *Land*, unifies desert science, arid environments, and development. Chapters identify land dynamics, address system risks and delineate human functions through original research in dryland regions.

Measurement and monitoring, field investigation, remote sensing, and data analysis are effective tools to investigate natural dynamics. Equally, inquiry into how policy and practice affect landscape sustainability is key to mitigating detrimental activity in deserts. Relations between socio-economic forces and degradation, ethnologies, agro-pastoral rangeland use, drought and disaster and resource extraction reflect land interactions. Contemporary themes of development and transition, food security, conflict and conservation are interlinked in arid environments. Through knowledge, documentation and evaluation, the book links academic disciplines to convey the complexities and possibilities existing across drylands.

The chapters draw together a great variety of global drylands. Research explores the Gobi Desert and China, Central Asia and the Middle East, as well as multiple African dimensions and Latin America. The aim is to understand the essential themes that arid and semi-arid regions encounter and engage with today. This interdisciplinary study reflects the vital links between social and environmental science in global deserts. Using mixed methodologies and varied techniques, the original work stresses today's topical themes to present novel analyses. The book offers much to dryland researchers, stakeholders and importantly, policy makers, who shape the world's vast desert landscapes.

Troy Sternberg, Ariell Ahearn
Special Issue Editors

land

MDPI

Review

Central Asian 'Characteristics' on China's New Silk Road: The Role of Landscape and the Politics of Infrastructure

Troy Sternberg *, Ariell Ahearn and Fiona McConnell

School of Geography, University of Oxford, Oxford OX1 3QY, UK; ariell.ahearn-ligham@ouce.ox.ac.uk (A.A.); fiona.mcconnell@ouce.ox.ac.uk (F.M.)

* Correspondence: troy.sternberg@geog.ox.ac.uk; Tel.: +44-(0)-186-528-5070

Received: 24 July 2017; Accepted: 18 August 2017; Published: 23 August 2017

Abstract: China's $1 trillion One Belt, One Road (OBOR) infrastructure project has significant landscape, socio-economic, and political implications in recipient countries. To date, investigation has focused on Chinese motivation and plans rather than OBOR impact in host nations. This paper examines the programme from the perspective of two Central Asian states—Kazakhstan and Kyrgyzstan—that are at the heart of OBOR. We identify geographical factors that constrain infrastructure, recognise geopolitical contestation between Russia and China, address historical and cultural factors, and consider issues of institutional capacity and marginality that may be impediments to China's initiative. The discussion then focuses on how OBOR may play out in Central Asian landscapes and suggests how to conceive and address the unprecedented transformation in the region's built environment. Critical issues are that OBOR has not been grounded in the physical geography, practical understanding of OBOR's impacts is missing, and the state-citizen-China nexus remains unexplored. As pivot nations, OBOR implementation in Kazakhstan and Kyrgyzstan will showcase the Chinese programme's strengths and highlight its weaknesses.

Keywords: Central Asia; landscape; One Belt; One Road; Kazakhstan; Kyrgyzstan; infrastructure; environment; New Silk Road

---

1. Introduction

The world's first $1 trillion infrastructure project has captivated global attention for its boldness, breadth, and transformative potential [1]. Presented as the New Silk Road, China's reach across Asia and beyond has no modern-day precedent. In May 2017, China's President Xi Jinping announced his 'One Belt, One Road' (OBOR) model to the world [2]. International attention has been focused on rail journeys from China to London, roads and pipelines across Asia, and rapid development trajectories. In China, the programme is presented as "a great undertaking that will benefit peoples around the world" [3] (p. 93). This mega-investment and infrastructure programme will have significant impact on the nations through which it traverses [4–6]. In the rush to construction, questions of sovereignty, cost, and obligations are yet to be resolved in recipient countries. Here we examine how building a silk road supernova resonates in two neighbouring nations at its Central Asian heart: Kazakhstan and Kyrgyzstan. Located on China's western border, they are pivot countries: how these states act and react will determine if the road paves a golden future or crumbles into a dusty dream.

OBOR is also known as Belt and Road Initiative (BRI), the Silk Road Economic Belt (SREB), and colloquially as the New Silk Road; use of OBOR in this paper reflects the most common and best-known acronym. A component programme is the Maritime Silk Road Initiative (MSRI) that expands China's interaction and investment throughout Southeast Asia, the Indian Ocean, and Arabian Sea through the Mediterranean to Europe [7]. In tandem, the land bridge and sea route comprise a grand vision of One

Belt, One Road that encompasses multiple economic and political aims across dozens of countries [8]. Narratives in the expanding literature on OBOR consider Chinese motivations, strategic implications, economic integration, and geopolitical forces [5,6,9,10]. Recent scholarship is starting to conceptualise the programme through a range of theoretical lenses, yet has limited detail or regional and national specificity [7]. Parsing meaning at state and sub-national levels in recipient countries has yet to feature; nor do narratives engage with the role that physical environments through which OBOR travels will have. This paper examines the overland Belt and Road Initiative from the perspective of two Central Asian 'frontier states' (from China's viewpoint) and differentiates Kazakh and Kyrgyz roles and abilities to shape or benefit from OBOR. Further, we identify geographical factors that constrain infrastructure, recognise geopolitical contestation between Russia and China, address historical and cultural factors, and consider issues of institutional capacity and marginality that may be impediments to China's OBOR.

Whilst the emergent discussion of OBOR presents a policy, economic, and international relations perspective on the programme [5–7], the lack of engagement with landscapes for the world's largest infrastructure programme is striking. As OBOR aims to cross the Himalaya, Pamir, and Tien Shan mountains, traverse the deserts and steppe of Central, South, and Southwest Asia, and pass through active conflict zones (notably Afghanistan), overcoming geography, culture, and sparse resources (think water or food) will be a major challenge and consume much time and resources. Creating infrastructure involves residents, communities, governments, and physical embodiments of the nation state; thus, how land and society interact with an expansive foreign hegemon is an essential research focus. Integrating notions of land and geography, this paper evaluates how OBOR may play out in Kazakhstan and Kyrgyzstan. We examine the limited body of literature on OBOR that incorporates Central Asia (see [5,11–13]), tease out specific Kazakh and Kyrgyz motivations and concerns within this assessment (see [14–16] and introduce an applied geography view to consider the implications when 'earth and infrastructure' meet [17] (p. 2). The paper first contextualises OBOR in Central Asia and then considers regional geo-politics and geo-economics. Next, the physical landscape and infrastructure are examined and relevant social forces highlighted. The discussion then focuses on how OBOR may play out in Central Asia and suggests how the paper's original research idea can address an unprecedented transformation in the region's built environment.

2. Contextualising OBOR in Central Asia: Regional Histories, Geopolitics, and Geoeconomics

Chinese presentation of OBOR is as an idea or theme for engagement and development rather than a firm plan or roadmap [6]. Whilst OBOR details are elusive to outsiders, from the Chinese perspective it is 'an abstract and metaphorical concept' with the ancient silk road as a touchstone [18] (p. 2). This obscures the realpolitik of the programme that is President Xi Jinping's 'one and only major foreign policy initiative' [6] (p. 4). Any major transnational infrastructure programme is laden with expectation and liabilities; one country's outreach and expansion is another's incursion and obligation. Perhaps the two sides find a shared space of differentiated but positive benefit for both; for that to happen requires clarity of intent, roles, politics, and economic versus social benefits for giver and receiver at the outset. Cooley [5] presents the New Silk Road as a contest between 'traders'—development and economic interests—versus 'gamers' that see geopolitical angles as primary. Sidaway and Woon [4] (p. 4) assimilate this to suggest OBOR is a 'geopolitical strategy, albeit dependent and anchored on economic rationale'. China's intentional ambiguity [9] feeds a 'net gain' approach as opposed to a more benign regional development and soft power effort [19]. Other interpretations of the motivations underpinning OBOR abound (Table 1), from it being a project to absorb excess Chinese labour and capacity to the internationalisation of the yuan [19,20]. Whatever socio-political paradigm is intended or dominates, the impact will be most directly felt in its immediate neighbours that control the gateway to the west—Kazakhstan and Kyrgyzstan.

Central Asia is the historic core of silk roads through time [21–25]. It is here that Alexander the Great died in his quest for empire; where Marco Polo searched for silk routes to Venice, and

Britain and Russia contested the Great Game that contributed to regional isolation [26]. Past silk roads between China and empires to the west presented Central Asia as 'mere transit stations' along the route [27] (p. 433). A century ago a British geographer and politician [28] identified Central Asia as the geographical 'pivot' of history, which all great land empires must control. This enduring saga is germane today, for this is where geography framed the past and economics aims to determine the future [29]. The Dzungarian Gate on the China-Kazakh border and the Torugart pass from Kashgar to Naryn, Kyrgyzstan are two of the few openings in the Tien Shan and Pamir mountains that allow past and present trade caravans to flourish (see Figure 1). Before China can access European markets overland, before goods can flow to West Asia and the Mediterranean and resources return to China, Kazakhstan and Kyrgyzstan must be crossed.

Table 1. Motivating factors for One Belt, One Road.

Reason	Citation
Internationalisation of the Yuan	Chatham House (2017) [20]
Infrastructure development in Asia	Lim 2016 [6]
Consume overproduction, overcapacity, excess products	Lim 2016 [6]
Reduce unskilled unemployment in China	Tang 2015 [30]
Improve regional transport links	Lain and Pantucci 2015 [29]
Gain access to natural resources, oil, gas	Tang 2015 [30]
Increase Chinese soft power and good will	Sidaway and Woon 2017 [4]
Foreign policy initiative	Lim 2016 [6]
Sino-centric unipolar Asia	Malik *in* Sulekha 2017 [31]
Road to empire	Financial Times 2015 [32]
Benefits of development and prosperity	Sidaway & Woon 2017 [4]
Enhance Chinese social stability and security	Rolland 2017 [3]
Central Asia as "key areas" for China's national energy security	Rolland 2017 [3]
Integrate Eurasian continent by 2050	Rolland 2017 [3]

Figure 1. Map of Central Asia with geographical features (mountains, desert, steppe) and existing major infrastructure (pipelines, railroads).

Based on their histories, politics and socio-cultural contexts, Central Asian states have their own distinctive interpretations of what a New Silk Road means [7]. Location makes Kazakhstan and

Kyrgyzstan the two key Central Asian players, though the countries have vastly divergent interests and needs. Kazakhstan presents an autocratic, rentier state that has extensive trade links and higher per capita income than China ($12,436 vs. $7616) [11,33]. Driven by energy resources and pipelines to western China, Kazakhstan can self-fund and self-direct infrastructure and development if desired; its energy hub is centred in the Caspian basin and is thus closer to Europe than China. Kyrgyzstan, with its struggling democracy, low living standard ($1267 GDP per capita) and mountainous terrain has limited economic resources; already 50% of its national debt is owed to China. Both nations fear Chinese stealth encroachment through OBOR, with hints at new 'fangs of imperialism' [31] (p. 1). At the same time, both states are members of the Russia-driven Eurasian Economic Commission (EEU) that cedes significant trade rights and favours political deference to Russia. This drives a bifurcation between two spheres, as economics and development is dominated by China [13], yet must exist within Russian military, social, and political hegemony. In fact, the term 'New Silk Road' first appeared in 2011 as a US State Department initiative for 'new North-South transit and trade routes that complement vibrant East-West connections across Eurasia' ([34] (p. 1); [35]). The American initiative built on past Soviet economic flows through Central Asia. Announced by the US State Department as part of its exit strategy from Afghanistan, its aim was to re-integrate the country into traditional trading routes. Though conceptual frameworks have varied, the initiative's key points appear to presage and to have become core elements of China's nearly identical 2015 OBOR outline (Table 2) [9].

Table 2. New Silk Road Initiatives.

2011	US-New Silk Road Initiative	2015	China-New Silk Road *
1	Regional energy markets	1	Connectivity-infrastructure, logistics
2	Trade and Transport		energy, communication
3	Customs and border operations	2	Trade and customs
4	Businesses and people-to-people	3	Financial integration
		4	People to people bond
		5	Policy coordination

* Document officially titled Visions and Actions on Jointly Building Silk Road Economic Belt and twenty-first century Maritime Silk Road [36].

With such a major infrastructure programme come a suite of ancillary factors that are influenced by domestic circumstances in host nations. These include indirect OBOR implications from reshaping border zones (e.g., Khorgos Special Economic Zone, Kazakhstan), state capacity to direct, absorb, and monitor projects, and manifestations of marginality (social, economic, ecological, political) to ethnic contestation, (dis)integration along transport routes, and engagement and benefit from OBOR at community levels [14,37]. The uncertainty and potential direction of OBOR implementation is reflected in the $46 billion Chinese-Pakistan Economic Corridor (CPEC) where the programme is most advanced [3]. The infrastructure crosses the Himalayas to Gwadar port on the Indian Ocean and involves tens of thousands of Chinese workers. Less direct are the routing of the pipeline and transport links within Pakistan, interaction with local populations, ethnic unrest, implications for fragile state institutions, and perceptions of benefit or exclusion. Issues of security and surveillance, rights of residents, water and environmental degradation, and wielding of de facto supra-state power raise questions about levers of control and for who's benefit is OBOR designed [38]. Resultant political infighting, grasping for infrastructure, volta face in nationalistic policy, and weak monitoring institutions point to the disruption major external infrastructure projects can have. Similar complexities can be expected as OBOR expands into Central Asia.

With an $8 trillion Asian infrastructure deficit, China's 'most significant and far reaching project ... ever' expects to find receptive hosts in need of what may be on offer [5] (p. 11). Yet in traversing a region, economic and political forces become entwined with realpolitik. The five post-Soviet Central Asian states are caught in a bind: they must reconcile Russia's historic role with China's expanding economic dominance [35]. Participation in OBOR may not be centred on local interest and consent as much as acceptance of projects sanctioned by the regional hegemons. Thus, Russia will favour train lines through Kazakhstan that bring it goods and tariffs over tracks to the Caspian Sea that benefit China or Europe instead [12]. With funding directed by China, benefits will accrue through opening new markets for goods, procuring resources, and increasing access and engagement in previously underserved areas. Presented as an economic corridor, an OBOR of pipelines, bridges, roads, etc. is imagined for Central Asia to enhance security and energy cooperation [4]. The concurrent launch of the Asian Infrastructure Investment Bank (AIIB) provides both a funding source and implied international backing for large scale building projects.

Whilst OBOR, BRI, and the New Silk Road are all about China [39], frontier states are not a vacuum waiting to be filled or controlled [7]. A large part of any regional infrastructure programme goes through former Soviet territory that remains Russia's near-abroad today. This geopolitical background has created a convoluted *ménage a trois* between China, Russia, and comparatively weak Central Asian states [35]. Russia combines its roles as former (and aspiring) political overlord, military guarantor of (or perhaps threat to) sovereignty, trade partner through the Eurasian Economic Union (EEU), and cultural (particularly language, alphabet (Cyrillic vs. Latin) and media) touchstone with its former republics [10,40]. Though there are existing tripartite relations through the ongoing Shanghai Cooperation Organisation security alliance, the supercedent EEU dominates structural interaction between Russia and Central Asian states without China's participation [41]. Separate from Russian-Chinese relations, the EEU sets out a protocol that Kyrgyzstan and Kazakhstan follow for multi-faceted relations with trade benefits and restrictions within which China-Central Asian relations must be balanced. Thus, the two states are situated between two superpowers with different agendas. This is most notable in economics, where Central Asian trade with China increased from $1 billion in 2002 to $45 billion in 2014 [42]. The political and economic divergence leaves these states facing two masters with different remits and agendas, and potentially conflicting rather than co-existing interests. Russian self-interest will favour infrastructure that will be beneficial through trade, resources, and transit fees [12]. Reflecting this dichotomy, both countries export more to Russia than China, yet China is the largest source of imports. The unequal stature of the two key states finds that >80% of China's regional trade is with Kazakhstan; benefits will thus accrue to Kazakhstan with limited support for disadvantaged Kyrgyzstan. Further, ethnic clashes have soured relations between Kyrgyzstan and Uzbekistan [43,44] with implications for China's envisioned shared projects (pipelines, railroads) through the Fergana Valley and onward to Afghanistan, Iran, and West Asia.

China has been instrumental in Kyrgyzstan's meagre development efforts whilst Russia maintains a military base. Vast amounts of new infrastructure is needed to replace aging and obsolete Soviet power plants, roads, bridges, dams, and hydro-electric power in Kyrgyzstan [45]. Yet the value of infrastructure-build for China in the country is not clear. Kyrgyzstan is central to the key planned rail route from Kashgar through Naryn and on to Uzbekistan [46], though currently China finds faster access traversing through Kazakhstan to Xinjiang Province. Such structural issues may limit and prescribe benefits to Kyrgyzstan. An unmentioned part of transport development in Kyrgyzstan is that its most valuable transit commodity is not minerals or oil but drugs. The state is on the main heroin route from Afghanistan to Europe through Kazakhstan and Russia where 25% of the supply is transported [47]. Any expansion of roads without concomitant increased drug interdiction will potentially facilitate criminal activity and trafficking, including to China's eastern heartland. Infrastructure also represents possible links between ethnic groups separated by borders, facilitates transnational networks, and represents pathways for migrants. For example, China claimed Uighur separatists were responsible for the 30 August 2016 car bomb attack on the Chinese Embassy in Bishkek, Kyrgyzstan [48].

In sum, OBOR is a project writ large with myriad intentions and possible implications. At the macro-scale, it is a presentation of China to the world in the 21st century. A Chinese government perspective stresses 'development and prosperity', that OBOR 'is not a tool of geopolitics' and 'growing power and investments should not be seen as threatening' [4] (p. 3). What is written on

OBOR is reflective of foreign perceptions of China's global engagement; circumspect articles decipher signals framed by external expectations and power relations vis-à-vis China. What is missing is the transition in thought to what OBOR tentacles may look like and engender on the ground in receiving nations. Existing narratives present OBOR as China's coming out event; now an interpretation and understanding of what this means for frontier states and host nations is imperative. As receivers of action, investment, and infrastructure, states are exposed to direct (physical, economic) and ancillary (social, environmental, political) benefits and costs.

Central Asia is largely overlooked in existing literature on OBOR, which has thus far largely attended to Chinese narratives about the initiative ([4], the Maritime Silk Road component (e.g., [7]), Chinese economic and political expansion in South East Asia [49], and the China-Pakistan Economic Corridor [3]. Research by Central Asian scholars and specialists has begun to turn attention to the potential impacts of OBOR in the region, focusing on narratives of state ideologies and development [13,16,43], the inability of Central Asian states to exert sovereignty [15], marginal borderlands [1,50], and expanding Chinese influence in the region [5,12]. These themes are important for OBOR, as they address the capacity of Kyrgyzstan and Kazakhstan to engage and absorb major infrastructure projects, and the existing literature suggests constraints and mixed outcomes. For example, ref. [10] (p. 376) views the New Silk Road as a series of mobilities through Central Asia that are corridors for 'hydrocarbons ... migrants ... drugs ... trade ... and military', a process matched by borders, immobility, and competition. Meanwhile, Reeves [50] examines how infrastructure brings hope as well as roads and questions territorial integrity. The varied discourses highlighted in this scholarship stresses Central Asian limitations in institutional capacity [51], environmental governance [52], and interaction with China [3] that are integral to a discussion on OBOR. Though land degradation and water stress are documented in the region [53,54], connections between landscape, infrastructure, and land use, critical to OBOR implementation, have not been investigated. In the remainder of this paper we seek to sketch out what such a research agenda might look like.

3. OBOR in Central Asia: Landscape, Infrastructure, and Socio-Political Dynamics

3.1. The Role of Landscape

Infrastructure does not exist in a vacuum; it is a very land-based, physical undertaking that conforms to or manipulates an environment. The connection between 'earth and infrastructure exhibits a series of temporalities' [17] (p. 2). Geography matters; altitude, temperature, climate, people, and governance all affect infrastructure projects. Without context roads are ephemeral and unrealised and if built may not be fit for purpose. As conceived, the New Silk Road transects landscapes to connect places, transit goods, and peoples and spread interaction. OBOR implementation places Central Asian geography at the crux of any Chinese westward endeavour (Figure 1); physical factors contrive against expansion from China's heartland within the Great Wall. The Tibetan Plateau and Himalayas obstruct southwestern routes, concentrating CPEC's efforts on the Karakoram Highway. At 4693 m, the Khunjerab Pass makes a treacherous choke point for roads, pipelines, and railroads making the long journey from Gwadar, Pakistan to Chinese territory. To the west, the Pamir and Tien Shan mountains are no less daunting; Peak Somoni (aka Stalin or Communism Peak) at 7495 m, Sina Peak (aka Lenin Peak) at 7134 m, and Kongur Tagh at 7649 m block physical access. To the north are the 2800 km-long Tien Shans featuring Jengish Chokusu (aka Victory Peak) at 7439 m and Khan Tengri (aka Lord of the Spirits) at 7010 m; they are breached by the Torugart Pass, at 3752 m, that enables transit between Kashgar in Xinjiang and Naryn on the Kyrgyz high plateau (see Figure 1).

To the north, the Dzungarian Gate is the geographical and historical mountain pass between China and Kazakhstan, described as the 'one and only gateway in the mountain-wall which stretches from Manchuria to Afghanistan' [55,56] (p. 266). Through the gate flows oil and gas pipelines, China Highway 312, the train to Europe, and Chinese workers to make OBOR come to life in the near abroad. Now on the Kazakh side Khorgos is being developed as a Special Economic Zone and logistics hub for trade with China and set to be the primary OBOR portal within the decade [10].

Physical geography adds complexity to infrastructure, and may catalyse in unexpected events and 'provoke social and political crises that may not be contained' [17] (p. 2). Land is but the obvious barrier between two regions. With great mountains come severe cold, glacial landforms, low precipitation, temperature extremes, high winds, and solid rock barriers that not only impede and isolate the region but can thwart and crush infrastructure (Table 3). This is married to barren desert landscapes, little water, and poor agricultural productivity [53]. Natural disasters are a further threat to infrastructure, as Almaty, the largest city in the region (1.7 m inhabitants), is on a series of fault lines radiating out from the Tien Shan Mountains. The city was destroyed in a 1911 earthquake when shocks reached Kyrgyzstan; it is at high risk for another quake [57]. Central Asia's 'appalling record of water management' due to unsustainable use, dysfunctional systems, and poor cooperation among states suggests that the potential economic benefits come with high environmental costs and barriers [52] (p. 976).

Table 3. Physical factors in Central Asia.

Factor	Kyrgyzstan	Kazakhstan
Border with China, km	1063	1765
Elevation in metres, mean	2988	387
Temperature, annual mean	5	2.2
Precipitation, annual mean	375	275
Deserts, %	10	>80
Arable land, %	6.7	8.9
Agriculture, % of water	93	66
Population density, km^2	27	6

Environmental management has become a contentious issue in Central Asia as former Soviet comrades have become independent, and interdependent, nations where water is linked to energy and security [54]. Concurrently, increasing land degradation, driven by unsustainable irrigation and farming practices and overgrazing, directly impacts land and livestock productivity, incomes, and rural livelihoods [58]. Soil and water erosion, salinization, pollutants, and poor land management have contributed to >60% of Kazakhstan and Kyrgyzstan's land identified as degraded and possibly desertified [53]. Land and water systems are based on Soviet structures, are not designed for transboundary cooperation, and are aggravated by historical legacies [58]. Existing land issues are exacerbated by China's interest in such resources, land-take for infrastructure, and increased bi-lateral assistance [53,54]. The transboundary nature of water adds complexity—Kyrgyzstan is water-rich yet infrastructure poor, Kazakhstan needs more water for agriculture, whilst China has plans to build dams in these states to harness hydropower. Further, climate variability and hazard risk are reflected in recent regional drought as harvest reduction, increased prices, and food insecurity reflect the fragility of land systems in the area [59]. Opaqueness in Chinese procurement of resource rights in the region points to contestation over land and water between current rural users and state-driven contracts. Acknowledgement and integration of environmental limitations within OBOR construction objectives is vital to minimise confrontation in this conflict-prone region [54].

3.2. *The Politics of Infrastructure*

The region's physical geography and geopolitical history both have significant influence on the existing and possible future infrastructure developments. Central Asia's Soviet-era railway gauge is 1.52 m whereas Chinese train track width is 1.435 m. Thus, no direct train transport can take place; all train cars need to have the undercarriage physically changed to wide-gauge supporting bogies (wheel framework). Customs processing into Central Asia is the least efficient and most time consuming in the world; goods take up to 79 days to clear the border [5]. This points to rent seeking behaviour, inefficient government, and lack of interest or motivation to work effectively with neighbours. In fact, if days/weeks/months are lost in customs the point of OBOR—increased connectivity and time savings—are negated and goods will continue to go by sea. Assessment by Cooley [5] identifies that Chinese officials expect 30% of investment to be lost to corruption in Central Asia and 80% lost in Pakistan. This does not include the layers of guanxi (connections) that filter through projects originating in China [19].

The list of infrastructure projects intended for Central Asia under OBOR is unclear; previous investment has stressed pipelines, refineries, highways, and power plants (Tables 4 and 5). Any such list has two component parts: what China wants to build, and what the countries want to have built. The decision-making process is opaque, leaving the project drivers unidentified. Here, Kazakhstan and Kyrgyzstan need to be differentiated; the former wants infrastructure that compliments exports and energy integration with minimal disturbance to society [11]. The latter needs roads and railroads, better energy supply, and jobs for its people (>one million Kyrgyz (16%) work in Russia) [60]. Both want benefits with an invisible, or moderated, Chinese presence due to distrust of Chinese intentions and perceived economic hegemony and land grabs by China [9]. Sinophobia pervades the region, perhaps inflamed by pervasive Russian media [14,61]. Can these differences be reconciled? One drawback of China's current 'all things to all people' stage is that there is no blueprint or document stating China's aims, standards, code of conduct, and treatment of local interests and sensitivities. Without a minimum framework trust remains low and arrangements are between states without community buy-in. In 2016, protests forced autocratic Kazakhstan to withdraw a law allowing foreigners to lease land in what was perceived as an imperial $1.9 billion Chinese grab for agricultural lands and an implied stranglehold on food production [62].

Table 4. Chinese investments in Central Asia.

Country	Project	Cost-US $	Year Signed/Implemented
Kazakhstan	Zhongfu Investment Group into oilseed processing	$1.2 billion	2016
Kazakhstan	MangistauMunaiGas (50%)	$2.6 billion	2009
Kazakhstan	Kazakh portion-Central Asia-China gas pipeline	$6.7 billion	2009
Kazakhstan	Ekibastuz GRES-2 Power Plant	$400 billion	2016
Kazakhstan	Kazakhstan-China Oil Pipeline	$3 billion	2006
Kyrgyzstan	Zhongda Oil Refinery	$430 million	2013
Kyrgyzstan	Kyrgyz portion-Turkmenistan-China gas pipeline	$1.4 billion	2016
Kyrgyzstan	North-South Highway	$400 million	2013
Kyrgyzstan	Thermal power plant	$386 million	2014
	Tang/FMSO [30]; see additional references		

Table 5. Proposed One Belt, One Road (OBOR) investments.

Country	Potential OBOR Investments	Project
Kazakhstan	China-Central Asian pipeline	Natural gas
Kazakhstan	Eurasian Land bridge	Railway corridor
Kazakhstan	China-Central Asia-West Asia corridor	China to Iran rail link
Kazakhstan	Khorgos-Aktau railway	Caspian to China link
Kyrgyzstan	China-Kyrgyzstan-Uzbekistan railway	High-speed rail
		Cooley [5] 2016

That such a turn of events happened in Kazakhstan bodes poorly for Chinese endeavours in Kyrgyzstan, Central Asia's most democratic-leaning state. Dissatisfaction with an ineffective government, elections, voting on regional and ethnic lines, and enduring poverty contribute to Kyrgyz discontent. State instability, intercommunal violence, a lack of jobs, and corruption continue to unsettle the country [44]. Since 2005 this has led to a series of 'Tulip Revolutions' through to street protests in

2017 [63]. Having borrowed heavily from China, 38% of its $3.7 billion external debt is to China's ExIm Bank alone [45]. For example, the $390 million Datka-Kemin north-south power line and $400 million north-south road are part of the existing Chinese debt package [13], highlighting how decisions against China's wishes may become problematic. What Kyrgyzstan most wants—jobs—are the very thing that Chinese infrastructure projects do not provide. Instead, Chinese labourers are brought in by the (tens of) thousands, as Pakistan's CPEC experience shows [38].

Discussions in Central Asia with civil society and academics stress the high awareness of OBOR and equally the lack of clarity and uncertainty about intent, timeframe, and expected outcomes. The local perspective focuses on what will be imposed without apparent consultation or meaning for communities. In Kyrgyzstan, past examples and a perceived influx of Chinese create a negative perception of what OBOR will mean domestically. Comments on a railway from Kashgar to Uzbekistan drily note few Kyrgyz will be on the train, as it will not pass through Bishkek. These observations reflect the work of Alff [14] (p. 440) on Kyrgyzstan as a Chinese 'vassal state'. Transport and potential subsequent development corridors are also viewed as aiding expansion of drug trafficking [10]. Kazakhstan is more favourable to OBOR, as the programme may increase exports and thus income. Infrastructure will be far removed from the business capital of Almaty, yet the mooted possible increased Chinese involvement in food production raises nationalistic concern. In both states the programme is perceived as externally directed with minimal domestic interaction.

To set some context for how OBOR is and will be received on the ground in Central Asia, it is important to note key socio-cultural dynamics in the region. Kyrgyz and Kazakhs are Turkic nationalities with shared linguistic, religious, and cultural roots from across what was once called Turkestan, a vast area stretching from the Tibetan Plateau and Himalayas to the Caspian Sea and Turkey. Identity, kinship, spiritual practices, and lifestyles remain vastly different from that of the Han Chinese. In the Soviet era, there was a massive influx of other ethnic groups into the region, including millions of Russians and significant numbers of Ukrainian, German, and Korean immigrants. More than 3.5 million Russians remain in Kazakhstan and Kyrgyzstan, further differentiating the people and orientation of the region from China. Today this contributes to separation and animosity between local residents and incoming Chinese with fear of de facto economic appropriation expressed in major cities [14,61]. Government to government relations are strong, but as 2016 protests of perceived Chinese land-grabs in Kazakhstan demonstrated, these may not be attuned to local sentiments [62]. Thus, as OBOR projects are initiated, citizen interest and acceptance needs to be considered and addressed.

This raises important subsidiary issues that come with mega-projects and investments. Are institutions, capacity, and expertise in place for consensual development programmes? Are adequate legal structures, customs unions, institutional oversight, and security measures set up to handle new flows and volume of goods and people? In a traditionally autocratic region, what role do citizens and communities have in recommending, shaping, and deciding on investments, locating routes, demarcating impact zones, and monitoring social and environmental impacts and liabilities? These currently unaddressed issues are key to OBOR's success. Without citizen engagement, economic relevance, and adequate planning, what happens once the concrete dries will remain an unknown. China's $430 million Zhongda oil refinery in northern Kyrgyzstan exemplifies how quickly Chinese mega-projects can go awry. Only a fraction of the 850,000 ton annual capacity is being used, as China failed to secure crude oil supply before building the facility. The investor has asked for Russia's assistance; Russia does not have interest in supplying raw materials to new Chinese players in their near abroad [64]. Without better understanding of local context and rational business planning development in Central Asia risks becoming at best a muddled story or infrastructure folly, and at worst igniting regional tensions. Currently several major Chinese projects are scheduled for Central Asia spanning roads, railroads, airports, power plants, and dams (Table 5). Like the Zhongda refinery, they are grand investments that are presented with limited scrutiny and uncertain economic justification. The in-country decision-making process is not transparent and whether projects are requested by or aligned with local interests is not ascertained. In this way OBOR reads less a partnership of friends as cogs in a Chinese mega-development plan. Thus, how new works address and incorporate local needs is obfuscated by fait-accompli announcements and state relations.

The presentation of OBOR in Central Asia provides little evidence of planning for environmental impacts or engaging with residents. This may include project siting, adequate resources such as water for both the project build and ongoing functioning, and land degradation and rehabilitation [52]. Roles for local communities, jobs, livelihood changes, and security are unstated. Further, ancillary processes such as implications for culture, ethnicity, and migration are unaddressed. Sinophobia in Central Asia is well known and inhibits rational assessment of infrastructure projects, a process clouded by the lack of transparency and limited role for host communities [11,42]. The sentiment may in part be a response to inflammatory Chinese government comments such as that Central Asia was a "rich piece of cake given to today's Chinese people by heaven" [3] (p. 118). As one local Kyrgyz commented, 'first they started sending their goods, then they started moving in, and finally they marry our local women." [30] (p. 18).

Self-perception and socio-economic factors colour trade and thus OBOR relations with China. Autocratic Kazakhstan views itself as the Central Asian power and with a positive balance of trade (>$38 billion annually) [33] is not dependent on China for physical development. This is reflected in Kazakhstan's trade surplus with China, a major buyer of the country's energy resources [65]. Conversely, Kyrgyzstan has few alternate methods to fund major infrastructure. Whilst the government encourages Chinese investment, citizens are not welcoming to China's expanded presence. As a struggling democracy, the Kyrgyz political class can in part attract voters by appearing to keep China at an arm's length. In the 2017 presidential elections, only a small number of candidates support OBOR in their campaign rhetoric.

Soft power themes ask if benefit will be accomplished through trans-cultural interaction or by receiving unreciprocated Chinese business and tourism. The term soft power implies a welcome, an invitation to exchange people, goods, ideas, and opportunity. Is this in the nature of a road or a pipeline? Or are the Chinese asking citizens to focus on second and third generation benefits delayed through oil contracts, debt payments, and rentier bureaucracies? The real possibility exists that China starts below zero on a soft power scale in Central Asia. Pantucci [61] identifies Sinophobia as the biggest regional problem that China faces with its Silk Road Economic Belt policy in Central Asia. This is nowhere in Silk Road plans and agendas. Without an OBOR charter of local principles or explanation of the Kazakh or Kygyz rights and responsibilities in a way that can be monitored and transform behaviour, residents have little justification to expect transparency or the prospect of participation in unfolding OBOR stories.

To reconcile the potential value of OBOR with its possible adverse consequences will take active acknowledgement of challenges and the engagement of principal players. Greater transparency in decision-making, clarity in project ownership and control, channels for local input and consultation, and a collaborative approach amongst diverse stakeholders is essential. The process would start with Chinese willingness to develop a Code of Conduct stating project rights and remedies—with governments informing citizens about OBOR plans and expectations and explanation of how communities are involved, geography accommodated, and monitoring enabled. Two mega-projects suggest ways forward: China National Petroleum Corp mentions Corporate Social Responsibility efforts in its report on Kazakhstan, suggesting nascent community awareness and a way to improved relations. The Kumtor Gold Mine in Kyrgyzstan (owned by Centerra of Canada) is embroiled in long-running disputes with the government over royalties and alleged environment damage. This may, in fact, be a positive process, as it reflects the impact of citizen protest in making infrastructure development accountable. Yet the lack of constructive mechanisms to resolve contested issues infers a tortuous development path awaits. Most OBOR interaction is stymied by perceived heavy-handed infrastructure-as-expansion, fissures between states and citizens, weakness of host nations vis-a-vis China, and an insensitive and overbearing Chinese agenda. Geographical challenges may have alternate routes, engineering solutions, or mitigation potential. As geo-political and social fractures solidify, discord may become both intractable and OBOR's major legacy.

4. Uncertainty, National Differences, Marginality, and Sino-Centrism

'Advancing westward was "a historical necessity for the Chinese nation, and it is also our destiny." ... Central Asia should thus be regarded "as territory to be recovered in our advance, not as a border region"' PLA General Liu Yazhou [3] (pp. 117–118).

A massive infrastructure programme, a major Chinese westward expansion, an offering of altruistic friendship, an example of manifest destiny—evaluation of OBOR offers a tabula rasa for Central Asia and the world. Interpretations abound yet facts are few on the ground. We stress four themes in the Central Asian context: uncertainty is an OBOR hallmark, events and infrastructure will play out differently in Kazakhstan and Kyrgyzstan, capacity and marginality will challenge infrastructure development, and ultimately all roads lead back to China (see [4]). A core tenet presented in the work of Barry [17] is that infrastructure is constructed on real geography through living communities, rather than in presidential speeches and grand edicts, is critical in a region of 7000+ metre mountains, deserts, extreme climates, and different cultures. The 'physical' will meet the 'theoretical' in OBOR's march westward. As Peyrouse and Raballand [35] (p. 413) state, 'a route on a map ... does not create traffic and trade'.

Interrogation of the uncertain, muddled human dimensions will provide a more robust evaluation of a New Silk Road in Central Asia. This proffers questions about institutional transformation or subordination, external annexation or domestic-driven collaboration, and benefit defined at the mega, national, or personal scale. Issues of livelihood enhancement and increased human opportunity, environmental sustainability, community development, and economic improvement are yet to be scrutinised and integrated into the grand rhetoric. The human element exposes its darker side—marginalisation of citizens, lost or foregone opportunity and jobs, roads as pathways to migration and trafficking, a diminution of rural prospects and populations, concentration of benefit in urban hubs, and degradation of environmentally based lives as agriculture, pastoralism, and livestock raising are isolated by infrastructure or overtaken by markets manifest in land grabs. Is organic growth on offer or is this a coercive multi-state project lacking civic relevance or nation-bonding through sacrifice of public space and engagement?

Within Central Asian dynamics Kazakhstan is well-placed to profit from OBOR as it is both politically and economically close to Russia whilst a major oil exporter to China. Once over the Tien Shan Mountains geography favours infrastructure across the flat steppe and established rail routes link the three countries. Solid state finances offer Kazakhstan a buffer; it does not need OBOR for development or domestic (e.g., voting) exigencies. In comparison, Kyrgyzstan is mountain-bound, on the proposed train line to (unfriendly) Uzbekistan, without oil, and in need of what China may decide to offer. This weak hand gives little leverage for the country and few alternate development sources. Where Kazakhstan perceives itself as an equal in negotiation, Kyrgyzstan is an OBOR supplicant.

The idea of marginality runs through the rural territories that will host infrastructure. Such contexts are on the fringe of political and economic loci of control that initiate projects, are spatially isolated and often on the edge of borders, and are often further separated by poverty, social group, and livelihood opportunity. High speed rail and pipelines designed to circumvent locals may lack relevance and result in failed projects (such as the Zhonga refinery) or, opposite to development, promote further disintegration within fraying social-political contexts. Whilst citizens in Kazakhstan or Kyrgyzstan cannot derail OBOR, they can, and have, changed regimes in Kyrgyzstan and may again protest against the aging autocracy in Kazakhstan. If infrastructure becomes corridors without context, alters or degrades an environment, or segregates development trajectories the 'soft power' path will be missed whilst inequality, Sinophobia, and dispossession remain. As Diener [10] (p. 380) states, 'roads may separate as much as they connect'. The tantalising rewards of OBOR will come with a foundation that engages with national and local discourses. In coming years more can be done by

OBOR proponents to identify standards and procedures that encourage domestic participation and acceptance rather than an external, top-down imposition of Chinese will on frontier states. This raises the spectre of nations' abilities to digest OBOR against a host of domestic factors from stability, ethnicity, and marginality to control and external influence.

Finally, procedures, announcements, efforts at persuasion, and maps show OBOR as a China-directed, optimistic, multi-faceted expansion effort (see Table 1). It is thus self-reflective—how can infrastructure benefit China and Chinese in state, business, employment, and citizen realms—reflected in General Liu's presumption of 'destiny'. Whilst the effort shows China going out into the world, in truth all roads lead back to China [39]. This is where benefit will be determined, success claimed, and new markets or resources conquered. Central Asia is the first stop on the parade. With small populations, malleable governments, and an inability to create and drive mega-projects, Kazakhstan and Kyrgyzstan are prime candidates to accommodate China's largess and attention. Success by flexible and undefined measures belongs to China. Failure of physical structure, economic profit, or citizen disapproval (even damage) will rest with the Central Asians. This is where grand integration meets with uncontrolled forces. These could be in the form of climate, earthquakes, droughts, famine, terrorism, conflict, or revolution—all have happened in Central Asia. Any of these can change Chinese hubris into humbleness as the New Silk Roads careen their way through Central Asia.

5. Conclusions

Tracking OBOR into Central Asia frames the possibilities and challenges the programme encounters in Kazakhstan and Kyrgyzstan. Historical, cultural, and economic factors separate the nations from China whilst development differentiates the two states. Borders and geography constrain infrastructure, superpower geopolitics impede state autonomy, corridors favour Chinese interests (resources, business) over local agendas, whilst OBOR objectives and outcomes in the region remain opaque [31]. Grounding the narrative identifies Central Asian characteristics that will ameliorate or mitigate the grand intentions of the New Silk Road.

Within OBOR several competing strands present themselves. Foremost are considerations regarding what is built and for whom. The implications of pipelines, roads, airports, etc. are tangible manifestations of Chinese prowess, development, and interstate collaboration. Secondary physical expressions—changes in water, land, and environment—are also part of OBOR. Less concrete but equally OBOR related are the suite of social, institutional, economic, and governance benefits and disruptions, with values assigned by perspective. To China, building infrastructure is a clear public good, a mantra often repeated in the Tibet and Xinjiang contexts. To Kazakhstan, infrastructure enabling trade and economic gain is positive; in Kyrgyzstan, any new construction is a boon, filling key gaps the state is unable to address. These macro-level representations extol the 'development and prosperity' the literature promises. These are the voices and vistas that will be shared for global consumption.

Given the scale, complexity, and inherent uncertainties regarding OBOR, identifying potential mechanisms for addressing challenges faced in Central Asian states is difficult if not unachievable. Based on our review of existing literature and analysis of geographical, social, and political contextual factors, the following key issues for further discussion emerge.

The lack of transparency regarding planned or proposed OBOR projects needs to be rectified. A first step would be ensuring that channels of communication are in place between Chinese planners/developers and Central Asian governments and, crucially, local communities that will be affected by the infrastructure developments.

Communication and coordination across China's 'frontier states', including Pakistan, where OBOR projects are more advanced (e.g., CPEC), can be facilitated and strengthened.

Clear OBOR objectives, standards, and rights should be established and expressed at local, state, and transnational levels.

Whilst current research considers manifestations of OBOR in Central Asia, this paper identifies three critical research ideas for future investigation. First, OBOR has not been grounded in Central Asia's physical landscape. Understanding the infrastructure footprint can be done through fieldwork and remote sensing to capture land area sacrificed and demands on water, and land use vis-a-vis dependent communities; Second, improved knowledge of the practical implications of OBOR on rural residents, livelihoods, and communities presents a gaping hole in current project understanding. Third, the host state institutions-citizens-Chinese outreach nexus awaits exploration. In combination, these significant research themes will offer insight, astuteness, and realistic knowledge about China's infrastructure expansion into Central Asia.

Acknowledgments: The research was funded by the John Fell Fund, University of Oxford and encouraged by Oxford's School of Geography. The University of Central Asia, American University of Central Asia, KIMEP University, and Kazakh National University were supportive of research and the Aga Khan Development Network added perspective and insights. We thank anonymous reviewers for their recommendations that improved the manuscript.

Author Contributions: Troy Sternberg conceived, researched and wrote the paper; Ariell Ahearn and Fiona McConnell contributed research framing, ideas, context and wordsmithing.

Conflicts of Interest: The authors declare no conflict of interest.

References

1. NY Times behind China's $1 Trillion Plan to Shake Up the Economic Order. 2017. Available online: https://www.nytimes.com/2017/05/13/business/china-railway-one-belt-one-road-1-trillion-plan.html?_r=0 (accessed on 13 May 2017).
2. Xinhua. President Xi's Speech at Opening of Belt and Road Forum. 2017. Available online: news.xinhuanet.com/english/20USD17-05/14/c_136282982.htm (accessed on 14 May 2017).
3. Rolland, N. *China's Eurasian Century? Political and Strategic Implications of the Belt and Road Initiative*; NBR: Seattle, WA, USA, 2017.
4. Sidaway, J.; Woon, C. Chinese Narratives on "One Belt, One Road" in Geopolitical and Imperial Contexts. *Prof. Geogr.* **2017**, 1–13. [CrossRef]
5. Cooley, A. The Emerging Political Economy of OBOR: The Challenges of Promoting Connectivity in Central Asia and Beyond. Center for Strategic and International Studies. Available online: www.csis.org/analysis/emerging-political-economy-obor (accessed on 16 March 2017).
6. Lim, T.W.; Chan, H.H.; Tseng, K.H.; Lim, W.X. *China's One Belt One Road Initiative*; Imperial College Press: London, UK, 2016.
7. Blanchard, J.; Flint, C. The Geopolitics of China's Maritime Silk Road Initiative. *Geopolitics* **2017**, *22*, 223–245. [CrossRef]
8. Blanchard, J. Probing China's Twenty-First-Century Maritime Silk Road Initiative (MSRI): An Examination of MSRI Narratives. *Geopolitics* **2017**, *22*, 246–268. [CrossRef]
9. Summers, T. China's 'New Silk Roads': Sub-National regions and networks of global political economy. *Third World Quart.* **2017**, *37*, 1628–1643. [CrossRef]
10. Diener, A. Parsing mobilities in Central Eurasia: Border management and new Silk Roads. *Eurasian Geogr. Econ.* **2015**, *56*, 376–404. [CrossRef]
11. Kassenova, N. China's Silk Road and Kazakhstan's Bright Path: Linking Dreams of Prosperity. *Asian Policy* **2017**, *24*, 110–116. [CrossRef]
12. Pantucci, R.; Lain, S., III. The Plan: Infrastructure and Integration. *Whitehall Pap.* **2016**, *88*, 30–46. [CrossRef]
13. Toktomushev, K. Central Asia and the Silk Road Economic Belt. Policy Brief 2016-1. 2016. Available online: http://ucentralasia.org/Resources/Item/1218 (accessed on 11 August 2017).
14. Alff, H. Flowing goods, hardening borders? China's commercial expansion into Kyrgyzstan re-examined. *Eurasian Geogr. Econ.* **2016**, *57*, 433–456. [CrossRef]
15. Doolot, A.; Heathershaw, J. State as resource, mediator and performer: Understanding the local and global politics of gold mining in Kyrgyzstan. *Cent. Asian Sur.* **2015**, *34*, 93–109. [CrossRef]
16. Sullivan, C. State-Building in the Steppe: Challenges to Kazakhstan's Modernizing Aspirations. *Strateg. Anal.* **2017**, *41*, 273–284. [CrossRef]

64. Rickleton, C. Kyrgyzstan: Chinese Investor Struggling without Russian Help. 2015. Available online: www.eurasianet.org/node/71776 (accessed on 7 May 2017).
65. Trading Economics Kazakhstan Balance of Trade. 2017. Available online: www.tradingeconomics.com/kazakhstan/balance-of-trade (accessed on 24 May 2017).

land

MDPI

Article

Plantation Development in the Turkana Basin: The Making of a New Desert?

Edward G. J. Stevenson

Department of Anthropology, University College London, London WC1H 0BW, UK; e.stevenson@ucl.ac.uk

Received: 30 November 2017; Accepted: 20 January 2018; Published: 25 January 2018

Abstract: Since the early 20th century, "desert reclamation" has been synonymous with large-scale waterworks and irrigation. These techniques have made it possible to produce abundant crops in arid or semi-arid environments. The costs have often been externalized, with increased environmental productivity in the new croplands counterbalanced by increased aridity elsewhere. In this paper I consider whose interests are served by such projects, and what kinds of social constructions of the natural and human environment make them possible. I focus on the Turkana basin, a watershed spanning the Ethiopian and Kenyan borders, where large dams and irrigation projects are currently being established with the goal of producing cash crops and hydro-electricity. In the narratives of the projects' proponents, the schemes are represented as part of a tradition of development stretching back to the American West. In the discourse of critics, the Aral Sea of Central Asia is frequently invoked. Considering Turkana in relation to these cases sheds light on the political and ecological gambits involved in desert reclamation, and helps us to understand the costs and benefits of such projects.

Keywords: desert reclamation; desertification; river basin development; political ecology; water

1. Introduction

In the late 1970s and early 1980s, in the wake of the Sahel drought, the concept of desertification gained wide currency. A key idea associated with desertification was that the practices of farmers and herders in the world's drylands were leading to rapid degradation of dryland ecosystems [1,2]. This narrative resonated with rising international concerns about the fragility of the natural environment, but the empirical basis for it was shaky. As Swift observes, "dry years (especially in the 1970s and early 1980s) were often compared with wet years (especially in the 1960s); this was then interpreted as a secular decline in productivity, rather than as a variation in the response of the natural vegetation or crops to soil moisture availability" [1] (p. 83). Nevertheless, the idea took hold, and was boosted by the endorsement of the UN Conference on Desertification (UNCOD) in 1978. The prescriptions for preventing desertification included large-scale afforestation and terracing projects, resettlement schemes, and promotion of alternative livelihoods. Such projects, led by governments and supported by international donors and non-government organizations, involved the mobilization of virtual armies, and absorbed vast amounts of capital. The short-term impacts on soil conservation were modest or even negative [3], and in some longitudinal studies of particular 'at-risk' landscapes, forest cover was observed to increase despite rises in population density (e.g., [4,5]).

Why did the desertification concept come to be so widely accepted? One answer is that the desertification narrative "met a need" and provided "a useful discourse" for powerful actors including governments, international agencies, and non-governmental organizations [1] (p. 84). Hoben characterized this discourse as a neo-Malthusian narrative of Africa as a "spoiled Eden", burdened by overpopulation, and in breach of carrying capacity—a situation remediable only through "massive investments in environmental reclamation" [6] (p. 11). The narrative appealed to influential groups who saw in it a way of extending their own power (cf. [7]).

Inspired by the literature on desertification, in this paper I take a critical perspective on the phenomenon of 'desert reclamation', and investigate whose needs are served by this idea. In some ways desert reclamation is a mirror image of desertification: it represents a conviction that land degradation is not an inevitable process, but can be reversed through the application of proven techniques of resource management. In other ways, it closely resembles the desertification narrative. Just as desertification was used as a justification for large-scale projects that often had negative impacts for the livelihoods of rural people, so desert reclamation has led to manifestly inequitable outcomes. Since the early 20th century 'reclamation' has been synonymous with a particular set of techniques involving large-scale waterworks and irrigation. These techniques have made it possible to produce abundant crops in arid or semi-arid environments. But the costs of this model have often been 'externalized', in the economist's terms, with increased environmental productivity in the new croplands counterbalanced by increased aridity elsewhere. Given that the areas targeted for such schemes have usually been previously inhabited and worked by smallholder farmers or herders, their effects have also been to produce increased hunger and food insecurity.

I focus on the Omo-Turkana basin, a watershed spanning the Ethiopian and Kenyan borders, where large dams and irrigation projects are currently being established, with the goal of producing cash crops and hydro-electricity. Controversy clings to this project, and a highly polarized debate has emerged between proponents and critics. In the narratives of the projects' proponents, the schemes are represented as part of a tradition of economic development stretching back to the American West—large-scale interventions, such as the Hoover Dam, which are seen as yielding large benefits for national progress [8]. In the discourses of critics, on the other hand, the Aral Sea of Central Asia is frequently invoked as a reference point, and the inequitable distribution of costs and benefits between 'national' and local parties is emphasized (e.g., [9,10]). Despite the frequency with which these two comparison cases are mentioned, the similarities and differences between the Omo-Turkana situation and the American and Central Asian experiences have rarely received close attention. In what follows I consider the Omo-Turkana case in light of these two comparison cases, and ask what each suggests about the calculus of social and environmental planning around river basin engineering schemes. In particular, I consider the historical contexts in which these three interventions took place, and the narratives that accompanied them (see Table 1).

Table 1. Characteristics of the river basin case studies.

River Basin	Location *	Time Period	Area Irrigated (ha) †	Principal Crops	Reference
Colorado	USA-Mexico	1902-present	2.4 m	alfalfa, wheat	[11]
Aral Sea	Central Asia	1954-present	7.5 m	cotton	[12]
Turkana	Ethiopia-Kenya	2005-present	175 k	sugar	[13]

* The Aral Sea basin includes parts of Uzbekistan, Kazakhstan, Tajikistan, Kyrgyzstan, Afghanistan, and Turkmenistan.
† Data are for circa 2000–2010 for the Colorado and circa 1988 for the Aral Sea basin. Data for the Turkana basin is the projection of the government of Ethiopia (Kuraz Sugar Development Project) (cf. [10]). In the lower Omo as of February 2016, sugar had been planted on approx. 11 kha, and an additional 13 kha had been cleared [13]. (Units: mha = million hectares; kha = thousand hectares.).

One consideration that guides me in comparing these cases is the social construction of the natural environment that is implied in the planning process. Relevant questions include: Who is making claims to land and water? And whose prior claims are affected, or placed in jeopardy? One of the similarities that emerges from these comparisons is the location of significant hydrological engineering schemes on the borders or peripheries of states. The downstream impacts can be more easily ignored or written off if they occur 'over the border', beyond the planners' purview. The second consideration concerns historical and spatial scale. The degree to which ecosystems are improved or degraded by irrigation schemes can look very different depending on whether the question is considered on a small or large geographical scale, and over a short or long time horizon. At the time of writing, the Turkana basin schemes remain relatively small, and the major irrigation ventures there date back less than a decade. One of the advantages of comparison is to provide a longer timescale over which to consider the social and ecological processes involved.

2. Dams and Development

The scale and scope of dam-building and irrigation schemes in the 20th century is difficult to overstate. Between 1950 and 2000, on average, two large dams were built per day, with the number of large dams worldwide reaching approximately 45,000 by the end of the century [14,15]. Irrigation systems associated with large dams were first built by British colonists in Egypt in the late 19th century [16] and in the Sudan in the early 20th century [17]. In the wake of the Tennessee Valley Authority project, river-basin-wide systems of dams became a template for export after the Second World War, promoted by the Bretton Woods institutions and by European and American engineers and development advisors [18,19]. Post-independence governments in Africa, such as the administration of Kwame Nkrumah in Ghana, promoted dams as a centrepiece of national economic strategy [20]. In Ethiopia after the restitution of Haile Selassie by British forces, the government promoted a similar model, with the first river basin engineering scheme launched in the Awash Valley in the 1960s [21,22].

From the vantage point of the early 21st century, scholars can appraise the costs and benefits of this hydro-irrigation explosion. At a global level, there have been major benefits: an estimated 15% of food supply comes from land irrigated from large dams, and 19% of electricity is generated by hydro-electric turbines [14]. However, planners have systematically understated the financial costs of large dams, which are, on average, almost twice as high as initially estimated [23]. When other kinds of losses are taken into account, e.g., farmland inundated by reservoirs, and reduced productivity of riverine and delta fisheries, the costs rise further. Social costs include the direct displacement of 40–80 m people [14], with at least 450 m more people affected by changes in river flows and environmental conditions [24]. These changes, the effects of which may be felt over multiple generations, erode people's sense of control over their lives, and undermine their psychosocial well-being [25,26].

From a political ecology perspective, the social costs listed above might be counted as part of the attraction of hydro-electricity and irrigation projects. At the same time as they present an appearance of economic rationality and national advancement, mega-projects may function as projects of social engineering that place vital resources at the disposal of elites [16,17]. By undermining or sweeping away one socio-ecological order and replacing it with another, they present experts with the opportunity to redesign landscapes and impose new kinds of discipline on people and places they deem recalcitrant or unproductive. In the following section I introduce the landscape of the Turkana basin, and explore the processes that led to the construction of dams and plantations in the region.

3. The Turkana Basin: Wetland, Dryland, or Desert?

The Turkana Basin includes a wide range of ecosystems, from highland forests in southwest Ethiopia to drylands around Lake Turkana, the world's largest permanent desert lake. In terms of rainfall, the lower Omo (i.e., the area downstream of the River Mago) is semi-arid, with mean annual rainfall in the region of 250–500 mm; Turkana County, Kenya, receives <250 mm/year. But the geographical classification of this region must also take account of the river that runs through it, which includes riverine swamps, thick forests in the delta region, and (until recently) one of the most potentially productive fisheries on the continent [10]. In these terms, the Omo/Turkana region invites comparison with Africa's other great wetlands, including the Okavango and Zambezi [27].

The River Omo rises in Ethiopia's south-west highlands, fed by the rains that wash over the region each year, part of a great climate system that links the Indian Ocean and the Horn of Africa. The countless streams charged by these rains feed major rivers that in turn feed into the Omo [28]. The largest of them, the Gibe, is itself a major river. In its watershed rose the kingdoms and states of the Macha Oromo, which until the late 19th century constituted a political system independent of Ethiopia [29]—part of an archipelago of cities and states that marked precolonial Africa. Swelled by the Gibe, the Omo rolls through the territory of peoples speaking a dozen languages; in its lower course, the old river meanders and then fans out into a great inland delta before emptying into the salty waters of Lake Turkana. The flood of the Omo is the ecological pulse of the entire region, each seasonal inundation providing a cue to grow. For the aquatic life of the lake, it provides nutrients that prime the reproductive systems of fish to spawn [30]. For the people of the river's lower reaches—one of the most culturally and linguistically diverse regions of Africa—the flood, along with the appearance of the Pleiades in the night sky, marks the new year [31,32]. The flood is vital for their agro-pastoralist livelihoods [33].

The lower Omo was formally incorporated in Ethiopia through military campaigns under Emperor Menelik II in the late 19th century. After these campaigns, the southern marches, including the lower Omo, were economically important for highland Ethiopia, exploited for ivory and slaves, but remained politically peripheral to the Ethiopian state [34]. The principal impediment to settler colonialism was the semi-arid climate, which made the land unsuitable for plough agriculture as practiced by the highland Ethiopian colonists. The exceptions prove the rule: by the early 20th century, Maale and Ari—pockets of highlands where rainfall was higher—were readily incorporated into the system of tenant/serf farming and revenue extraction [35].

The first intensive farming operations in the lower Omo were attempted by North Korean state conglomerates in the early 1990s, in the last years of the military socialist Derg regime. Irrigation systems were installed to grow cotton, but a combination of unsuitable soils and salinity build-up led the venture to be abandoned. Nonetheless investors remained interested in the idea of plantation agriculture in the lower Omo, and soon after the takeover of power by the Ethiopian People's Revolutionary Democratic Front (EPRDF), plans to develop the Omo were drawn up. Woodroofe and Associates, under the auspices of the Government of Ethiopia and with external support from the African Development Bank, produced a 'Master Plan' for hydro-electricity generation and irrigated agriculture on the Omo in 1996 [36]. The scale of the designs that the state pursued on the Omo in the following years, however, was far greater than the Master Plan suggested.

The transformation of the basin began relatively modestly, with the construction of the Gilgel-Gibe I dam in the upper catchment of the basin (completed in 2004) and the Gilgel-Gibe II hydropower plant (completed in 2010). Gilgel-Gibe III (hereafter 'Gibe III') was a far greater undertaking. Two-hundred and forty meters high, and with an installed capacity of approximately 1870 Megawatts, it would more than double Ethiopia's electricity-generating capacity. Italian engineering firm Salini (which had previously built Gilgel-Gibe I and II) was handed the contract without open competition, and work on the dam began in 2006, before any environmental or social impact assessments (ESIAs) had been carried out. Unable to attain funding from other sources (the World Bank, the European Development Bank, or the African Development Bank declined their requests), the Ethiopian government eventually secured financing for the dam from the Chinese state. Although ESIAs were subsequently commissioned [37], they overlooked some of the most important potential impacts of the schemes, which would stem from the irrigation projects to be established in conjunction with the dam. The largest of them, the Kuraz Sugar Development Project, was projected to cover 175,000 ha [13].

In 2015, when the sluice gates of the Gibe III were closed and the reservoir behind the dam began to fill, conflicting predictions of the project's implications were afloat. Proponents of the dam and plantation schemes saw prosperity, urbanization: a desert reclaimed. Critics saw imminent ecological disaster and heightened economic marginalization for the people of the lower Omo and Lake Turkana. In the narratives of the project's proponents, Gibe III and Kuraz were sometimes explicitly, sometimes implicitly, placed as part of a tradition of economic development stretching back to the American West—large-scale interventions yielding large benefits for national progress [8,38]. The abundant waters of the lower catchment of the Omo, flowing south into Lake Turkana, were construed as a hydro-agricultural resource that demanded to be developed for the good of the nation. The narrative took the form: "This is how you (Europeans and Americans) got where you are. How dare you deny it to us?" In the discourse of the critics, the resources invested in the building of mega-dams—estimated at €1.5bn for Gibe III [39]—would have been better spent on small-scale development projects that provided immediate benefits to local people, and gave them a say in planning their futures [40]. Critics also pointed to the incalculable environmental and biocultural costs of replacing a diverse landscape managed by indigenous stewards with commercial estates growing crops for export [41]. The narrative took the form: "Wherever this has been done, it has led to disaster." A recurring reference point was the Aral Sea (e.g., [9,10]).

In the following sections I review the cases of the Colorado and the Aral Sea and ask what each suggests about the calculus of social and ecological planning involved in hydro-agricultural projects. In the final part, I consider the shorter time series of evidence for ecological change in the Omo-Turkana since 2015, and identify common themes that emerge from consideration of the three case studies.

4. The Colorado Basin

> "We have in the Colorado [River] an American Nile awaiting regulation, and it should be treated in as intelligent and vigorous a manner as the British government has treated its great Egyptian prototype."
>
> Lippincott, 1912, cited by Worster [42] (p. 155)

Located in the American Southwest, and watered principally by snowmelt from the Rocky Mountains, the Colorado River runs through seven U.S. states, traversing the Sonoran Desert and northern Mexico, where the river historically terminated in a delta and emptied into the Gulf of California. It was in this region that high dams attained global prominence, and the damming and tapping of the Colorado is directly implicated in the rise of the economy of southern California and massive demographic growth in the U.S. Southwest since the mid-20th century. In recent years, with decreasing precipitation in the upper catchment, and increasing abstraction in the lower catchment of the Colorado, barely a drop of the river's water crosses the border to Mexico [43].

The story of the Colorado is crucial for understanding the history of desert reclamation—indeed, the term "reclamation" had its origin here, in the 1902 Reclamation Act by which the U.S. federal government mandated major works on the region's rivers in the interest of agricultural development and settlement of the West. This legislation, and the administrative apparatus that it spawned—including the Bureau of Reclamation, which went on to design and construct dams and irrigation systems across the region—was but one phase in a longer history of competition over natural resources, a drama played out among Native Americans, Mexicans, and descendants of European settlers, among others. In the reading of legal scholar Douglas J. Kenney, the "economic paradigm" that governed the usage of the Colorado from the nineteenth century onwards privileged the initiatives of European settlers over others, with Native Americans and Mexicans placed at the back of the queue for water rights [44]. In the Colorado Compact of 1944, 10% of the total annual flow of the Colorado (1.5 million acre feet, or approx. 2 km^3) was allocated to Mexico, with the remaining 90% shared equally between the U.S. states of the upper and lower basin [45]. Over the course of the 20th century, large federal subsidies made possible the irrigation of millions of acres of land, supporting beef and dairy cattle, alfalfa, wheat, and (on a smaller scale, but much more profitably) fruit, nuts, and vegetables [11].

The development model devised by the Bureau of Reclamation involved the building of large dams that would serve the dual purpose of generating electricity and facilitating irrigation, with the high costs of dam building and irrigation works offset in part by electricity sales after the projects' completion [42]. The promise of irrigated land made available to settlers at bargain prices was supposed to lure white homesteaders from the more densely populated eastern states, and the Reclamation Act made provisions for this scenario by limiting the size of plots in its initial offer to 160 acres, and by instituting a requirement that owners should live on the land for at least five years [42]. In the event, few settlers took up the offer—in large part because of the hardship of living in a newly irrigated desert—while speculators acquired as much land as they could for resale. The largest beneficiaries of these projects were those who already had substantial capital, including established landowners whose property increased in value as a result of irrigation. Large landowners in the vicinity of cities such as Phoenix, Arizona, for example, benefitted from public funds spent on the Roosevelt Dam (completed in 1911), which provided increased water for their private estates [42]. The region's cities benefitted too, with increased quantities of water and electricity supporting rapid urban growth [46].

For the elites who came to control the majority of the new irrigated lands, a major challenge was obtaining enough workers to tend their estates. The solution they arrived at was to hire migrant workers, a majority of them Mexicans [42]. Meanwhile, as water abstractions for the expanding estates on the U.S. side of the border grew, the quality and quantity of water available for Mexico deteriorated. By the 1960s, high salinity rendered the water flowing into Mexico practically useless for agriculture. Mexican politicians complained that the U.S. government was in breach of the 1944 agreement [42,47].

While Mexico had been allocated 10% of the river, and found itself with water that was too salty to use, the position of Native Americans in relation to the irrigation projects was more ambiguous. For most of the twentieth century, de facto exclusion of Native Americans from the profits accruing from industrial agriculture coexisted, albeit uneasily, with de jure rights of those living on reservations to "the water resources needed to sustain [their] new tribal homelands." As Kenney observes:

> Translating this principle into actual water management in the Colorado River basin is an ongoing process, subject to considerable debate .... Arizona, in particular, features several tribes with Colorado River rights of great seniority, as these rights are defined as originating with the dates of the Indian treaties ... that typically took place before widespread homesteading by Anglos [descendants of European settlers]. [...] By some estimates, large reservations such as the Navajo reservation in north-eastern Arizona—could conceivably by awarded the entire flow of the Colorado under this calculus. Politically this outcome is unacceptable to the non-Indians that would be displaced, so the 'solution' has been to withhold from tribes the financial resources needed to develop water projects until they agree to settlements that dramatically scale-back the size of their rights.
>
> [44] (p. 132)

While Native Americans were effectively denied access to the benefits of the irrigation boom in the USA, those with the most attractive land and the greatest access to capital increasingly came to dominate, notably the agro-industrialists of Arizona and southern California [11,42]. Considering the basin as a whole, however, the highest price was paid by Mexicans. The logical end result of the economic paradigm—treating the river as a commodity—was to raise competition for water among the basin partners; but rather than competing on a level playing field, those upstream held a strong advantage. Since the 1960s, the Colorado River has rarely reached the sea. Its delta in Mexico disappeared, as did the forests and fisheries it used to support, and the riverine farms of the indigenous Cucapá people who lived there [15] (p. 29).

The drying-up of the Colorado delta roughly coincided with the emergence in the USA of an alternative paradigm that meaningfully challenges the economic paradigm, and highlights the human and environmental costs of river-basin development [44]. While communities seeking the removal of existing dams and the restitution of landscapes to their former states face strong opposition, some of these campaigns have paid off. The success stories include California's Echo Park (site of the first environmental preservation campaign in U.S. history) and Mono Lake [48]. As the region's ageing stock of dams is increasingly threatened by reservoir sedimentation, more dams are being decommissioned than are being built in the USA [15] (p. lxi). Judging from the experience with the Colorado, the timescale of a hydraulic civilization's rise to maturity and possible decline would appear to be roughly a century.

The experience of the Aral Sea, however, suggests it may be shorter still.

5. The Aral Sea

The Aral Sea was in the early twentieth century the world's fourth-largest inland water body. Located on the borders of the Central Asian states of Kazakhstan and Uzbekistan, it was for most of the 20th century part of the Soviet Union, with economic and political control exercised from Moscow. Beginning in the late 1950s, large-scale irrigation works serving cotton estates led to major abstractions from the principal rivers that fed the Aral Sea, the Amu Dar'ya and Syr Dar'ya, which have their sources in the snow and glaciers of the Pamir and Tian Shan mountain ranges [49]. The construction of major canals diverting water from these rivers, and concurrent increases in land area devoted to cotton in the 1970s and 1980s led to the complete drying up of the Amu Dar'ya and occasional drying up of the Syr Dar'ya. As a consequence of the reduced inflow, the depth and surface area of the Aral Sea declined precipitously, and split into two or more bodies of water. Exposed areas of former lake bed turned to salt pans; the maritime climate of the region turned dryer; biodiversity plummeted. Endemic fish species, such as the Aral salmon, became extinct, and by the early 2000s scientists surveying terrestrial biodiversity reported fewer than half of the species of mammals and birds previously observed [50].

The negative impacts of the disappearance of the sea were felt most of all by the people who had formerly lived along its shores, especially the Karakalpak fishers and herders of Uzbekistan. Fisheries that once sustained one of the Soviet Union's largest fish-processing factories were destroyed; grazing lands dried out; and entire communities were marooned as navigation routes between coastal towns and villages were lost. Initially Karakalpakstan shared in the wealth produced by the new schemes, growing cotton, rice, wheat, and vegetables with irrigation infrastructure. But being situated at the end of the line in terms of access to the irrigation system meant that the water they received grew increasingly saline. Production peaked in the 1980s and has since been in decline, with increasing volumes of water required to flush previously accumulated salt out of the soil [51].

Other effects were felt further afield. Windblown salt and dust from the desiccated lake bed had negative effects on agriculture in a wide zone downwind of the basin. Impacts on human health (in addition to those experienced via loss of employment in the fishery and transport sectors) included respiratory infections exacerbated by windblown salts, as well as an increased incidence of cancers and other conditions potentially related to exposure to the heavy metals and pesticides associated with industrial farming operations [52].

Although the most critical phases of the Aral Sea disaster played out in the later years of the Soviet Union, relations between Muscovite planners and the region can be traced to the conquest of Central Asia by imperial Russia in the 19th century. Economically important first as a source of furs, and later for its mineral wealth, the relationship between the eastern and western parts of the empire was long characterized by extraction. In the vision of modernist planners, this frontier region "was to be reclaimed from wilderness with the help of modern technology and railroads, colonized and transformed from a cold and unwelcoming wasteland roamed by backward savages into a populous and prosperous part of Russia itself" [53] (p. 282). The historical experience of the Aral Sea region renders this civilizing narrative bitterly ironic, as—far from reclaiming the region from "wilderness"—the very projects that were projected to bring wealth and civilization produced in their turn a new and distinctive kind of desert.

6. Discussion: Omo and Lake Turkana Revisited

When any given large dam or irrigation project is launched, it is often claimed that it is too early to say how productive or destructive the transformed landscape of the basin will be. The value of historical comparison is that it helps us consider probable trajectories of change, and, in light of them, to think anew about the possibilities for course correction or mitigation of harms. In the literature on river basin development, two types of historical comparison are commonly made: country- or region-specific comparisons (e.g., [22]); and global surveys that attempt to discern modal patterns in terms of the contributions of large hydro-irrigation projects to development (e.g., [14,23,24]). In this article I have taken a different approach, focusing not on the experience of a single world area, nor on global patterns, but on those comparison cases that recur most frequently in the rhetoric of proponents and critics.

What can we glean from these two cases that is of relevance to Turkana? In both cases, two narratives can be discerned: a heroic narrative of feats of engineering and a tragic narrative of desiccation (the Colorado as "a river no more" [44]; the Aral Sea as a cartographic error). In the case of the Colorado, it is the heroic story—accompanied by a myth of equitably shared benefits—that prevails; in the case of the Aral Sea, it is the tragic one. But despite the rhetorical uses to which they have been put, the experiences of the Colorado basin and the Aral Sea are in many ways similar. The similarities include the conceptual importance of regional or international boundaries for externalizing ecological costs, and the ethnic "othering" or simple erasure of the people whose lands and livelihoods were affected by the rerouting or repurposing of rivers. The existence of political boundaries and gradients of power facilitate what Steve Rayner calls "institutionalized forgetfulness" [54] (p. 110): they help to hold at bay knowledge that would otherwise be problematic for the achievement of institutional objectives.

Since the Gibe III reservoir began filling in 2015, the shoreline at Ferguson's Gulf, the most productive fishery on the lake, has receded by 1.5 km [55]. Abstraction of increasing quantities of water for irrigation upstream would cause the lake levels to drop further, potentially splitting into two or more smaller lakes, like the Aral Sea [10]. Indeed, even if irrigation were suspended, fisheries biologists predict that the regulation of the river by Gibe III will lower fish yields from Lake Turkana by more than two-thirds, since the lake's fish depend on the seasonal flood pulse (the nutrients and fresh water that the Omo delivers) as a cue for breeding [30]. This is "uncomfortable knowledge," in Rayner's terms, and those whose interests are served by the Gibe III dam and Omo plantations can be expected to deny it [54]. So far, the Kenyan government has not forced its Ethiopian counterpart to acknowledge the threats these developments pose for those whose livelihoods depend on the lake. Despite attempts by the United Nations Environment Programme to broker an agreement between the governments, the key problem of water-sharing remains unresolved [55–57].

And what of the people of the lower Omo valley? For centuries they have relied on a three-pillar strategy to make a living in the semi-arid climate: combining herding, farming on rain-fed land, and farming on land inundated by the annual rise of the river [58]. The river flood, since it carried water from a large catchment in highland Ethiopia, dependably provided the water and silt needed for farming. With the Gibe III dam, the flood has ended, and the people of the lower Omo are therefore deprived of one crucial component of their livelihood. Some of the most valuable agricultural lands have been annexed by the expanding sugar plantations, and the availability of vital fall-back resources, in the form of game animals and wild foods, is severely constrained, as scrubland has been razed for plantation development [59]. At the same time as thousands of new settlers and migrant labourers converge on the plantations of the lower Omo, a food insecurity crisis looms for the region's indigenous people [60]. The safety net provided by the Ethiopian government in the form of a "villagization" programme has been woefully inadequate [61]. To avert famine, there is a need for a comprehensive and well-funded livelihood reconstruction programme [62].

In most scholarly work and media reports on this unfolding crisis, the voices and opinions of the people of the lower Omo and Turkana have been conspicuously absent. Ironically, the most prominent opponents of the Omo dam and irrigation schemes—advocacy groups such as International Rivers and the Oakland Institute—are based not in Africa but in the western USA. But there is a logic in this. The region of the world with the longest history of massive hydraulic engineering has given birth to some of its most forceful critics. If there is one common denominator in the history of large dam and irrigation schemes, it is the unfair distribution of costs and benefits. The people of the American West have certainly not benefitted equally from the experiments carried out on the region's rivers, any more than the peoples of the Soviet Union or Central Asia benefitted equally from the sacrifice of the Aral Sea. The critics' refrain—"Wherever this has been tried, it has been a disaster"—might be amended: Wherever this has been tried, it has benefitted some at the expense of others.

Acknowledgments: I have benefitted from conversations with Jennifer Hodbod, Benedikt Kamski, Dan Mains, Ruth Mandel, Selam Shiferaw, David Turton, and members of the Omo-Turkana Research Network. I am grateful to Troy Sternberg and Ariell Ahearn for their invitation to present some of the ideas in this paper at the Oxford Desert Conference, in June 2017. Errors that remain are my own.

Conflicts of Interest: The author declares no conflict of interest.

References

1. Swift, J. Desertification: Narratives, winners and losers. In *The Lie of the Land: Challenging Received Wisdom on the African Environment*; Leach, M., Mearns, R., Eds.; International African Institute/James Currey: Oxford, UK, 1996; pp. 73–90.
2. Mortimore, M. *Roots in the African Dust: Sustaining the Sub-Saharan Drylands*; Cambridge University Press: Cambridge, UK; New York, NY, USA, 1998.
3. Herweg, K.; Ludi, E. The performance of selected soil and water conservation measures—Case studies from Ethiopia and Eritrea. *CATENA* **1999**, *36*, 99–114. [CrossRef]
4. Nyssen, J.; Haile, M.; Naudts, J.; Munro, N.; Poesen, J.; Moeyersons, J.; Frankl, A.; Deckers, J.; Pankhurst, R. Desertification? Northern Ethiopia re-photographed after 140 years. *Sci. Total Environ.* **2009**, *407*, 2749–2755. [CrossRef] [PubMed]
5. Tiffen, M.; Mortimore, M.; Gichuki, F. *More People, Less Erosion: Environmental Recovery in Kenya*; Wiley-Blackwell: Chichester, UK; New York, NY, USA, 1993.
6. Hoben, A. Paradigms and politics: The cultural construction of environmental policy in Ethiopia. *World Dev.* **1995**, *23*, 1007–1021. [CrossRef]
7. Ferguson, J. The Anti-Politics Machine: Development, Depoliticization, and Bureaucratic Power in Lesotho. University of Minnesota Press: Minneapolis, MN, USA, 1994.
8. Pottinger, L. A Tale of Two Dams: Comparing Ethiopia's Grand Renaissance to Hoover. International Rivers Blog. Post date: 9 July 2013. Available online: https://www.internationalrivers.org/blogs/229/a-tale-of-two-dams-comparing-ethiopia%E2%80%99s-grand-renaissance-to-hoover (accessed on 20 November 2017).
9. International Rivers. *The Downstream Impacts of Ethiopia's Gibe III Dam: East Africa's "Aral Sea" in the Making?* International Rivers: Berkeley, CA, USA, 2013. Available online: https://www.internationalrivers.org/resources/gibe-iii-s-impacts-on-lake-turkana-7773 (accessed on 20 November 2017).
10. Avery, S. *What Future for Lake Turkana? The Impact of Hydropower and Irrigation on the World's Largest Desert Lake*; African Studies Center, University of Oxford: Oxford, UK, 2013. Available online: http://www.africanstudies.ox.ac.uk/lake-turkana-and-lower-omo-hydrological-impacts-major-dam-and-irrigation-developments (accessed on 20 November 2017).
11. Cohen, M.; Christian-Smith, J.; Berggren, J. *Water to Supply the Land: Irrigated Agriculture in the Colorado River Basin*; Pacific Institute: Oakland, CA, USA, 2013.
12. Saiko, T.A.; Zonn, I.S. Irrigation expansion and dynamics of desertification in the Circum-Aral region of Central Asia. *Appl. Geogr.* **2000**, *20*, 349–367. [CrossRef]
13. Kamski, B. The Kuraz Sugar Development Project (KSDP) in Ethiopia: Between 'sweet visions' and mounting challenges. *J. East. Afr. Stud.* **2016**, *10*, 568–580. [CrossRef]
14. World Commission on Dams. *Dams and Development: A New Framework for Decision-Making*; Earthscan: London, UK, 2000.
15. McCully, P. *Silenced Rivers: The Ecology and Politics of Large Dams*; Enlarged & Updated Edition; Zed Books: London, UK, 2001.
16. Mitchell, T. *Rule of Experts: Egypt, Techno-Politics, Modernity*; University of California Press: Berkeley, CA, USA, 2002.
17. Bernal, V. Colonial Moral Economy and the Discipline of Development: The Gezira Scheme and "Modern" Sudan. *Cult. Anthropol.* **1997**, *12*, 447–479. [CrossRef]
18. Sneddon, C. *Concrete Revolution: Large Dams, Cold War Geopolitics, and the US Bureau of Reclamation*; University of Chicago Press: Chicago, IL, USA, 2015.
19. Sneddon, C.; Fox, C. The Cold War, the US Bureau of Reclamation, and the technopolitics of river basin development, 1950–1970. *Political Geogr.* **2011**, *30*, 450–460. [CrossRef]

20. Miescher, S.F. "Nkrumah's Baby": The Akosombo Dam and the dream of development in Ghana, 1952–1966. *Water History* **2014**, *6*, 1–26. [CrossRef]
21. Kloos, H. Development, drought, and famine in the Awash Valley of Ethiopia. *Afr. Stud. Rev.* **1982**, *25*, 21–48. [CrossRef]
22. Kloos, H.; Legesse, W.; McFeeters, S.K.; Turton, D. Problems for pastoralists in the lowlands: River basin development in the Awash and Omo valleys. In *Water Resources Management in Ethiopia: Implications for the Nile Basin*; Kloos, H., Legesse, W., Eds.; Cambria Press: New York, NY, USA, 2010.
23. Ansar, A.; Flyvbjerg, B.; Budzier, A.; Lunn, D. Should we build more large dams? The actual costs of hydropower megaproject development. *Energy Policy* **2014**, *69*, 43–56. [CrossRef]
24. Richter, B.; Sandra, P.; Carmen, R.; Thayer, S.; Bernhard, L.; Allegra, C.; Morgan, C. Lost in Development's Shadow: The Downstream Human Consequences of Dams. *Water Altern.* **2010**, *3*, 14–42.
25. Colson, E. *The Social Consequences of Resettlement*; Manchester University Press: Manchester, UK, 1971.
26. Scudder, T. *The Future of Large Dams: Dealing with Social, Environmental, Institutional, and Political Costs*; Routledge: London, UK, 2006.
27. Adams, W.M. *Wasting the Rain: Rivers, People and Planning in Africa*; University of Minnesota Press: Minneapolis, MN, USA, 1992.
28. Butzer, K.W. *The Recent History of an Ethiopia Delta: The Omo River and the Level of Lake Rudolf*; University of Chicago Press: Chicago, IL, USA, 1971.
29. Hassen, M. *The Oromo of Ethiopia: A History, 1570–1860*; Red Sea Press: Trenton, NJ, USA, 1994.
30. Gownaris, N.J.; Pikitch, E.K.; Aller, J.Y.; Kaufman, L.S.; Kolding, J.; Lwiza, K.M.M.; Obiero, K.O.; Ojwang, W.O.; Malala, J.O.; Rountos, K.J. Fisheries and water level fluctuations in the world's largest desert lake. *Ecohydrology* **2017**, *10*, e1769. [CrossRef]
31. Girke, F. *Homeland, Boundary, Resource: The Collision of Place-Making Projects on the Lower Omo River, Ethiopia*; Working Papers/Max-Planck-Institute for Social Anthropology No. 148; Max Planck Institute for Social Anthropology: Halle/Saale, Germany, 2013.
32. Ruggles, C.L.N. (Ed.) Mursi and Borana Calendars. In *Handbook of Archaeoastronomy and Ethnoastronomy*; Springer: New York, NY, USA, 2015; pp. 1041–1050. [CrossRef]
33. Turton, D. The Downstream Impact. In Proceedings of the Gibe III Dam, School of Oriental and African Studies (SOAS), London, UK, 10 November 2010. Available online: http://www.mursi.org/pdf/RAS%20Talk%20-%20Copy.pdf/view (accessed on 20 November 2017).
34. Donham, D.L.; James, W. (Eds.) *The Southern Marches of Imperial Ethiopia: Essays in History and Social Anthropology*; Cambridge University Press: Cambridge, UK, 1986.
35. Donham, D.L. Old Abyssinia and the new Ethiopian empire: Themes in social history. In *The Southern Marches of Imperial Ethiopia: Essays in History and Social Anthropology*; Cambridge University Press: Cambridge, UK, 1986; pp. 3–50.
36. Woodroofe, R. *Omo-Gibe River Basin Development Master Plan Study: Final Draft Report*; Federal Democratic Republic of Ethiopia, Ministry of Water Resources: Addis Ababa, Ethiopia, 1996.
37. Agriconsulting S.p.A.; Mid-Day International. Gibe III Hydroelectric Project: Environmental and Social Impact Assessment. Additional Study on Downstream Impact; Electric Power Corporation: Addis Ababa, Ethiopian, 2009.
38. Tadesse, K. Ethiopia: Big Nile Dam Could Ease Africa Power Failures. Christian Science Monitor. Available online: https://www.csmonitor.com/Environment/Latest-News-Wires/2013/0703/Ethiopia-Big-Nile-dam-could-ease-Africa-power-failures (accessed on 3 July 2013).
39. Salini Impregilo. Ethiopia Inaugurates Tallest RCC Dam in World Built by Salini Impregilo. Press Release. Available online: http://www.salini-impregilo.com (accessed on 17 December 2016).
40. Fong, C. *The Scramble for Water, Land, and Oil in the Lower Omo Valley*; International Rivers: Berkeley, CA, USA, 2015.
41. Survival International. *Progress Can Kill: How Imposed Development Destroys the Health of Tribal Peoples*; Survival International: London, UK, 2007.
42. Worster, D. *Rivers of Empire: Water, Aridity, and the Growth of the American West*; Oxford University Press: Oxford, UK, 1985.

43. Owen, D. Where the River Runs Dry: The Colorado and America's Water Crisis. The New Yorker. Available online: http://www.newyorker.com/magazine/2015/05/25/the-disappearing-river (accessed on 25 May 2015).
44. Kenney, D.J. The Colorado River: What prospect for "a river no more"? In *River Basin Trajectories: Societies, Environments and Development*; Molle, F., Wester, P., Eds.; CABI: Wallingford, UK, 2009; pp. 123–146.
45. Hundley, N. *Water and the West: The Colorado River Compact and the Politics of Water in the American West*; University of California Press: Berkeley, CA, USA, 1975.
46. Reisner, M. *Cadillac Desert: The American West and Its Disappearing Water (Revised)*; Penguin: London, UK, 1993.
47. Brownell, H.; Eaton, S.D. The Colorado River salinity problem with Mexico. *Am. J. Int. Law* **1975**, *69*, 255–269. [CrossRef]
48. Horowitz, H. Aral sea analogs in the American West. In *Disaster by Design: The Aral Sea and Its Lessons for Sustainability*; Edelstein, M.R., Cerny, A., Gadaev, A., Eds.; Emerald: Bingley, UK, 2012; pp. 89–103.
49. Vostokova, E.A. Ecological disaster linked to landscape composition changes in the Aral Sea basin. In *Creeping Environmental Problems and Sustainable Development in the Aral Sea Basin*; Glantz, M., Ed.; Cambridge University Press: Cambridge, UK; New York, NY, USA, 1999; pp. 26–46.
50. Micklin, P. The Aral Sea Disaster. *Annu. Rev. Earth Planet. Sci.* **2007**, *35*, 47–72. [CrossRef]
51. Pearce, F. *When the Rivers Run Dry*; Transworld: London, UK, 2006.
52. O'Hara, S.L.; Wiggs, G.F.; Mamedov, B.; Davidson, G.; Hubbard, R.B. Exposure to airborne dust contaminated with pesticide in the Aral Sea region. *Lancet* **2000**, *355*, 627–628. [CrossRef]
53. Glebov, S. Siberian ruptures: Dilemmas of ethnography in an imperial situation. In *An Empire of Others: Making Ethnographic Knowledge in Imperial Russia and the USSR*; Cvetkovski, R., Hofmeister, A., Eds.; Central European University Press: Budapest, Hungary, 2013; pp. 281–310.
54. Rayner, S. Uncomfortable knowledge: The social construction of ignorance in science and environmental policy discourses. *Econ. Soc.* **2012**, *41*, 107–125. [CrossRef]
55. Mutambo, A. Ethiopia Denies Forcing through Gibe Dams Project. *Daily Nation (Newspaper)*; Kenya. Available online: http://www.nation.co.ke/news/Gibe-Dams-project-on-Kenya-and-Ethiopia/1056-3816882-10d7oky/index.html (accessed on 17 February 2017).
56. Arnold, D. Kenya, Ethiopia Mediating Omo River Water Controversy. Voice of America (News Agency), USA. Available online: https://www.voanews.com/a/kenya-ethiopia-mediating-omo-river-water-controversy/1770973.html (accessed on 16 October 2013).
57. Human Rights Watch. Ethiopia: Dams, Plantations a Threat to Kenyans. 2017. Available online: https://www.hrw.org/news/2017/02/14/ethiopia-dams-plantations-threat-kenyans (accessed on 20 November 2017).
58. Turton, D. Mursi Response to Drought: Some Lessons for Relief and Rehabilitation. *Afr. Aff.* **1985**, *84*, 331–346. [CrossRef]
59. Buffavand, L. 'The land does not like them': Contesting dispossession in cosmological terms in Mela, south-west Ethiopia. *J. East. Afr. Stud.* **2016**, *10*, 476–493. [CrossRef]
60. Carr, C.J. *River Basin Development and Human Rights in Eastern Africa: A Policy Crossroads*; Springer: Cham, Switzerland, 2017.
61. Stevenson, E.G.J.; Buffavand, L. 'Do our bodies know their ways?' Villagization, food insecurity, and ill-being in Ethiopia's Lower Omo valley. *Afr. Stud. Rev.* **2018**, in press.
62. Turton, D. Hydropower and irrigation development in the Omo valley: Development for whom? In Proceedings of the International Conference on Ethiopian Studies, Warsaw, Poland, 28 August 2015.

land

MDPI

Brief Report

Dust Storms from Degraded Drylands of Asia: Dynamics and Health Impacts

Shinji Otani [1,*], Yasunori Kurosaki [2], Youichi Kurozawa [3] and Masato Shinoda [4]

[1] International Platform for Dryland Research and Education, Tottori University, Tottori 680-0001, Japan
[2] Arid Land Research Center, Tottori University, Tottori 680-0001, Japan; kuro@alrc.tottori-u.ac.jp
[3] Division of Health Administration and Promotion, Faculty of Medicine, Tottori University, Yonago 683-8503, Japan; kurozawa@med.tottori-u.ac.jp
[4] Graduate School of Environmental Studies, Nagoya University, Nagoya 464-8601, Japan; shinoda.masato@g.mbox.nagoya-u.ac.jp
* Correspondence: otanis@alrc.tottori-u.ac.jp; Tel.: +81-857-30-6316

Received: 16 October 2017; Accepted: 22 November 2017; Published: 24 November 2017

Abstract: Asian dust events are massive meteorological phenomena during which dust particles from Chinese and Mongolian deserts are blown into the atmosphere and carried by westerly winds across Northeast Asia. Recently, there has been steady increase in both the frequency and the severity of Asian atmospheric dust events. Concern has been expressed regarding the potential health hazards in affected areas. The principal nature of the damage associated with Asian dust events differs between the emission (sandstorm) and downwind (air pollution) regions. In the emission region, the health impacts of dust storms are reflected in the high prevalence of respiratory diseases and severe subjective symptoms. Extreme dust storm events may cause a disaster to happen. In downwind regions such as Japan, analysis of Asian dust particles has shown the presence of ammonium ions, sulfate ions, nitrate ions, and heavy metal compounds that are considered not to originate from soil. Asian dust particles have been thought to adsorb anthropogenic atmospheric pollutants during transport. Therefore, Asian dust events coincide with increases in daily hospital admissions and clinical visits for allergic diseases such as asthma, allergic rhinitis, and conjunctivitis. Although the effect of Asian dust on human health in each region is influenced by a variety of different mechanisms, human activities are partly responsible for such negative effects in many situations. We therefore need to address these environmental problems.

Keywords: Asian dust; human health; Mongolia; Japan

---

1. Introduction

Dust storms originate in many dryland areas. Estimates of the relative magnitudes of dust emissions in different parts of the world indicate the Sahara contributes >50% of the global total, followed by China and Central Asia (about 20%) [1]. Dust events often affect human life and health not only in the originating drylands but also in downwind regions. Although the Sahara is the most important source of dust globally, approximately 60% of windblown Saharan dust moves southward toward the Gulf of Guinea across areas where few people live [2]. Asian dust events involve the long-range transport of atmospheric particulate matter originating from dryland areas, which is carried by westerly winds across Northeast Asia (Figure 1). Such events can affect large numbers of people because East Asia is the one of the most densely populated areas of the world. According to Kurosaki and Mikami [3], both the frequency and the severity of Asian atmospheric dust events have been increasing, raising concern regarding consequent health hazards in affected areas. The principal nature of the effects of Asian dust events on human health differs between the emission and downwind regions [1]. Here, we describe the health effects in each of these regions.

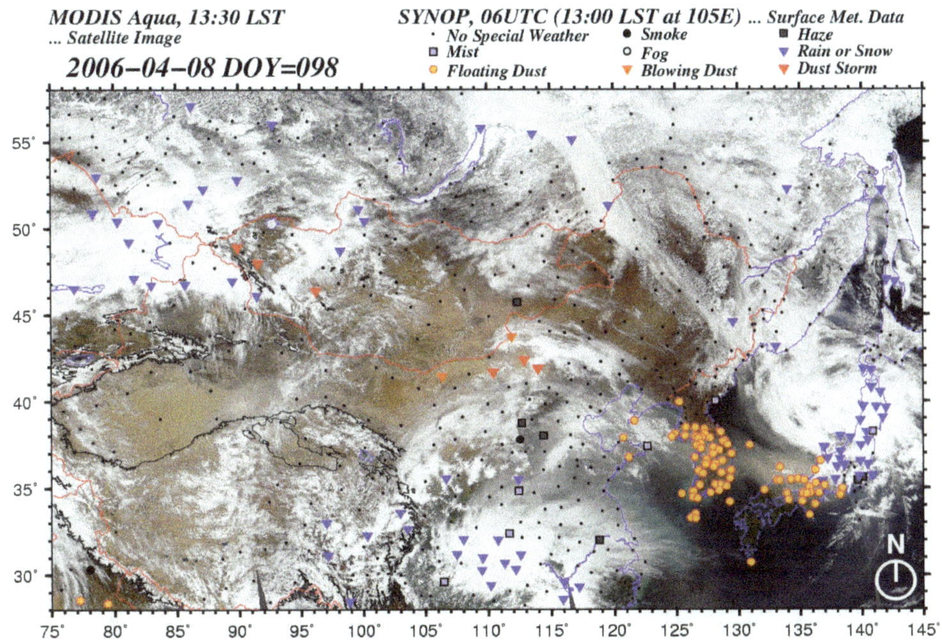

Figure 1. Typical Asian dust event. MODIS: Moderate Resolution Imaging Spectroradiometer true-color images with surface weather conditions on 8 April 2006 (NASA: National Aeronautics and Space Administration/GSFC: Goddard Space Flight Center, Rapid Response).

2. Impacts of Asian Dust on Health in Emission Regions

In Mongolia, during 17–20 April 1980, and on 5–6 May 1993, between nine and 16 people and 100,000–675,000 head of livestock died because of severe snow and dust storms [4]. Most recently, because of an intense dust storm that occurred across a broad area of Mongolia on 26–27 May 2008, 52 people lost their lives and 320,000 animals were killed. In an emission region such as Mongolia, severe dust storm events constitute major disasters.

We have previously documented subjective symptoms of the eyes and of the respiratory system, as reported by inhabitants (n = 123) in urban and desert areas of Mongolia immediately after this latest dust storm. The data collection method adopted in that study involved face-to-face interviews with a questionnaire. A detailed description of the survey has been published elsewhere [5]. The prevalence of eye symptoms was higher among those subjects living in the desert area than among the urban area subjects. Dust storms have been associated with a high prevalence of eye symptoms in Mongolia. We also performed a cross-sectional survey on Health-Related Quality of Life (HRQOL) using a 36-item short-form health survey (i.e., an index of health condition) and livestock loss data one year after this dust storm event. Our results provide preliminary evidence that livestock loss has long-term effects on HRQOL; some HRQOL markers (general health and vitality) were lower among people who had lost livestock than among those who had not. Therefore, saving lives, maintaining animal husbandry, and ensuring medical health support after dust storm disasters (e.g., medical care patrols and psychological consultations) are primary requirements. Moreover, the development of an appropriate early warning system to prevent dust storm damage is also needed [6].

Exposure to dust particulates irritates the respiratory tract and is associated with respiratory disorders such as asthma, pneumonia, and nonindustrial silicosis [1]. In fact, the outpatient morbidity associated with diseases of the respiratory system in provinces with high frequency of occurrence of dust events has remained stably high [3,7] (Figure 2). For example, the morbidity rate of respiratory diseases in Omnogovi Province, located in the Gobi Desert, was 2157.61 per 10,000 inhabitants in 2011, i.e., about twice the national average (1048.17) [7]. It is suggested that populations particularly vulnerable to airborne particulate matter and associated respiratory diseases are children, the elderly, and people with pre-existing heart and lung diseases [1].

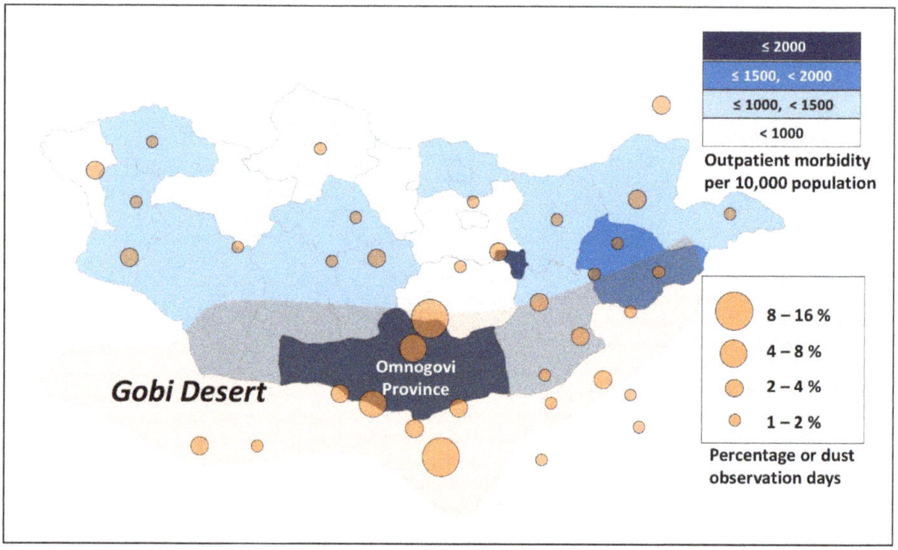

Figure 2. Morbidity associated with diseases of the respiratory system in each province of Mongolia (2011) and observations of dust event (1988–2004). Dust days were based on Code Table 4677 of the World Meteorological Organization (WMO) and dust outbreak frequency was defined as the percentage of the number of dust outbreaks to the total number of observations. The elements of surface meteorological data were the 3-hourly present weather and the surface wind speed at a height of 10 m in East Asia from March 1988 to May 2004 from WMO (see detail in Reference [3]). This dust outbreak map is a modified version of Figure 3 in Reference [3].

3. Impacts of Asian Dust on Health in Downwind Regions

Using trajectory analysis, confirmed by satellite observation data, we have clarified that Asian dust events transport both soil-derived and anthropogenic metals [8]. Additionally, based on aerosol sampling in both Mongolia and Japan, it has been established that "bioaerosols" (i.e., airborne microorganisms) carried by dust events include fungi and bacteria [9,10]. Therefore, it is necessary to consider not only the physical impacts of dust but also the physiological-biochemical responses when we evaluate the health effects of Asian dust events.

In downwind regions of Asian dust events such as Japan, South Korea, and Taiwan, recent epidemiological studies have shown that such events have coincided with increases in daily hospital admissions and clinical visits for allergic diseases such as asthma, allergic rhinitis, and conjunctivitis [1,11]. Children have been found to be particularly vulnerable. Globally, exposure to dust particles transported by desert storms is associated with increased hospital admissions for childhood asthma [12]. Worsening asthma symptoms caused by Asian dust might be attributed to the combination of particulate matter from soil and anthropogenic pollutants [13]. In relation to cardiovascular diseases, it has been suggested that Asian dust is a potential trigger of acute myocardial infarction [14].

As we have previously reported, it has been highlighted that Asian dust can induce symptoms such as itchy eyes and skin, nasal congestion, and sore throats in otherwise healthy subjects [15].

We evaluated the association between daily symptoms and dust events in 54 healthy subjects. In this study, scored symptoms were significantly higher on Asian dust days than on non-Asian dust days, and the skin symptom scores were correlated positively with the levels of suspended particulate matter [15].

4. Allergic Reactions and Asian Dust in Downwind Regions

The association between skin symptoms and Asian dust events has been reported previously [16]. Analysis of Asian dust particles in Japan has shown the presence of ammonium ions, sulfate ions, nitrate ions, and heavy metal compounds that are considered not to originate from soil. During transport, Asian dust particles are thought to adsorb atmospheric pollutants caused by anthropogenic activities. To compare skin symptoms on an Asian dust day with metal allergic reactions, we performed patch testing, which is a useful method for the estimation of contact dermatitis. In this previous study [16], we performed tests on nine subjects with skin symptoms and 11 subjects without skin symptoms on an Asian dust day. It was found that 89% of subjects with skin symptoms reacted to nickel samples. Conversely, 82% of subjects without skin symptoms had no reaction to nickel samples [16]. Nickel is a common cause of contact allergies, and it has been found that skin symptoms on Asian dust days are significantly associated with the levels of atmospheric nickel in corrected total suspended particulates [17]. Skin symptoms during Asian dust events might reflect allergic reactions to Asian dust particle-bound metals.

Recent studies have shown that microbes such as bacteria and fungi can migrate vast distances during Asian dust events by attaching themselves to dust particles. In Korea, springtime air contains a large variety of fungi and potentially high levels of fungal allergens including *Penicillium* [18]. In Japan, the possibility of bacterial attachment to aeolian dust particles has been identified [19]. We have previously published a relationship between serum immunoglobulin E (IgE) levels and subjective symptoms on Asian dust days with 25 healthy subjects [19]. Significant association was found between IgE levels of microbial allergens and nasal symptom scores. Asian dust events might trigger some kind of hypersensitivity to fungal allergens [19].

5. Conclusions

Although the effects of Asian dust on human health differ between the emission and downwind regions, they are closely connected to each other by human activities leading to desertification. Moreover, industrial development makes the issue more complicated. An implication for public policy in East Asia is that to protect public health, anthropogenic sources of particulate pollution must be controlled more rigorously in areas highly impacted by Asian dust events.

Acknowledgments: This work was supported by JSPS KAKENHI (Grant Number 25220201). We thank James Buxton MSc from Edanz Group (www.edanzediting.com./ac) for editing a draft of this manuscript.

Author Contributions: Shinji Otani conceived, researched, and wrote the paper; Yasunori Kurosaki, Youichi Kurozawa, and Masato Shinoda contributed research framing, ideas, context, and wordsmithing.

Conflicts of Interest: The authors declare no conflict of interest.

References

1. Goudie, A.S. Desert dust and human health disorders. *Environ. Int.* **2014**, *63*, 101–113. [CrossRef] [PubMed]
2. Karanasiou, A.; Moreno, N.; Moreno, T.; Viana, M.; de Leeuw, F.; Querol, X. Health effects from Sahara dust episodes in Europe: Literature review and research gaps. *Environ. Int.* **2012**, *47*, 107–114. [CrossRef] [PubMed]
3. Kurosaki, Y.; Mikami, M. Regional Difference in the Characteristic of Dust Event in East Asia: Relationship among Dust Outbreak, Surface Wind, and Land Surface Condition. *J. Meteorol. Soc. Jpn.* **2005**, *83A*, 1–18. [CrossRef]

4. Dulam, J. Discriminate Analysis for Dust Storm Prediction in the Gobi and Steppe Regions in Mongolia. *Water Air Soil Pollut. Focus* **2005**, *5*, 37–49. [CrossRef]
5. Mu, H.; Battsetseg, B.; Ito, T.Y.; Otani, S.; Onishi, K.; Kurozawa, Y. Health effects of dust storms: Subjective eye and respiratory system symptoms in inhabitants in Mongolia. *J. Environ. Health* **2011**, *73*, 18–20. [PubMed]
6. Mu, H.; Otani, S.; Shinoda, M.; Yokoyama, Y.; Onishi, K.; Hosoda, T.; Okamoto, M.; Kurozawa, Y. Long-term effects of livestock loss caused by dust storm on mongolian inhabitants: A survey 1 year after the dust storm. *Yonago Acta Med.* **2013**, *56*, 39–42. [PubMed]
7. State Implementing Agency of Health, Government of Mongolia. *Health Indicators 2011*; State Implementing Agency of Health: Ulaanbaatar, Mongolia, 2012; p. 96.
8. Onishi, K.; Kurosaki, Y.; Otani, S.; Yoshida, A.; Sugimoto, N.; Kurozawa, Y. Atmospheric transport route determines components of Asian dust and health effects in Japan. *Atmos. Environ.* **2012**, *49*, 94–102. [CrossRef]
9. Maki, T.; Kurosaki, Y.; Onishi, K.; Lee, K.C.; Pointing, S.B.; Jugder, D.; Yamanaka, N.; Hasegawa, H.; Shinoda, M. Variations in the structure of airborne bacterial communities in Tsogt-Ovoo of Gobi desert area during dust events. *Air Qual. Atmos. Health* **2017**, *10*, 249–260. [CrossRef] [PubMed]
10. Maki, T.; Puspitasari, F.; Hara, K.; Yamada, M.; Kobayashi, F.; Hasegawa, H.; Iwasaka, Y. Variations in the structure of airborne bacterial communities in a downwind area during an Asian dust (Kosa) event. *Sci. Total Environ.* **2014**, *488–489*, 75–84. [CrossRef] [PubMed]
11. Shepherd, G.; Terradellas, E.; Baklanov, A.; Kang, U.; Sprigg, K.A.; Nickovic, S.; Boloorani, A.D.; Al-Dousari, A.; Basart, S.; Benedetti, A.; et al. *Global Assessment of Sand and Dust Storms*; United Nations Environment Programme (UNEP): Nairobi, Kenya, 2016; pp. 40–42. ISBN 978-92-807-3551-2.
12. Kanatani, K.T.; Ito, I.; Al-Delaimy, W.K.; Adachi, Y.; Mathews, W.C.; Ramsdell, J.W.; Toyama Asian Desert Dust and Asthma Study Team. Desert dust exposure is associated with increased risk of asthma hospitalization in children. *Am. J. Respir. Crit. Care Med.* **2010**, *182*, 1475–1481. [CrossRef] [PubMed]
13. Watanabe, M.; Yamasaki, A.; Burioka, N.; Kurai, J.; Yoneda, K.; Yoshida, A.; Igishi, T.; Fukuoka, Y.; Nakamoto, M.; Takeuchi, H.; et al. Correlation between Asian dust storms and worsening asthma in Western Japan. *Allergol. Int.* **2011**, *60*, 267–275. [CrossRef] [PubMed]
14. Matsukawa, R.; Michikawa, T.; Ueda, K.; Nitta, H.; Kawasaki, T.; Tashiro, H.; Mohri, M.; Yamamoto, Y. Desert dust is a risk factor for the incidence of acute myocardial infarction in Western Japan. *Circ. Cardiovasc. Qual. Outcomes* **2014**, *7*, 743–748. [CrossRef] [PubMed]
15. Otani, S.; Onishi, K.; Mu, H.; Kurozawa, Y. The effect of Asian dust events on the daily symptoms in Yonago, Japan: A pilot study on healthy subjects. *Arch. Environ. Occup. Health* **2011**, *66*, 43–46. [CrossRef] [PubMed]
16. Otani, S.; Onishi, K.; Mu, H.; Yokoyama, Y.; Hosoda, T.; Okamoto, M.; Kurozawa, Y. The Relationship between Skin Symptoms and Allergic Reactions to Asian Dust. *Int. J. Environ. Res. Public Health* **2012**, *9*, 4606–4614. [CrossRef] [PubMed]
17. Onishi, K.; Otani, S.; Yoshida, A.; Mu, H.; Kurozawa, Y. Adverse health effects of Asian dust particles and heavy metals in Japan. *Asia Pac. J. Public Health* **2015**, *27*, 1719–1726. [CrossRef] [PubMed]
18. Oh, S.Y.; Fong, J.J.; Park, M.S.; Chang, L.; Lim, Y.W. Identifying airborne fungi in Seoul, Korea using metagenomics. *J. Microbiol.* **2014**, *52*, 465–472. [CrossRef] [PubMed]
19. Otani, S.; Onishi, K.; Mu, H.; Hosoda, T.; Kurozawa, Y.; Ikeguchi, M. Associations between subjective symptoms and serum immunoglobulin E levels during Asian dust events. *Int. J. Environ. Res. Public Health* **2014**, *11*, 7636–7641. [CrossRef] [PubMed]

land

MDPI

Article

Bequest of the Norseman—The Potential for Agricultural Intensification and Expansion in Southern Greenland under Climate Change

Chatrina Caviezel, Matthias Hunziker and Nikolaus J. Kuhn *

Department of Environmental Sciences, Physical Geography and Environmental Change, University of Basel, Klingelbergstrasse 27, CH-4056 Basel, Switzerland; c.caviezel@unibas.ch (C.C.); matthias.hunziker@unibas.ch (M.H.)

* Correspondence: nikolaus.kuhn@unibas.ch; Tel.: +41-61-207-0741

Received: 20 October 2017; Accepted: 4 December 2017; Published: 7 December 2017

Abstract: The increase of summer temperatures and a prolonged growing season increase the potential for agricultural land use for subarctic agriculture. Nevertheless, land use at borderline ecotones is influenced by more factors than temperature and the length of the growing season, for example soil quality, as the increasing lengths of dry periods during vegetation season can diminish land use potential. Hence, this study focuses on the quality of the soil resource as possible limiting factor for land use intensification in southern Greenland. Physical and chemical soil properties of cultivated grasslands, reference sites and semi-natural birch and grassland sites were examined to develop a soil quality index and to identify the suitability of soils for a sustainable intensification and expansion of the agriculture. The study revealed that soils in the study area are generally characterized by a low effective cation exchange capacity (CEC_{eff}) (3.7 ± 5.0 meq 100 g^{-1}), low pH $CaCl_2$ (4.6 ± 0.4) and low clay and silt content (3.0 ± 1.0% and 38.2 ± 4.7%, respectively). Due to the high amount of coarse fraction (59.1 ± 5.8%) and the low amount of soil nutrients, an increasing threat of dry spells for soils and yield could be identified. Further, future land use intensification and expansion bears a high risk for concomitant effects, namely further soil acidification, nutrient leaching and soil degradation processes. However, results of the soil quality index also indicate that sites which were already used by the Norseman (980s–1450) show the best suitability for agricultural use. Thus, these areas offer a possibility to expand agricultural land use in southern Greenland.

Keywords: subarctic agriculture; Greenland; soil quality index; farming at its limits; air temperature increase; increase of growing season

1. Introduction

Agriculture in subarctic latitudes is characterized by a cool and short growing season. Thus, the agricultural use consists mostly of cool season forage and vegetable crops, small grains, raising cattle, sheep, goats, pigs and poultry, as well as reindeer herding [1]. Consequently, agriculture in these areas is of a relatively small economic relevance. Nevertheless, climate change is projected to improve the potential of growing forage crops and thus raising animals [1].

Since 1960, the rate of warming in northern latitudes has been two to three times higher than the global average [1] and future climate change is expected to be more pronounced than in other regions [1]. For southern Greenland, climate models predict a temperature increase of 3–6 °C for the next 100 years [1]. The Danish Meteorological Institute (DMI) modeled a lengthening of the growing season of approximately two months until 2100 [2]. Despite the ecological restrictions, Greenlandic agriculture has already experienced an enormous intensification during the last decades, e.g., by tillage of fodder fields and application of industrial fertilizers [3,4]. This trend is expected to grow further over the next century as climate warming is anticipated to increase the length of the growing season and productivity [4].

Southern Greenland offers a historic perspective on land use and climate change. During the medieval climate optimum, the Norse settlers (985–1450) practiced animal husbandry in this region. The settlement by the Norse farmers, however, ended ultimately with their complete disappearance. Literature suggests different reasons for the abandonment of the Norse settlements, for example the worsening of climatic conditions [5] or the overexploitation of land resources as grazing pressure exceeded the resilience of the landscape and pasture economy failed [6,7]. The at least partially environment-induced failure of the Norse highlights the importance of developing sustainable land use today.

"Re-colonialization" of Greenland took place in 1771, when Denmark–Norway sent a missionary expedition to Greenland [8] and permanent settlement and land use in South Greenland was resumed. During the 18th and 19th century, land use was practiced only for self-sufficiency with few domestic animals kept close to settlements [9]. Modern sheep farming in southern Greenland began after 1915 [9] and can be subdivided into two management phases. Sheep were first kept grazing outside for the whole year [10]. After some harsh winters in the 1960s and 1970s, when many sheep starved to death, a new management plan was introduced in 1982, following a pattern of seven months of barn feeding and 5 months grazing on outlying fields [9]. Barn feeding required hay production for winter fodder and the area used for haying increased steadily from 70 ha in 1980 [10] to 780 ha in 2001 to currently 1081 ha [11]. At the same time, the number of sheep owners decreased from 95 in 1982 to 48 in 2008 [12].

Agriculture in Greenland actually encompasses primarily livestock breeding of sheep. According to the Greenlandic Statistics for 2013 [13], 43 farms operated with agriculture as their primary source of income based on sheep farming with roughly 46,000 sheep, about 20,000 ewes and 26,000 lambs, which graze during the summer and are slaughtered in autumn. Grazing land included 242,000 ha, while areas used for hay production accounted for 1081 ha in 2013 [13]. The agricultural sector accounts for only about 0.1% of total export of goods [11]. To augment the agricultural exports, Greenland aims at increasing the number of sheep by fostering hay production and grazing capacity [3]. In addition, the future potential for commercial vegetable production in southern Greenland has recently been celebrated in several popular media and functions as a symbol for climate change. Carrying headlines include "Global Warming a Boon for Greenland's Farmers" [14] or "Farming to the fore as Greenland ice thaws" [15]. However, Greenlandic agriculture is far from self-sufficiency. In 2007, Greenland imported 1000 t sheep fodder, corresponding to 40–50% of the winter fodder [16]. An additional 600 ha of hayfields would be necessary to reach self-sufficiency in feeding the current number of ewes during winter [17]. This corresponds to an increase of 55% of the area that is used nowadays. The Agricultural Consulting Service states that efforts should therefore focus on the expansion of cultivated land in the coming years [17]. Another possibility for increasing the autonomous sheep food supply would be a further raise in the yield per area using fertilizers. According to the Greenlandic government, Greenlandic agriculture already applied 600 to 700 t of industrial fertilizer on an area of about 1000 ha in 2007 [16]. Literature indicates a yield range between 2000 and 2500 FE ha^{-1} (1 FE = nutrition value of 1 kg barley [18]) in the years 2000–2007 [17,19,20]. Comparing the yield in 1994 (1420 FE ha^{-1}) to the yield in 2007 (2100 FE ha^{-1}) [19], the increase can be interpreted as a reaction to the increasing application of fertilizers. However, newer numbers of the Greenlandic Agriculture Commission [18] highlight a significant decrease in yields during the last decade with 2100 FE ha^{-1} in 2007, 1000 FE ha^{-1} in 2010, 1800 FE ha^{-1} in 2011 and 1500 FE ha^{-1} in 2012, respectively. According to the Greenlandic Agriculture Commission [18], the decrease in yield is caused by the increasingly dry summers, especially in the more protected areas of the inner fjords. Thus, the question arises of whether the prospected increase in temperature and growing season has the potential to strengthen the agricultural sector in Greenland.

Westegaard-Nielsen et al. [21] modeled the potential aboveground biomass production for sheep farming in southern Greenland until 2100 based on predicted climate change. Their results indicate an increase of biomass, suitable for increasing livestock breeding. However, the modeled increase in future biomass production is based on using air temperatures only. The authors stress that soil nutrients and plant available water are two main components that need to be included in precise assessments of future plant growth. Thus, the question arises whether the increase of temperature and the prolonged growing season actually contribute to the expansion of farming. Furthermore, climate change does not necessarily imply steadily increasing temperatures and length of growing season. On the contrary, variability in temperature, but also in precipitation patterns, for example an increased length of summer dry spells, have to be considered [2].

Finally, agriculture in the pristine landscape of southern Greenland is not without consequences for ecosystem stability [4]. Unsustainable land use can affect ecosystem functions in several ways, as for example soil degradation by loss of organic matter, salinization, compaction, structural destruction, sealing, contamination or acidification [22]. Consequently, inappropriate land use can compromise the maintenance of further productivity [22]. According to Doran and Parkin [23], soil quality assessment is therefore essential to determine the sustainability of land management systems in the near and distant future. The definition of soil quality and thus the suitability of soils for sustainable agricultural use has long been a challenging issue, since soils are highly variable in properties, characteristics and functions [22,23]. To define the capacity of soil systems for a sustainable intensification and for preventing concomitant effects of agricultural use, indicators for soil quality are needed [24]. Indicators should designate the resilience of soil against physical, chemical and biological disturbances [24] and thus provide information about the ability of the soil to provide essential environmental services [22]. They should be easy to measure and provide evident information on ecosystem functioning and productivity [22]. Thus, soil quality indicators should not only identify the condition of the soil resource, but also define the economic and environmental sustainability of land management practices [25].

The main goal of this study is to answer the question whether the proposed intensification of agricultural use following climate warming in southern Greenland is possible and sustainable considering the soil resources. Specifically, the study will examine the hypothesis whether plant available nutrients and plant available water limit the potential for a sustainable intensification and expansion of agriculture in southern Greenland. To achieve this aim, the study will examine the impact of agriculture on soil and landscape ecology by identifying concomitant effects on the soil resource. A key outcome is the identification of areas which are most suitable for a sustainable land use intensification and expansion based on an integrated soil quality index (SQI).

2. Material and Methods

2.1. Study Area

The study on the potential for land use expansion and intensification in southern Greenland took place in the area of Igaliku (60°59′27″ N, 45°25′9″ W) between the head of Igalikup Kangerlua (Einar's Fjord) and Tunulliarfik fjord (Erik's Fjord) (Figure 1). The study area was chosen due to its long land use history, its good accessibility and due to existing contacts to local farmers. The study area is one of 15 regions in southern Greenland where sheep farming is practiced. Hay making area, as well as sheep number in the study region account to approximately 10% of the total area and sheep number in southern Greenland. Due to its location in the inner fjords, the region is climatically more favorable than other regions; however, precipitation during vegetation period can be scarce compared to other regions [26]. In addition, soils conditions on selected protected areas of the inner fjords show to be more favorable for agricultural use than in the outer fjords and less protected areas [3].

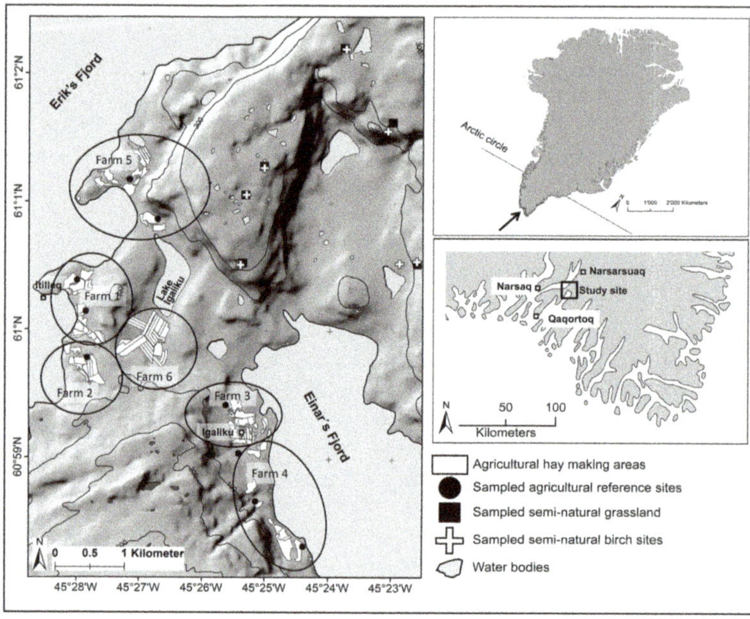

Figure 1. Study area and hay making fields in the area (114 ha), fields were grouped according to the farmer managing the areas and its allocation. Farm 6 (27 ha) did not participate in the study, however it is still shown in the figure as it represents almost a fourth of the total hay making area, and as its vicinity to Lake Igaliku poses a severe risk for the eutrophication of the water body.

Geologically, the Ketilidian basement, which was formed 1850–1725 million years ago, consists of granitic rocks of the Julianehåb batholith, which is partly overlain by Gardar sandstone sediments and lavas with outcrops of basaltic magmas and the intrusion of early dyke swarms [27]. Mean annual precipitation for the climatological standard normal 1961–1990 at closest meteorological station in Narsarsuaq is 615 mm. Mean annual temperature is 0.9°, with means of 10.3 °C in July and −6.8 °C in January [28]. The climate is thus subarctic because the mean temperature of the warmest month is above 10 °C [29]. On average, the number of growing days (≥5 °C) in Narsarsuaq was 138 days during the standard period 1961–1990. The mean number of growing days increased to 146 days for the period 1991 to 2010. The occurrence of dry winds from the Icecap called foehn [26], as well as the high variability in summer precipitation [2,26] can cause prolonged dry periods during vegetation period [2,26]. According to the Greenlandic Agricultural Consulting Service [3], both prolonged dry periods and increasing number of intensive rainfall events have been observed during the last decade.

Vegetation is characterized by local willow and birch copses at lower elevations. On less protected areas dwarf-shrub willow heath (*Salix glauca*) and an understorey of herbs as clustered lady's mantle (*Alchemilla glomerulans*) can be found [29]. Above the altitudinal limit of willow heath, depending more on microclimatic and soil conditions than on defined altitudes, vegetation is characterized by wind exposed, largely open fjell fields. Species include dispersed and low growing patches of lichens, dwarf shrubs like arctic willow (*Salix arctica*) and common juniper (*Juniperus communis*), and grasses, such as wood-rush (*Luzula confusa*) and moss campion (*Silene acaulis*) [29,30]. Natural grassland slopes are dominated by different species (*Anthoxanthum odoratum*, *Deschampsia flexuosa* and subordinated *Alchemilla alpina*), moist fens with sedges and different species of peat mosses (*Carex rariflora*, *Eriophorum scheuchzeri*, and *Sphagnum* sp.) are frequent but cover limited area due to topography [29,30]. Furthermore, cultivated grassland can be found. According to Daniëls [31], vegetation of cultivated hay fields and pastures in southern Greenland most likely belongs to the class of Molinio-Arrhenatheretea, a.o. red fescue (*Festuca rubra*), common sorrel (*Rumex acetosa*), dandelion (*Taraxacum officinale*) and meadow buttercup (*Ranunculus acris*). The weed vegetation of arable fields around Igaliku consists mainly of seedmixes with high amount of timothy grass (*Phleum pratense*). Daniëls [31] concludes, that vegetation communities on farmed land in southern Greenland mainly correspond to the typical communities for the European cultural landscapes.

Soils in southern Greenland are generally described as acidic with low amount of nutrients and limited capacity to store water [3]. However, on protected areas in the inner fjords, brown earth can be found [3]. According to Feilberg [29], generally Brown earth and Podzols occur in densely vegetated areas, while shallow Syrosems and Rankers developed where vegetation is scattered or constitutes just a continuous cover.

Igaliku has long been recognized as the historic site of Gardar, the seat of the bishop of the Norse settlements in Greenland during the Middle Ages (986 to 1450) and has already experienced intensive agricultural use during this period [4,32,33]. The current settlement of Igaliku has been established in 1782 by the Norwegian colonist Anders Olsen, whose family subsisted on few domestic cattle and sheep [32]. Population ranged between 35 and 50 inhabitants during the last 20 years [34]. In 2011, 2100 sheep were kept at five farms in and around Igaliku [34]. For winter fodder production, a total area of about 115 ha was managed by the five farmers [3]. The areas used for hay production are common land, but assigned for management to individual farmers. They are located at a relatively low elevation (max. 75 m a.s.l., mean 28 m a.s.l.) between hills reaching 300 m a.s.l. and surrounded by mountains with elevation of up to 1800 m a.s.l.

2.2. Sampling Setup

To assess the potential of agricultural intensification and expansion, as well as the impact of land use, chemical and physical soil parameters of agricultural fields around Igaliku, were sampled. In total, 87 ha of agricultural land, farmed by four participating farmers, was studied (Figure 1). In 2013, the participating famers kept 1400 mother sheep, which are fed through the winter. The fields of Farms 3 and 4 are managed by the same farmer, but the fields of Farm 4 are allocated more south. For each Farm, 1–2 reference sites (9 in total) were sampled next to a representative agricultural field (in the following called agricultural reference sites). Furthermore, seven birch areas and six grassland areas in the northeast of the study site were sampled as additional reference sites (Figure 1). These areas represent the natural vegetation associations at the subarctic ecotone border [29]. However, sites were characterized as semi-natural as there is evidence for pastoral activities. The characteristics of the sampling sites are summarized in Table 1.

Table 1. Overview of the sampled areas and its site characteristics.

Site	Sampling Site Characteristics		Size of Sampling Area
Farm 1 n = 10	Mean elevation (m a.s.l.)	21 ± 14 Stdv.; Range 38	Parcel size
	Mean slope (°)	6 ± 3 Stdv.; Range 9	
	Aspect	N: 0, NE: 0, E: 0; SE: 0, S: 0, SW: 2, W: 6, NW: 2	
	Slope position	CR: 0, UP: 0, MS: 6, LS: 4, TS: 0, BO: 0	
	Slope form	SS: 8, SV: 0, SC: 1, VS: 1, VV: 0, VC: 0, CS: 0, CV: 0, CC: 0	
	Soil texture	Sandy loam: 10, Loamy sand: 0, Sand: 0	
Farm 2 n = 11	Mean elevation (m a.s.l.)	53 ± 15 Stdv.; Range 40	Parcel size
	Mean slope (°)	7 ± 3 Stdv.; Range 9	
	Aspect	N: 2, NE: 0, E: 0; SE: 0, S: 0, SW: 0, W: 7, NW: 2	
	Slope position	CR: 0, UP: 0, MS: 3, LS: 5, TS: 3, BO: 0	
	Slope form	SS: 6, SV: 0, SC: 2, VS: 0, VV: 0, VC: 0, CS: 3, CV: 0, CC: 0	
	Soil texture	Sandy loam: 11, Loamy sand: 0, Sand: 0	
Farm 3 n = 8	Mean elevation (m a.s.l.)	22 ± 9 Stdv.; Range 24	Parcel size
	Mean slope (°)	3 ± 2 Stdv.; Range 4	
	Aspect	N: 0, NE: 0, E: 7; SE: 0, S: 1, SW: 0, W: 0, NW: 0	
	Slope position	CR: 0, UP: 0, MS: 4, LS: 4, TS: 0, BO: 0	
	Slope form	SS: 5, SV: 0, SC: 2, VS: 0, VV: 1, VC: 0, CS: 0, CV: 0, CC: 0	
	Soil texture	Sandy loam: 8, Loamy sand: 0, Sand: 0	

Table 1. *Cont.*

Site	Sampling Site Characteristics		Size of Sampling Area
Farm 4 $n = 5$	Mean elevation (m a.s.l.) Mean slope (°) Aspect Slope position Slope form Soil texture	34 ± 12 Stdv.; Range 24 5 ± 2 Stdv.; Range 4 N: 1, NE: 0, E: 4; SE: 0, S: 0, SW: 0, W: 0, NW: 0 CR: 0, UP: 0, MS: 5, LS: 0, TS: 0, BO: 0 SS: 2, SV: 0, SC: 0, VS: 1, VV: 0, VC: 1, CS: 0, CV: 0, CC: 1 Sandy loam: 4, Loamy sand: 1, Sand: 0	Parcel size
Farm 5 $n = 10$	Mean elevation (m a.s.l.) Mean slope (°) Aspect Slope position Slope form Soil texture	28 ± 14 Stdv.; Range 45 6 ± 3 Stdv.; Range 10 N: 1, NE: 0, E: 0; SE: 0, S: 0, SW: 5, W: 1, NW: 3 CR: 0, UP: 0, MS: 5, LS: 2, TS: 0, BO: 3 SS: 3, SV: 0, SC: 2, VS: 1, VV: 1, VC: 0, CS: 2, CV: 0, CC: 1 Sandy loam: 10, Loamy sand: 0, Sand: 0	Parcel size
Agricultural reference sites $n = 9$	Mean elevation (m a.s.l.) Mean slope (°) Aspect Slope position Slope form Soil texture	33 ± 12 Stdv.; Range 34 7 ± 3 Stdv.; Range 9 N: 0, NE: 1, E: 3; SE: 0, S: 3, SW: 0, W: 2, NW: 0 CR: 0, UP: 0, MS: 9, LS: 0, TS: 0, BO: 0 SS: 5, SV: 1, SC: 2, VS: 1, VV: 0, VC: 0, CS: 0, CV: 0, CC: 0 Sandy loam: 7, Loamy sand: 2, Sand: 0	10 m × 10 m
Semi-natural birch sites $n = 7$	Mean elevation (m a.s.l.) Mean slope (°) Aspect Slope position Slope form Soil texture	199 ± 48 Stdv.; Range 180 12 ± 5 Stdv.; Range 16 N: 0, NE: 0, E: 0; SE: 3, S: 1, SW: 2, W: 1, NW: 0 CR: 0, UP: 0, MS: 4, LS: 2, TS: 1, BO: 0 SS: 5, SV: 0, SC: 1, VS: 0, VV: 0, VC: 0, CS: 1, CV: 0, CC: 0 Sandy loam: 3, Loamy sand: 4, Sand: 0	10 m × 10 m
Semi-natural grassland sites $n = 6$	Mean elevation (m a.s.l.) Mean slope (°) Aspect Slope position Slope form Soil texture	221 ± 48 Stdv.; Range 136 14 ± 6 Stdv.; Range 13 N: 0, NE: 0, E: 0; SE: 2, S: 3, SW: 0, W: 1, NW: 0 CR: 0, UP: 0, MS: 5, LS: 1, TS: 0, BO: 0 SS: 42, SV: 0, SC: 1, VS: 1, VV: 0, VC: 0, CS: 0, CV: 0, CC: 0 Sandy loam: 2, Loamy sand: 2, Sand: 1, Missing: 1	10 m × 10 m

Abbreviations. For slope positions: CR, Crest (summit); UP, upper slope (shoulder); MS, Middle slope (back slope); LS, Lower slope (foot slope); TS, Toe slope; Bo, bottom. Abbreviations for slope forms of plan and profile curvature: SS, Straight straight; SV, Straight convex; SC, straight concave; VS, Convex straight; VV, Convex convex; VC, convex concave; CS, Concave straight; CV, Concave convex; CC, Concave concave [35].

2.3. Soil Sampling and Analysis

Sampling on the agricultural fields was performed on the resolution of field size according to the map of agricultural fields provided by the Agricultural Consulting Service in Qaqortoq (Figure 1). The fields were selected to ensure that all variabilities in topographic parameters and crops were represented. Sampling took place within or shortly after the yearly harvest at the end of July 2014. Soil samples were taken on 44 agricultural fields as composite sample of 15 to 25 cores, which were randomly taken all across the field by a hand auger reaching rooting depths of 30 cm. On reference sites, soils were predominantly very shallow and reached only around 30 cm. Same sampling depth for all the soils was chosen to compare the results. Additionally, three soil samples were taken by a metal cylindrical core of 100 cm^3 volume and 5 cm length at 12–17 cm on each field to determine soil bulk density and coarse fraction. Reference sites were sampled likewise within a 10 m × 10 m area (see Table 1).

Soil analysis results refer to concentration per area using the volumetric basis calculated by bulk density of the fine earth fraction <2 mm and investigated soil depth. Precise information on the soil processing and analysis, as well as the reasons for parameter selection is given in Table 2. To extrapolate the measurements of the 44 agricultural fields to the surrounding fields, the topographic parameters (angle, orientation, position and form of the slope), parent material and crop were mapped according to the FAO Guidelines for soil description [35] for all sampling sites and on further 76 agricultural fields. Soil quality of the agricultural fields was evaluated considering single indicating parameters as well as using an overall soil quality index (SQI).

Table 2. Measured soil properties, its significance as soil quality indicator and applied analysis methods.

Parameter	Significance As Index for Soil Fertility	Analytical Method for Data Determination
C_{org} (%)	Index for soil fertility and soil structure. Increases the CEC and the water-holding capacity [36], thus prevents nutrient leaching.	Sieved (<2 mm), ball-milled, total carbon content was determined by the dry combustion technique Leco CN 628 Elemental Determinator, (C_{tot} was set as C_{org} as there was no evidence for inorganic carbon)
N_{tot} (%)	Initial source for N mineralization—>N_{min}-N (NO^{3-}-N + NH^{4+}-N)	Sieved (<2 mm), ball-milled, nitrogen content was determined by the dry combustion technique by Leco CN 628 Elemental Determinator
Macronutrients K, P, Mg, Ca (kg ha^{-1})	Index for the capacity to support plant growth, environmental quality indicator [36].	CO_2 extraction 1:2.5 soil solution FAL CO_2-EX [37], ionchromatography, IC 761
N_{min} (NO_3-N + $NH_4{}^+$-N) (kg ha^{-1})	Index plant available soil inorganic nitrogen, leaching potential, mineralization [36].	Sieved (<2 mm), cooled until soil extraction; extraction with 0.01M Calciumchlorid (1:4); FAL NM-Ex [37], soil nitrate and ammonium determination with QUANTOFIX® rapid test (Machery & Nagel)
pH ($CaCl_2$) (-)	Index for soil acidity/alcalinity. Reduced availability of nutrients in acidic soils [36]. Soil acidity influences soil microbes, reduces bacterial activity and thus decomposition [38].	1:2.5 0.01 M $CaCl_2$ suspension
Cation exchange capacity (CEC) (meq 100 g^{-1})	Index for soil capacity to supply exchangeable nutrient cations Mg_{ex}, K_{ex}, Ca_{ex} to the soil solution for plant uptake [39].	For soils with pH (H_2O) ≤ 5.9; FAL KUK1-Ex with HCl + H_2SO_4; KUK1-H; for soils with pH (H_2O) > 5.9; KUK2-EX with Bariumchlorid KUK2-H [37]; Measured on CIROS VISION – ICP-OES Spectrometer, $CEC_{pot} = (Ca^{2+}) + (Mg^{2+}) + (K^+) + (Na^+) + (H^+)$, $CEC_{eff} = (Ca^{2+}) + (Mg^{2+}) + (K^+) + (Na^+)$
Exchangable sodium (Na_{ex}) percentage (ESP) (%)	Addresses potential hazards posed by sodium soils, as for example a decrease in permeability, soil dispersion, poor water infiltration, and possible sodium toxicity to plants [40].	$ESP = [(Na^+)/CEC_{pot}] \times 100$
Base saturation (BS) (%)	Indicates whether a soil should be limed or not [41].	$BS = [CEC_{eff}/CEC_{pot}] \times 100$
C/N (-)	Indicates mineralization/immobilization of soil nitrogen [42].	Sieved to 2 mm, ball-milled, the organic carbon content was determined by the dry combustion technique by Leco CN 628 Elemental Determinator; Ratio C/N
Bulk density (BD) < 2 mm (g cm^{-3})	Index for plant rooting, porosity and soil organisms [43].	Undisturbed soil cores 100 cm^3, air dried, sieved to 2 mm; BD = weight dried fine earth/cylinder volume Volume > 2 mm was measured by water replacement of the material > 2 mm, volume < 2 mm was calculated by total volume of the cylinder minus the volume > 2 mm
CF (Vol%)	Index for water and nutrient retention potential [40].	CF = Vol > 2mm/(Vol > 2 mm + Vol < 2 mm)
Texture (%)	Index for retention and transport of water and chemicals [36].	Treatment with H_2O_2 (15%) to exclude organic material, Mastersizer 2000 (Malvern, Germany)

2.4. Soil Quality Index (SQI)

To identify the most significant variables to calculate the soil quality index (SQI) that best represents the soil functions associated with sustainable land use intensification, a minimum dataset (MDS) of parameters is needed to avoid redundancies of the parameters [44]. According to a review by Zornoza et al. [22], the most common parameters used to calculate an index for soil quality were pH, electric conductivity (EC), soil organic carbon (C_{org}), total nitrogen (N_{tot}) and available P. Other indicators such as nitrate (NO_3^--N), ammonium (NH_4^+-N), exchangeable cations (Na_{ex}, K_{ex}, Ca_{ex}, Mg_{ex}), bulk density (BD), texture and available water content have, according to Zornoza et al. [22], also been used by various authors. The MDS used to calculate the SQI in this study was selected using principal component analysis (PCA). PCA is a way to reduce the dimensionality of a data-set while limiting the loss of information [44]. The PCA for the presented study was performed on the analyzed 16 chemical and physical soil parameters, using SPSS statistic package (IBM 24.0). The mineralized Nitrogen (N_{min}-N) was not included, because values of N_{min}-N cannot be used as soil quality index for crop growth for the following growing period. The reduction of the number of components was obtained using the eigenvalue-one criterion; only PCA's with an eigenvalue >1 [45], that explained at least 5% of the total variance and had loadings <0.4 [46] were selected. The PCA output is summarized in Table 3. To avoid inter-related variables among the indicators in one PCA, that could have remained within the variables explaining the 5% in the total variance, correlation among the indicators was examined [47]. Variables were determined as redundant if the correlation coefficients in between the variable of the PCA were higher than $R^2 = 0.65$; given the case, the higher ranked variable was kept in the MDS.

Table 3. Output of principal component analysis (PCA) of the soil quality indicators and their classification into functional groups. Parameters are ordered according to their loading. Bold component-loadings correspond to the indicators included in the MDS.

	PC 1	PC 2	PC 3	PC 4	Functional Group
Eigenvalue	7.549	2.543	1.423	1.166	
Percent	47.18	15.894	8.893	7.285	
Cumulative percent	47.18	63.073	71.966	79.251	
ESP (%)	**0.933**	0.166	0.009	0.138	The less the better [48]
Na_{ex} (meq 100 g^{-1})	0.852	0.344	0.093	0.157	
C:N ratio (-)	−0.763	−0.224	−0.245	0.039	
BS (%)	0.665	0.229	0.651	0.059	
Mg_{ex} (meq 100 g^{-1})	**0.609**	0.483	0.243	−0.115	The more the better [43]
C_{org} (%)	0.343	**0.799**	0.179	0.317	The more the better [43]
N_{tot} (%)	0.477	0.755	0.219	0.246	
P (mg 100 g^{-1})	0.343	**0.739**	−0.032	−0.133	Optimal value [49]
BD > 2mm (g cm^{-3})	−0.088	**−0.711**	−0.133	0.37	The less the better [43]
K_{ex} (meq 100 g^{-1})	0.45	**0.561**	0.33	−0.051	The more the better [43]
Ca_{ex} (meq 100 g^{-1})	0.194	0.429	**0.837**	0.097	The more the better [43]
pH ($CaCl_2$)	0.036	−0.149	**0.82**	−0.414	Optimal value [50]
CEC_{eff} (meq 100 g^{-1})	0.533	0.528	0.644	0.062	
Sand (%)	0.064	−0.298	0.074	**−0.897**	The less the better [38]
Clay (%)	0.237	−0.074	0.015	0.844	
CF (Vol-%)	0.005	0.277	0.078	**−0.53**	The less the better [38]

The SQI is used for synthesizing the information of the measured indicators based on scoring different soil indicators to estimate the agricultural suitability of the sampled area. Measured values are transformed into unit-less values, so that scores may be combined to form a single value. Thus, the selected parameters were subdivided into different groups to assign different functions representing the best soil functionality of the parameter (Table 3). The groups are: (i) the more the better; (ii) the less the better; and (iii) optimal value. The data for the chosen parameters were ranked by linear scoring technique [43,51]. All the values of one variable were ranked with a score ranging from 0 to 1 applying the three functions. If the best soil functionality was associated with high values, highest score was given for the highest value, if best soil functionality was associated with low values, the highest score was given for the lowest value, remaining values were scored based on their percentage of the highest value. If the best soil functionality was associated with an optimal value highest score was given for the value closest to the optimal value. After ranking, SQI was calculated by summing scores of each indicator and dividing the result by the total number of indicators [43,52]. Ranking of available P was performed using the data not based on volumetric recalculation, but on measured concentrations to apply the classification system developed by Agroscope for the Swiss agricultural research agency [49], which also takes into account humus and clay content of the sample. The nutrients classification of Agroscope was chosen because it is also applicable to alpine soils, whose development is comparable to southern Greenlandic soils. As thresholds depend on geographic regions, the soil quality index can only be used locally to compare the quality differences within the fields of different farmers, differences to agricultural reference sites, semi-natural birch areas and grasslands. Visual classification of the scores was performed by geometrical intervals (ArcGIS 10.2).

3. Results and Discussion

3.1. Suitability of the Sampled Fields for Agricultural Use

The major concern for the intensification and expansion of agricultural use can be identified in the low nutrient status of the soils (Table 4). The sandy texture (Table 5) and the acidic parent material of the soils (Table 4) limit nutrient retention, thus soils cannot buffer natural and management induced pH changes, which leads to further nutrient leaching and a further decrease of the pH-value. A low pH affects plant availability of macronutrients, especially P, Ca and Mg, but also K and N [50]. The process of nutrient leaching due to acid soil conditions is also indicated by the significant lower Ca and Mg levels on agricultural sites compared to the semi-natural reference sites (Table 4). The unsuitable nutrient status is also shown by the low effective cation exchange capacity (CEC_{eff}) and the low base saturation (BS) (Table 4), both indicating a low nutrient saturation. The analysis of CEC_{eff} revealed very low values on Farms 1, 2 and 5, as well as on the semi-natural birch and grass sites. Farms 3 and 4 showed a median CEC_{eff} of 12.4% and 6.5%, respectively. However, on Farm 4, the exchangeable sodium percentage (ESP), expressed in Na_{ex}, contributed with median 11.5% to the CEC_{eff}, while on Farm 3, 6.2% of the CEC_{eff} derived from Na-cations. According to Scheffer et al. [39], the percentage of Na_{ex} on the CEC_{eff} on slightly acidic soils in humid climate normally accounts to only <2%. Thus, the measured Na_{ex} on Farms 3 and 4 show overproportional percentages on CEC_{eff}. However, excluding Na_{ex}-cations for the CEC_{eff} calculation, the CEC_{eff} was still significantly higher on Farms 3 and 4 than on the remaining farms.

According to the nutrient classification of Swiss Agroscope [49], the amount of P (Table 4) is particularly poor for most of the sampled sites, illustrating that P represents a limiting factor. K levels (Table 4) showed generally a higher nutrient status, and all agricultural fields were classified as "sufficient", "stocked" and "enriched", except for Farm 5 where only 20% of the fields showed "sufficient" K concentrations. No significant K differences were found between the reference and farmed sites.

Table 4. Chemical parameters. Parameters marked with (*) show normally distributed data, thus mean and standard deviation is given, significances were calculated with *t*-test. Parameters marked with (**) show not normally distributed data, thus median and interquartile range is given, significances were calculated with Mann—Whitney rank sum test. Ag. Ref. = Agricultural reference sites; Birch = Semi-natural birch sites; Grass = Semi-natural grassland sites.

Site	$pH(CaCl_2)$ * (-)	P ** (kg ha^{-1})	K ** (kg ha^{-1})	Ca ** (kg ha^{-1})	Mg ** (kg ha^{-1})	CEC_{eff} ** (meq 100 g^{-1})	ESP ** (%)	BS ** (%)	N_{min}-N ** (kg ha^{-1})	C_{org} Stock ** (tha^{-1})	N_{tot} Stock ** (tha^{-1})	C:N Ratio * (-)
Farm 1	4.4 ± 0.2^{abc}	2.1 ± 2.1^{bc}	107.0 ± 29.7	128.7 ± 49.9^{bc}	24.4 ± 7.7^{bc}	2.6 ± 1.1^{ad}	0.8 ± 0.3^{ab}	16 ± 7^{ad}	25.1 ± 34.6	116 ± 23^{b}	6.8 ± 1.1^{ab}	16.3 ± 1.4^{ad}
Ag. Ref 1	4.7 ± 0.0^{bc}	0.0^{f}	58.5 ± 13.1^{b}	50.7 ± 61.1^{b}	32.4 ± 24.5	2.4 ± 1.4	1.2 ± 0.5	18 ± 9	$0.0\ b^{e}$	92 ± 41	5.5 ± 2.8	16.9 ± 1.4^{bg}
Farm 2	4.3 ± 0.3^{abcd}	4.7 ± 2.8^{bc}	48.5 ± 22.8^{abcd}	75.0 ± 64.1^{bc}	22.8 ± 10.4^{bc}	2.8 ± 2.3	0.8 ± 0.4^{abd}	18 ± 8	2.4 ± 1.7^{bcd}	105 ± 20^{ab}	6.7 ± 1.7^{ab}	15.6 ± 2.0^{ad}
Ag. Ref 2	4.4	0.0	84.0	57.7	28.9	2.4	1.1	14	3.0	140	8.2	17.1^{g}
Farm 3	4.7 ± 0.5	88.1 ± 54.7^{abcd}	156.2 ± 224.0	113.0 ± 224.0^{c}	40.8 ± 153.1^{d}	12.4 ± 4.1^{abcd}	6.2 ± 4.6^{bcd}	38 ± 11^{bcd}	16.5 ± 9.5^{ad}	183 ± 87^{bc}	15.4 ± 6.8^{bcd}	12.5 ± 0.9^{bcd}
Ag. Ref 3	5.1	13.1^{b}	96.2	38.2	12.2	10.1	10.3^{b}	36	7.7	133	10.7	12.5^{c}
Farm 4	4.8 ± 0.3^{d}	5.9 ± 27.3^{bc}	91.2 ± 50.4	19.8 ± 32.1^{bcd}	16.6 ± 36.0	6.5 ± 6.1^{bcd}	11.5 ± 2.3^{abcd}	40 ± 17^{bcd}	5.4 ± 1.8^{bc}	93 ± 48	7.5 ± 4.4^{b}	11.6 ± 0.6^{abcd}
Ag. Ref 4	5.1 ± 0.2^{e}	3.9 ± 9.5	123.8 ± 56.2	24.0 ± 100.4^{b}	32.8 ± 59.6	8.7 ± 9.0^{b}	9.3 ± 7.4^{b}	39 ± 20^{b}	14.2 ± 13.5	167 ± 162^{b}	16.2 ± 14.5^{be}	11.0 ± 0.6^{bceg}
Farm 5	4.8 ± 0.2^{bc}	0 ± 0	80.8 ± 28.1	163.7 ± 152.1	31.3 ± 12.5^{bc}	3.5 ± 0.9	1.0 ± 0.1^{b}	22 ± 14	17.6 ± 10.3	112 ± 37^{b}	6.6 ± 2.5^{b}	15.6 ± 1.2^{d}
Ag. Ref 5	5.1 ± 0.3	1.3 ± 2.6	95.4 ± 72.1	577.6 ± 369.8^{e}	84.8 ± 64.1	$10.6 \pm 1.2b^{f}$	0.5 ± 1	37.5 ± 17^{f}	12.0 ± 12.8	129 ± 37^{b}	9.7 ± 2.1^{b}	13.3^{bcf}
Farms 1–5	4.6 ± 0.4^{abc}	3.6 ± 11.1^{bc}	86.8 ± 54.8	103.9 ± 112.7^{bc}	25.9 ± 17.5^{bc}	3.7 ± 5.0	1.0 ± 1.3^{b}	22 ± 17	6.0 ± 19.1^{b}	112 ± 35^{bc}	7.0 ± 2.1^{bc}	14.7 ± 2.2^{b}
Ag. Ref. 1–5	4.9 ± 0.3^{e}	2.0 ± 3.9^{c}	84.0 ± 58.8	57.7 ± 98.3^{b}	32.8 ± 32.6	8.2 ± 6.9	1.4 ± 8.3^{b}	34 ± 17^{b}	5.6 ± 11.2^{bc}	133 ± 35^{b}	9.9 ± 2.6^{b}	13.7 ± 2.7^{bc}
Birch	5.2 ± 0.4^{e}	0.0 ± 0.0^{e}	88.9 ± 29.5	273.7 ± 295.0^{e}	66.6 ± 33.0^{e}	3.4 ± 2.5	0 ± 0^{ae}	14 ± 13^{a}	27.3 ± 19.5^{ae}	74 ± 19^{ae}	4.6 ± 1.8^{ae}	16.4 ± 1.3^{ae}
Grass	5.1 ± 0.4^{e}	0.0 ± 0.0^{ae}	103.6 ± 33.1	326.5 ± 177.1^{e}	63.8 ± 17.0^{e}	2.9 ± 1.8	0 ± 1.9^{e}	17 ± 6	21.8 ± 16.9^{a}	67 ± 69	4.2 ± 3.4	16.2 ± 0.9^{a}

[a] Sign. differences ($p < 0.05$) to agricultural reference sites (1–5); [b] sign. differences ($p < 0.05$) to birch sites; [c] sign. differences ($p < 0.05$) to grassland sites; [d] sign. differences ($p < 0.05$) to the remaining agricultural sites; [e] sign. differences ($p < 0.05$) to all agricultural sites (Farms 1–5); [f] sign. differences ($p < 0.05$) to the corresponding farm; [g] sign. differences ($p < 0.05$) to the remaining agricultural reference sites.

Table 5. Physical soil parameters. Parameters marked with (*) show normally distributed data, thus mean and standard deviation is given, significances were calculated with *t*-test. Parameters marked with (**) show not normally distributed data, thus median and interquartile range is given, significances were calculated with Mann–Whitney rank sum test. Ag. Ref = Agricultural reference sites; Birch = Semi-natural birch sites; Grass = Semi-natural grassland sites.

Site	BD < 2 mm * (gcm^{-3})	CF ** (Vol%)	Clay ** (%)	Silt ** (%)	Sand ** (%)
Farm 1	1.06 ± 0.11^{bd}	5.5 ± 5.3	3.2 ± 0.8^{b}	38.8 ± 3.8^{c}	58.2 ± 4.0^{c}
Ag. Ref 1	1.16 ± 0.08^{b}	2.5 ± 1.0	2.7 ± 0.7	36.8 ± 13.2	60.5 ± 13.9
Farm 2	0.98 ± 0.13^{b}	5.0 ± 2.7	2.3 ± 0.7^{b}	36.4 ± 3.3^{c}	60.3 ± 3.9^{c}
Ag. Ref 2	1.13^{b}	4.3	2.8	41.0	56.2
Farm 3	0.68 ± 0.17^{acd}	8.0 ± 4.5^{d}	2.7 ± 0.5^{b}	36.7 ± 7.3^{c}	60.7 ± 8.5
Ag. Ref 3	0.87	4.8	2.7	39.9	57.4
Farm 4	0.93 ± 0.11	3.9 ± 1.8	2.1 ± 2.6	33.0 ± 2.6^{d}	64.8 ± 3.8
Ag. Ref 4	0.97 ± 0.11^{b}	7.0 ± 6.6	3.2 ± 1.6	45.4 ± 22.7	51.4 ± 24.3
Farm 5	1.10 ± 0.19^{bd}	4.7 ± 5.0	3.6 ± 0.7^{bd}	38.5 ± 3.6^{bc}	57.6 ± 3.5^{bc}
Ag. Ref 5	0.96 ± 0.10	6.5 ± 10.3	3.2 ± 0.1	40.6 ± 12.0	56.2 ± 12.0
Farms 1–5	0.96 ± 0.21^{b}	5.2 ± 4.8	3.0 ± 1.0^{b}	38.2 ± 4.7^{bc}	59.1 ± 5.8^{bc}
Ag. Ref. 1–5	1.02 ± 0.13^{b}	4.3 ± 5.0	3.0 ± 0.5^{b}	41.0 ± 10.7^{c}	51.4 ± 10.8^{bc}
Birch	0.76 ± 0.13^{ace}	3.7 ± 8.0	0.9 ± 1.7^{ae}	28.9 ± 17.2^{e}	59.9 ± 19.8^{ae}
Grass	1.06 ± 0.16^{b}	10.8 ± 8.7	0.7 ± 2.7^{d}	22.2 ± 16.3^{ae}	63.3 ± 18.9^{ae}

[a] sign. differences ($p < 0.05$) to agricultural reference sites (1–5); [b] sign. differences ($p < 0.05$) to birch sites; [c] sign. differences ($p < 0.05$) to grassland sites; [d] sign. differences ($p < 0.05$) to the remaining agricultural sites; [e] sign. differences ($p < 0.05$) to all agricultural sites (Farms 1–5); [f] sign. differences ($p < 0.05$) to the corresponding farm; [g] sign. differences ($p < 0.05$) to the remaining agricultural reference sites.

Comparing the autumn nutrient status (P_2O_5, K_2O) to the economically optimal nutrient range for seeded grasslands, i.e., yield compared to the need of fertilizer application [53], we can conclude that only the fields of Farm 3 have an economically feasible nutrient condition (Figure 2).

A possibility to adapt land use to the unsuitable soil conditions could also be reached by cultivating plant species which are more adapted to the soil conditions. The Agricultural Consulting Service is currently running experiments with various perennial types of grasses for the production of hay and silage [3].

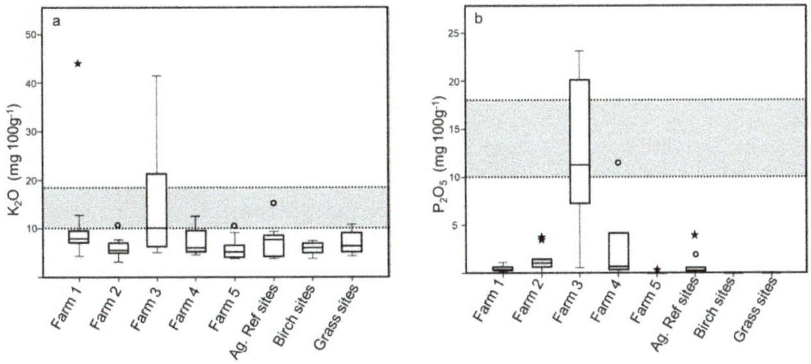

Figure 2. (**a**) After harvest values for the nutrients K_2O; and (**b**) P_2O_5 compared to economically optimal nutrient values for seeded grasslands [53]. Circles represent outliers that have values greater than 1.5 times the interquartile range, stars represent extreme outliers that have values greater than 3 times the interquartile range. Ag. Ref. sites = Agricultural reference sites; Birch sites = Semi-natural birch sites; Grass sites = Semi-natural grassland sites.

The second major concern for land use intensification is related to climate variability and can be found in reduced plant available water. Due to mostly sandy texture (Table 5) of the soils, with insignificant amounts of clay (Table 5), and due to the high coarse fraction in some soils (Table 5), the water holding capacity is generally low. Archeological evidence of irrigation channels in Igaliku illustrates that the Norseman faced the challenge of limited water availability already [32,54]. The risk of dry soils for agricultural production is also mentioned by Christensen [55] who also pointed out that agricultural use should concentrate on moist depression zones not prone to drought during periods without rain. According to the Agricultural Consulting Service [3], the high variability of annual weather conditions, including the lack of rainfall during the growing season, represents a major concern for expanding agriculture. Water availability during growing season (May–September) is already very limited because only approximately half of the mean annual rainfall of 615 mm (1961–1990) at Narsarsuaq station falls during the growing season [28]. According to Hansen [26], precipitation in the inner fjords can be as low as 110 mm during the growing season in dry years. Katabatic dry winds emanating from the ice cap are a characteristic of the region and can persist for many days [56]. The dry winds result in an evapotranspiration of up to 16 mm day^{-1} and can lead to drought stress for the vegetation [26]. Adderley and Simpson [56] modeled soil chemical, physical and soil-water hydraulic properties for two study sites near the eastern settlement, in Qassiarsuk and Tasersuaq about 20 km northwest of Igaliku and concluded that there is a frequent requirement for irrigation, both currently and during the Norse settlement period. Recent press communications underline the threat of droughts. In 2008, for example, the 50% reduction in yield of hay and silage for sheep husbandry caused by the drought in southern Greenland was seen as a national problem [57], severe yield reduction due to dry summers continued in the years 2010, 2011, 2012 and 2015 [18,58,59]. The threat of dry growing seasons for fodder availability can even be more pronounced in the future as the Danish Meteorological Institute (DMI) modeled an increase in the lengths of dry periods (consecutive days with less than 1 mm precipitation) during the vegetation period [2]. According to the DMI, dry periods will last up to 10 days longer until 2081–2100 compared to the period 1991–2010 with its longest dry period of 20 days [2]. Another risk related to climate variability could be found in the high discrepancies of summer precipitation in southern Greenland from year to year but also within one summer [2]. As not only the length of dry periods but also the total amount of summer precipitation is expected to increase for 10% until 2081–2100 compared to the period 1991–2010 [2], this would infer heavier rainfall events. Considering the poor nutrient retention potential of the soils, such events would impose a high risk of nutrient leaching.

3.2. Impact of the Agricultural Use on Soil and Landscape Ecology

The comparison between farmed areas and reference sites already revealed that agriculturally-used areas show a significantly lower pH ($CaCl_2$) compared to the reference sites (Table 4). Even though comparability of the values on farm level is limited, due to the low amount of reference sites for the single farms, a trend in the decrease of pH of the already acidic soils is still visible. The efficiency of applied fertilizers decreases with decreasing pH [60], highlighting another limitation of the soils for farming. The low pH ($CaCl_2$) on the studied agricultural fields indicates high nutrient leaching, and thus waste of approximately 50 to 70% of the applied fertilizer [60]. The N_{min}-N values after harvesting (Table 4) confirm the risk of nitrate leaching. Due to percolating water during autumn rainfalls, autumn N_{min}-N soil values typically leach up to 100% on sandy soils and 30–70% on more cohesive soils [61]. Thus, N_{min}-N after harvesting should not exceed 30 kg ha^{-1} [62] to prevent water bodies from eutrophication. Information on the amount of fertilizer added on the fields was not available, however the high N_{min}-N values after harvesting, especially on Farms 1 and 5, indicate an inappropriate use of fertilizers (Figure 3). Confirming the leaching, a nutrient enrichment in the nearby Lake Igaliku (see Figure 1) has recently been identified by Belle et al. [63]. The study concludes that Lake Igaliku is in a worrying ecological state because of the disturbed nitrogen cycle, a loss of littoral biodiversity, and alterations of its trophic functioning [63].

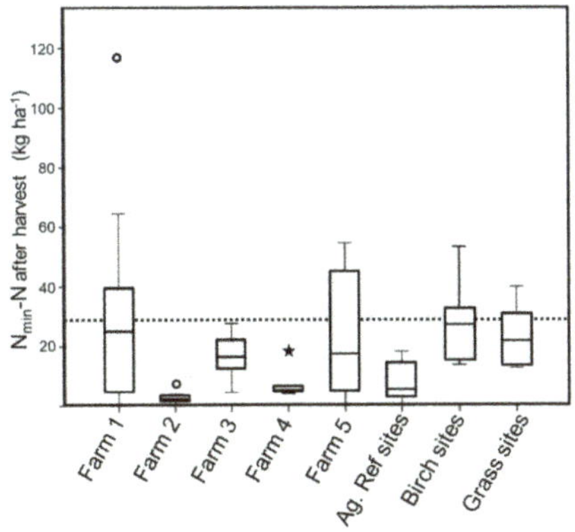

Figure 3. After harvest N_{min}-N values compared to advised N_{min}-N values after harvest to protect water bodies [62]. Circles represent outliers that have values greater than 1.5 times the interquartile range, stars represent extreme outliers that have values greater than 3 times the interquartile range. Ag. Ref. sites = Agricultural reference sites; Birch sites = Semi-natural birch sites; Grass sites = Semi-natural grassland sites.

The inappropriate use of fertilizer, especially if the fertilizer is not directly taken up by the plants, can increase the amount of H^+ and thus lead to further leaching of nutrients [48]. Especially ammonium based fertilizers are known to affect soil pH by increasing soil acidity as the process of converting ammonium to nitrate in the soil (nitrification) releases H^+ [48]. The high amount of H^+ is thereby indicated by the very low base saturation (BS) on the sampled agricultural fields of Farms 1, 2 and 5 compared to the agricultural reference sites. Krøjer [64] as well as Høegh et al. [19] advise a more careful use of artificial fertilizers in the region, to prevent nutrient leaching and soil acidification. Soil acidification affects several soil processes, for example a lowered buffering capacity leading to further decrease in pH and more nutrient leaching. The hampered decomposition of litter leads to poor soil structure, root growth and microbial activity [65], which potentially reduces pore volume and water retention capacity. A reduced pore volume is also shown by higher bulk density values, which was found on Farms 1 and 5 (Table 5).

As described above, plant water availability will be a crucial parameter for agricultural use in the future, as the lengths and frequency of dry periods will increase [2]. Besides the given soil conditions and the already mentioned management induced processes which can negatively affect plant available water, a further process affecting water retention potential was detected in this study. The high percentage of exchangeable sodium (Na_{ex}) observed on fields of Farms 3 and 5 can affect aggregate stability, cause a decrease in permeability and reduce water infiltration ability [36,40]. Most likely the high amounts of Na_{ex} on Farms 3 and 4 are caused as soils were affected by seawater. This process is particularly conceivable for the fields of Farm 4 (median Na_{ex} 11.5%) (Table 4) which are situated adjacent to the sea. Another process contributing to the Na_{ex} enrichment may have been the use of seaweed as fertilizer during the Norseman period [66]. This enrichment process may have affected the fields of Farm 3 (median Na_{ex} of 6.2%) because they are situated in the village of Igaliku and were already used by the Norseman [4,32,33,67].

The analyses of the soil characteristics revealed two major processes, namely soil acidification and nutrient leaching, which both pose a risk for landscape ecology under land use intensification and expansion. However, more detailed information on fertilizer application, yield, sheep stock number, nutrient leaching into water bodies and a higher amount of reference sites providing a better comparability of unused and used areas on farm level would be needed to make a distinct estimation on the concomitant effects of land use in southern Greenland.

3.3. Identifying the Most Suitable Areas

Agriculture in south Greenland already affects soil—and landscape ecology. Consequently, it is necessary to identify areas which show suitable soil conditions for expansion and intensification to guarantee sustainable land use in the future. Applying the SQI by integrating the measured physical and chemical soil properties into one index, enables an overall assessment of the soil quality. SQI values for both reference sites and farms are shown in Figure 4. It is noteworthy that the fields belonging to Farm 3 show significantly ($p < 0.01$) the best soil quality, while the remaining farms with lower quality do not differ significantly amongst each other, the agricultural and the semi-natural reference sites.

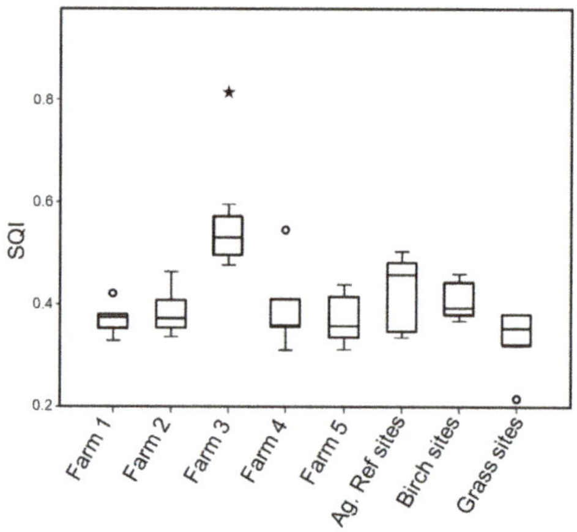

Figure 4. Boxplots indicating soil quality index for agricultural areas, agricultural reference sites and semi-natural reference sites. Circles represent outliers that have values greater than 1.5 times the interquartile range, stars represent extreme outliers that have values greater than 3 times the interquartile range. Ag. Ref. sites = Agricultural reference sites; Birch sites = Semi-natural birch sites; Grass sites = Semi-natural grassland sites.

Analyzing the pattern of the calculated SQI (Figure 5) reveals an interesting finding, which could help to identify suitable areas for agricultural use easily. According to the descriptions of Arneborg [32,68,69], who performed archeological studies in Igaliku, most of the fields of Farm 3 were already used by the Norseman. The fields represent the "inner fields" or "homefields" of the Norseman in Igaliku (Figure 5b) which were used for winter fodder production [32,68]. A few fields at Farm 3 (301/325 and 302) and the northernmost fields of Farm 4 (402 and 403), which lie outside the stonewall confining the "inner fields", are close to former animal pens and have presumably been used as pasture [32]. This area coincides with the fields showing the highest soil quality index.

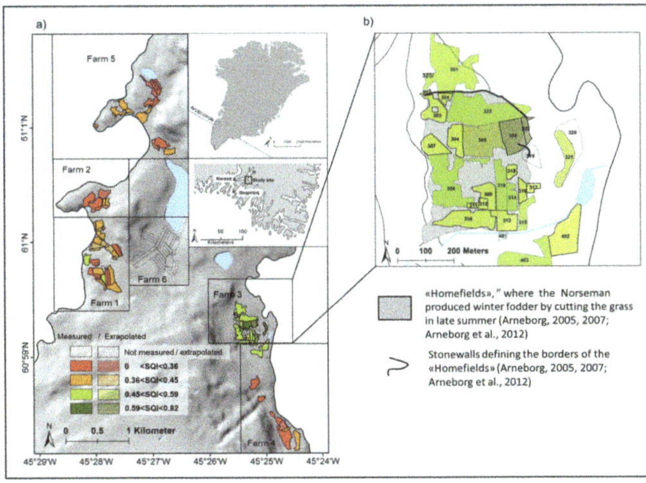

Figure 5. (**a**) Parcel based soil quality index map for the agricultural areas. The hatched fields correspond to the fields where the SQI was extrapolated, considering similarities in the mapped properties (angle, orientation, position and form of the slope, parent material and crop, mapped according to the FAO Guidelines for soil description [35]. (**b**) Areas used for haymaking during the Norseman period (980s to 1450), after [32,68,69].

No evidence for agricultural use such as irrigation channels or ruins has been found on the fields of Farms 1, 2, 5 and the remaining fields of Farm 4 [32,68,69]. These areas, which have presumably not been used by the Norse, show a significantly lower SQI (Figures 4 and 5). The SQI differences appear to provide evidence that the Norse settlements were situated on areas which were either suited best for agricultural use or were improved by the Norseman, e.g., by manuring or algae application. Thus, a possibility to expand agricultural land use in southern Greenland would be to concentrate the expansion on areas which were already used during the Norseman period, or at least on such sites that show similar geo-ecologic properties. Even though the present cultivated area far exceeds the area cultivated by the Norseman, there are still areas which were used by the Norseman that lie unused today [70].

3.4. Improvement of Soil Quality by Directed Measures

Soil acidification seems to be an obvious hindrance for a sustainable agriculture in southern Greenland. Liming acidic soils to balance acidification has already been recommended by Christensen, [55] and later by Krøjer, [64] and also by Høeg et al. [19]. Liming would also be indicated considering base saturation values (BS) (Table 4), which are classified, as "poor" or even "very poor" on all sampled fields [41]. Liming or the amendment of soluble soil calcium (gypsum) would also represent an accurate measure against sodic soils, which also affect the availability of plant water, replacing the sodium due to the more powerful electrical charge of gypsum [48]. Although the problem of soil acidification has apparently been identified several decades ago, our study as well as the literature [17,19,20] indicate that measurements to reduce it were not applied consequently.

The second problem, namely soil water supply during the growing season, could be addressed by irrigation systems, as used by the Norseman [32,54,68], to reduce the variability in yield production. Several irrigation channels were already reestablished for the farming area within Igaliku, corresponding to the fields of Farm 3, in the 1960s [68]. However, facing more pronounced dry periods in the future requires further measures like for example soil ameliorations, which enhance the ability to store plant available water for example by increasing soil organic carbon. Further, adaptions to the increasing length of dry periods could include cultivating plant species, which are more adapted to dry soil conditions.

A further issue concerning the adaption to climate variability and sustainable land use is the increasing size of the farms. According to Hayashi [12], the Greenlandic government introduced ordinances under the Home Rule Act of 1979, which made investments mandatory and thus made the small scaled sheep farming no longer viable. In the following, part-time farmers disappeared, while sheep flock size of the remaining farmers increased to keep their farms economically profitable. The number of sheep owners in Greenland decreased from 95 in 1982 (shortly after the Home Rule Act of 1979) to 48 in 2008 [12]. The Rambøll report states that a farmer needs at least 400 sheep to be able to make a living from it [11]. In contrast, in 1933, under self-sufficient side line farming, the number of sheep owners was 206, while flock size was mostly fewer than 100 sheep, the total number of sheep reached approximately 5000 [12]. The increasing farm size bears several risks. First, a bigger flock size infers higher use intensity than several small farms, because small farms are less centralized than one big farm. Big flock sizes also increase the risk of soil erosion, as land use, especially grazing may concentrate on smaller areas. Soil erosion due to overgrazing has been documented for several Norse settlements [7,10,33,67,71] and has also been mentioned as one of the reasons for the disappearing of the Norseman in 1450 [7,33]. Today, the increased sedimentation rate in Lake Igaliku (see Figure 1) during the 20th century indicates an increased soil erosion rate within the study area [67]. Small scale, decentralized farming, grazing regulations and smaller flock size could help minimizing soil erosion, but would require political action aimed at soil protection and subsidies providing incentives for a sustainable agriculture. Re-cultivation and expansion of agricultural use may also infer the readiness of farmers to live and establish themselves isolated and far away from the next inhabited area, and to adapt agricultural use on the given natural resources and climate variability by running a more diverse and small scaled agriculture.

Figure 6 schematically summarizes the presented hindrances for future land use intensification and expansion given by the soil and climate conditions. Possibilities to increase the potential for an agricultural development as well as the adaptions needed to guarantee that the agricultural development is sustainable are also outlined in the diagram. Finally, it highlights the risks for landscape and soil ecology.

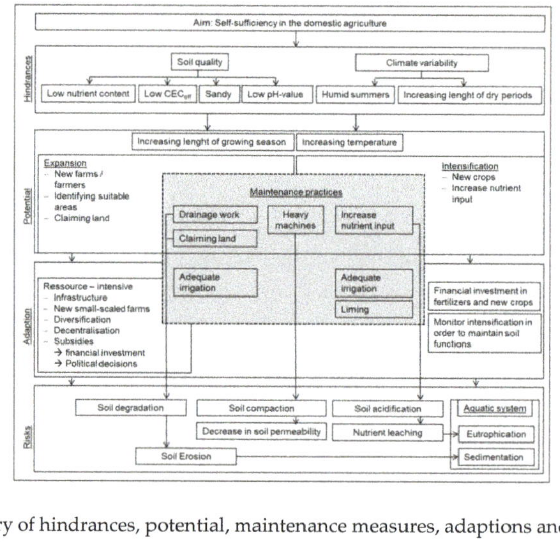

Figure 6. Summary of hindrances, potential, maintenance measures, adaptions and risks concerning the expansion and intensification of southern Greenlandic agriculture.

4. Conclusions

The study evaluated the conditions for agricultural land use in southern Greenland, focusing on the soil as the natural resource for food production. The results of this study illustrate three major findings. First, even though the study area around Igaliku is characterized by more favorable climate and partly more favorable soil conditions than other regions in southern Greenland, an increase of summer temperatures and the length of the growing season, does not automatically increase the potential for land use intensification and expansion. The main hindrance to expand or intensify agricultural use is given by the low nutrient status and the unsuitable soil texture, which affects the ability of soil to store water and nutrients. On the contrary, climate change with increased lengths of dry periods poses new risks for yield production, especially due to the mentioned unfavorable soil texture.

However, the SQI shows that, even though soil characteristics in general are not favorable, there still can be found areas that provide better possibilities for land use expansion. In the study area, these areas corresponded to those sites which were already used by the Norseman (980s–1450 AD). In the context of sustainable landscape ecology, it seems to be essential to focus land use expansion on areas that were used in medieval times.

Finally, the study revealed two major risks which can affect landscape ecology. The analysis showed a trend for soil acidification on the already acidic soils. Furthermore, the unfavorable soil texture poses a risk for nutrient leaching. To ensure a future sustainable land use, these two major problems should be addressed.

Acknowledgments: The authors want to express appreciation for the help of the students Helena Barth, Tabea Baumgartner, Alexandra Diesslin, Guy Doerfel, Lisette Kaufmann, Nicole Notter and Oliver Zoller, participating in the field course in southern Greenland, the help of the participating farmers as well as the help of Ruth Strunk assisting the laboratory analysis. Furthermore, the authors want to thank the Tomcik foundation and the "Freiwillige Akademische Gesellschaft Basel", for financially supporting the field campaign. The authors also thank the "Departementet for Erhverv, Råstoffer og Arbejdsmarkt" of the Greenlandic Government for providing research permission.

Author Contributions: C.C., M.H. and N.J.K. conceived and designed the experiments; C.C. and M.H. performed the field research and the data analysis; N.J.K. contributed materials and analysis tools; and C.C. wrote the paper. All authors critically commented on the manuscript.

Conflicts of Interest: The authors declare no conflict of interest. The founding sponsors had no role in the design of the study; in the collection, analyses, or interpretation of data; in the writing of the manuscript, and in the decision to publish the results.

References

1. Hassol, S.J. *Impacts of a Warming Arctic: Arctic Climate Impact Assessment*; Cambridge University Press: Cambridge, UK; New York, NY, USA, 2004; ISBN 978-0-521-61778-9.
2. Christensen, J.; Olesen, M.; Boberg, F.; Stendel, M.; Koldtoft, I. *Fremtidige Klimaforandringer i Grønland: Kujalleq Kommune*; Videnskabelig Rapport 15-04 (1/6); Danmarks Meteorologiske Institut: København, Denmark, 2016.
3. Greenland Agricultural Consulting Service. Available online: http://www.nunalerineq.gl/english/landbrug/index-landbrug.htm (accessed on 2 March 2016).
4. Bichet, V.; Gauthier, E.; Massa, C.; Perren, B.; Richard, H.; Petit, C.; Mathieu, O. The history and impacts of farming activities in south Greenland: An insight from lake deposits. *Polar Rec.* **2013**, *49*, 210–220. [CrossRef]
5. Dugmore, A.J.; Keller, C.; McGovern, T.H. Norse Greenland Settlement: Reflections on Climate Change, Trade, and the Contrasting Fates of Human Settlements in the North Atlantic Islands. *Arct. Anthropol.* **2007**, *44*, 12–36. [CrossRef]
6. Edwards, K.J.; Schofield, J.E.; Kirby, J.R.; Cook, G.T. Problematic but promising ponds? Palaeoenvironmental evidence from the Norse Eastern Settlement of Greenland. *J. Quat. Sci.* **2011**, *26*, 854–865. [CrossRef]
7. Fredskild, B. Erosion and Vegetational Changes in South Greenland Caused by Agriculture. *Geogr. Tidsskr.-Dan. J. Geogr.* **1992**, *92*, 14–21. [CrossRef]
8. Sørensen, A.K. *Denmark-Greenland in the Twentieth Century (Vol. 341)*; Museum Tusculanum Press: Copenhagen, Denmark, 2007; ISBN 978-87-635-3070-5.

9. Austrheim, G.; Asheim, L.J.; Bjarnason, G.; Feilberg, J.; Fosaa, A.M.; Holand, Ø.; Høegh, K.; Jonsdottir, I.S.; Magnusson, B.; Mortensen, L.E.; et al. Sheep grazing in the North-Atlantic region—A long term perspective on management, resource economy and ecology. *Rep. Zool. Ser.* **2008**, *3*, 1–82.
10. Jacobsen, N.K. Studies on soils and potential for soil erosion in the sheep farming area of south Greenland. *Arct. Alp. Res.* **1987**, *19*, 498–507. [CrossRef]
11. Rambøll Management Consulting. *Where Can Development Come From? Potentials and Pitfalls in Greenland's Economic Sectors towards 2025*; Rambøll Management Consulting: Copenhagen, Denmark, 2014; p. 65. Available online: http://download.ramboll-environ.com/ramboll/Greenpeace-report.pdf (accessed on 29 February 2016).
12. Hayashi, N. Incorporating Climate Change Remedies into Community Development in Greenland. In *Cultural Dynamics of Climate Change and the Environment in Northern America*; Sommer, B., Ed.; Climate and Culture; Brill: Leiden, The Netherlands; Boston, MA, USA, 2015; Volume 3, pp. 283–303.
13. Ministry of Fisheries, Hunting & Agriculture, Greenland. *Grønlands Statistik. 2014 Statistisk Årbog*; Fiskeri, Fangst og Landbrug: Nuuk, Greenland, 2014; p. 36.
14. Taufetter, G. Arctic Harvest: Global Warming a Boon for Greenland's Farmers. *Spieg. Online*. 2006. Available online: http://www.spiegel.de/international/spiegel/arctic-harvest-global-warming-a-boon-for-greenland-s-farmers-a-434356.html (accessed on 2 March 2016).
15. Pleitgen. *Farming to the Fore as Greenland Ice Thaws.* 2012. Available online: http://edition.cnn.com/2012/12/04/world/greenland-secrets/ (accessed on 11 July 2016).
16. Ministry of Fisheries, Hunting & Agriculture, Greenland. *Økonomisk. Analyse af det Grønlandske Landbrug*; Ministry of Fisheries, Hunting & Agriculture: Nuuk, Greenland, 2008. Available online: http://naalakkersuisut.gl/~/media/Nanoq/Files/Publications/Fangst%20og%20fiskeri/DK/Udgivelser_EL_%C3%98konomisk%20analyse%20af%20det%20gr%C3%B8nlandske%20landbrug_april%202008_DK.pdf (accessed on 2 March 2016).
17. Høeg, K.; Isaksen, E.; Frederiksen, A.; Hansen, H.; Poulsen, K.; Iversen, A.; Devisme, A. Årsberetning 2006/2007 fra Konsulenttjenesten for Landbrug. 2007. Available online: http://www.nunalerineq.gl/dansk/raad/aarsberet/Aarsberetning%202006%202007%20DK.pdf (accessed on 16 May 2016).
18. Landbrugskommisionen. *Landbrugskommissionens Betænkning*; Landbrugskommisionen: Nuuk, Greenland, 2014; p. 110. Available online: http://naalakkersuisut.gl/~/media/Nanoq/Files/Attached%20Files/Fiskeri_Fangst_Landbrug/DK/2016/Final_Rapport%20landbrug%202014_Chair_DK_pdf.pdf. (accessed on 13 November 2017).
19. Høegh, K.; Lund Stidsen, P.; Frederiksen, A.; Hansen, H.; Motzfeldt, N.; Gard, J.; Petersen, M. Årsberetning 2007/2008 fra Konsulenttjenesten for Landbrug. 2008. Available online: http://www.nunalerineq.gl/dansk/raad/aarsberet/Aarsberetning%202007%202008%20DK.pdf (accessed on 16 May 2016).
20. Høegh, K.; Isaksen, G.; Frederiksen, A.; Hansen, H.; Poulsen, K.; Iversen, A.; Devisme, A. Årsberetning fra Konsulenttjenesten for Landbrug. 2006. Available online: www.nunalerineq.gl/dansk/raad/aarsberet/Aarsberetning%202005%202006%20DK.pdf (accessed on 16 May 2016).
21. Westergaard-Nielsen, A.; Bjørnsson, A.B.; Jepsen, M.R.; Stendel, M.; Hansen, B.U.; Elberling, B. Greenlandic sheep farming controlled by vegetation response today and at the end of the 21st Century. *Sci. Total Environ.* **2015**, *512–513*, 672–681. [CrossRef]
22. Zornoza, R.; Acosta, J.A.; Bastida, F.; Domínguez, S.G.; Toledo, D.M.; Faz, A. Identification of sensitive indicators to assess the interrelationship between soil quality, management practices and human health. *Soil* **2015**, *1*, 173–185. [CrossRef]
23. Doran, J.W.; Parkin, T.B. Defining and assessing soil quality. *Defin. Soil Qual. Sustain. Environ.* **1994**, *35* 3–21.
24. Schiefer, J.; Lair, G.J.; Blum, W.E. H. Indicators for the definition of land quality as a basis for the sustainable intensification of agricultural production. *Int. Soil Water Conserv. Res.* **2015**, *3*, 42–49. [CrossRef]
25. Doran, J.W. Soil health and global sustainability: Translating science into practice. *Agric. Ecosyst. Environ.* **2002**, *88*, 119–127. [CrossRef]
26. Hansen, B.U. Satellite Monitoring of the Biomass Production in Southern Greenland. *Geogr. Tidsskr.-Dan. J. Geogr.* **1988**, *88*, 94–101. [CrossRef]
27. Brooks, K. A tale of two intrusions—Where familiar rock names no longer suffice. *Geol. Today* **2012**, *28*, 13–19. [CrossRef]
28. Hanna, E.; Cappelen, J. Recent climate of southern Greenland. *Weather* **2002**, *57*, 320–328. [CrossRef]

29. Feilberg, J. *A Phytogeographical Study of South Greenland, Vascular Plants*; Commission for Scientific Research in Greenland: Copenhagen, Denmark, 1984.
30. Madsen, C.K. Pastoral Settlement, Farming, and Hierarchy in Norse Vatnahverfi, South Greenland. Ph.D. Thesis, University of Copenhagen, Copenhagen, Denmark, 2014.
31. Daniëls, F.J. A geobotanical impression of South. Greenland with some remarks on its "Boreal Zone". In Proceedings of the Fifth International Workshop: Conservation of Arctic Flora and Fauna (CAFF) Flora Group. Circumboreal Vegetation Mapping (CBVM) Workshop, Helsinki, Finland, 3–6 November 2008; CAFF Technical Report No. 21. Talbot, S., Charron, T., Barron, T., Eds.; CAFF International Secretariat: Helsinki, Finland, 2010; pp. 85–94.
32. Arneborg, J. *Saga Trails-Brattahlid, Gardar, Hvalsey Fjord's Church and Herjolfnesnes: Four Chieftain's Farmsteads in the Norse Settlement of Greenland*; The National Museum of Denmark: Copenhagen, Denmark, 2007.
33. Perren, B.B.; Massa, C.; Bichet, V.; Gauthier, É.; Mathieu, O.; Petit, C.; Richard, H. A paleoecological perspective on 1450 years of human impacts from a lake in southern Greenland. *Holocene* **2012**, *22*, 1025–1034. [CrossRef]
34. Kommune Kujalleq Kommuneplanen-Igaliku, Redegørelse og Delområder. Available online: http://kujalleq.odeum.com/download/pdf/da/14_igaliku_redegoerelse_og_delomraader.pdf (accessed on 20 July 2016).
35. FAO. *Guidelines for Soil Description*, 4th ed.; Food and Agriculture Organization of the United Nations: Rome, Italy, 2006; ISBN 978-92-5-105521-2.
36. Arshad, M.A.; Martin, S. Identifying critical limits for soil quality indicators in agro-ecosystems. *Agric. Ecosyst. Environ.* **2002**, *88*, 153–160. [CrossRef]
37. Agroscope FAL Reckenholz. *Schweiz. Referenzmethoden der Eidg. Landwirtschaftlichen Forschungsanstalten*; Eidgenössische Forschungsanstalten FAL, RAC, FAW: Zürich, Switzerland, 1996.
38. Sponagel, H. *Bodenkundliche Kartieranleitung: Mit 41 Abbildungen, 103 Tabellen und 31 Listen*; Bundesanstalt für Geow. und Rohstoffe: Hannover, Germany, 2005; ISBN 978-3-510-95920-4.
39. Scheffer, F.; Schachtschabel, P.; Blume, H.-P. *Soil Science*, 1st ed.; Springer: Berlin, Germany, 2016; ISBN 978-3-642-30941-0.
40. Amacher, M.C.; O'Neil, K.P.; Perry, C.H. *Soil Vital Signs: A New Soil Quality Index (SQI) for Assessing Forest Soil Health*; Department of Agriculture, Forest Service, Rocky Mountain Research Station: Fort Collins, CO, USA, 2007; pp. 1–12.
41. Flisch, R.; Sinaj, S.; Charles, R.; Richner, W. Grundlagen für die Düngung im Acker- und Futterbau-Kapitel 7-10. *Agrar. Schweiz* **2009**, *16*, 32–49.
42. Stevenson, F.J.; Cole, M.A. *Cycles of Soil: Carbon, Nitrogen, Phosphorus, Sulfur, Micronutrients*, 2nd ed.; Wiley: New York, NY, USA, 1999; ISBN 978-0-471-32071-5.
43. Marzaioli, R.; D'Ascoli, R.; De Pascale, R.A.; Rutigliano, F.A. Soil quality in a Mediterranean area of Southern Italy as related to different land use types. *Appl. Soil Ecol.* **2010**, *44*, 205–212. [CrossRef]
44. Armenise, E.; Redmile-Gordon, M.A.; Stellacci, A.M.; Ciccarese, A.; Rubino, P. Developing a soil quality index to compare soil fitness for agricultural use under different managements in the Mediterranean environment. *Soil Tillage Res.* **2013**, *130*, 91–98. [CrossRef]
45. Kaiser, H.F. The Application of Electronic Computers to Factor Analysis. *Educ. Psychol. Meas.* **1960**, *20*, 141–151. [CrossRef]
46. Wander, M.M.; Bollero, G.A. Soil Quality Assessment of Tillage Impacts in Illinois. *Soil Sci. Soc. Am. J.* **1999**, *63*, 961–971. [CrossRef]
47. Qi, Y.; Darilek, J.L.; Huang, B.; Zhao, Y.; Sun, W.; Gu, Z. Evaluating soil quality indices in an agricultural region of Jiangsu Province, China. *Geoderma* **2009**, *149*, 325–334. [CrossRef]
48. Western Plant Health Association. *Western Fertilizer Handbook*; Waveland Press: Long Grove, IL, USA, 2010; ISBN 978-1-57766-679-0.
49. Flisch, R.; Sinaj, S.; Charles, R.; Richner, W. Grundlagen für die Düngung im Acker- und Futterbau-Kapitel 3-6. *Agrar. Schweiz.* **2009**, *16*, 6–31.
50. Prasad, R.; Power, J.F. *Soil Fertility Management for Sustainable Agriculture*; CRC/Lewis Publishers: Boca Raton, FL, USA, 1997; ISBN 978-1-56670-254-6.
51. Liebig, M.A.; Varvel, G.; Doran, J. A Simple Performance-Based Index for Assessing Multiple Agroecosystem Functions. *Agron. J.* **2001**, *93*, 313–318. [CrossRef]
52. Andrews, S.S.; Karlen, D.L.; Cambardella, C.A. The soil management assessment framework. *Soil Sci. Soc. Am. J.* **2004**, *68*, 1945–1962. [CrossRef]

53. Landwirtschaftskammer Nordrhein-Westfalen. *Grunddüngung. mit Phosphat., Kali, Magnesium*; Landwirtschaftskammer Nordrhein-Westfalen: Münster, Germany, 2015; pp. 1–4. Available online: https://www.landwirtschaftskammer.de/landwirtschaft/ackerbau/pdf/phosphat-kalium-magnesium-pdf.pdf (accessed on 10 July 2016).
54. Edwards, K.J.; Schofield, J.E. Investigation of proposed Norse irrigation channels and dams at Garðar/Igaliku, Greenland. *Water Hist.* **2013**, *5*, 71–92. [CrossRef]
55. Christensen, K.N. Grønlands jord under kultur. *Tidskr. Grønl.* **1953**, *11*, 407–413.
56. Adderley, W.P.; Simpson, I.A. Soils and palaeo-climate based evidence for irrigation requirements in Norse Greenland. *J. Archaeol. Sci.* **2006**, *33*, 1666–1679. [CrossRef]
57. De Neergaard, E.; Stougaard, P.; Høegh, K.; Munk, L. Climatic changes and agriculture in Greenland: Plant diseases in potatoes and grass fields. In Proceedings of the IOP Conference Series: Earth and Environmental Science, Copenhagen, Denmark, 10–12 March 2009.
58. Dollerup-Scheibel, M. Borgmester Jørgen Wæver Johansen Efterlyser Hjælp til Trængte Landmænd. *SermitsiaqAG Online*. 2015. Available online: http://knr.gl/da/nyheder/landbruget-ramt-af-t%C3%B8rke (accessed on 7 July 2016).
59. Hannestad, A. Tørke Hærger Grønlands Landbrug. *Politiken*. 2015. Available online: http://politiken.dk/oekonomi/gloekonomi/ECE2888287/toerke-haerger-groenlands-landbrug/ (accessed on 7 July 2016).
60. Myers, R. *The Basics of Chemistry (Basics of the Hard Sciences)*; Greenwood Press: Westport, CT, USA, 2003; ISBN 978-0-313-31664-7.
61. Niedersächsisches Landesamt für Ökologie (NLÖ). *Anwenderhandbuch für Die Zusatzberatung Wasserschutz: Grundwasserschutzorientierte Bewirtschaftungsmaßnahmen in der Landwirtschaft und Methoden zu Ihrer Erfolgskontrolle; Hildesheim*. 2001. Available online: http://www.nlwkn.niedersachsen.de/download/95902/NLWKN_2015_Anwenderhandbuch_fuer_die_Zusatzberatung_Wasserschutz_Grundwasserschutzorientierte_Bewirtschaftungsmassnahmen_in_der_Landwirtschaft_und_Methoden_zu_ihrer_Erfolgskontrolle_Band_21_.pdf (accessed on 15 July 2016).
62. Chambre d'Agriculture. Das Info.-Blatt des Wasserschutzberatungsdienstes der Landwirtschaftskammer. Chambre d'Agriculture Wasserschutzberodung: Strassen, Luxenburg, 2015; pp. 1–4. Available online: http://www.lwk.lu/download/838/lwk-waasserschutzberodung-info-2015-4.pdf (accessed on 7 July 2016).
63. Belle, S.; Rius, D.; Bichet, V.; Massa, C.; Mavon, C.; Millet, L. Combining limnology and paleolimnology to assess the influence of climate change on two lakes in Southern Greenland. *Polar Biol.* **2017**, 1–13. [CrossRef]
64. Krøjer, A.K. Landbrug i Sydgrønland-Svære produktionsvilkår i det arktiske miljø. *Tidskr. Grønl.* **2000**, *8*, 283–289.
65. Galler, J. *Kalkdüngung Basis für Fruchtbare Böden*; Landwirtschaftskammer Salzburg, Betriebsentwicklung und Umwelt: Salzburg, Austria, 2008; pp. 1–27.
66. Retamero, F.; Schjellerup, I.; Davies, A. (Eds.) *Early Agricultural Remnants and Technical Heritage (EARTH): 8000 Years of Resilience and Innovation*; Oxbow Books: Oxford, UK; Philadelphia, PA, USA, 2016; ISBN 978-1-84217-359-6.
67. Massa, C.; Bichet, V.; Gauthier, É.; Perren, B.B.; Mathieu, O.; Petit, C.; Monna, F.; Giraudeau, J.; Losno, R.; Richard, H. A 2500 year record of natural and anthropogenic soil erosion in South Greenland. *Quat. Sci. Rev.* **2012**, *32*, 119–130. [CrossRef]
68. Arneborg, J. Greenland irrigation systems on a West Nordic background: An overview of the evidence of irrigation systems in Norse Greenland c. 980-1450 AD. *Pamatky Archeol. Suppl.* **2005**, *17*, 137–145.
69. Arneborg, J.; Lynnerup, N.; Heinemeier, J.; Møhl, J.; Rud, N.; Sveinbjörnsdóttir, Á.E. Norse Greenland dietary economy ca. AD 980-ca. AD 1450: Introduction. *J. North Atl.* *39*. [CrossRef]
70. Kielsen, K.; Frederiksen, S. *Muligheder for Klimatilpasning i Landbrugserhvervet—Status og Handlemuligheder*; Aarhus Universitet and Government of Greenland: Aarhus, Denmark, 2017; pp. 1–50.
71. Schofield, J.E.; Edwards, K.J.; Mighall, T.M.; Martínez Cortizas, A.; Rodríguez-Racedo, J.; Cook, G. An integrated geochemical and palynological study of human impacts, soil erosion and storminess from southern Greenland since c. AD 1000. *Palaeogeogr. Palaeoclimatol. Palaeoecol.* **2010**, *295*, 19–30. [CrossRef]

land

MDPI

Article

Re-Placing the Desert in the Conservation Landscape: Charisma and Absence in the Gobi Desert

Stephen Lezak

School of Geography and the Environment, University of Oxford, South Parks Road, Oxford OX1 3QY, UK; stephen.lezak@gmail.com

Received: 1 December 2018; Accepted: 28 December 2018; Published: 29 December 2018

Abstract: Across the Gobi Desert in China and Mongolia, millions of newly planted trees struggle to survive amid adverse ecological conditions. They were planted by a wide variety of actors in an attempt to protect, restore, or modify the local environment, despite evidence of their negative consequences upon local ecosystems. This paper investigates how these afforestation projects both challenge and affirm recent theoretical work on conservation, while also providing key insights into the decision-making framework of land management across the world's third largest desert region. This analysis, supported by evidence from corporate practice, government policy, and participant observation, builds primarily on the work of Jamie Lorimer and other authors who identify the charisma of certain species as a primary driver of contemporary conservation. But the case of afforestation in the Gobi is inadequately explained by a desire to protect individual species; rather, I show how the charisma at the level of the landscape influences conservation practice. I extend this analysis to suggest that the management of deserts worldwide may be mediated by their perception as absent or empty spaces, thus explaining projects like afforestation which seem to re-place rather than conserve. Using the framework of absence and presence to better understand land use and environmental governance could have implications extending well beyond the Gobi Desert.

Keywords: absence; afforestation; charisma; China; conservation; desertification; Gobi; Mongolia

---

1. Introduction

In recent decades, the practice of environmental conservation has become the subject of sustained ethnographic inquiry. Researchers writing in the traditions of anthropology, environmental history, and science and technology studies have examined how land management and its constituent sciences are shaped not only by objective biophysical referents, but also by cultural inputs and human bias. This paper builds on this literature by examining these recent theoretical developments in the arid and semi-arid landscape of the Gobi Desert in Mongolia and China, a markedly different setting to the bucolic countryside of Europe and the dense forests of North America from which these theories originated. In particular, I examine the widespread practice of drylands afforestation, which is supported by individual, corporate, and public actors across the Gobi region. This analysis suggests that environmental conservation in the world's third-largest desert is inexplicable by recent critical scholarship that focuses rather narrowly on the role of charisma in individual species as the drivers of conservation management practice [1–3]. In the Gobi, where afforestation efforts are largely unconcerned with inter-species distinctions, what underlies conservation decision-making instead is an epistemology of ecological wellbeing situated at the scale of the landscape itself. Put differently, conservation actors are engaged in projects to promote landscape-wide flourishing, with little regard for the welfare of constituent species. In this context, the role of charismatic organisms—usually upheld as central to the "conservation assemblage" [1]—is secondary at best. Perhaps critiques thus far have paid so much attention to the proverbial trees that they've failed to see the forest.

2. Background

Contemporary critiques of conservation practice arose in part from the wider cultural turn in anthropology and philosophy in the late 20th Century. As early as the 1990s, the perceived marginalization of non-Western perspectives in land management led to sustained calls for integrating more diverse voices into the politics and policies of environmental governance [1]. "Fortress conservation" [2]—the exclusion of human inhabitants from protected natural areas—became the flashpoint of a latent conversation within the conservation movement: what does it mean to protect and care for the unbuilt environment? Were competing visions of conservation owed to information deficits and inappropriate biases, or could there exist multiple conflicting but equally legitimate answers to the questions at hand?

Traditionally, the answers to these questions were provided by conservation biologists, who represent themselves as simultaneously positivist and normative in their practice [3]. The conservationist is at once the impartial scientist and also the pro-environmental activist [4,5]—like Dr. Seuss' Lorax, who "speak[s] for the trees" [6]. But critics have challenged the orthodoxy that conservation biology practitioners can simultaneously lay claim to ethical truth and natural fact. Without ever doubting good intentions, questions arose as to whether the positivist enterprise of conservation could really achieve the detached impartiality to which it lay claim [4].

William Cronon's essay, "The Trouble with Wilderness" [7], has become a cornerstone of this perspective. Writing in the mid-1990s, Cronon argues that the "natural"—as distinct from the "human"—is a cultural invention, made possible by industrialization and urbanization, when people first began to build environments that looked and felt radically different from farmland or forests. It was in precisely this vein that authors began to raise pressing questions about whether conservation biology itself is a culturally and politically mediated practice. These critiques suggest that a concept such as *biodiversity* or *wilderness* may not be "a true object that science progressively uncovers, but ... an historically produced discourse" [4] (p. 54). Jamie Lorimer [3], in particular, has written at length about how conservation science and policy emerge from an assemblage of human and nonhuman interactions [8]; the scientific knowledge and management decisions that result are contingent, dynamic, political, and disquietingly subjective.

The way in which society, political interests, and aesthetic bias come to bear upon these discourses is not simply a matter of intellectual inquiry; the role of conservationists as experts of environmental management brings real-world consequence to their practice [9]. What's more, the backdrop of global environmental change raises the stakes significantly. How landscapes transform in the Anthropocene depends directly on how humans deliberate what constitutes good conservation and what is healthy or harmful to nonhuman life. Yet even as many environmentalists embrace a cross-disciplinary perspective (and as "The Trouble with Wilderness" enters the canon of environmental studies reading lists), some conservationists bristle at the question of whether their own scientific methods are ensnared in cultural entanglements. It seems that many of these researchers and practitioners perceive the science of ecology as relatively insulated from the silty waters of culture. They assert that at a time of global environmental destruction, critical theorists who question the epistemological underpinnings of conservation can inadvertently arm anti-environmentalism or otherwise sew disunity among an already fragmented movement [10,11]. This sustained divide between traditional ecology and a conservation practice that acknowledges contingency [3,12] continues to frustrate efforts to assimilate theory into practice. This paper focuses on the theory-practice nexus of one ecosystem in particular: the arid and semi-arid drylands of the Gobi Desert.

In the following pages, I focus on the practice of afforestation (planting trees where they did not previously grow) in the Gobi Desert in Northern China and Southern Mongolia (see Figure 1). I present three miniaturized "scenes", each an independent case study of an afforestation site, ordered by increasing scale. I begin with a firsthand anthropological account from fieldwork conducted in the Noyon administrative district in the summer of 2017; the second scene is a study of corporate environmental remediation; the third examines national afforestation policies in Mongolia and

China. While this regional analysis affirms much of the scholarship that identifies human bias in conservation practice, it also poses a challenge to a key theoretical tenet: that conservation management is explicable by a species-level ontology of the *charisma* of particular organisms in combination with their broader ecological role [3,13]. Jamie Lorimer defines nonhuman charisma "as the distinguishing properties of a nonhuman entity or process that determine its perception by humans and its subsequent evaluation" [13] (p. 915). At the level of the species—"the basic unit of conservation biology" [14] (p. 52)—environmental protection exhibits a tension between protecting a group of designated organisms and using alternative epistemologies to determine desirable ecological outcomes. The former tends to prevail, generating a species-centered practice that Ernest Small calls "the New Noah's Ark" [15] for its necessary exclusion of a large share of organisms. Those species that are easier to identify and observe, aesthetically remarkable, and thrilling to encounter receive a disproportionate share of attention from the global assemblage of conservation actors [15–17]. One analysis found that the "flagship species" of environmental groups are best explained in terms of marketing theory rather than ecological significance [18].

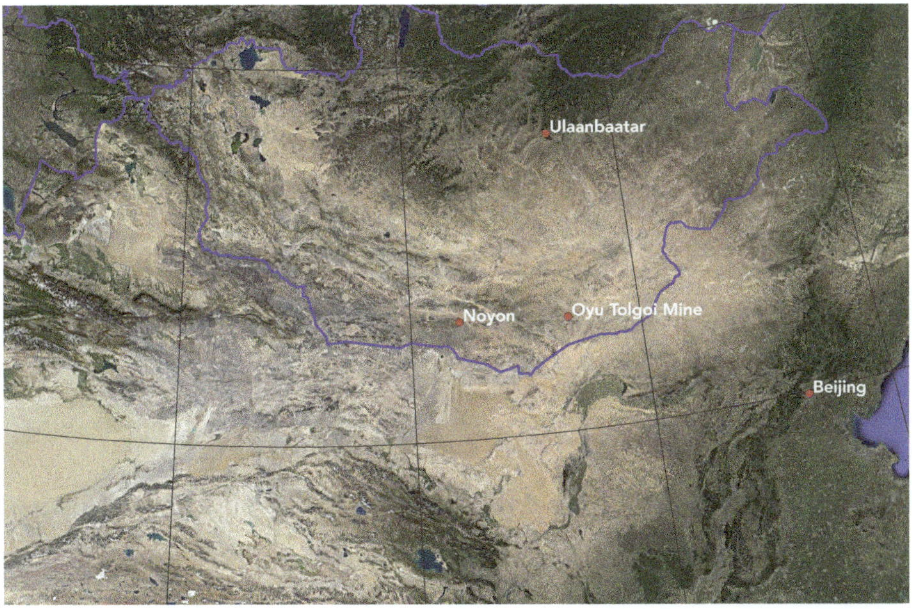

Figure 1. The Gobi Desert, with Noyon and Oyu Tolgoi. Imagery: Landsat/Copernicus; Data SIO; NOAA; US Navy; NGA; GEBCO (2018). Data: Google (2018).

The case of Gobi afforestation, however, refuses to conform to this model of conservation. In this regional instance, the planting of trees as an environmental intervention is insoluble in a species-level ontology of environmental conservation. As shown below, afforestation is undertaken with a desire to protect the environment and justified in scientific (rather than aesthetic) vernacular, yet the region-wide effort is neither ecologically rational nor explained by a desire to protect one or more particular species. The following scenes suggest that the mobilization of private industry and government resources to plant trees in the Gobi is responding instead to nonhuman charisma at the level of the *landscape*. Specifically, the aesthetic perception of deserts as barren, lifeless spaces leads actors to associate afforestation with environmental protection. Lacking in "natural" charisma, the protection of the desert is realized as its re-placement. In this instance, conservationists are filling the metaphorical ark with whatever charismatic biomass is closest at hand—many of the scraggly native scrub species are evidently of minimal interest. The discussion section of this paper addresses this in further detail by considering whether the landscape-level charisma of the Gobi is an exceptional case, or whether it might generalize across more regions and biomes, particularly those landscapes perceived of as absent before human eyes.

3. The Champion Herder

Despite the relative lack of plant agriculture [19], trees have always had a precious place in Gobi society, and deliberate tree-planting in the region dates at least to the 13th Century [20]. But preindustrial silviculture was made extremely difficult by the region's climate. The Gobi is the world's third largest desert, with a current size over 2 million km^2 [21]. Its edges are semi-arid, with annual rainfall of less than 200 mm, but in its driest regions, precipitation is well below 100 mm annually [22]. On the Mongolian (north) side of the desert, cultivated agriculture is a rare sight, prohibited by climate and the relative lack of groundwater wells. The landscape accommodates scrub vegetation, including large areas of grassland on which nomadic and semi-nomadic herders graze a variety of large herbivores [23]. Scattered trees are occasionally found in canyons and ephemeral streambeds, well-positioned to capture a torrent of summer rainfall [22,23]. Marco Polo, who traveled across Asia and met with the Emperor, Kublai Khan, noted that tree planting was a common policy in the Mongolian Empire. Perhaps anticipating nine-hundred years of environmental theory, the Venetian observed that both nature and culture were written into the Empire's afforestation policy: a hybrid of rural land planning and spiritual appeasement.

> *There is another regulation adopted by the Great Khan, equally ornamental and useful. At both sides of the public roads he causes trees to be planted, of a kind that become large and tall ... to point out the road—when the ground is covered with snow.*
>
> *... It may be added that the Great Khan is the more disposed to plant trees because astrologers tell him that those who plant trees are rewarded with long life.* [20] (pp. 228–229)

This sort of hybrid land planning—partly practical, partly spiritual—remains common today. I found it epitomized by one "champion herder" who lives just beyond the dozen mirrored canyons of the Noyon Syncline (see Figure 1). He had achieved public recognition for having over 1000 animals in his herd. His economic success enabled him to plant an unruly grove of trees amid a landscape otherwise dominated by coarse gravel. In the summer of 2017, the elderly grandfather toured me around his miniature forest. I followed him into a small depression below the hill on which he made his seasonal camp (like many Mongolian herders, he moved with his animals several times each year). At the bottom of the basin we passed the pump house which drew up ancient rainfall into a maze of irrigation hoses. A lush grove of young trees lay just beyond, the tallest reaching about four meters in height. The old man explained that he had planted exactly 108 trees: an "auspicious number" for Mongolians, he noted.

When I asked why he planted the grove, he told me he had a desire to "protect nature". He was aware that his small forest was a local biodiversity hotspot for birds and insects. He also gave away seeds to his neighbors and expressed his pride that South Gobi Province was leading the tree planting movement in Mongolia. When we sat down in a tiny gazebo in the center of the grove our conversation turned to more spiritual matters. Three benches were arranged in the shade around a table made from a tree trunk too wide to have come from his relatively young plantation. The herder spoke with sadness about the local mining activity and the decline of nomadic herding. He said that the rise of one and the tapering off of the other had both contributed to local land degradation. I was reminded of similar conversations I've had with cattle and sheep ranchers in Western Wyoming, USA. Our conversation was cut short by the arrival of the local shaman—he had come to conduct ceremonies on the herder's behalf, that he might better understand what was troubling the local land spirits.

Not unlike Kublai Khan's, this champion herder's tree planting expressed an epistemology of ecological well-being that drew upon observed ecological phenomena but relied equally upon culture, spirituality, and aesthetics in order to assess and draw conclusions from objective (i.e., "out there") referents. What's more, the herder was not especially concerned with the pre-interventionist state of the landscape—he was improving rather than conserving. This may seem strange, given that the Gobi is not short on threatened charismatic species; for instance, the elusive snow leopard is the focus of extensive conservation efforts [24]. Yet afforestation at any scale does not appear to figure into a larger plan to protect these large mammals, or to protect any particular species for that matter. Rather, the champion herder wanted to "protect Nature", writ-large. He looked out onto the steppe and saw a lack of flourishing. The sparse desert scrub did not seem worth cultivating—it did not convey the sort of vibrancy that a healthy environment would feature. In response, the herder aimed to create a livelier landscape: most relevant was not the charisma of one species or another, but the cumulative *absence* of aesthetic charisma from the ecosystem he called home.

A similar sentiment can be found in the discourse of Mongolian Buddhist religious leaders. The proceedings from the 2005 Northern Buddhist Conference on Ecology and Development, hosted in Ulaanbaatar, features a summary of recent reforestation efforts in the Gobi. Whilst described as "conservation", the planting of elm trees is advocated as "an organic means of commemorating the lives lived" whilst making "an important contribution to fighting desertification" [25]. Land management, in this case, belongs to the realms of both culture and nature, science and spirituality—without pretense of pure rationality. While the herder's grove is hardly an example of modern conservation biology, it offers an archetype for how a landscape-level epistemology of an ecosystem motivates environmental action that is entirely unconcerned with the welfare of already-existing charismatic species. I open with this scene to hold a mirror to more modernist conservation practices in the Gobi, examining differences and similarities alike. The following two scenes explore these parallels.

4. The Transnational Mining Company

Roughly 200 km east of the champion herder's summer camp is Oyu Tolgoi, until recently the world's largest undeveloped combined copper and gold mine [26]. The mine is jointly owned by the Mongolian government and Rio Tinto, one of the world's largest mining companies [27]. Expected to someday account for one-third of Mongolia's GDP [28], Oyu Tolgoi has become a point of constant reference in national political conversations. The mine is renowned for voluntarily pursuing an ambitious target of "net-positive impact", aimed at more-than-offsetting any ecological harm caused in the course of its operations. To meet this goal, Oyu Tolgoi is employing "biodiversity offsets", detailed in an expansive environmental and social impact assessment (ESIA) [29]. The design and implementation of the Oyu Tolgoi remediation plan has been praised for setting a new standard in Mongolian extractive industry—a welcome contrast to the routinely poor practices that are commonplace in the sector. Twice-annual audits for ESIA compliance have been undertaken since 2013 by independent consultancies in Canada and Italy, with results published online alongside shareholder reports and an encyclopedic environmental management plan. Oyu Tolgoi was routinely praised in my conversations with Mongolian NGO employees, who felt it set a new standard for extractive industry in the country.

But even Oyu Tolgoi is planting trees in the Gobi where before there were none. In 2014, Oyu Tolgoi planted over 10,000 saxaul trees, using irrigated plots to ensure the trees' survival [30]. Curiously, Oyu Tolgoi's own ESIA states that the mine expects "no certain losses to offset" [29] on the local saxaul groves. That is, Oyu Tolgoi is planting trees not as a like-for-like remediation, but rather as an offset to other ecological harm, such as water contamination, airborne dust, and habitat destruction. The harm that Oyu Tolgoi has set out to remedy is not concerned with one species over another; like the champion herder, the perceived damage exists on a landscape-level ontology that is taxonomically irreducible. Within that framework, the planting of saplings is not about the trees *per se*; it is a socioecological compensation for other types of environmental harm. Such practices evoke what Mette High, in her research on Mongolian artisanal and small-scale (*ninja*) miners, calls *cosmoeconomics*. High documents the practice of corporate and small-scale miners employing lamas to conduct "appeasement ceremonies" at the groundbreaking of new mines, partly to placate angered land and water spirits [31] (p. 101).

There is an additional strangeness worth noting about Gobi afforestation. An abundance of evidence suggests that tree plantations in the Gobi Desert do more to exacerbate ecological degradation than repair it. Ecologists studying Gobi afforestation have written at length about the foolishness of planting trees in this arid and semi-arid region [32–34]. Rather than stabilizing topsoil and preventing desertification, newly-planted trees increase soil erosion by upsetting the surface water equilibrium due to increased evapotranspiration. Particularly in windy areas such as the Gobi, this increases evaporation and depletes soil moisture. The resulting topsoil is drier and more susceptible to wind- and waterborne sediment transport, further degrading soils and increasing desertification. Post-perturbation recovery in dry climates is a slow and difficult process, possibly requiring hundreds of years for an ecosystem to resemble its pre-disturbance state [35]. What passes as a biodiversity offset is more like a damage multiplier.

A report by the World Bank's East Asia and Pacific Environment and Social Development Department investigated several tree-planting initiatives in the Mongolian Gobi. In 2004, thirty-three million seedlings were planted in Mongolia across nearly ten-thousand hectares [22]. A partial survey of these planting sites yielded striking results: among those sites in the Gobi, none had a survival rate higher than eight percent; a majority of surveyed plots had no living trees at all. The reasons for this lack of success were numerous, but most often stemmed from inadequate management such as degraded fencing and insufficient irrigation. The World Bank report also highlighted poor site selection as a confounding factor: none of the Gobi sites were located in areas that could support trees without irrigation. The authors note that one site was located in an area where the water requirements of the saplings were 150% of annual rainfall. The report also noted a strong bias toward planting species that were particularly ill-suited to the Gobi climate. Even as native saxaul plantations struggled, much costlier Gobi sites were developed with elm, poplar, willow, and tamarix. These sites, roughly thirty times more cost-intensive per hectare than the saxaul sites, reveal a generous bias toward non-native or rarer tree species. Perhaps unsurprisingly, these trees that have received such a disproportionate share of resources bear little resemblance the scruffy demeanor of saxaul; they are archetypes of the forest: tall, lush aliens in the desert landscape.

In Oyu Tolgoi's 2017 shareholder report, a photo shows rows of smiling employees holding saplings beneath a vast nursery canopy [36]. Is it possible that this act of ecological charity is not so different from the champion herder's grove or from the trees that lined the avenues of 13th-Century Mongolia? Even this multinational mega-mine's practice of conservation maneuvers between biophysical referents and subjective preferences. At the end of the day, the irrigated saxaul grove near Oyu Tolgoi, exponentially larger than that of the champion herder, also represents a vision of ecosystem wellbeing that is neither wholly "cultural" nor entirely "natural", and is clearly invested in the flourishing of the landscape rather than the welfare of a subset of its charismatic inhabitants. The planting of trees at mining sites like Oyu Tolgoi shows landscape conservation taking the form of landscape substitution. Actors stage a recuperation by replacement, inserting organisms that did not and would not have existed otherwise. That this landscape-level conservation is practiced by a giant of global capitalism suggests that the champion herder mentioned above may be more of an exemplification than an exception.

5. The National Policies

Perhaps Marco Polo would not have been surprised to know that nearly a millennium after his visit, the Mongolian government once again mandates the planting of trees. The 2012 Minerals Law introduced a policy of exponential afforestation:

> *18.1 The central state administration in charge of geological and mining affairs shall exercise the following authority:*

...

18.2.11 to develop a method, a methodology and a procedure for planting ten trees for every cut down tree and foresting [sic]

...

114. The mandatory rehabilitation of environment affected by prospecting, exploration, mining and processing activities:

...

114.4 A license holder shall be obliged to plant 100 trees in place of each tree that was cut down in the course of prospecting, exploration, mining and processing activities. [37]

The central government has passed these laws despite the inability to sustain many tree plantations without groundwater irrigation, depleting fossil reserves that hydrogeologists classify as "non-renewable" [38].

This contemporary practice of state-sponsored afforestation also echoes 20th Century policies across Europe and Asia, in which high-modernist statecraft frequently turned to tree-planting. Nazi Germany, Mussolini's Italy, and the Soviet Union under Stalin all sponsored ambitious afforestation programs [39]. Of these, the earliest and longest-lived was the Soviet Union's "Great Stalin Plan for the Transformation of Nature", which sought to create nearly six million hectares of new forest to serve as shelterbelts along the southern border, intended to stop dry winds coming from Central Asia and increase precipitation in Southern Russia [39]. While these practices were concerned with environmental engineering and resource management, their proponents also saw landscape modification and conservation as a means to foster national identity and ecological nativism [40]. Dr. Konrad Guenther, a German conservationist, asserted in the early 1930s that "the chords of the German soul are tuned to nature. Let us allow nature to speak, and let us be happy to be German!" (qtd. in [39]). In all of these programs, the image of the forest seems to serve as nature's archetype: healthy, resilient, and worthy of protection.

The history of afforestation in the Gobi is relatively more recent. Qing-era China showed little interest in forest management, especially compared to the regenerative forestry practiced by Japan beginning in the 18th Century [41]. In the earth 20th Century, Chinese leaders including Mao Zedong showed little interest in environmental issues [39], but his successors have pursued an ambitious afforestation project for the past five decades. A US $7 billion tree planting initiative has been underway since 1978 [42]. The Three-North Shelterbelt, also known as the Great Green Wall, is the world's largest tree planting project [43], an attempt to slow the northern winds coming across the Gobi and prevent soil erosion and airborne dust [44]. Fifty-billion trees [45] have been planted in the hopes of slowing the expansion of the desert and reducing the severity of regional dust storms that carry as far as Beijing [46]. It appears that this instance of afforestation—"the dominant ecological engineering program in China" [44] (p. 14)—has emerged without the same Tibetan Buddhist roots as its Mongolian counterparts, but persists despite ecologists' continued claim of its perverse outcomes. While China's National Forestry and Grassland Administration describes the Shelterbelt as a "great success" [47], Chinese ecologists have decried these afforestation efforts at length, for all the same reasons listed above [48]. As of 2005, the survival rate of trees planted as part of the Three-North Shelterbelt was 15% [33]. Ironically, these efforts may be counteracting a decade-long upward trend in regional precipitation that led the Gobi to contract by as much as 18% between 2000 and 2012 [21].

Is the Shelterbelt relevant to other instances of Gobi afforestation? The Great Green Wall is undoubtedly a conservation program, meant to "protect" non-arid landscapes from a desert which is perceived as encroaching. Nowhere in this discourse is there a consideration that desert itself might need protecting—it is as if the desert is separate from Nature in a way that the forest is not. Furthermore, the aims and methods of the Shelterbelt seem largely unconcerned with protecting vulnerable species. What seems to drive this ecological mega-project is the fear of an infectious *absence* in the Gobi. Chinese engineers have been tasked with its containment. The resulting conservation plan seems to embrace the ethic, *don't just stand there—plant something!* Evidently native species don't count.

Jiang Gaoming, an ecologist at the Chinese Academy of Sciences, astutely observes that the government is "trying to make forests, that is, planting trees in areas that did not have forests distributed, or are at least treeless in the last several thousand years" (quoted in [49]). The making of forests exhibits the characteristics of a completely secular, modernist, landscape-level ontology of conservation, but still vulnerable to the aesthetics of charisma. In this particular case, government land managers set out to do precisely what took place at Oyu Tolgoi and at the champion herder's home: to substitute one landscape with another, gardening in the name guardianship. It was as though the mere insertion of these saplings, with their own vital energy, would enliven a landscape threatened by infertility. In these acts of conservation, the Gobi is being re-placed with forest.

6. Discussion: Geographies of Absence

In these case studies, conservation practice is not fully reducible to ecological rationality and the allure of charismatic species. The planting of non-native trees across this desert expanse is irreducible in mainstream ecological and theoretical accounts of conservation. Rather, Gobi afforestation seems to involve a rowdy cast of characters that are rarely present in contemporary literatures of conservation planning and practice. In some instances (e.g., Section 3), environmental protection is primarily concerned with religiously-inflected aesthetics. In other instances, a modernist, interventionist vision of human-nature relations structures land conservation as engineering projects. But a common thread connects these: afforestation-as-conservation poses an important counternarrative to the conventional wisdom that conservation is primarily driven by species-level charisma.

To a large extent this paper builds upon the work of Lorimer [3] and others, showing how conservation is indeed a messy assemblage. But the context of the Gobi reveals an analytic container that is not fully sealed. To date, the dominant scholarship that frames conservation as a hybrid of natural science and cultural inputs has focused primarily on the charisma of plant and animal life [13], and on species in particular, as the drivers of a hybrid conservation assemblage that is more-than-human and more-than-science. Much of this is framed by discussions of "charismatic megafauna"—most often birds or large mammals—that disproportionately occupy the attention of conservation biologists' scientific practice.

In a telling illustration, a review of over 63.5 million UK biological records indicates that 65% of all entries document bird species. A further 22% concern vascular plants. The remaining 13% of total records spans all non-avian vertebrates, lower plants, fungi, and invertebrate life forms [50]. It is no surprise that these data focus on the most charismatic species of the British Isles and neglect those that less effectively captivate human interest. Jamie Lorimer, noting this genomic injustice, writes that "UK conservation is guided less by the panoptic logic of biodiversity and more by a taxonomy of nonhuman charisma" [3] (p. 75).

Yet in the context of Gobi afforestation, charismatic organisms do not seem to figure prominently in the logic of conservation. Indeed, biologists hired by mining companies and governments seem relatively uninterested in protecting indigenous life forms. Conservationists instead insert unthreatened species into an alien ecology, suggesting that a verdant, charismatic, synthetic landscape is preferable to a "natural" one. In this regional study, conservation management is not driven by the presence of charismatic species but rather by their absence—a quality that can only exist at the level of the landscape.

I close this paper with a conjecture: there is something about drylands that disrupts the role of species-level charisma in motivating land management. If true, this particular phenomenon could have important implications for how humans across cultures manage desertscapes. Conservation biology in landscapes with high net primary productivity usually takes the form of protecting what is (or was) already there, often framed around keystone or flagship species. By contrast, attempts to protect the Gobi take the form of inserting the largest possible organisms into a landscape frequently described as "barren" or "empty".[1] Is this furnishing fullness to eliminate absence? Does an "empty" landscape connote sickness and a "full" one indicate health? I propose that *absence* itself may be a worthwhile object of study, to better understand the dynamics of landscape conservation in the deserts of the world.

Perhaps a more appropriate conservation strategy for the Gobi would not be invested in creating more life or filling a void. This would require disposing of our usual metaphors for understanding ecosystem health, e.g., as a body or a population [51]. Indeed, the naming of landscape-level charisma calls for a return to certain biophysical basics to counter this bias. Such an approach would embrace absence and shift focus to species diversity, resilience, equilibria, or geologic indicators such as soil transport and groundwater reserves. It would be critical of species-level ontologies with the architecture of flagships and keystones. It would be skeptical of top-down human interventions, particularly those that re-place the landscape with non-native critters. Above all, this approach would be cognizant of the discursive mode in which land managers represent—or misrepresent—the flourishing of landscapes.

A final question lingers: could this same geography of absence present itself in other ecosystems? There are parallels waiting to be drawn: consider the relatively high rate of alpine conservation compared with lowland conservation, or the precipitous state of global blue-water marine habitat—an altogether different type of desert but with an undeniable aura of absence [52]. Even further afield, we may consider how absence adds context to aerospace engineers' increasingly urgent warnings about the growing dangers of orbital space debris [53]. Taken as a whole, a much broader critique emerges, suggesting that the bulk of conservation practice is dedicated not just to *charisma* but to reinforcing *presence*—of mountains, large mammals, deep canyons, striking butterflies, mighty rivers, and forests. How might environmentalists better harness their resources and knowledge to protect those spaces that are, by nature, *absent* before our human eyes?

7. Conclusions

This paper traces the hybrid ecologies of the Gobi along the roots of its newly planted groves. How and why these afforestation programs emerged ties together ecology, charisma, modernity, and the geographies of absence. While the present analysis does not necessarily generalize beyond the Gobi, I see no compelling reason to suggest that it is limited to this one region. Jamie Lorimer has already illustrated the hybridity of conservation practice on a variety of scales.

In 2015, the conservation and ecology journal *Oryx* published a special issue on trees. Peter Crane writes in the opening editorial that "The charismatic megaflora are a gift to the whole of humanity bequeathed to us by millions of years of evolution" [54] (p. 378). It is difficult to imagine the same adulatory words written about seaweed, scrub, or the biotic communities in soils—those less charismatic flora and bacteria. It seems even less probable that the same could be written about "the desert" at large. This paper suggests that when charisma, presence, or cultural affinity cambers the perception of landscapes as healthy or ailing, land management can veer toward ecological irrationality—species need never enter the conversation. It appears that the afforestation efforts in the Gobi are exemplary of this. Identifying this phenomenon asks environmental managers to hold a mirror to their practice and interrogate biases—however large or small—that might influence how we shape and inhabit the deserts of the world.

Funding: The research was funded by the Economic and Social Research Council and the Global Challenges Research Fund, grant number ES/S000798/1; and by the University of Oxford John Fell Fund, grant number 00005620.

---

[1] It bears noting that, more than most plants, trees have particularly anthropomorphic qualities (linguistically evidenced by their *limbs*, *trunks*, and *crowns*). In this sense, desert afforestation could be read as *populating* an absent landscape.

Acknowledgments: I am deeply grateful to Ariell Ahearn and Troy Sternberg; this research would not have been possible without their generous mentorship. Olivia Ashmoore, Cavin Wilson, Michaela Korodimou, Daniel Lezak, and two anonymous reviewers provided comments on previous drafts. S. Soyolbold and M. Tserennadmid gave crucial support and guidance during fieldwork.

Conflicts of Interest: The author declares no conflict of interest.

References

1. *Conservation and Mobile Indigenous Peoples: Displacement, Forced Settlement, and Sustainable Development*; Chatty, D., Colchester, M., Eds.; Studies in Forced Migration; Berghahn Books: New York, NY, USA, 2002; ISBN 978-1-57181-841-6.
2. Brockington, D. *Fortress Conservation: The Preservation of the Mkomazi Game Reserve, Tanzania*; African Issues; International African Institute in Association with James Currey; Indiana University Press: Oxford, UK; Bloomington, IN, USA, 2002; ISBN 978-0-253-34079-5.
3. Lorimer, J. *Wildlife in the Anthropocene: Conservation after Nature*; University of Minnesota Press: Minneapolis, MN, USA, 2015; ISBN 978-1-4529-4428-9.
4. Escobar, A. Whose Knowledge, Whose nature? Biodiversity, Conservation, and the Political Ecology of Social Movements. *J. Political Ecol.* **1998**, *5*, 53–82. [CrossRef]
5. Robertson, D.P.; Hull, R.B. Beyond Biology: Toward a More Public Ecology for Conservation. *Conserv. Biol.* **2001**, *15*, 970–979. [CrossRef]
6. Geisel, T.S. *The Lorax*; Random House: New York, NY, USA, 1971; ISBN 978-0-394-82337-9.
7. Cronon, W. The Trouble with Wilderness: Or, Getting Back to the Wrong Nature. In *Uncommon Ground: Rethinking the Human Place in Nature*; Norton: New York, NY, USA, 1996; pp. 69–90.
8. Latour, B. *Reassembling the Social: An Introduction to Actor-Network-Theory*; Clarendon Lectures in Management Studies; Oxford University Press: Oxford, NY, USA, 2007; ISBN 978-0-19-925605-1.
9. Rutherford, S. Green governmentality: Insights and opportunities in the study of nature's rule. *Prog. Hum. Geogr.* **2007**, *31*, 291–307. [CrossRef]
10. Chaloupka, W. Jagged Terrain: Cronon, Soule', and the Struggle over Nature and Deconstruction in Environmental Theory. *Strateg. J. Theory Cult. Politics* **2000**, *13*, 23–38. [CrossRef]
11. Soulé, M.E. The Social Siege of Nature. In *Reinventing Nature? Responses to Postmodern Deconstruction*; Soulé, M.E., Lease, G., Eds.; Island Press: Washington, DC, USA, 1995; pp. 137–170, ISBN 978-1-55963-310-9.
12. Biro, A. Towards a Denaturalized Ecological Politics. *Polity* **2002**, *35*, 195–212. [CrossRef]
13. Lorimer, J. Nonhuman Charisma. *Environ. Plan. D Soc. Space* **2007**, *25*, 911–932. [CrossRef]
14. Braverman, I. Conservation without nature: The trouble with in situ versus ex situ conservation. *Geoforum* **2014**, *51*, 47–57. [CrossRef]
15. Small, E. The new Noah's Ark: Beautiful and useful species only. Part 2. The chosen species. *Biodiversity* **2012**, *13*, 37–53. [CrossRef]
16. Clucas, B.; McHugh, K.; Caro, T. Flagship species on covers of US conservation and nature magazines. *Biodivers. Conserv.* **2008**, *17*, 1517–1528. [CrossRef]
17. Minin, E.D.; Moilanen, A. Improving the surrogacy effectiveness of charismatic megafauna with well-surveyed taxonomic groups and habitat types. *J. Appl. Ecol.* **2014**, *51*, 281–288. [CrossRef]
18. Home, R.; Keller, C.; Nagel, P.; Bauer, N.; Hunziker, M. Selection criteria for flagship species by conservation organizations. *Environ. Conserv.* **2009**, *36*, 139–148. [CrossRef]
19. Sneath, D. *Mongolia Remade: Post-Socialist National Culture, Political Economy, and Cosmopolitics*; Amsterdam University Press: Amsterdam, The Netherlands, 2018; ISBN 978-90-485-4213-0.
20. Polo, M. *The Travels of Marco Polo, the Venetian*; Wright, T., Ed.; Bohn: London, UK, 1854.
21. Sternberg, T.; Rueff, H.; Middleton, N.; Sternberg, T.; Rueff, H.; Middleton, N. Contraction of the Gobi Desert, 2000–2012. *Remote Sens.* **2015**, *7*, 1346–1358. [CrossRef]
22. Mühlenberg, M.; Batkhishig, T.; Dashzeveg, T.; Drößler, L.; Neusel, B.; Tsogtbaatar, J. *Lessons From Tree Planting Initatives in Mongolia*; Mongolia Discussion Papers, East Asia and Pacific Environment and Social Development Department; World Bank: Washington, DC, USA, 2006.
23. Wesche, K.; Ronnenberg, K.; Retzer, V.; Miehe, G. Effects of large herbivore exclusion on southern Mongolian desert steppes. *Acta Oecol.* **2010**, *36*, 234–241. [CrossRef]

24. Johansson, Ö.; McCarthy, T.; Samelius, G.; Andrén, H.; Tumursukh, L.; Mishra, C. Snow leopard predation in a livestock dominated landscape in Mongolia. *Biol. Conserv.* **2015**, *184*, 251–258. [CrossRef]
25. Robinson, J.; Verboom, G.; Urantsatsral, C.; Thrift, E. Northern Buddhist Conference on Ecology and Development, Summary of Sessions. Ulaanbaatar, Mongolia, 2005. Available online: http://www.arcworld.org/downloads/UB_conference.pdf (accessed on 26 November 2018).
26. Anonymous. Halfway to Where? *The Economist*, 8 October 2011.
27. Kapelus, P. Mining, Corporate Social Responsibility and the "Community": The Case of Rio Tinto, Richards Bay Minerals and the Mbonambi. *J. Bus. Ethics* **2002**, *39*, 275–296. [CrossRef]
28. Anonymous. Mine, all Mine. *The Economist*, 21 January 2012.
29. The Biodiversity Consultancy. *Fauna & Flora International ESIA Appendix 5: Net Positive Impact forecast for the Oyu Tolgoi Project*; The Biodiversity Consultancy: Cambridge, UK, 2012.
30. OT Plants Saxaul Trees to Restore Gobi Landscape. Available online: http://en.mongolianminingjournal.com/content/57143.shtml (accessed on 26 November 2018).
31. High, M.M. *Fear and Fortune: Spirit Worlds and Emerging Economies in the Mongolian Gold Rush*; Cornell University Press: Ithaca, NY, USA, 2017; ISBN 978-1-5017-0754-4.
32. Cao, S. Why Large-Scale Afforestation Efforts in China Have Failed to Solve the Desertification Problem. *Environ. Sci. Technol.* **2008**, *42*, 1826–1831. [CrossRef] [PubMed]
33. Cao, S.; Tian, T.; Chen, L.; Dong, X.; Yu, X.; Wang, G. Damage Caused to the Environment by Reforestation Policies in Arid and Semi-Arid Areas of China. *Ambio* **2010**, *39*, 279–283. [CrossRef]
34. Cao, S.; Chen, L.; Shankman, D.; Wang, C.; Wang, X.; Zhang, H. Excessive reliance on afforestation in China's arid and semi-arid regions: Lessons in ecological restoration. *Earth-Sci. Rev.* **2011**, *104*, 240–245. [CrossRef]
35. Bailey, R.M. Spatial and temporal signatures of fragility and threshold proximity in modelled semi-arid vegetation. *Proc. R. Soc. B Biol. Sci.* **2011**, *278*, 1064–1071. [CrossRef]
36. Turquoise Hill Resources. *Made by Mongolia: Annual Report 2017*; Turquoise Hill Resources, Ltd.: Vancouver, BC, Canada, 2017.
37. Law of Mongolia: Minerals Law. Available online: http://en.mongolianminingjournal.com/content/38041.shtml (accessed on 26 November 2018).
38. Ma, J.; Ding, Z.; Edmunds, W.M.; Gates, J.B.; Huang, T. Limits to recharge of groundwater from Tibetan plateau to the Gobi desert, implications for water management in the mountain front. *J. Hydrol.* **2009**, *364*, 128–141. [CrossRef]
39. Brain, S. The Great Stalin Plan for the Transformation of Nature. *Environ. Hist.* **2010**, *15*, 670–700. [CrossRef]
40. Williams, J.A. "The Chords of the German Soul are Tuned to Nature": The Movement to Preserve the Natural Heimat from the Kaiserreich to the Third Reich. *Cent. Eur. Hist.* **1996**, *29*, 339. [CrossRef]
41. Saito, O. Forest history and the Great Divergence: China, Japan, and the West compared. *J. Glob. Hist.* **2009**, *4*, 379. [CrossRef]
42. Li, M.; Liu, A.; Zou, C.; Xu, W.; Shimizu, H.; Wang, K. An overview of the "Three-North" Shelterbelt project in China. *For. Stud. China* **2012**, *14*, 70–79. [CrossRef]
43. Hanjie, W.; Hao, Z. A simulation study on the eco-environmental effects of 3N Shelterbelt in North China. *Glob. Planet. Chang.* **2003**, *37*, 231–246. [CrossRef]
44. Wang, X.M.; Zhang, C.X.; Hasi, E.; Dong, Z.B. Has the Three Norths Forest Shelterbelt Program solved the desertification and dust storm problems in arid and semiarid China? *J. Arid Environ.* **2010**, *74*, 13–22. [CrossRef]
45. Ahrends, A.; Hollingsworth, P.M.; Beckschäfer, P.; Chen, H.; Zomer, R.J.; Zhang, L.; Wang, M.; Xu, J. China's fight to halt tree cover loss. *Proc. R. Soc. B Biol. Sci.* **2017**, *284*, 20162559. [CrossRef] [PubMed]
46. Wang, X.; Dong, Z.; Zhang, J.; Liu, L. Modern dust storms in China: An overview. *J. Arid Environ.* **2004**, *58*, 559–574. [CrossRef]
47. Three-North Shelterbelt Program. Available online: http://english.forestry.gov.cn/ (accessed on 26 November 2018).
48. Jiang, G.; Han, X.; Wu, J. Restoration and Management of the Inner Mongolia Grassland Require a Sustainable Strategy. *Ambio J. Hum. Environ.* **2006**, *35*, 269–270. [CrossRef]
49. Jiang, H. China's Great Green Wall Proves Hollow. *Epoch Times* **2009**, *29*, 2009.
50. Burnett, J.; Copp, C.; Harding, P. *Biological Recording in the United Kingdom: Present Practice and Future Development*; UK Department of the Environment: Oxford, UK, 1994.

51. Larson, B. *Metaphors for Environmental Sustainability: Redefining Our Relationship with Nature*; Yale University Press: New Haven, CT, USA, 2014; ISBN 978-0-300-20581-7.
52. Jackson, S.E. The water is not empty: Cross-cultural issues in conceptualising sea space. *Aust. Geogr.* **1995**, *26*, 87–96. [CrossRef]
53. Shan, M.; Guo, J.; Gill, E. Review and comparison of active space debris capturing and removal methods. *Prog. Aerosp. Sci.* **2016**, *80*, 18–32. [CrossRef]
54. Crane, P. Can we save the charismatic megaflora? *Oryx* **2015**, *49*, 377–378. [CrossRef]

land

Article

Assessing Riparian Vegetation Condition and Function in Disturbed Sites of the Arid Northwestern Mexico

Lara Cornejo-Denman [1], Jose Raul Romo-Leon [1,*], Alejandro E. Castellanos [1], Rolando E. Diaz-Caravantes [2], Jose Luis Moreno-Vázquez [2] and Romeo Mendez-Estrella [1]

1 Departamento de Investigaciones Científicas y Tecnológicas, Universidad de Sonora, Hermosillo 83000, Mexico; laracornejodenman@gmail.com (L.C.-D.); acastell@guaymas.uson.mx (A.E.C.); mendezromeo@hotmail.com (R.M.-E.)

2 El Colegio de Sonora, Hermosillo 83000, Mexico; rdiaz@colson.edu.mx (R.E.D.-C.); jmoreno@colson.edu.mx (J.L.M.-V.)

* Correspondence: joser2@email.arizona.edu; Tel.: 11+52-662-259-2169

Received: 27 November 2017; Accepted: 18 January 2018; Published: 22 January 2018

Abstract: Transformation or modification of vegetation distribution and structure in arid riparian ecosystems can lead to the loss of ecological function. Mexico has 101,500,000 ha of arid lands, however there is a general lack of information regarding how arid riparian ecosystems are being modified. To assess these modifications, we use eight sites in the San Miguel River (central Sonora) to analyze (1) riparian vegetation composition, structure and distribution using field sampling and remote sensing data from Unmanned Aerial Vehicles (UAV); (2) productivity (proxies), using vegetation indices derived from satellite data; and (3) variability posed by riparian vegetation and vegetation adjacent to riparian habitats. The development of a simple yet informative Anthropogenic-disturbance Index (ADI) allowed us to classify and describe each study site. We found sharp differences in vegetation composition and structure between sites due to the absence/presence of obligate-riparian species. We also report significant difference between *EVI* (Enhanced Vegetation Index) values for the dry season among vegetation types that develop near the edges of the river but differ in composition, suggesting that land cover changes form obligate-riparian to facultative-riparian species can lead to a loss in potential productivity. Finally, our tests suggest that sites with higher disturbance present lower photosynthetic activity.

Keywords: riparian ecosystems; Sonoran desert; remote sensing; land cover/land use

1. Introduction

It has been thoroughly documented that human activities, which modify water availability and promote drastic habitat modifications, have caused dramatic changes in riparian vegetation of arid ecosystems in the American southwest [1–4]. Due to the similarities between geographic and ecological conditions of this region and the Mexican northwest, it is reasonable to expect that the disturbance processes could be similar, although fewer studies of riparian ecosystems in the Mexican northwest have been conducted.

Water availability in the Mexican northwest is critical, since many of its watersheds and groundwater sources are overexploited, due to intensive use by economic activities such as agriculture and cattle ranching [5–7]. Moreover, previous work suggests that water management practices could potentially modify the structure and function of key ecosystems in arid environments [3,8–10]. In this sense, recent studies describe how groundwater depth restricts the establishment of riparian vegetation in watersheds located in arid environments of central Sonora (northwestern Mexico), particularly in places where high levels of groundwater depth coincide with a denser distribution of agricultural wells [8]. Groundwater levels can also affect surface runoff, possibly narrowing the river channel; a negative correlation has been found between channel width and the percentage of agricultural land use along the river sides, in several rivers in Sonora [11]. Alteration of groundwater levels and surface runoff can prevent the establishment of obligate-riparian species such as *Salix gooddingii* and *Populus fremontii* [12], sometimes causing the complete loss of riparian vegetation or its replacement by drought tolerant desert species like *Prosopis* sp. or *Acacia* sp.; these changes in composition can have several consequences on the ecosystem: loss of local biodiversity [13], increase the susceptibility to invasive species [3,14], loss of riparian native corridors [15,16] and decrease in vegetation indices values [17].

Riparian ecosystems are often considered hotspots for ecosystem services [18–20]. In arid environments, this is evident since most economic, social and cultural activities are closely tied to this systems [21–23]. Therefore, riparian ecosystems in arid environments are highly dynamic in terms of changes on land use and land cover [24,25]. These changes often result in the modification of the structure and function of riparian vegetation due to active change (e.g., the clearing of riparian vegetation to establish agricultural areas) or changes introduced by constant pressure posed by human activities (e.g., Cattle Ranching) [11,26,27].

For arid environments, there is a lack of information regarding the assessment of disturbance, however, it is of utmost importance to understand how different disturbance-causing activities are changing riparian habitats in Sonora, and how these changes could be compromising the provision of ecosystem services. Since agriculture and cattle ranching are the main subsistence and commercial activities in the region and because these activities are registered to be the ones that cause the greatest changes in ecosystems all over the state [28,29], we consider them as playing an essential role when assessing disturbance in riparian habitats. Also, for a region with critical water availability issues it is important to have a better understanding of how human activities are changing riparian ecosystems, thus, the monitoring of different vegetation parameters and the generation of a disturb-based classification method can help assess the ecological condition of these ecosystems in Sonora.

Different remote sensing techniques can be used to identify areas where obligate-riparian vegetation is being transformed in terms of its composition, aerial images can be especially useful to discriminate vegetation types along rivers [30,31]; at the same time satellite image analysis can be used to derive vegetation indices [32–34] which can provide important information about the behavior of the ecosystem in terms of its productivity. Previous studies using remote sensing suggest a trend towards the modification of riparian vegetation, due to the establishment on facultative species in north western Mexico [8,35]. However, assessments using high spatial resolution imagery to evaluate the previous (using UAV's) are scarce.

The objective of the present study was to analyze disturbance and its effects on function and structure of riparian habitats in arid regions of northwestern Mexico. For the previous we explore the combined use of satellite imagery, Unmanned Aerial Vehicles (UAV) photography and field vegetation sampling to assess the differences and similarities within and between 8 disturbed sites along the San Miguel River in central Sonora. Also, we evaluated several characteristics of riparian vegetation and the adjacent desert scrub: composition, diversity, cover and photosynthetic activity (represented by the Enhanced Vegetation Index, *EVI*). Finally, we assessed the disturbance degree by the development and application of the ADI.

By following the suggested approach, we will be able to assess how disturbance at the local level affects riparian vegetation variables (composition, structure, diversity and cover) and how changes in these relate to shifts in function (photosynthesis) at the landscape level.

2. Materials and Methods

In order to assess disturbance of riparian habitats we developed specific qualitative parameters and applied them during site visits. Following this, we evaluated structure, composition and cover of vegetation on different sites along the river by field sampling. To complement site description, we used aerial imagery to define which vegetation types were present in each site and determine plant cover. Finally, satellite imagery was used to derive the *EVI*.

2.1. Study Area

The San Miguel river (SMR) is located in the central part of the northern Mexican state of Sonora (Figure 1), between 30°45.75′N–29°6.5′N latitude and 111°4.11′W–110°21.1′W longitude and it is a sub watershed of the much larger Sonoran river watershed. It has an extension of 3845 km^2. These two rivers (San Miguel and Sonora) are the main sources of water for the capital city of Hermosillo. The mean annual temperature is of 21 °C and the mean annual precipitation is of 421 mm. The 8 study sites are located along the San Miguel River, two in the northern part, two in the northern-central, two in the southern-central and two in the southern part. According to Shreve and Wiggins [36] classification, the SMR is located in the Plains of Sonora, which is designated as Arbosuffrutescent Desert and it is characterized by the presence of *Olneya tesota* trees and *Encelia farinosa* bushes. Classification made by the National Institute of Geography and Statistics establishes the following types of vegetation and land uses in the area: oak forest, oak-pine forest, microphyllous desert scrub, mesquite forest, xerophilous mesquital, subtropical scrub, native and introduced grasslands, riparian vegetation and agriculture. The main economic activities in the region are agriculture and cattle ranching. There is a total of 8266 hectares of irrigated lands, 75% of which correspond to forage production [21]. These activities have developed in the region over hundreds of years, however in the past decades, due to the technological advances and the intensive use of riparian habitat, the system has become very dynamic in terms of land cover shifts. On top of this, it has been reported that the SMR aquifer shows signs of overexploitation based on recharge-extraction analyses; official data reported an annual deficit of 1,500,000 m^3 [7], although the most recent update reports an availability of 17,508,107 m^3 [5], which arises some confusion and concern due to the fact that this last number was based on a 2008 hydrogeological study elaborated for a mining project. Given the above, riparian habitats show different scenarios where water availability has become an economical [22], social [21,23] and ecological [8,37] issue that needs to be addressed by managers and land owners.

The study sites where selected through satellite image observation and exploration field trips. Final selection was based on the presence of several land uses of interest: agriculture, livestock grazing, human settlements, roads, recreational activity and the presence of riparian vegetation.

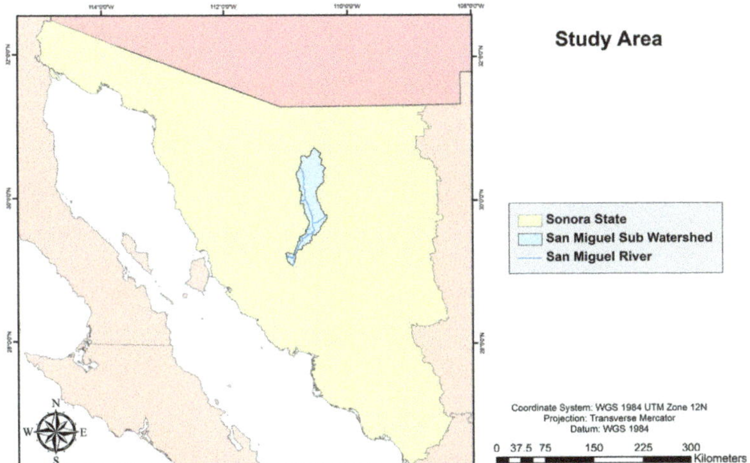

Figure 1. Location of study area.

2.2. Disturbance Assessment

To establish the degree of disturbance on each site, we developed a series of qualitative criteria regarding (1) the intensity and type of human activities; (2) the state of the riparian vegetation; (3) the presence/absence of surface water. Classification of sites was made through applying the ADI, which evaluates the sites according to several criteria (Table 1). A number was assigned to indicate the presence, absence, evidence or intensity of the features observed: superficial water flow, *Populus* and/or *Salix* species, cattle, agriculture, roads and/or human settlements and recreational activity.

Although the San Miguel River is an intermittent river [38,39], superficial water flow can be continuous during rainy season. Average precipitation in the region during the sampling year was of 382 mm [40]. Since field sampling was done through the rainy season, superficial water flow was expected to be present and, thus, it was considered among the classification criteria. *Populus* and *Salix* species are particularly susceptible to groundwater fluctuations and their absence could be an indicator of groundwater scarcity. Livestock and agriculture are the most common economic activities in the area, so their presence and closeness to the river was considered for site classification. Human settlements and recreational activity can also be sources of disturbance so they were also considered for site classification.

The selection of the criteria used to develop this index is based on literature that evaluates human water use and its impacts on river flow in the American southwest; these studies report that overexploitation of rivers has caused a reduction in the presence of riparian trees such as *Populus fremontii* and *Salix gooddingii* [12,13]. Agriculture, livestock grazing, human settlements and recreational activity were also considered as criteria since these are the main disturbance types reported for riparian ecosystems [4,24,41]. Criteria selection and structure for the index was also based on the environmental condition of riparian areas, as a methodology reported by González and García [42]. The criteria and interpretation of the index values are explained on Tables 1 and 2.

Table 1. Criteria used to determine the ADI.

	Yes	VALUE	Little/Few	VALUE	No	VALUE
Superficial water flow	Continuous flow	1	Intermittent flow	2	Dry riverbed	3
Presence of *Populus* and/or *Salix* species	High frequency of these species	1	Lower frequency than other trees	2	Absence of these species	3
Presence or evidence of cattle	Absence of cattle	1	Evidence of grazing, feces, or buffel grass	2	Cattle was seen near or around the river	3
Agriculture	No agricultural areas were seen	1	Presence of agricultural areas smaller than 100 sq. meters	2	Agricultural areas less than 500 m from the river	3
Roads and/or human settlements	Study site located 500 m away from any road or house	1	Presence of roads but no houses	2	Houses and roads near the river	3
Recreational activity	No human presence or infrastructure	1	Human presence but no infrastructure	2	Evidence of human presence (parking places, grills, trash cans, trash)	3

Table 2. Interpretation of the ADI values.

ADI	Disturbance Category
6–8	Low: there's no observable evidence of anthropogenic disturbance. Superficial water flow is continuous (during rainy season); vegetation is dominated by obligate-riparian species.
9–11	Low-medium: there's an observable evidence of human activities in low intensity; vegetation is dominated by obligate-riparian species.
12–14	Intermediate: superficial water flow may be intermittent, human activities diversify; vegetation is characterized by the presence of facultative-riparian species along with obligate-riparian species.
15–17	Medium-high: superficial water flow is intermittent or the riverbed is dry, human activities diversify, absence of obligate-riparian species and dominance of non-riparian species.
18	High: dry riverbed, human activities intensify, absence of obligate-riparian species and dominance of non-riparian species.

2.3. *Field Measurements and Derivation of Diversity Index*

Vegetation description on each site was made through the registry and derivation of several variables (Table 3). Field sampling was done following the *relevé* method and protocol [43]. Two major vegetation types where sampled on each site (riparian vegetation and adjacent desert scrub), the purpose of this vegetation description was to compare the variation of several vegetation characteristics of riparian sites and of the adjacent desert scrub sites.

Table 3. Variables used for the vegetation description.

Variable	Description	How Was It Measured?
Composition	The list of species found in each stand.	Field identification and registry, herbarium identification.
Cover	Vertical crown or shoot-area projection for each vegetation strata (herbs, shrubs and trees) per plot.	Continuous (65%) = crown touching. Intermittent (30–64%) = interlocking or touching crowns interrupted by openings. Open (29%) = crowns not touching or infrequently touching.
Richness	Total number of species per stand (for tree and shrub strata only).	Count the different species present on each stand.
Diversity	Diversity of each stand (for tree and shrub strata only).	Represented by the Shannon-Weiner diversity index.
Dominance	Tree and shrub species with the highest number of individuals per stand.	Count the number of individuals per species.

Two sampling units (stands) were established on each site, one for riparian vegetation and another for the adjacent desert scrub located approximately 200 m from the first stand and in the opposite direction of the river, this distance was considered enough to go past the transition area between riparian vegetation and the adjacent desert vegetation. San Juan Ranch was the only site where desert scrub was not sampled, due to time and climate restrictions.

Field sampling consisted in identifying vegetation strata and registering composition and cover for each. First, a 30 × 30 m square was delineated, in which tree species where registered along with the number of individuals per species, cover was estimated for the whole tree strata. Then, a second square of 10 × 10 m was delineated inside the first one and shrub species, number of individuals and strata cover were registered. Finally, a third square of 1 × 1 m was delineated inside the second square and the number of different herbaceous species and strata cover were registered. Two repetitions were made, resulting in a 30 × 90 m (three repetitions per vegetation strata in each sample) stand with a total of 2700 sq. meters sampled for each vegetation type on each site.

For each sampling site we calculated diversity of the tree and shrub strata by the derivation of the Shannon-Weiner diversity index [44]. Diversity of the shrub and tree strata for each stand was estimated using the Shannon-Weiner diversity index:

$$H' = -\sum_{i=1}^{s} p_i log_2 p_i$$

where S is the total number of species, p_i the proportion of individuals of each species based on the total of individuals (ni/N), ni is the number of individuals of each species and N the total of all individuals of all species.

2.4. Aerial Imagery Sampling and Mosaicking

Aerial imagery acquisition requires two processes. First, obtaining aerial images in the field and second, processing those images for the orthomosaic generation. A total of 13 flights were performed in the different study sites, using a UAV (Unmanned Aerial Vehicle). See Supplementary Materials for spatial and temporal specifications of each flight (Table S1). The following subsections describe the process of the image acquisition and orthomosaic generation.

2.4.1. Aerial Imagery Sampling

For the acquisition of aerial images several flights were performed at each study site. See Supplementary Materials for flight requirements and image specifications (Table S2). All flights were performed during growing season (August, September and October 2016) with the purpose of recording vegetation cover at its greatest; and during the day (between 11 a.m. and 2 p.m.) to ensure appropriate light conditions.

2.4.2. Mosaicking process

The image processing is based on pattern recognition, which establishes a relationship between a pattern (vector of features describing an object) and a class label; those features can be spectral reflectance, texture, emittance values from optical imagery, or geographical features and the object can be a single pixel or a set of adjacent pixels forming a geographical entity [45].

The mosaicking process is performed by photogrammetry software. The process consists on finding common points (keypoints) between the images. When the same keypoint is found on 2 different images they are matched and this will generate a 3D point (Pix4D). These 3D points are needed to generate a Point Cloud from which a Digital Surface Model and an Orthomosaic will derive. This process requires a high overlap between images, which in most cases is at least 75% (Pix4D).

Since the UAV has a built-in GPS, all images taken were automatically georeferenced and no ground control points were needed. Georeferenced images from each flight where added to the photogrammetry software for the stitching process, resulting in a total of 13 orthomosaics with an approximate area of 5 hectares each. The spatial resolution expected from the orthomosaic is of 3 cm.

2.5. Satellite Imagery

Satellite data was obtained as a land surface reflectance (LASRC) from the Landsat 8 OLI (Operational Land Imager)/TIRS (Thermal Infrared Sensors) Pre-Collection L1T data type, directly from the United States Geological Survey 'Earth Explorer' website; these products are radiometrically calibrated and orthorectified using ground control points and digital elevation model data to correct for relief displacement [46]. Landsat 8 data includes 11 bands, from which 7 of them (1–7) were selected and processed in ERDAS IMAGINE 9.2 to create the final stack. Bands 8, 9, 10 and 11 were not used.

Since the study area comprises two Landsat 8 scenes and because we wanted to compare photosynthetic activity variations between two seasons, a total of 4 scenes were chosen for two different dates on 2016: May and September. See Supplementary Materials for scene specifications (Table S3).

The vegetation index was derived for the complete four scenes but the only pixels involved in the analysis were those corresponding to the digitalized areas from the orthomosaics, obtained from the UAV analysis previously described.

2.6. Vegetation Index Derivation

Vegetation indices (VI) are designed based on the spectral response to light of each type of vegetation; this is how indices can provide qualitative and quantitative measurements that can be used to describe photosynthetic activity, structure characteristics and canopy density, among other variables [47,48]. As explained by the previous references, several VI have been related to structural aspects of vegetation, such as Leaf Area Index (LAI); this parameter is related to the amount of available photosynthesizing biomass. The previous can be used as a theoretical basis to derive Net Primary Production from satellite data, as explained by Running et al. [49].

The Normalized Difference Vegetation Index (NDVI) is commonly used in arid regions because of their strong seasonality change but when vegetation density increases during the growth season NDVI can reflect high values with very little variation, which can be interpreted as a spectral signal saturation [50]. Thus, another vegetation index was selected for this study, the Enhanced Vegetation Index (*EVI*), which presents higher sensitivity in areas where vegetation is dense and it is less prone to light saturation.

Enhanced Vegetation Index (*EVI*) is derived by using the following equation:

$$EVI = G\frac{\rho_{NIR} - \rho_{red}}{\rho_{NIR} + C_1 \times \rho_{red} - C_2 \times \rho_{blue} + L}$$

where ρ is the surface reflectance with atmospheric correction, L is the canopy background adjustment; C_1 and C_2 are the aerosol resistance coefficients, as described by [50].

EVI values range from 0 to 1, those closer to 1 indicate a greater density and photosynthetic activity of vegetation, those closer to 0 indicate vegetation is scarce or less productive. Although *EVI* values do not reflect a direct amount of Net Primary Productivity, it is considered as an accurate measure of photosynthetic activity, which is related to the amount of biomass.

2.7. Land Cover Assessment

Once obtained, we used orthomosaics to identify the different vegetation types in the sample areas. Our class scheme (Table 4) was developed using observations from the field (regarding community composition and stand physiognomy) and previous classification studies conducted in the area [8]. In order to simplify the classification scheme, all desert vegetation types adjacent to the river is referred to as "Desert Scrub", even though the formal designation for the Plains of Sonora region (as described by Shreve and Wiggins 1964 classification) is Arbosuffrutescent Desert.

Division of riparian classes (Riparian Vegetation and Riparian Mesquite Woodland) is based mainly on the differences in composition between two classes. Considering that Riparian Vegetation is composed mainly by obligate-riparian species and Riparian Mesquite Woodland by facultative-riparian species [51].

Land cover distinction was done through field observation and comparing the field sampling sites with what was shown on the orthomosaic, thus the field sampling sites served as ground control points. Differentiation between land cover types, on the aerial photography was possible through (1) extensive training on the field and (2) training on the analysis and management of each orthomosaic generated. Therefore, our land cover classification was conducted via "heads up digitizing" analysis through expert knowledge. Some of the main elements of photo interpretation used for this analysis were: color (different tree species vary in color), cover (e.g., riparian mesquite woodland tends to be denser than riparian vegetation), distribution, frequency and pattern (e.g., bare ground patches dominated by columnar cacti like *Stenocereus thurberi* and *Lophocereus schottii*).

Table 4. Land cover descriptions.

Land Cover Class	Description	Literature
Riparian Vegetation	The most important tree species found in riparian ecosystems in Sonora are *Salix goodingii* and *S. bonplandiana*, *Prosopis velutina* and *Populus fremontii*. Common shrubs include *Baccharis salicifolia*, *Hymenoclea monogyra*, *Acacia constricta* and *A. farnesiana*, *Celtis pallida*, *Nicotiana glauca*.	Scott et al. 2009 [11]
Riparian Mesquite Woodland	Common vegetation found on streamways includes *Prosopis*, *Parkinsonia florida*, *Olneya*, *Acacia greggii* and *Acacia occidentalis*. Common shrubs include *Celtis pallida*, *Baccharis sarothroides* and *Lycium*.	Shreve and Wiggins, 1964 [36]
Desert Scrub	Open scrubland with small, low-branching trees, with irregular colonies of shrubs and widely spaced columnar cacti (*Stenocereus thurberi*, *Lophocereus schottii*). *Parkinsonia*, *Olneya* and *Prosopis* find their optimum conditions, alternated with colonies of *Larrea*. Common shrubs include *Phaulothamnus spinescens*, *Mimosa laxiflora* and *Celtis pallida*.	Shreve and Wiggins, 1964 [36]
Herbaceous cover	This cover was considered as an independent class, due to the amount of cover variation between seasons, which has a considerable influence on *EVI* values.	All herbaceous cover identified from the orthomosaic was integrated in this class.

2.8. Cover Estimations from the Combination of Aerial and Satellite Imagery

Cover estimations were derived from each orthomosaic. Orthomosaics were displayed on top of the satellite scenes (Figure 2) and the image pixel was used as the measuring unit for the vegetation types seen on the orthomosaic, thus, every orthomosaic was divided in a grid of 30 × 30 m cells. Orthomosaic area selection consisted in selecting those pixels with more than 90% of visible coverage and excluding pixels with more than 10% of visual errors such as dark parts (generally pixels at the edges of the mosaic) or pixels with too much shadow.

Figure 2. Left: Orthomosaic showing the river section in Galera site. **Right**: Image showing the over-positioning of orthomosaic on Landsat scene, each square represents a Landsat pixel.

Estimations were made for each of the 900 square meter Landsat 8 pixels by direct digitalization using GIS, starting in the inferior left corner of each pixel and continuing clockwise. Measures were registered for each vegetation type found on each pixel and then added to have a total sum for each cover class; this sum was converted into a percentage based on the total area used for digitalization.

2.9. Photosynthesis Analysis on Land Cover Classes

Using *EVI* values as a variable for vegetation function, a comparison was made among all land cover classes, among particular land cover classes of interest and between photosynthetic activity and disturbance. *EVI* values were selected as described below.

Each pixel of the Landsat 8 scenes has a unique *EVI* value. The interposition of the orthomosaic over the Landsat scene allowed us to discern the vegetation types of each pixel, thus having a precise distinction of the photosynthetic activity for each class, which facilitated its comparison. Not all pixels of the orthomosaics were used for the photosynthesis analysis, only those where the land cover class covered more than 50% of the pixel. This pixel selection was done to guarantee that the photosynthetic activity signal corresponds to a dominant land cover class.

2.9.1. Seasonal Photosynthetic Activity per Land Cover Class

Photosynthetic activity variation within each vegetation type was analyzed in order to determine if there's a significant difference among them. *EVI* values for both seasons (May and September) and for each land cover class were selected and residuals for change were obtained by subtracting May *EVI* values from September *EVI* values. A Kruskal-Wallis test was applied to the residues of each land cover class. This analysis was applied to all land cover classes present in all of the study sites.

2.9.2. Photosynthetic Activity of Riparian Vegetation and Riparian Mesquite Woodland

The first step for this analysis was to identify which sites represented Riparian Vegetation and Riparian Mesquite Woodland along the edge of the river. The objective is to compare the physiological response (measured as photosynthetic activity) of two land cover classes that develop in the same area and supposedly have the same resource availability (in this case, water).

In order to assess the differences in photosynthetic activity between the two riparian land cover classes (Riparian Vegetation and Riparian Mesquite Woodland), *EVI* values for both seasons (May and September) were selected and analyzed using two Mann-Whitney tests for each site, one for May-data and another for September-data.

2.9.3. Photosynthetic Activity and Disturbance

Photosynthetic activity represented by the vegetation index is an important parameter to assess vegetation condition. Thus, it is important to know how this parameter varies in relation to disturbance.

Since the ADI was applied to evaluate the condition of riparian sites, this analysis only includes data from land cover classes developed near the river (Riparian Vegetation and Riparian Mesquite Woodland). In order to establish if there's a relation between photosynthetic activity values of riparian vegetation and disturbance, a linear regression was applied between *EVI* average values for September and the ADI values for each site.

3. Results

3.1. Site Classification According to the ADI

Results of the Anthropogenic-disturbance Index application are presented on Table 5. Distribution and ranking of study sites are presented on Figure 3.

Table 5. Criteria intensity (**Y** yes, **L/F** low or few, **N** no) registered on each site for derivation of the Anthropogenic-disturbance Index.

	Superficial Water Flow			Presence of *Populus* and/or *Salix* Species			Presence or Evidence of Cattle			Agriculture			Roads and/or Human Settlements			Recreational Activity			ADI	Disturbance Category
	Y	L/F	N	Y	L/F	N	Y	L/F	N	Y	L/F	N	Y	L/F	N	Y	L/F	N		
Aguilar Ranch	X			X				X		X			X					X	11	Low-medium
Cienega		X			X		X					X	X					X	12	Intermediate
Living Fences	X			X			X			X			X					X	12	Intermediate
El Cajon	X					X	X			X			X			X			16	Medium-high
Nogalera			X			X		X		X			X					X	15	Medium-high
Galera	X			X				X		X			X					X	11	Low-medium
San Juan Ranch	X			X				X			X		X					X	10	Low-medium
Tomas Ranch	X			X				X			X		X					X	10	Low-medium

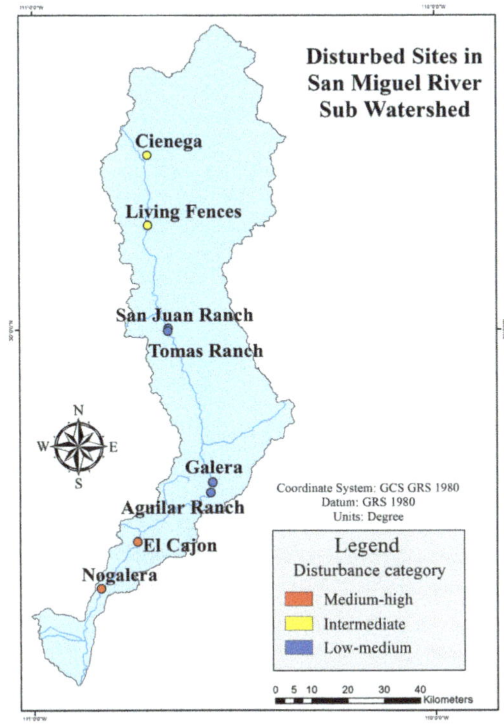

Figure 3. Distribution of sites along the sub watershed.

The two northern sites have an intermediate level of disturbance—they are approximately at a 10 km distance from the largest town (Cucurpe) and share the presence of cattle near the river. Both sites present some type of management. Cienega stands out due to the presence of cattle all over the site. On the other hand, the Living Fences site presents living fence rows made of *Salix* an *Populus* species in several river sections of the site, all of them in different stages of growth.

All central sites present the lowest level of disturbance, low-medium, which was unexpected because most of the sites are ranches were different types of management are definitely present. In spite of this, the vegetation on these sites did not show signs of great disturbance, their main common characteristics are: the presence and dominance of obligate-riparian species such as *Populus fremontii* and *Salix gooddingii*, the presence of a constant superficial water flow and the absence of recreational activities.

The two southern sites are the ones with the highest level of disturbance (Cajon and Nogalera) and have certain similarities such as the absence of obligate-riparian trees and the presence of at least three human activities developed around the river. Commercial agricultural activity in Nogalera stands out, due to the presence of extensive pecan tree (probably *Carya illinoinensis*) fields about 500 m away from the river. El Cajon stands out due to the presence of intense recreational activity.

3.2. Cover, Composition and Diversity from Field Estimations

Field measurements are presented for Riparian Vegetation and Desert Scrub stands on each site. Cover is presented as a percentage and it was estimated for each vegetation strata (herbs, shrubs and trees). Composition was determined for the bush and tree strata on each site. Diversity was derived from field data applying the Shannon-Wiener diversity index.

3.2.1. Vegetation Strata Cover for Riparian Vegetation and Desert Scrub Stands

Riparian Vegetation in all sites shows an herbaceous cover of 50% or more, this can be due to the fact that field sampling was done during the growing season. Shrub cover is variable among all sites, the lowest percentages in Aguilar Ranch, Cienega and San Juan Ranch can be due to the constant presence of cattle and the negative effect it has on shrubs [4,11,26,52–55]. Trees cover more than 30% of the area in most sites (Figure 4), which indicates certain stability among the canopy. Although tree composition in all sites varies, canopy stability is important to maintain riparian habitat conditions.

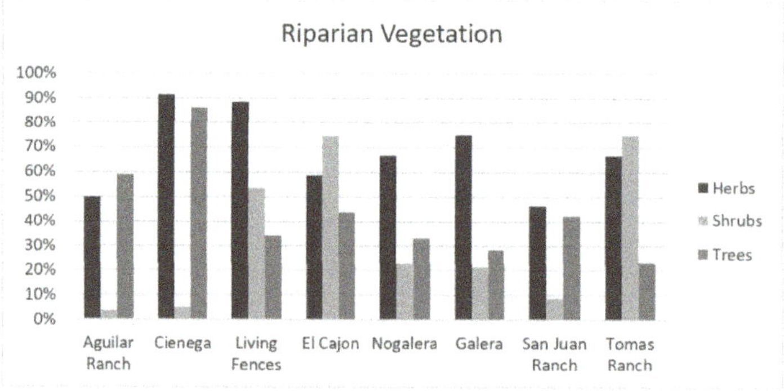

Figure 4. Cover percentage for vegetation strata on the riparian vegetation stands.

Desert Scrub cover (Figure 5) varies greatly among sites, this makes sense given the distinct terrain characteristics of each site. The two most disturbed sites (El Cajon and Nogalera) differ greatly in herbaceous and shrub cover. Shrubs have the highest cover in El Cajon (sampling site was a rocky hill) and the lowest in Nogalera (sampling site was a plain). Herbaceous cover on the Nogalera site stands out as the highest of all sites. Tree strata is higher in the northern site Cienega (63%) and lowest in the other northern site Living Fences (7%), sampling stand in this last site was a steep rocky hill. In contrast with riparian sites, covers for desert scrub are considerably low.

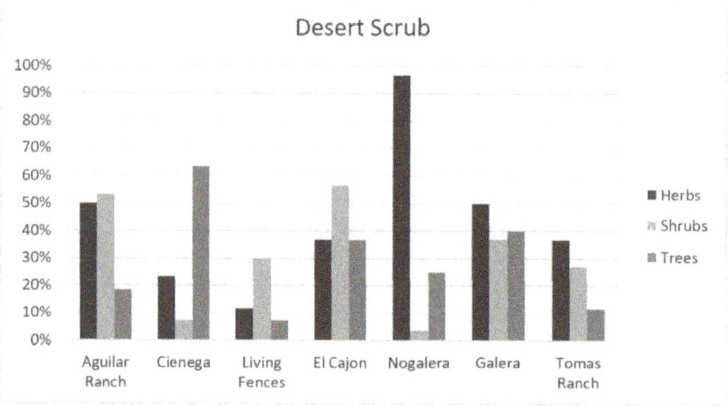

Figure 5. Cover percentage for vegetation strata on the desert scrub stands.

3.2.2. Shrub and Tree Composition for Riparian Vegetation and Desert Scrub Stands

Riparian Vegetation stands represented a dominance of obligate-riparian trees such as *Populus fremontii*, *Fraxinus velutina* and *Salix gooddingii*, for sites Aguilar Ranch, Cienega, Living Fences,

Galera, San Juan Ranch and Tomas Ranch (Table 5). El Cajon and Nogalera sites were dominated by facultative-riparian and non-riparian tree species such as *Prosopis velutina* and *Parkinsonia florida*.

The bush strata in half of the sites were dominated by naturalized/disturb-indicator species such as *Ricinus communis* and *Nicotiana glauca* (Table 6). It is important to note that buffel grass (*Cenchrus ciliaris*) was present in both southern sites, very close to the river, in low frequency but higher in Nogalera than in Cajon. Nogalera site was the only site where a desert tree (*Parkinsonia florida*) presented the highest number of individuals in the Riparian Vegetation stand, other Desert Scrub species were also found near the dry riverbed such as *Cylindropuntia leptocaulis* and *Lophocereus schottii*, although in low frequency.

Desert Scrub stands are dominated by facultative-riparian and desert species (Table 7). Some bushes such as *Celtis pallida* share their distribution between stands and are found along almost every site.

Facultative-riparian tree *Prosopis velutina* was the most common tree found in both the riparian and desert habitats, most of the time having a higher number of individuals on the Desert Scrub stands, except for sites El Cajon and Nogalera, where mesquite trees were present in a much higher frequency in the Riparian Vegetation stands than in the Desert Scrub stands.

Table 6. Dominant bush and tree species by number of individuals in Riparian Vegetation stands for each site.

	Aguilar Ranch	Cienega	Living Fences	El Cajon	Nogalera	Galera	San Juan Ranch	Tomas Ranch
Bushes	*Ricinus communis* *Baccharis salicifolia*	*Celtis pallida*	*Baccharis salicifolia* *Nicotiana glauca*	*Celtis pallida* *Vallesia glabra*	*Lycium berlandieri*	*Ricinus communis* *Nicotiana glauca* *Ambrosia ambrosioides*	*Nicotiana glauca* *Baccharis salicifolia* *Ricinus communis*	*Baccharis salicifolia*
Trees	*Populus fremontii* *Prosopis velutina* *Acacia farnesiana*	*Fraxinus velutina*	*Salix gooddingii*	*Prosopis velutina*	*Parkinsonia florida* *Prosopis velutina*	*Populus fremontii* *Parkinsonia aculeata*	*Salix gooddingii*	*Salix gooddingii*

Table 7. Dominant bush and tree species by number of individuals in Desert Scrub stands for each site.

	Aguilar Ranch	Cienega	Living Fences	El Cajon	Nogalera	Galera	San Juan Ranch	Tomas Ranch
Bushes	*Celtis pallida*	*Lycium berlandieri* *Celtis pallida*	*Mimosa laxiflora*	*Mimosa laxiflora*	*Jathropha cardiophylla*	*Celtis pallida*	n/a	*Celtis pallida*
Trees	*Prosopis velutina*	*Prosopis velutina*	*Prosopis velutina*	*Stenocereus thurberi* *Olneya tesota*	*Olneya tesota* *Parkinsonia florida*	*Prosopis velutina*	n/a	*Prosopis velutina*

3.2.3. Shrub and Tree Diversity and Richness for Riparian Vegetation and Desert Scrub Stands

Shannon-Weiner diversity index results (Table 8) shows that, for Riparian Vegetation, one of the least disturbed sites (Galera) has the higher value and one of the intermediate disturbed sites (Cienega) has the lowest value. In general, half of the sites had higher diversity values for the Desert Scrub and the other half had higher diversity values for Riparian Vegetation. Both of the southern sites (El Cajon and Nogalera) present high levels of shrubs and trees diversity.

Table 8. Shannon-Weiner diversity index and richness values for Riparian Vegetation and Desert Scrub stands, for each site (shrubs and trees).

	Riparian Vegetation		Desert Scrub	
	Shannon-Weiner Diversity Index	**Richness**	**Shannon-Weiner Diversity Index**	**Richness**
Aguilar Ranch	1.5821	8	1.6510	11
Cienega	0.4957	4	1.1365	9
Living Fences	1.8725	13	1.8230	11
El Cajon	1.8535	11	1.7179	7
Nogalera	1.3557	12	1.9141	14
Galera	1.9389	11	0.9598	7
San Juan Ranch	1.8029	11		
Tomas Ranch	0.5909	3	1.0350	11

3.3. *Land Cover Assessment from Aerial Imagery*

A total of 15 orthomosaics were derived from UAV flights, 8 correspond to Riparian Vegetation sites and 7 to Desert Scrub sites (Figures 6 and 7).

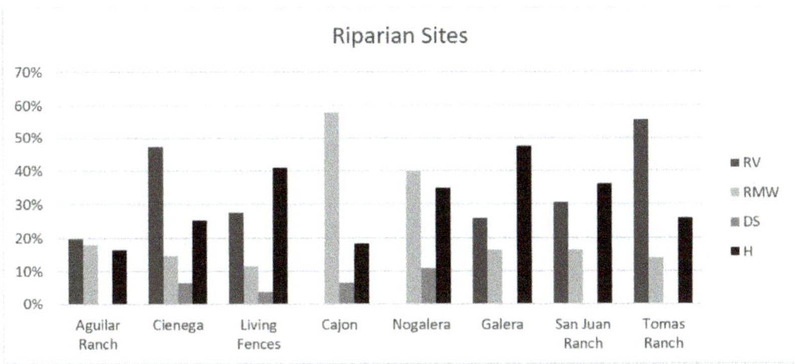

Figure 6. Cover percentage for Land Cover Classes for each site, derived from orthomosaics of Riparian sites. RV (Riparian Vegetation), RMW (Riparian Mesquite Woodland), DS (Desert Scrub), H (Herbaceous cover).

Tomas Ranch has the highest cover of riparian vegetation, which may include trees as well as shrubs and when compared with field measurements it can be established that most of the riparian vegetation in this site is represented by shrubs. Cienega is the next site with the highest cover of riparian vegetation, which is dominated mostly by ash trees (*Fraxinus velutina*). Aguilar Ranch stands out to be the site with less vegetation coverage for all land cover classes, this could be due to the closeness of agriculture fields to the riparian study site. Cajon and Nogalera sites differ from the rest of the sites because riparian vegetation here is represented by mesquite woodland. Cajon has the highest cover of Riparian Mesquite Woodland and most of it is composed by shrubs. This high percentage of cover could be related to the physiography of the place, since its surrounded by low hills which could lower species dissemination to adjacent places and promote species gathering in the same area [44], this could also explain the high richness and diversity of this site. The high percentages of herbaceous cover also mean there are more open spaces between the canopy, along with the fact that none of the classes cover more than 60% of the terrain, this can be a sign of landscape modifications and species dominance shifts [56].

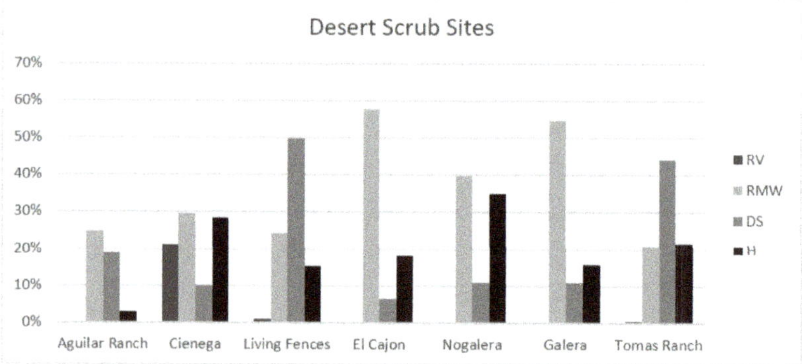

Figure 7. Cover percentage for Land Cover Classes for each site, derived from orthomosaics of Desert Scrub sites. RV (Riparian Vegetation), RMW (Riparian Mesquite Woodland), DS (Desert Scrub), H (Herbaceous cover).

Living Fences and Tomas Ranch have the highest cover for Desert Scrub class. The steep hill where Desert Scrub was sampled in Living Fences site can be a cause of its cover reaching 50%, preventing the place to have continuous disturbance such as the one caused by cattle, since it would not be as easy to graze on the hills as it is on flatter terrains. Even though disturbance was not assessed on the Desert Scrub sites, in some places management practices were evident, such as in Tomas Ranch, were half of the site was covered with buffel grass, which comprises at least half of the Desert Scrub cover percentage.

3.4. Photosynthetic Activity Analysis

3.4.1. Seasonal Photosynthetic Activity per Vegetation Type

The analysis of variance performed with the residual *EVI* values of May and September shows a clear seasonal difference between the Desert Scrub class and the rest of the cover classes and minor differences among the other classes (Riparian Vegetation, Riparian Mesquite Woodland and Herbaceous Cover), this could be due the relative stability of water availability for riparian vegetation and the opposite for desert scrub, thus showing a drastic increase on photosynthetic activity during the rainy season in some of our sites (Figure 8).

Desert Scrub class values in Aguilar Ranch show to be higher than any other class, meaning that this class has a higher seasonal variation. Riparian Vegetation and Herbaceous cover show the lowest residual values, indicating a low seasonal variation in its photosynthetic activity.

Desert Scrub residual values indicate a strong seasonal photosynthetic response. The other two classes present values lower than zero indicating that May's photosynthetic activity is higher or similar to September's photosynthetic activity.

Desert Scrub and Herbaceous residual values are the highest in this site, indicating a higher seasonal variation than the other classes. Riparian Vegetation shows values below zero, indicating that in some cases May's photosynthetic activity is higher or similar to September's photosynthetic activity.

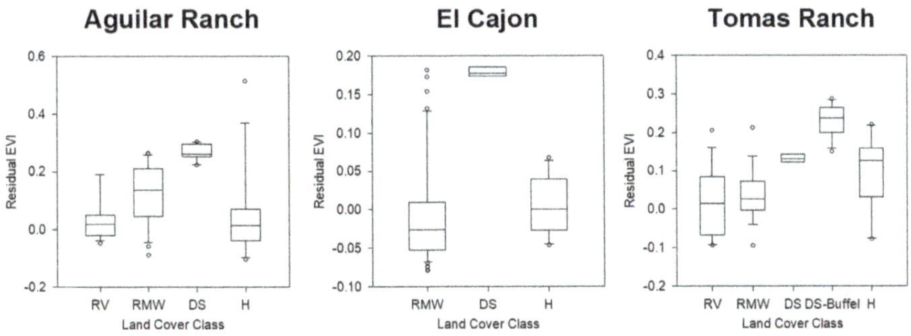

Figure 8. Differences among residual values of each land cover class in Aguilar Ranch, El Cajon and Tomas Ranch. RV (Riparian Vegetation), RMW (Riparian Mesquite Woodland), DS (Desert Scrub), H (Herbaceous cover). Aguilar Ranch: H = 28.23, $P \leq 0.001$, N = RV(16) RMW(24) DS(11) H(13). El Cajon: H = 14.76, $P \leq 0.001$, N = RMW(44) DS(5) H(14). Tomas Ranch: H = 33, $P \leq 0.001$, N = VR(17) MR(15) MD(5) MD-Buffel(11) HE(17).

3.4.2. Photosynthetic Activity of Riparian Vegetation and Riparian Mesquite Woodland

There were five sites with Riparian Vegetation and Riparian Mesquite Woodland present along the edges of the river. September *EVI* values show no significant difference for most of the sites but two (San Juan Ranch and Tomas Ranch). These two sites are ranches close to each other, they have a low-medium disturbance and riparian tree species are dominant in both. On the other hand, May *EVI* values show a significant difference between the two classes on four of the sites, with Riparian Vegetation maintaining higher photosynthetic activity than Riparian Mesquite Woodland (Table 9). This suggests that even when Riparian Mesquite Woodland establishes near the river it stills requires having precipitation to maintain photosynthetic activity values similar to those of Riparian Vegetation.

Table 9. *EVI* averages for May and September for sites with Riparian Vegetation (RV) and Riparian Mesquite Woodland (RMW). Significance level (*P*). Values with * show a significant difference.

	EVI-May		*P*	*EVI*-September		*P*
	RV	**RMW**		**RV**	**RMW**	
Aguilar Ranch	0.408	0.372	0.132	0.432	0.408	0.228
Living Fences	0.364	0.221	0.003 *	0.361	0.425	0.145
Galera	0.364	0.406	0.009 *	0.484	0.469	0.132
San Juan Ranch	0.330	0.201	0.001 *	0.456	0.349	0.001 *
Tomas Ranch	0.388	0.211	0.008 *	0.436	0.354	0.001 *

3.4.3. Photosynthetic Activity and Disturbance

We performed a correlation analysis between *EVI* and ADI values of seven of our sites (we exclude Cienega since it presented unique conditions). Our finding suggests that disturbance, as measured by the ADI, could partially predict photosynthetic activity of a site (estimated by *EVI*) (Figure 9). Productivity decline in riparian ecosystems has been reported as one of the main consequences of disturbance caused by human activities, such as water deviation, agriculture, raise-cattle and mining [4,20,41]. More specifically, vegetation indices such as NDVI and *EVI* have been found to have a relationship with many climatic variables and also with surface and groundwater levels, finding that, in many cases, the indices decline with deeper levels of groundwater and with decreased flood flows [17,57].

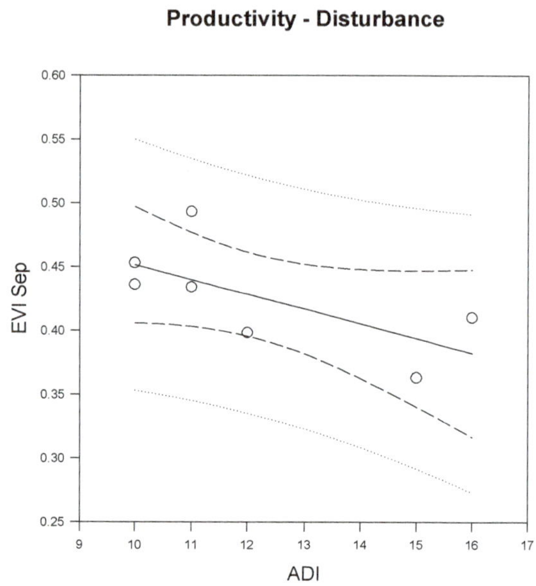

Figure 9. EVI-September values of Riparian Vegetation relation with the Anthropogenic Disturb Index (ADI). $R^2 = 0.448$, $P = 0.100$.

4. Discussion

4.1. Disturbance in the SMR Watershed

There is no previous study in the SMR that analyses disturbance in riparian ecosystems. Previous land use and vegetation classification studies in the area have found that agriculture and cattle ranching are the activities that contribute the most to land cover dynamics [8], which is consistent with the present study given that those activities were present in all of the study sites and are determinant to the disturbance degree of riparian habitats. Different indices that evaluate ecological parameters in riparian habitats have been used in many places in Mexico [15,58–60], all of them use vegetation parameters such as composition, cover and species density to estimate riparian habitat condition. The ADI proposed in this study includes composition and human activities mainly but not cover. Even though cover is not included in the ADI, our results show that this parameter could be related to disturbance.

The Riparian Quality Index (RQI) developed by Gonzalez and Garcia [42] uses several vegetation characteristics (e.g., continuity, regeneration and composition) and hydro-morphological factors to evaluate the ecological condition of rivers, even though this index was designed for a particular European riparian environment, most of its elements can be used in other environments. Even though the ADI shares some similarities with the RQI, our main focus is the assessment of disturbance to evaluate the ecological condition of rivers.

Distribution of disturbed sites along the river shows that those with the highest ADI are located in the south, closer to the city of Hermosillo. These results are consistent with previous studies [23] that demonstrate how peri-urban communities experience changes in land use due to urban water transfer, these changes include the loss of agricultural land and habitat transformations due to recreational sites establishment. The two most disturbed sites in the present study (El Cajon and Nogalera) have been modified by non-typical activities for the region: commercial agriculture and the other is recreational activity. These activities require a different management and resource administration compared to other more traditional activities (such as subsistence agriculture). Therefore, ecological modification of the riparian habitat by either of the activities mentioned could represent new scenarios for local people.

Sites located in the central part of the sub watershed have a low-medium disturbance, most of them are ranches. This stands out due to the fact that ranches are places with frequent human presence and most of the times include some type of resource management. In this case, it could be that the effect of disturbance is not so strong due to the relatively small size of the ranches, where human activities are present but not necessarily intense.

The northern sites present an intermediate disturbance degree; they both include some type of management. In the case of Cienega there's an intense presence of cattle, which could be related to the low shrub cover in both Riparian and Desert habitats, which was less than 10% in both cases. Living Fences site stands out due to the presence of cottonwood and willow fences along the edges of the river, it is important to notice that most of the riparian tree cover was composed by these living fences, that is, almost no other riparian trees were found that were not managed. These living fences are a main characteristic of the traditional type of agricultural activity in the region. Considering the high disturbance degree found in these riparian habitats, the present study could be an update on what Nabhan and Sheridan [61] called a "stable agroecosystem in the upper San Miguel," stating that it is not necessarily stable if disturbance is so high and riparian vegetation so scarce, although living fences might still hold some additional benefits to the riparian habitat.

4.2. Cover and Composition of Riparian habitats and Desert Habitats

Field estimations show that Riparian Vegetation cover is higher than Desert Scrub cover, for all vegetation strata; which is most likely due to resource availability and physical characteristics in riparian habitats which enable vegetation growth, such as soil moisture, organic matter deposition and groundwater [62].

Herbaceous cover was the strata with the highest values for both habitats but it did not reach 100% in any site. Since herbs and shrubs are the most affected by cattle [4], it is interesting to note that one of the places with greatest cattle presence (Cienega) shows a 90% herbaceous cover and a 5% shrub cover, in the riparian habitat. In this particular site, cattle might be having a greater effect on shrub removal and the high percentage of herbaceous cover may be a response to the climatic conditions at the time of sampling (rainy season). The same situation can be seen in the desert habitat for the same site, where shrub cover is only 7%.

Compositions of both habitats present some similarities, such as *Prosopis velutina*, *Parkinsonia florida*, *Celtis pallida* and *Lycium berlandieri*. Riparian habitats show the presence of exotic species *Ricinus communis* and *Nicotiana glauca* in most of the low-medium disturbed sites. *Ricinus communis* is considered as an indicator of disturbed habitats [63]. *Nicotiana glauca* is widely common in Sonoran rivers [64] and high dominant values in its presence have been found in many places where half of the land has been transformed to agriculture [11]. It stands out that these species were found in the sites with the lowest disturbance degree.

Even though many desert species can be found on riparian habitats, this study considers that the abundance and high presence of these species can be a sign of change in the riparian ecosystem. This can be seen in the two sites with the highest disturbance degree (Cajon and Nogalera) where *Prosopis velutina* and *Parkinsonia florida* dominated the tree strata. The presence of cacti such as *Cylindropuntia leptocaulis* and *Lophocereus schottii*, along with the presence of the invasive buffel grass in riparian habitats of these two sites, could also be an indicator of a gradual vegetation transformation or an increase of aridity in the riparian habitat.

Diversity of shrub and tree strata in Cienega is the lowest for riparian habitats, this might be due to the fact that most of the canopy was dominated by a single tree species *Fraxinus velutina* and almost no other tree species were present. Something similar can be seen in Galera, where the desert habitat has the lowest diversity value and the composition shows a dominance of mesquite in the sampling site.

Contrary to what was expected, vegetation class covers derived from orthomosaics (UAV) show similar averages for both habitats. Riparian habitat shows a 25% cover for Riparian Vegetation and a 23% for Riparian Mesquite Woodland and Desert habitat shows a 21% cover for Desert Scrub and a 27% for Riparian Mesquite Woodland. These similarities can also be a sign of degradation of riparian habitats, where vegetation cover is expected to be higher than in desert habitats. Although when riparian habitat covers are compared within each site, Riparian Vegetation presents higher covers than Riparian Mesquite Woodland in all sites, except for those where it is not present (Cajon and Nogalera). In the desert habitat, Riparian Mesquite Woodland has a higher cover than Desert Scrub in most of the sites.

4.3. *Photosynthetic Activity of Riparian Habitats and Desert Habitats*

Differences in *EVI* averages between riparian and desert habitats are consistent with other findings [8,17,35] where a higher productivity is attributed to riparian vegetation when compared to its adjacent vegetation and with the fact that water availability promotes vegetation productivity in riparian habitats [55,65,66]. Also, the seasonal difference in *EVI* values for desert habitat is much greater than in riparian habitat, showing that riparian habitats maintain more stable productivity rates than desert habitats, which respond drastically during the growth season when water is available through precipitation.

Riparian vegetation shows higher *EVI* average values than Riparian Mesquite Woodland in most of the 5 sites where these two classes are present; this was consistent for both seasons. This photosynthetic activity variation is important and could mean that, even when both classes share the same environment and water availability, Riparian Vegetation remains more productive than Riparian Mesquite Woodland. Soil characteristics between these two classes might explain part of the productivity differences, since, as registered by Trujillo [67], Riparian Vegetation in the San Miguel river has a greater amount of carbon and nitrogen and also greater mycorrhizal activity, than Riparian Mesquite Woodland.

The productivity-disturbance inverse relationship shows to be significant and consistent with what has been reported before [41], in terms of productivity decline being one of the main consequences of disturbance (according to ADI scores). Based on *EVI* analysis, our results suggest that riparian sites with high disturbance due to human activities (which include changes in water availability, as suggested by previous studies [8,23,35] and changes in the composition and structure of vegetation) have lower photosynthetic activity (and hence productivity) than less disturbed areas.

The use of aerial photography in this study was very helpful to perform a detailed classification analysis since different vegetation types can be easily identified with high spatial resolution imagery. Also, digitalization of the different vegetation types allows for very accurate cover estimation, due to the clear visualization of tree and shrub canopies. In addition, when combined with a coarser resolution sensor (OLI) we were able to identify the exact productivity values for each vegetation type, which was an essential part of this study.

5. Conclusions

According to our analysis, the increase of land-cover-change associated human activities can increase disturbance on the vegetation of riparian ecosystems in the San Miguel River. As disturbance increases, photosynthetic activity of Riparian Vegetation decreases.

The two sites with the highest degree of disturbance according to the ADI are characterized by the development of particular activities (non-traditional) that were not found in any other site, such as commercial agriculture and recreational activity. Riparian composition in these two sites differs greatly from riparian vegetation in the rest of the sites, due to the almost total absence of obligate-riparian species. Composition in places with the highest disturbance is dominated by facultative-riparian species.

The presence of non-riparian species and facultative-riparian species in riparian habitats is not uncommon, in this study we suggest that the dominant presence of such species could be related to the intensity of human activities developed in the site. Further studies need to be made to sustain or dismiss this suggestion.

Disturbance as measured by the Anthropogenic-disturbance Index shows to have no significant effect on shrub and trees richness and diversity. Further studies would need to assess herbaceous composition and diversity and its relation to disturbance.

Unmanned Aerial Vehicles constitute an efficient tool to assess and monitor riparian ecosystems. Further studies or management proposals should take this into account for the development of regular monitoring programs on arid and semi-arid watersheds. The remote sensing methods used in this study should be further explored for the study of riparian ecosystems. Especially the use of high efficiency drones could be applied for a more extensive mapping of riparian habitats. The coupling of aerial technologies along with free-access satellite data, is a feasible option for riparian characterization and monitoring.

In spite of their small extension and of being highly variable, riparian habitats in arid regions are of crucial importance, an evidence of this is the capacity of Riparian Vegetation to maintain higher photosynthetic activity levels than other vegetation types such as Riparian Mesquite Woodland and Desert Scrub. Since riparian habitats demonstrate high variability in terms of land use, we believe that it is necessary to increase the number of studies regarding their composition and the effects/threats posed by human disturbance. For the previous, the present study provides potential tools to evaluate the effects of different management practices on riparian habitats in arid environments.

Supplementary Materials: The following are available online at http://www.mdpi.com/2073-445x/7/1/13/s1, Table S1: Spatial and temporal location of each flight performed. (R) and (D) after site name indicate flight performed over riparian vegetation (R) and over adjacent desert scrub (D). SMH is the abbreviation for San Miguel de Horcasitas municipality, Table S2: Flight requirements and image specifications, Table S3: Characteristics of Landsat 8 scenes.

Acknowledgments: The present work was supported by the following projects: (1) "Strengthening Resilience of Arid Region Riparian Corridors Ecohydrology and Decision–Making in the Sonora and San Pedro Watersheds" (funded by the National Science Foundation's Dynamics of Coupled Natural and Human (CNH) Systems Program); (2) Apoyo a la incorporacion de nuevos PTC (funded by SEP-PRODEP) and (3) grant support CB2013-223525-R (CONACYT). L.C.-D. would also like to thank the National Council for Science and Technology of Mexico (CONACYT) for their support through a postgraduate scholarship.

Author Contributions: Lara Cornejo-Denman was the primary author and all authors contributed to the final paper. Jose Raul Romo-Leon contributed to all steps of the analysis and commented on the manuscript. Alejandro E. Castellanos, Jose Luis Moreno-Vázquez and Rolando E. Diaz-Caravantes made important contributions during the development of the research project and commented and revised the manuscript. Romeo Mendez-Estrella contributed with discussion and field work. All authors reviewed and approved the final manuscript.

Conflicts of Interest: The authors declare no conflict of interest.

References

1. Stromberg, J.; Tellman, B. *Ecology and Conservation of the San Pedro River*; The University of Arizona Press: Tucson, AZ, USA, 2009.
2. Stromberg, J.; Briggs, M.; Gourley, C.; Scott, M.; Shafroth, P.; Stevens, L. Human alterations of riparian ecosystems. In *Riparian Areas of the Southwestern United States. Hydrology, Ecology, and Management*; Baker, M.B., Ffolliott, P.F., DeBano, L., Neary, D.G., Eds.; Lewis Publishers: Boca Raton, FL, USA, 2004; p. 429.
3. Obedzinski, R.A.; Shaw, C.G.; Neary, D.G. Declining woody vegetation in riparian ecosystems of the western united states. *West. J. Appl. For.* **2001**, *16*, 169–181.
4. Zaimes, G. *Understanding Arizona's Riparian Areas*; Zaimes, G., Ed.; Arizona Cooperative Extension—The University of Arizona: Tucson, AZ, USA, 2007.
5. CONAGUA. *Actualización de la Disponibilidad Media Anual de Agua en el Acuífero Río San Miguel (2625), Estado de Sonora*; Secretaría de Medio Ambiente y Recursos Naturales—Comisión Nacional del Agua: México, 2015.

6. Moreno, J.L.; Marañón, B.; López, D. Los acuíferos sobreexplotados: Origen, crisis y gestión social. In *El Agua en México: Cauces y Encauces*; Academia Mexicana de Ciencias: Tlalpan, México, 2010; p. 702.
7. CONAGUA. *Disponibilidad Media Anual de Agua Subterránea. 2625 Río San Miguel*; Secretaría de Medio Ambiente y Recursos Naturales—Comisión Nacional del Agua: México, 2009.
8. Mendez-Estrella, R.; Romo-Leon, J.; Castellanos, A.; Gandarilla-Aizpuro, F.; Hartfield, K. Analyzing landscape trends on agriculture, introduced exotic grasslands and riparian ecosystems in arid regions of mexico. *Remote Sens.* **2016**, *8*, 664. [CrossRef]
9. Williams, C.A.; Cooper, D.J. Mechanisms of riparian cottonwood decline along regulated rivers. *Ecosystems* **2005**, *8*, 382–395. [CrossRef]
10. Petrakis, R.E.; van Leeuwen, W.J.D.; Villarreal, M.L.; Tashjian, P.; Dello Russo, R.; Scott, C.A. Historical analysis of riparian vegetation change in response to shifting management objectives on the middle rio grande. *Land* **2017**, *6*, 29. [CrossRef]
11. Scott, M.L.; Nagler, P.L.; Glenn, E.P.; Valdes-Casillas, C.; Erker, J.A.; Reynolds, E.W.; Shafroth, P.B.; Gomez-Limon, E.; Jones, C.L. Assessing the extent and diversity of riparian ecosystems in Sonora, Mexico. *Biodivers. Conserv.* **2009**, *18*, 247–269. [CrossRef]
12. Lite, S.J.; Bagstad, K.J.; Stromberg, J.C. Riparian plant species richness along lateral and longitudinal gradients of water stress and flood disturbance, san pedro river, arizona, USA. *J. Arid Environ.* **2005**, *63*, 785–813. [CrossRef]
13. Stromberg, J.; Tiller, R.; Richter, B. Effects of ground water decline on riparian vegetation of semiarid regions: The san pedro, arizona. *Ecol. Appl.* **1996**, *6*, 113–131. [CrossRef]
14. Shafroth, P.B.; Cleverly, J.R.; Dudley, T.L.; Taylor, J.P.; Van Riper, C., III; Weeks, E.P.; Stuart, J.N. Control of tamarix in the western united states: Implications for water salvage, wildlife use, and riparian restoration. *Environ. Manag.* **2005**, *35*, 231–246. [CrossRef] [PubMed]
15. Villaseñor, J.F. *Habitat Use and the Effects of Disturbance on Wintering Birds Using Riparian Habitats in Sonora, Mexico*; University of Montana: Missoula, MT, USA, 2007.
16. DeBano, L.F.; DeBano, S.J.; Wooster, D.E.; Baker, M.B. Linkages between riparian corridors and surrounding watersheds. In *Riparian Areas of the Southwestern United States. Hydrology, Ecology, and Management*; Baker, M.B., Ffolliott, P.F., DeBano, L., Neary, D.G., Eds.; Lewis Publishers: Boca Raton, FL, USA, 2004; p. 429.
17. Nguyen, U.; Glenn, E.; Nagler, P.; Scott, R. Long-term decrease in satellite vegetation indices in response to environmental variables in an iconic desert riparian ecosystem: The Upper San Pedro, Arizona, USA. *Ecohydrology* **2014**, *8*, 610–625. [CrossRef]
18. Jones, K.B.; Slonecker, E.T.; Nash, M.S.; Neale, A.C.; Wade, T.G.; Hamann, S. Riparian habitat changes across the continental united states (1972–2003) and potential impications for sustaining ecosystem services. *Landsc. Ecol.* **2010**, *25*, 1261–1275. [CrossRef]
19. Chan, D. *Using Ecosystem Services to Understand the Impact of Land Cover Change: A Case Study of the Upper San Pedro Watershed*; University of Arizona: Tucson, AZ, USA, 2013.
20. Patten, D. Riparian ecosytems of semi-arid north america: Diversity and human impacts. *Wetlands* **1998**, *18*, 498–512. [CrossRef]
21. Navarro-Navarro, L.A.; Moreno-Vazquez, J.L.; Scott, C.A. Social networks for management of water scarcity: Evidence from the san miguel watershed, sonora, mexico. *Water Altern.* **2017**, *10*, 41–64.
22. Lutz, A.N. *Human Adaptation to Social and Environmental Change in Rural Communities of the San Miguel Watershed in Arid Northwest Mexico*; The University of Arizona: Tucson, AZ, USA, 2017.
23. Díaz-Caravantes, R.E.; Sánchez-Flores, E. Water transfer effects on peri-urban land use/land cover: A case study in a semi-arid region of mexico. *Appl. Geogr.* **2011**, *31*, 413–425. [CrossRef]
24. House-Peters, L.; Scott, C.A. Assessing the impacts of land use change on water availability, management, and resilience in arid region riparian corridors: A case study of the san pedro and rio sonora watersheds in southwestern USA and northwestern mexico. In Proceedings of the XIV World Water Congress of the International Water Resources Association, Porto de Galinhas, Brazil, 25–29 September 2011.
25. Ely, C.C.; Kepner, W.G.; Goodrich, D.C.; Nash, M.S. *Investigating Historic Parcel Changes to Understand Land Use Trends—A Methodology and Application for the San Pedro River Watershed*; United States Environmental Protection Agency: Washington, DC, USA, 2013.
26. Webb, R.H.; Leake, S.A. Ground-water surface-water interactions and long-term change in riverine riparian vegetation in the southwestern united states. *J. Hydrol.* **2006**, *320*, 302–323. [CrossRef]

27. Villarreal, M.; Leeuwen, W.V.; Romo, J.R. Mapping and monitoring riparian vegetation distribution, structure and composition with regression tree models and post-classification change metrics. *Int. J. Remote Sens.* **2012**, *33*, 4266–4290. [CrossRef]
28. Castellanos-Villegas, A.; Bravo, L.; Koch, G.; Llano, J.; López, D.; Méndez, R.; Rodríguez, J.; Romo, R.; Sisk, T.D.; Yanes-Arvayo, G. Impactos ecológicos por el uso del terreno en el funcionamiento de ecosistemas áridos y semiáridos. In *Diversidad Biológica de Sonora*; Freaner, F.M., Devender, T.V., Eds.; UNAM-CONABIO: Hermosillo, Mexico, 2010.
29. Búrquez, A.; Martínez-Yrízar, A. Conservation and landscape transformation in sonora, méxico. *J. Southwest* **1997**, *39*, 371–398.
30. Akasheh, O.Z.; Neale, C.M.U.; Jayanthi, H. Detailed mapping of riparian vegetation in the middle rio grande river using high resolution multi-spectral airborne remote sensing. *J. Arid Environ.* **2008**, *72*, 1734–1744. [CrossRef]
31. Jensen, A.; Hardy, T.; McKee, M.; Chen, Y. Using a multispectral autonomous unmanned aerial remote sensing platform (aggieair) for riparian and wetlands applications. In Proceedings of the International Geoscience and Remote Sensing Symposium, Vancouver, BC, Canada, 24–29 July 2011; pp. 3413–3416.
32. Box, E.O.; Fujiwara, K. Vegetation types and their broad-scale distribution. In *Vegetation Ecology*, 2nd ed.; John Wiley & Sons, Ltd.: New York, NY, USA, 2013; p. 578.
33. Lambers, H.; Chapin, F.S., III; Pons, T.L. *Plant Physiological Ecology*, 2nd ed.; Springer: New York, NY, USA, 2008.
34. Cohen, W.B.; Goward, S.N. Landsat's role in ecological applications of remote sensing. *BioScience* **2004**, *54*, 535–545. [CrossRef]
35. Mendez-Estrella, R.; Romo-Leon, J.; Castellanos, A. Mapping changes in carbon storage and productivity services provided by riparian ecosystems of semi-arid environments in northwestern mexico. *Int. J. Geo-Inf.* **2017**, *6*, 298. [CrossRef]
36. Shreve, F.; Wiggins, I.L. *Vegetation of the Sonoran Desert*; Stanford University Press: Palo Alto, CA, USA, 1964.
37. Galaz, O. *Estructura y Diversidad de la Comunidad de Plantas Del Hábitat Ribereño de la Parte Baja Del Río San Miguel, Sonora*; Universidad de Sonora: Hermosillo, Mexico, 2016.
38. Vivoni, E.; Gutiérrez-Jurado, H.; Aragón, C.; Méndez-Baroso, L.; Rinehart, A.; Wyckoff, R. Variation of hydrometeorological conditions along a topographic transect in northwestern Mexico during the north american monsoon. *J. Clim.* **2007**, *20*, 1792–1809. [CrossRef]
39. CONAGUA. *Programa Detallado de Acciones de Gestión Integral Para la Restauración Hidrológica Del Río Sonora*; Secretaría de Medio Ambiente y Recursos Naturales—Comisión Nacional del Agua: México, 2013.
40. SIAFESON, R.D.E.M.A.D.S. Remas. Available online: http://www.siafeson.com/remas/index.php/ (accessed on 15 January 2018).
41. Poff, B.; Koestner, K.A.; Neary, D.G.; Henderson, V. Threats to riparian ecosystems in western North America: An analysis of existing literature1. *J. Am. Water Resour. Assoc.* **2011**, *47*, 1241–1254. [CrossRef]
42. González, M.; García, D. Riparian Quality Index (RQI): A methodology for characterising and assessing the environmental conditions of riparian zones. *Limnetica* **2011**, *30*, 235–254.
43. Mueller-Dombois, D.; Ellenberg, H. *Aims and Methods of Vegetation Ecology*; John Wiley & Sons: New York, NY, USA, 1974.
44. Gurevitch, J.; Scheiner, S.M.; Fox, G.A. *The Ecology of Plants*; Sinauer Associates, Inc.: Sunderland, MA, USA, 2006; p. 574.
45. Mather, B.T.A.P.M. *Classification Methods for Remotely Sensed Data*, 2nd ed.; CRC Press—Taylor & Frances Group: Boca Raton, FL, USA, 2009; p. 357.
46. USGS. Landsat Missions. Available online: https://landsat.usgs.gov/geometry (accessed on 15 January 2018).
47. Bannari, A.; Morin, D.; Bonn, F.; Huete, A.R. A review of vegetation indices. *Remote Sens. Rev.* **1995**, *13*, 95–120. [CrossRef]
48. Glenn, E.; Huete, A.; Nagler, P.; Nelson, S. Relationship between remotely-sensed vegetation indices, canopy attributes and plant physiological processes: What vegetation indices can and cannot tell us about the landscape. *Sensors* **2008**, *8*, 2136–2160. [CrossRef] [PubMed]
49. Running, S.W.; Nemani, R.R.; Heinsch, F.A.; Zhao, M.; Reeves, M.; Hashimoto, H. A continuous satellite-derived measure of global terrestrial primary production. *BioScience* **2004**, *54*, 547–560. [CrossRef]
50. Huete, A.; Didan, K.; Miura, T.; Rodriguez, E.P.; Gao, X.; Ferreira, L.G. Overview of the radiometric and biophysical performance of the modis vegetation indices. *Remote Sens. Environ.* **2002**, *83*, 195–213. [CrossRef]

51. McLaughiln, S.P. Riparian flora. In *Riparian Areas of the Southwestern United States. Hydrology, Ecology, and Management*; Baker, M.B., Ffolliott, P.F., DeBano, L., Neary, D.G., Eds.; Lewis Publishers: Boca Raton, FL, USA, 2004; pp. 127–140.
52. Clary, W.P.; Kruse, W.H. Livestock grazing in riparian areas: Environmental impacts, management practices and management implications. In *Riparian Areas of the Southwestern United States. Hydrology, Ecology, and Management*; Baker, M.B., Ffolliott, P.F., DeBano, L., Neary, D.G., Eds.; Lewis Publishers: Boca Raton, FL, USA, 2004; p. 429.
53. Stromberg, J. Riparian mesquite forests: A review of their ecology, threats, and recovery potential. *J. Ariz-Nev. Acad. Sci.* **1993**, *27*, 111–124.
54. Stromberg, J.C.; Setaro, D.L.; Gallo, E.L.; Lohse, K.A.; Meixner, T. Riparian vegetation of ephemeral streams. *J. Arid Environ.* **2017**, *138*, 27–37. [CrossRef]
55. Malanson, G.P. *Riparian Landscapes*; Cambridge University Press: Cambridge, UK, 1993; p. 306.
56. Fernandes, M.R.; Aguiar, F.C.; Ferreira, M.T. Assessing riparian vegetation structure and the influence of land use using landscape metrics and geostatistical tools. *Landsc. Urban Plan.* **2011**, *99*, 166–177. [CrossRef]
57. Fu, B.; Burgher, I. Riparian vegetation ndvi dynamics and its relationship with climate, surface water and groundwater. *J. Arid Environ.* **2015**, *113*, 59–68. [CrossRef]
58. Mendoza, M. *Vegetación Ribereña: Indicador de la Salud Del Río de la Reserva de la Biósfera de la Barranca de Metztitlán*; Colegio de Postgraduados: Texcoco, Mexico, 2014.
59. Ramírez, J.E. *Caracterización Del Ecosistema de Ribera y su Valor Indicador Del Estado Ecológico en la Subcuenca Valle de Bravo-Amanalco, Estado de México*; Universidad Nacional Autónoma de México: Mexico City, Mexico, 2010.
60. LANCIS. *Evidencias de Las Afectaciones a la Integridad Funcional de Los Ecosistemas Del Río Bacanuchi y el Río Sonora Por el Derrame de la Mina Buena Vista Del Cobre: Avances Del Diagnóstico Ambiental*; Universidad Nacional Autónoma de México—Laboratorio Nacional de Ciencias de la Sostenibilidad: Mexico City, Mexico, 2016.
61. Nabhan, G.; Sheridan, T. Living fencerows of the Rio San Miguel, Sonora, Mexico: Traditional technology for floodplain management. *Hum. Ecol.* **1977**, *5*, 97–111. [CrossRef]
62. DeBano, L.F.; Schmidt, L.J. Definitions and classifications. In *Riparian Areas of the Southwestern United States. Hydrology, Ecology, and Management*; Baker, M.B., Ffolliott, P.F., DeBano, L., Neary, D.G., Eds.; Lewis Publishers: Boca Raton, FL, USA, 2004; p. 429.
63. CONABIO. Malezas de México. Available online: http://www.conabio.gob.mx/malezasdemexico/2inicio/home-malezas-mexico.htm (accessed on 15 January 2018).
64. Van Devender, T.; Felger, R.; Fishbein, M.; Molina-Freaner, F.; Sánchez-Escalante, J.; Reina-Guerrero, A. Biodiversidad de las plantas vasculares. In *Diversidad Biológica de Sonora*; Molina-Freaner, F., Van Devender, T., Eds.; UNAM—CONABIO: Hermosillo, Mexico, 2010; pp. 229–262.
65. Stromberg, J.; Beauchamp, V.; Dixon, M.; Lite, S.; Paradzick, C. Importance of low-flow and high-flow characteristics to restoration of riparian vegetation along rivers in arid south-western united states. *Freshw. Biol.* **2007**, *52*, 651–679. [CrossRef]
66. Richardson, D.; Holmes, P.; Esler, K.; Galatowitsch, S.; Stromberg, J.; Kirkamn, S.; Pysek, P.; Hobbs, R. Riparian vegetation: Degradation, alien plant invasions, and restoration prospects. *Divers. Distrib.* **2007**, *13*, 126–139. [CrossRef]
67. Trujillo, C. *Interacciones Bióticas y de Nutrientes en el Suelo de la Vegetación Ribereña y Mezquital en el río San Miguel, Sonora*; Universidad de Sonora: Hermosillo, Mexico, 2014.

land

MDPI

Article

Dirty Water, Muddied Politics: Hybridisation of Local and National Opposition to Kumtor Mine, Kyrgyzstan

Joseph Horrocks-Taylor

School of Geography, University of Oxford, Oxford OX1 3QY, UK; joseph.horrocks-taylor@st-hildas.ox.ac.uk; Tel.: +44-(0)-770-259-8604

Received: 6 March 2018; Accepted: 31 March 2018; Published: 3 April 2018

Abstract: From a Mongolian 'super mine' to China's One Belt One Road, rapid infrastructural development is reforging Central Asia as an economic pivot of the future. Such development offers enticing economic benefits, but threatens fragile environments and local livelihoods. Due to the weakness of the state, the emphasis will be on citizens to hold developers accountable to their social and environmental pledges. Reports of political elites influencing the demands of popular protests call into question the ability of citizens to fulfil this function. This paper examines protest authenticity in Kyrgyzstan, focusing on an environmental social movement against Kumtor gold mine. We trace the emergence and evolution of the social movement, identifying the flexible discursive and scalar strategies it uses to achieve emphasis of the local level and relevance on the national scale. The discussion focuses on how national political saliency may incentivise elite involvement with social movements. This involvement can mask the local demands of the social movement, fixing the environmental problem as a national issue. It is crucial to understand the scalar dynamics of elite-protest interaction if Central Asian civil society is to hold future infrastructural developments to account.

Keywords: Central Asia; Kyrgyzstan; infrastructure; environment; mining; social movements; protest; environmental justice; subversive clientelism

---

1. Introduction

In 1904, Halford Mackinder identified Central Asia as the geographic "pivot of history" [1] (p. 1). Today, this region is being reforged as an economic pivot of the future. The world's longest gas pipeline and two of the world's largest open-pit mines have been constructed, and the world's tallest dam is on track to come online in 2018 [2–4]. These developments are reliant on foreign capital looking to exploit the mineral reserves and geostrategic positioning of the region. However, these large infrastructure projects can have myriad environmental impacts, including land degradation, biodiversity loss, dust generation, water pollution and water depletion, which threaten the health and livelihoods of communities living in Central Asia's steppe drylands [5,6]. The significant risks and rewards of development projects mean that the extractive sector in the region is a 'battleground', with local, national and international actors all fighting for their interests [7]. This is a dynamic interface, with discourses constantly being rejected, recycled and reproduced to suit the shifting strategies of this diverse set of actors. Here, we examine how the changing landscape of this 'battleground' influences the evolution of a local social movement against Kumtor gold mine in Kyrgyzstan. While Kumtor demonstrates how existing infrastructure in Central Asia is heavily contested and has largely failed to live up to local expectations, future developments like China's $1 trillion One Belt, One Road (OBOR) project continue to progress apace [8]. It is essential to understand how specific discourses enable or prevent citizen protest from holding infrastructure development to account in Central Asia when faced with "repression from a state-capital symbiosis" [9] (p. 16).

While there are a variety of definitions of Central Asia, this paper will use the broader delimitation which included "Afghanistan, northeastern Iran, Pakistan, northern India, western China, Mongolia and the former Soviet Central Asian republics" [10] (p. 24). The mineral resources of this region are "a geologist's paradise", but "a pain" to develop [11,12]. The landforms of the region are restrictive of infrastructural development, from the KyzylKum and KaraKum deserts of the west to the Tian Shan and Pamir mountains of the southeast. This landscape can "thwart and crush infrastructure" [8] (p. 62). However, much of the blame for slow development has been attributed to the weakness of the state. The role of the state has been the focus of much of the expanding literature on infrastructure projects in Central Asia. Scholars have identified infrastructure as a tool of nation-building [4,13,14], a question of sovereignty [15–17] and a source of rent-seeking [18]. From this state-centric viewpoint, the emergence of environmental social movements opposing infrastructural developments in Central Asia occurs due to the absence of the state. The failure of the state to meet society's basic economic needs allows elites to capture community support by providing material or symbolic investments [19,20]. Elites can then use 'public' protests as "weapons of the wealthy" to achieve their private political goals [19] (p. 1). Alternatively, the contrast between the financial strength of foreign investors and the perceived weakness of the state can lead to questions of sovereignty, with the potential to trigger social movements demanding resource nationalisation [15].

In these theoretical discourses, the local scale is subordinate to the national scale, with local people's actions dictated by political elites. Recent scholarship on anti-mining social movements in Mongolia [9,21] and Kyrgyzstan [22,23] has attempted to offset the supremacy of the state as the preeminent scale of analysis. This work highlights local communities' concerns about the environmental impacts of infrastructure projects, especially risks to water supply in this arid region. Infrastructural development in mountainous areas threatens the "primary source of fresh water for Central Asia's rivers and aquifers", while infrastructural demands for water in drier areas often endanger the health and environment-dependent livelihoods of isolated pastoral communities [24] (p. 979). In addition to posing a threat to water supplies, infrastructural development can exacerbate natural dust generation, which is a significant health risk to communities in Central Asian steppe drylands [5]. This health risk is aggravated by the potential for infrastructural accidents, such as dam failure or toxic spillages. These perceived threats can generate an 'environmentalism of the poor', causing local 'eco-mobilisations' to arise largely independently of other scales and actors [22].

In order to mediate between these two contrasting bodies of literature, there is a need to engage directly with the scalar and temporal dynamism of environmental protest in the Central Asian context. We attempt to understand how protests evolve in time and space, using a 'politics of scale' approach to provide a nuanced understanding of environmental mobilisation against Kumtor gold mine in Kyrgyzstan. This paper first examines the literature on the emergence and evolution of social movements in Central Asia. We endeavour to tease out the spatial limitations of current theories and introduce 'politics of scale' as a theoretical lens. Next, the origin and development of local opposition to Kumtor is explored, with an emphasis on the strategies the movement uses to set 'scales of meaning' and to scale-jump. The discussion then focuses on the tensions that arise when social movements attempt to position the local scale as the legitimate scale of meaning for environmental conflicts, while simultaneously forming coalitions with other actors to scale-jump local claims to national and global levels. The majority of the academic literature suggests that these local and scale-jumping strategies are complementary [25,26], but this paper will argue that once social movements scale-jump to the national level they are more vulnerable to being co-opted by national actors to achieve alternate political ends.

2. Theories of Environmental Activism in Central Asia

2.1. Activism and the State

Environmental movements against infrastructure developments are on the rise in Central Asia. There have been anti-mining movements in Mongolia [9,21], environmental justice movements in Kazakhstan [27] and Tajikistan [28], and a variety of environmental protests in Kyrgyzstan [22]. This rise of environmental activism has been interpreted in academia largely through the lens and at the scale of the state. Academic analysis of the role of the state in Central Asia has a distinguished pedigree. Following their delimitation by Soviet "nation-makers" in the 1920s [29,30], the Central Asian nations underwent a reluctant "catapult to independence" initiated by the collapse of the Soviet Union in 1991 [31,32]. As Central Asian political elites focused on constructing a "common-sense of belonging" in these artificial and multi-ethnic nations, academic scholarship on the region began to concentrate on their strategies of state and nation building [30,33–36].

This statist and elite-centric focus is pervasive in contemporary scholarship on Central Asian infrastructure. Fillippo Menga [4] demonstrates how large-scale construction projects intersect with elite-led nation-building in the case of Rogun dam in Tajikistan. The symbolic value of the hydraulic project enables the ruling elite to disseminate its own idea of the nation, progress and development. Similarly, the Turkmenistan–Afghanistan–Pakistan–India (TAPI) natural gas pipelines project is understood as a 'virtual pipeline'; its primary value exists in its generation of domestic discourses of progress [37] (p. 1). While the construction of border roads in Kyrgyzstan is conceptualised as an unruly process, often escaping the political visions that initiate them, the state is still the actor with primary political agency [17]. Infrastructure projects have considerable political potency in this region.

Academic accounts of the state and infrastructure in Central Asia have also highlighted the endemic problem of corruption. In Uzbekistan, the state is experienced as organised crime [38], while in Kyrgyzstan the state has become "an investment market" for members of the elite [39]. Much of this "criminalisation of the state" in Kyrgyzstan is centred around the Kumtor gold mine [40]. Allegedly, members of the Akaev regime were involved in the laundering of up to $15 million using offshore vehicles [41]. In addition, a Kumtor executive estimated that half of the mining-funded Issyk Kul development fund disappears annually [42]. Political elites in Central Asia have maintained a "façade of compliance with international anticorruption standards" while consistently exploiting the state for private economic gain [43] (p. 26).

The dominance of the state-centric viewpoint has meant that analysis of resistance to infrastructure projects is often conducted through the lens and at the scale of the state. A primary way that statist literature understands social movements against infrastructure is as resource nationalism—the demand that the government increase its control over national resources [15]. Resource nationalism has been implicated in Kazakhstan [44], Mongolia [15] and Kyrgyzstan [7]. Its origin is often traced to the disparity between the wealth created by resources and levels of national poverty, contributing to a "growing sense amongst the citizens of mineral-rich countries that they have not received an appropriate share of the benefits" of resource extraction [45] (p. 10). This distrust of foreign investment is enhanced by the cultural changes that can come with it. In the case of Oyu Tolgoi mine in Mongolia, the decline of local herding coupled with the influx of Chinese mine workers trigger fears of Chinese colonization and threats to Mongolian national identity [13,15]. Resource nationalism does not emerge from the local level but is a strategy that "domestic elites employ in order to increase their control of natural resource" and gain political advantages [16,46] (p. 38).

State-centric accounts have also attributed the emergence of environmental social movements to subversive clientelism. Subversive clientelism is the theory that citizens may participate in environmental protests because elites ask them to [19]. This 'protest inauthenticity' has been implicated across Central Asia, from the state-managed 'Astroturf' protests in Russia [47] (p. 33) to the 'rent-a-mob' phenomenon in Kyrgyzstan [48]. In Kyrgyzstan, the failure of the state to meet society's basic economic needs has allowed elites to capture community support by providing material or symbolic investments [19,20]. When it is politically expedient for them, elites use mass support obtained through this subversive clientelism to mobilise against the state. In addition to developing their own support base, elites develop ties with other autonomous elites to expand their informal patronage networks [19]. The existence of a 'rent-a-mob' phenomenon is supported by a survey-based study in Kyrgyzstan, which found that 5.13% of respondents were offered some kind of compensation to participate in an environmental protest [22].

Theorisations of resource nationalism and subversive clientelism as triggers for environmental protests in Central Asia are heavily contested in the literature. These theories are considered to be reductionist, as they rely on material incentives, rational choice and cultural norms to interpret the behaviour of non-elites [40]. These assumptions constrain the political agency of ordinary citizens, providing a "spectator's view of action where non-elites are viewed in instrumental and passive terms" [49] (p. 15). Furthermore, this framing marginalised the real socio-environmental threats posed to locals by infrastructure projects. It is essential to analyse environmental protests from the viewpoint of the ordinary citizen, not just the elites [22].

In Central Asia, infrastructure development poses real environmental, material and cultural threats to local populations. These communities are often highly reliant on the environment for their livelihoods and wellbeing, and thus have a heightened understanding of what they stand to lose by not managing it carefully [50,51]. Threats to the environment, whether they materialise or not, can create 'relative deprivation' for local communities, generating local and specific grievances, which provide a general spur to activism [52,53]. Despite lacking meaningful capital, poor groups can undertake resistance ranging from overt, confrontational behaviour to subtle, everyday forms [40,54,55]. This 'environmentalism of the poor' is closely linked to the 'environmental justice' movement. Environment justice can be sub-divided into recognition, distributional and procedural components. Recognition refers to the acknowledgement of collective social identities and their specific needs, concerns and livelihoods [56]. Distributional environmental justice demands the equitable distribution within society of environmental benefits and burdens [57]. Procedural environmental justice calls for the just political and social functioning of society and its institutions, with an emphasis on participation in decision-making [57]. Several environmental mobilisations in Central Asia have been interpreted through an environmental justice lens [9,21].

Political opportunity structures can also determine if, when and where a community protests [58,59]. Following the collective interest model, potential protestors evaluate their political environment and calculate the likelihood of success of their collective action [60]. If a protestor believes the action they take will likely result in the outcome they desire, they may be willing to risk the costs of acting [22]. Transient political opportunities may combine with long-standing environmental grievances to enhance the "eco-mobilizeability" of an individual at a specific time [22].

Political opportunity structures do not just influence the actions of the poor, they may also influence the actions of elites. When locally originating social movements achieve a certain level of popular support and political saliency, a political opportunity arises for elites to exploit. However, there has been limited analysis of how "community concerns about the environment may drive elite involvement and co-optation" [22] (p. 340). This is possibly because of a focus in the literature on how environmental movements emerge and the spectacular one-off protest events they engender. There is a need for a more prolonged assessment of environmental movements to understand how communities' grievances and elites' political aspirations intersect throughout the evolution of the movement. Similarly, there has been a lack of interaction with scale in the literature on environmental movements in Central Asia. The theories of eco-mobilisation and elite co-optation conceptualise environmental protests as static scalar entities predominantly operating on either the local or national scale. There is a dearth of analysis on how actors socially construct the scale of environmental movements to suit their own political interests.

2.2. Politics of Scale

'Politics of scale' is a theory that allows for a more nuanced understanding of the dynamic scalar relations of environmental conflicts. This theory suggests that continuous social processes of socio-political and economic struggle occur between actors aiming to strengthen power and control of a certain scale while disempowering other scales [61,62]. Thus, scales are both socially constructed and politically contested, and are constantly in the process of being reorganized [63,64]. Actors engage strategically with scale to construct, hierarchize and reorganise territorial units in relation to each other [63,65].

As a global business based on place specific mineral deposits, mining has a rich politics of scale. Mining is often described as a 'glocal' phenomenon: international networks of production and consumption are realised in local communities and environments [66,67]. The concept of 'glocalisation' was coined to nuance the homogenising power of globalisation by showing that the effects of globalising forces are contingent upon place-based characteristics [62,66]. With regards to resource extraction, 'glocalisation' is now used to describe the scale-creation and scale-adjustment ability of both extractive industries and movements resisting extraction [68].

Both 'politics of scale' and 'glocalisation' are useful concepts to understand how governments and corporations use specific geographical scales to define mining projects. Mining projects are largely negotiated at a bilateral level between governments and corporations, converting these extractive places into national spaces of development generation [69,70]. This creates a gap between the 'scales of meaning' at which a mining project is experienced or perceived, and the 'scales of regulation' at which the project is commercially approved and politically defined [71,72]. Anti-mining movements challenge the existing hegemonic power structures that define the environmental conflict by restructuring the scales of meaning and regulation [63]. This is achieved by mounting a defence of local place, while simultaneously strategically rescaling the movement in a 'glocal' fashion, often through allegiances with national and supra-national networks [25,26].

Anti-mining movements jump scales from local to national to global in attempts to target concerns on certain geographical scales in order to challenge national decisions [26,64]. Scale-jumping can be achieved by creating broader networks of social actors from different geographical locations, expanding the geographical scale at which the anti-mining discourse operates [73]. These actors support local movements by providing experience, information, enhancing media attention and enabling a broader 'repertoire of contention' to include legal challenges [69,74]. A recent study found that when an environmental movement was connected to more civil society organisations, and especially when it was connected to more nationally or transnationally central civil society organisations, its outcome was more likely to be perceived as a success [75].

Including a wider range of actors and organisations in this broad network can result in initial discourses of local environmental justice being joined by broader discourses such as climate change, anti-neoliberal ideals and resource nationalism [76–78]. For instance, in the Pascua Lama conflict in Chile, local resistance movements referenced wider global claims such as climate change, glacier protection, democracy and participation [78]. In some cases, this can result in a loss of autonomy for local organisations, a need to alter strategies, ideological conflicts and identity loss [79]. Local movements need to be careful not to lose the grassroots while establishing links with supralocal organisations [75]. If there are inequities in coalitions, contradictory opinions on the appropriate values and arguments to use may mean that local environmental justice claims lose out to discourses introduced by more powerful supralocal organisations [80]. In Kyrgyzstan, where there are only 600–700 active national non-governmental organisations (NGOs) and few international organisations, local resistance movements may have to be less scrupulous about the social actors that they form advocacy networks with, increasing the likelihood of this discourse displacement [81].

3. The Case of Kumtor

3.1. Introducing Kumtor

The Kumtor gold deposit is the largest in Central Asia, and has been central to Kyrgyzstan's domestic economy since production started in 1997. In 2014, it accounted for 23.1% of Kyrgyzstan's industrial output and 7.4% of its GDP [7]. The economic importance of Kumtor to Kyrgyzstan has meant that the mine has been closely linked to national politics. The agreement between Cameco and the government in 1992 gave the Kyrgyz government about 60% of shares in the Kumtor Operating Company (KOC) [82]. However, the Kyrgyz government periodically sold shares to Cameco to fund President Akaev's election campaigns [83]. By 2004 the Kyrgyz government only had 15.6% of shares in Kumtor, while Cameco largely sold out to Centerra, another Canadian firm [83]. In 2009, Centerra and President Bakiev negotiated a New Terms Agreement that granted Kyrgyzstan a 33% share of Centerra [84]. This agreement was still in place at the time of writing.

The local environmental movement originated to claim compensation for a major industrial accident. In May 1998, a truck travelling to the mine lost 1.7 tonnes of sodium cyanide, which dissolved into the Barskoon River (see Figure 1) [84]. Despite villagers not being informed of the spill for five hours, an international commission concluded that 'no carcinogenic, mutagenic, teratogenic, reproductive or neurotoxological effects to people are expected' [85] (p. iii). However, local residents and NGOs claimed that more than 2500 people were affected by poisoning, 800 went to hospital and four died as a result of the spill [86]. The Kumtor Operating Company (KOC) provided $25 per adult as compensation for the accident, but local residents argued that this was insufficient compensation for the long-term health impacts and agricultural losses [86,87].

Figure 1. Map of Kyrgyzstan with Kumtor mine.

Local resident Erkingul Imankojoeva set up an NGO called Karek in 1998 to pursue locals' claims for compensation [88]. In addition to the preeminent demand for compensation, Karek also called for enhanced local economic benefits, ecological monitoring and improved safety measures [23]. Karek's 'repertoire of contention' was mostly limited to using protests and road blockades to pressurise Centerra and the government [23,59]. Local residents blockaded the road from Bishkek to the mine on at least five separate occasions between 1998 and 2006. In the July 2006 protest, more than 3000 locals blocked the main road leading to the mine in round-the-clock shifts [88]. In addition, Karek generated public awareness through conducting press meetings, accident victim photo exhibits and public showings of Zolotaya Avariya (Golden Accident), a film about the accident [23]. After advocating for compensation for eight years, in 2006, $3.7 million was distributed to five villages in the area, mostly for harvest damages [23].

The initial claims articulated by local anti-Kumtor activists closely fit the thesis of the environmentalism of the poor. Karek framed the Kumtor issue using local place-based claims, articulating a discourse of environmental justice and human health. The considerable impacts of the Barskoon spill provided a clear grievance that local communities could rally around, and the movement was clearly contained to those impacted by the spill. The use of protests and roadblocks in the villages within Issyk-Kul province cemented this local scale as the legitimate 'scale of meaning' that should be taken into account in the decision-making around Kumtor. Indeed, subsequently, it has been shown that Kumtor may also be polluting the river Naryn that runs through several villages in the Naryn province [89]. However, this region has not received any compensation due to Karek's success in positioning the villages in Issyk-Kul province as the location of Kumtor's impacts.

Anti-Kumtor activists pursued several strategies to scale-jump their concerns to a national and global level. Karek created links with Bishkek-based NGOs such as Tree of Life through an 'eco-safety network' [23]. These national NGOs articulated a unified discourse that situated the local scale as the dominant scale of the anti-Kumtor lobby, which added emphasis to Karek's public awareness campaigns. This successfully raised the local, place-based compensation claims of Karek to the national level. Additionally, the local anti-Kumtor lobby allied with a small number of global organisations to scale-jump the issue to the global level. Dutch NGO Milieukontakt and the Organisation for Security and Cooperation in Europe organised workshops to train local NGOs, while the Urgent Action Fund for Women's Human Rights conducted an online campaign to generate financial support for Karek [23].

Supra-national networks have been widely acknowledged as crucial to the scale-jumping ability of local social movements [73–75]. However, in this example of anti-Kumtor activism, the roadblocks and protests were a successful scale-jumping strategy in and of themselves. Infrastructural sabotage and public demonstrations often halted production, reducing Centerra's profits, which adversely affected company share price and government tax revenue. In the five days after the initiation of the July 2005 protests Centerra's share price fell by 8%, while in the five days after the initiation of the July 2006 protests, it fell by 28% (Table 1). Despite scale-jumping, the anti-Kumtor movement to the national and global level, the symbolic locus of the protests was clearly at the local point of disruption. Social movement theory has previously suggested that forming trans-national allegiances was critical to articulating a 'glocal' environmental conflict [26]. Local scale jumping strategies of this kind may negate the need for social movements to build trans-national allegiances in order to 'glocalise'. In fact, in the absence of these allegiances, environmental movements may be better positioned to retain the 'local' element of 'glocalisation'. This successful scale-jumping usage of protests, road blocks and NGO networks provided a template for other anti-Kumtor activists to follow [23].

Table 1. Response of Centerra Gold share price to Kumtor protests over a five-day period following protest onset (2005–2006).

Date	Percentage Change
18 July 2005	−8.03
3 August 2005	0.98
1 November 2005	−1.26
17 May 2006	−4.60
13 July 2006	−28.36

3.2. *Diversifying Discourses*

Since the success of the movement to claim compensation for the Barskoon spill, the demands of the anti-Kumtor movement have diversified. Demands have shifted from demanding distributional environmental justice to demanding procedural environmental justice. These procedural justice demands have taken a variety of forms. A key argument of local NGOs is that Issyk-Kul province bears the environmental costs of Kumtor, but due to procedural and institutional inequity receives an incommensurate lack of economic benefits [23]. Between 1994 and 2012, KOC paid $2.15 billion in tax and contributed 1% of its gross revenue to the Issyk-Kul Region Development Fund [84]. However, only a fraction of this money benefited local communities for two reasons. First, local governments receive 'shared revenue', taxes collected by the state and distributed equally to local administrations [90]. New legislation in 2011 meant that 2% of this total tax accrued directly to the local government of the mine, but Karek argues that this investment is inadequate compensation for the environmental issues associated with Kumtor [83]. Second, corruption prevents locals from benefitting from this investment. Karek has aligned with national anti-corruption NGOs Citizens Against Corruption and Taza Tabiyat to draw attention to how Kumtor has been used as 'an investment market' by political elites [39,91].

In addition to seeking procedural environmental justice in the form of equitable benefit-sharing, NGOs are leading the movement towards a more equal decision-making process. Currently, the Extractive Industries Transparency Initiative (EITI) in Kyrgyzstan is a top-down process, with all actions, decision-making and implementation centralised in Bishkek, "excluding regions altogether" [92] (p. 469). EITI meetings are conducted in Bishkek with no travel reimbursement for attendees, and reports are rarely translated into Kyrgyz [92]. 'Tree of Life' has been working to publicise and improve access to the EITI process in Kyrgyzstan [92].

Karek has aligned with national NGOs to criticise the Government and KOC on environmental grounds. The 'Tree of Life' argued that assigning part of the neighbouring Sarychat-Eertash nature reserve to a concession area of Kumtor contravened Kyrgyz regulations on environmental protection [22,90]. This NGO also organised protests in Bishkek in 2017 opposing glacier destruction by Kumtor [93]. The local anti-Kumtor campaign has also allied with national human rights NGOs Kylym Shamy (Torch of the Century) and Human Rights Bureau to publicise the issue of anti-Kumtor activists being beaten and tortured [23,94]. These NGOs have videotaped conversations with KOC officials and shared photographs of activists beaten in police custody [95].

After the success of the movement for compensation, the localised anti-Kumtor movement jumped scale to the national level by deepening allegiances with a diverse range of national NGOs. This mirrors Robbins' [96] observation that environmental movements emerge in response to material threats, but subsequently build 'bridges' to broader struggles. This bridge building has resulted in initial discourses of local environmental justice being joined by broader discourses introduced from the national and global sphere, constructing "completely entwined arguments and values in a glocal process" [78] (p. 225). For example, Tree of Life introduced post-materialist conservationist values to argue for the protection of nature from Kumtor, but directly related this back to the local level by arguing that glacier destruction presents a threat to local water supplies [93].

3.3. Overpowering Discourses

The initial success of the original localised anti-Kumtor movement, coupled with the formation of the supralocal allegiance network, succeeded in raising the political prominence of Kumtor at the national level. Ordinary people began engaging with Kumtor as an informal symbol of resistance to the state vision of development. The political prominence and symbolic nature of Kumtor made it an attractive political prospect for certain political elites to associate themselves with. Following the revolution of 2010, the Ata-Jurt political party has capitalised on this political opportunity, promoting nationalisation of Kumtor to undermine the Government.

Political elites do not passively join environmental movements in Kyrgyzstan, they actively shift the discourses and scale of the movement to suit their own ends. In 2012, Ata-Jurt MP Sadyr Japarov formed a parliamentary commission to assess the mine's environmental impact [42]. The 300-page report was published in June 2012, and alleged that the open cast mine had negatively impacted on the nearby glacier and rare plant species, along with poisoning streams feeding into the Barskoon river [97]. In an interview, Japarov stated that "all works at Kumtor have been implemented with violations of ecological regulations" [98]. Many of the environmental issues highlighted by this commission were similar to those broadcast by the local socio-environmental movement. However, these issues were framed as a failure of the government to protect the national legacy of Kyrgyzstan's environment, as opposed to the environmental justice framing of the local movement, which emphasised the impacts on local livelihoods. Although Japarov's call to nationalise Kumtor was defeated in parliament, he was successful in bringing the environmental issues of Kumtor back to the forefront of Kyrgyz politics.

On 3 October 2012, 1000 people attended a nationalisation rally in Bishkek organised by Ata-Jurt [84]. Party leader Kamchibek Tashiyev gave an inflammatory speech in which he proclaimed that "nobody but the nation should run the Kumtor (mine)" [98]. Following the speech, Tashiyev unsuccessfully led protestors in attempting to break into the parliament building, which resulted in his arrest [99]. Afterwards, Tashiyev claimed that Ata-Jurt "wanted the Kumtor problem to be raised to the state level" as "Kumtor is something that affects all Kyrgyz" [100,101]. This rally constituted another shifting of the 'scale of meaning' of the Kumtor problem to the national level. By arguing that Kumtor should be run by "the nation" as it "affects all Kyrgyz" and symbolically locating the protest outside the parliament building, this rally shifted the 'scale of meaning' of the anti-Kumtor movement to the national level, while simultaneously divorcing it from the local anti-Kumtor campaign [98,101].

These demands for the nationalisation of the mine spread to local anti-Kumtor campaigns (see Table 2). Although some NGOs spoke out against the Ata Jurt led calls for nationalisation, describing them as an attempt to "score some political points", the majority of locals supported this discursive shift [101]. From 28–31 May 2013, around 3000 protestors blocked the road to the mine demanding that environmental standards be upheld and that Kumtor be nationalised [102,103]. On 30th May, they seized an electricity station, disrupting power to the mine [103]. Clashes between protestors and police resulted in 92 people being arrested and 55 people being wounded [83]. Two months later a video appeared on state television showing two informal leaders of the May Barskoon protest demanding $3 million from a KOC director in exchange for a guarantee not to orchestrate protests [96]. In the video, both men claim to have the support of Ata-Jurt politicians Kamchybek Tashiyev and Sadyr Japarov, leading to claims in the Kyrgyz media that the Kumtor protests are another 'rent-a-mob' incident [42,104,105]. While the demands and the timing of the May protest certainly suited Ata-Jurt politically, the connection between Ata-Jurt and the local environmental movement does not fit with Radnitz' [19] theory of subversive clientelism. In subversive clientelism, elites utilise local support bases, but Ata-Jurt politicians Kamchybek Tashiyev and Sadyr Japarov originate from Southern Kyrgyzstan, and thus are geographically and politically removed from the Issyk-Kul oblast. Furthermore, local people have mobilised against Kumtor on environmental grounds for over a decade, well before nationalisation of the mine was politicised by Ata-Jurt. Additionally, subversive clientelism describes the mechanism by which elites initiate protests, not how they co-opt pre-existing movements.

We argue that the calls for resource nationalism in local Kumtor protests are indicative of a more passive elite co-optation of popular protests. The strong public profiles of Ata-Jurt politicians allowed them to monopolise media attention, facilitating an elite domination of anti-Kumtor discourse production and scalar framing. This discursive dominance is augmented by the potential of resource nationalism as a framing strategy. Resource nationalism is potent in Kyrgyzstan due to the extensive nation-building strategies implemented by the state, while depicting the sovereignty of Kyrgyzstan as threatened was particularly compelling at the time due to the perception that the sovereignty of the nation was vulnerable following the riots of 2010 [84,106]. Finally, the 'infrastructural hope' of resource nationalism is tempting in Kyrgyzstan where 32.1% of the population is below the poverty line [17,107].

The popularity of the resource nationalism discourse espoused by Ata-Jurt created a political opportunity for local protesters to enhance the political saliency of the local anti-Kumtor movement and scale-jump to the national scale. In this model of elite co-optation, local protestors are not passive victims, stripped of agency. They are active collaborators opportunistically pursuing a political opportunity.

Utilising the discourse of resource nationalism has enrolled national support for the anti-Kumtor movement, but has consolidated the Kumtor problem as a national rather than a local issue. This has undermined local claims for procedural and distributional environmental justice, preventing a decentralisation of the power, decisions and benefits of Kumtor. By articulating Kumtor as a problem of national relevance, and advocating for nationalisation as a solution, anti-Kumtor activists have empowered the national scale as the 'scale of meaning' of this environmental conflict. As a result, attempted 'solutions' to the dispute have been national in scale. In January 2014, the government approved a deal to swap its 32.7% stake in Centerra for a 50% stake in KOC, though this deal was subsequently rejected by parliament [108]. In 2017, the government reached another deal with Centerra, which involved KOC making one-time payments of US $7 million to a new, government-administered Cancer Care Support Fund and US $50 million to a new, government-administered Nature Development Fund [109]. These solutions pander to the calls for resource nationalism and environmental protection advocated by national actors, but fail to resolve local calls for environmental justice in the face of threats to their livelihoods.

Table 2. Location and demands of reported Kumtor protests.

Date	Location	Demands	Citation
10 July 1998	Local	Compensation	Wooden, 2017 [23]
18 July 2005	Local	Compensation	Institute for War & Peace Reporting (IWPR), 2006 [86]
3 August 2005	Local	Compensation	Radio Free Europe/Radio Liberty (RFE/RL), 2005 [87]
1 November 2005	Local	Compensation	New York Times, 2005 [110]
17 May 2006	Karakol	Compensation	Wooden, 2013 [22]
13 July 2006	Local	Compensation	Wooden, 2013 [22]
18 April 2007	Bishkek	Local environment and economy	Wooden, 2013 [22]
19 April 2007	Local	Local environment and economy	Wooden, 2013 [22]
7 May 2007	Local	Compensation	RFE/RL, 2007 [111]
25 January 2011	Local	Local environment and economy	Transitions (TOL), 2011 [42]
16 February 2011	Bishkek	Improve regional infrastructure,	Wooden, 2013 [22]
4 June 2012	Local	Local environment and economy	Reuters, 2012 [112]
3 October 2012	Bishkek	Nationalisation	Gullette, 2014 [84]
4 October 2012	Jalalabad	Nationalisation	Gullette, 2014 [84]
5 October 2012	Jalalabad	Nationalisation	RFE/RL, 2012 [99]
8 October 2012	Jalalabad	Nationalisation	RFE/RL, 2012 [113]
13 March 2013	Bishkek	Nationalisation	RFE/RL, 2013 [114]
24 April 2013	Bishkek	Nationalisation	RFE/RL, 2013 [100]
28 May 2013	Local	Environment, Nationalisation	RFE/RL, 2013 [103]
3 June 2013	Jalalabad	Nationalisation	IWPR, 2013 [115]
27 July 2013	Local	Cancel contract with Centerra	RFE/RL, 2013 [116]
28 August 2013	Local	Nationalisation	Gullette, 2014 [84]
7 October 2013	Karakol	Nationalisation	Gullette, 2014 [84]
8 October 2013	Local	Nationalisation	RFE/RL, 2013 [116]
11 October 2013	Local	Nationalisation	K-News, 2013 [117]
25 March 2014	Local	Distribute benefits locally	AKIpress, 2014 [118]
29 April 2014	Local	Distribute benefits locally	Gullette, 2014 [84]
8 November 2017	Bishkek	Glacier law change	IWPR, 2017 [48]

4. Discussion

Local social movements are often trapped in a scalar bind; they must choose between emphasising local demands (and risk remaining irrelevant on a national level) or appealing to national or global sympathies (and risk marginalising local concerns). The local environmental movement against Kumtor was initially able to escape this scalar bind. The lack of international NGOs operating in Kyrgyzstan meant that the local anti-Kumtor movement pursued a largely autonomous scale-jumping strategy. Protests and roadblocks in local villages produced the local scale as the legitimate 'scale of meaning', while disrupting the transport of goods to and from the mine, scale-jumping the issue to national and global prominence. Local NGOs also allied with national NGOs in an effort to publicise the plight of local communities. These allegiances were deepened following the success of the initial movement for compensation, resulting in the introduction of broader discourses to the anti-Kumtor movement such as post-materialist environmentalism, transparency and corruption, and human rights (Table 3). This trend of diversifying discourses as an environmental movement matures has been documented in Latin America, where environmental movements adjusted their terminology to reflect contemporary struggles, resulting in a hybridisation of local social justice and broader ecological concerns [119].

Table 3. Broad trajectory of the anti-Kumtor movement.

Timing	Main Demands	Key Actors	Scale of Meaning
1997–2006	Compensation	Local NGOs (Karek)	Local
2006–2012	Economic distribution Human rights Environmental Procedural justice	Local NGOs (Karek) National NGOs (Tree of Life, Taza Tabiyat, Human Rights Bureau)	Local/National
2012–2014	Nationalisation	National political party (Ata Jurt) Local activists	National

When supralocal actors introduce discourses to local environmental movements, the new discourses can hybridise with pre-existing environmental justice claims, or they may overpower local concerns. As Ata-Jurt's calls for the nationalisation of Kumtor grew in political prominence, there was a palpable shift towards nationalisation in the demands of the local environmental movement. This is in agreement with Central Asian literature, suggesting that resource nationalism does not originate at the local level, but is a political strategy used by national elites [16,46]. While adopting resource nationalism increased the national resonance of the local anti-Kumtor movement, it also positioned Kumtor as a national issue, undermining local claims for environmental justice. This contradicts claims in the literature that local environmentalism of the poor and glocal rescaling strategies are compatible strategies to escape the scalar bind [25,26]. Jumping scales may not always enable anti-mining movements to challenge hegemonic scalar power relations, and occasionally can lead to their reinforcement.

Forming advocacy allegiances with contentious political groups can transform perceptions of the causes and demands of environmental protests. Ata-Jurt's involvement was widely perceived as an attempt to "score some political points" [101] or a "shakedown scheme" [42]. In Mongolia, when the controversial Fire Nation, an ethno-nationalist group, allied with environmental NGOs, the organisation was described as having "rebranded itself as an environmentalist organisation" [120]. In these cases, forming an advocacy network with controversial organisations led to the real environmental issues being marginalised.

The discursive shift of the local environmental movement towards resource nationalism contradicts existing theorisations of subversive clientelism. The local movement against Kumtor originated as an autonomous ecomobilisation in response to a specific grievance, and showcased an environmentalism of the poor. Existing theories of structured elite co-optation describe how elites initiate protests, rather than how they co-opt pre-existing movements [19]. These reductionist theories of subversive clientelism struggle to elucidate the discursive shift of the local anti-Kumtor movement. Instead, this paper argues for a more nuanced co-optation, suggesting that the discursive shift of the local movement occurred due to the ability of Ata-Jurt to monopolise media attention and dominate production of national discourses regarding Kumtor. While some local activists resisted this discursive co-optation, the majority acquiesced to the resource nationalism framing of the Kumtor issue.

The decision of Ata-Jurt to discursively co-opt the Kumtor issue to suit its own political ends, and the acquiescence of local activists to the resource nationalism discourse can be understood through political opportunity structure theory [58,59]. There was a political opportunity for Ata Jurt to co-opt the Kumtor issue in 2012; as Kumtor was nationally prominent, protests organised by the local movement had reduced following the awarding of compensation in 2006 (Table 2), playing on concerns for national sovereignty powerfully following the recent 2010 riots [84], and Ata-Jurt needed to distract from criminal charges against several of its members [101]. Aligning with Ata-Jurt was a political opportunity for the local movement due to the extensive media presence of Ata Jurt politicians, which drew attention to Kumtor and the paucity of other international and national actors available to form a network to scale-jump the issue to supralocal levels. In this model of elite co-optation, local protestors are not passive victims, but active collaborators pursuing a political opportunity.

Transient elite involvement in locally originating environmental issues is not limited to Kyrgyzstan; it is widely prevalent across Central Asia. When the Mongolian Ongi River Movement became politically prominent, electoral candidates joined protest marches and advertised their support [21]. However, the River Movement adhered to its initial framing of mining as a local environmental issue, and refrained from engaging with "politically ambitious and predominantly urban-based" movements that emphasised the political and economic drawbacks of mining [21] (p. 94). This contrasts with the discursive Faustian bargain made by the local Kumtor environmental movement with Ata-Jurt. From this contrast, we suggest that the ability of a social movement to resist elite co-optation may depend on the strength and clarity of the movement's identity. Movements may be more vulnerable to elite co-optation when they become salient on the national level, but also when they start to embody a wider range of demands that locals do not relate to.

5. Conclusions

Kumtor is often framed as either a question of "the vision of the state and the role of strategic resources in the present and future of the country" [7] (p. 9), or local "social, economic and environmental issues combined" [3]. Tracing the trajectory of anti-Kumtor protests in Kyrgyzstan shows that environmental activists often engage with both of these framings, relating to theories that social movements must simultaneously legitimise the local level as the 'scale of meaning' while scale-jumping local claims to supralocal relevance through advocacy networks. However, due to the dearth of civil society actors in Kyrgyzstan and lack of international NGOs operating in the region, the onus is often on social movements to scale-jump independently of advocacy networks. The Kumtor movement achieved this by disrupting access to the mine, affecting production and generating fluctuations in the Canadian company's share price. This demonstrates that social movements are not dependent on extensive networks to scale-jump, but have the ability to do so autonomously.

When situated environmental claims become nationally prominent, they become attractive targets for elite co-optation. In the case of Kumtor, this elite co-optation is unlikely to have been through structured channels and material incentives proposed by subversive clientelism [19]. We theorise that a more passive elite involvement in anti-Kumtor protests occurred through discourse monopolisation and opportunistic local alignment. This discursive co-optation by elites can restrict the ability of local movements to scale-jump to the national level in Post-Soviet Kyrgyzstan. In order to scale-jump to the national level and escape the scalar bind, movements may need to undertake autonomous scale-jumping, or form extensive horizontal networks with other national and international local groups.

This paper identifies three critical research areas for further study. First, there has been limited analysis of the dynamic nature of environmental movements in Central Asia. Shifting focus from understanding the demands of emerging environmental movements to tracing the emergent demands of environmental movements would shed light on how a stable civil society could function in this region. Second, an improved understanding of more passive ways that elites interact with and influence environmental protests would nuance the current highly structured theorisations. There is need for further study to assess strategies that local activists use to resist discursive co-optation. Third, accounts of environmental threats in Central Asia need to engage with how actors socially construct scale. These broad research themes would facilitate a more reflexive understanding of environmental movements in Central Asia.

Acknowledgments: Ariell Ahearn and Troy Sternberg provided advice and insightful comments on earlier drafts of this manuscript. Three anonymous reviewers made discerning recommendations that improved the manuscript.

Conflicts of Interest: The author declares no conflict of interest.

References

1. Mackinder, H.J. The geographical pivot of history. *R. Geogr. Soc.* **1904**, *170*, 298–321. [CrossRef]
2. Resource Investor China Secures Gas from Turkmenistan: Who's the Real Winner? 2009. Available online: http://www.resourceinvestor.com/News/2009/12/Pages/China-secures-gas-from-Turkmenistan-Whos-the-real-winner.aspx (accessed on 16 February 2018).
3. The Guardian. Tensions Flare over Environmental Threat of Canadian Gold Mine in Kyrgyzstan. Available online: https://www.theguardian.com/global-development/2016/jan/05/environmental-threat-canadian-gold-mine-kyrgyzstan (accessed on 5 March 2018).
4. Menga, F. Building a Nation through a Dam: The Case of Rogun in Tajikistan. *Natl. Pap.* **2015**, *43*, 479–494. [CrossRef]
5. Sternberg, T.; Edwards, M. Desert Dust and Health: A Central Asian Review and Steppe Case Study. *Int. J. Environ. Res. Public Health* **2017**, *14*, 1342. [CrossRef] [PubMed]
6. Abdolvand, B.; Mez, L.; Winter, K.; Mirsaeedi-Gloßner, S.; Schütt, B.; Rost, K.; Bar, J. The dimension of water in Central Asia: Security concerns and the long road of capacity building. *Environ. Earth Sci.* **2015**, *73*, 897–912. [CrossRef]
7. Fumagalli, M. The Kumtor Gold Mine and the Rise of Resource Nationalism in Kyrgyzstan. Central Asian Economic Paper, No. 16. 2015. Available online: https://app.box.com/s/qeluit6p5qmg8sohl21qmx13gpt87pu5 (accessed on 31 March 2018).
8. Sternberg, T.; Ahearn, A.; McConnell, F. Central Asian 'characteristics' on China's new Silk Road: The role of landscape and the politics of infrastructure. *Land* **2017**, *6*, 55. [CrossRef]
9. Byambajav, D. Mobilizing against dispossession: Gold mining and a local resistance movement in Mongolia. *J. Cent. North. Humanit.* **2012**, *5*, 13–32.
10. Dani, A.H. *History of Civilizations of Central Asia: Towards the Contemporary Period: From the Mid-Nineteenth to the End of the Twentieth Century*, 1st ed.; UNESCO: Paris, France, 2005.
11. Hughes, G. Mining, Development and Environment in Central Asia: Toolkit Companion with Case Studies. 2012. Available online: http://www.zoinet.org/web/sites/default/files/publications/companion_ENG.pdf (accessed on 16 February 2018).
12. The Economist Gold in the Hills—Mining in Kyrgyzstan, 2013. Available online: https://www.economist.com/news/asia/21573615-attitudes-towards-foreign-investors-hold-troubled-country-back-gold-hills (accessed on 16 February 2018).
13. Jackson, S.L. Imagining the mineral nation: Contested nation-building in Mongolia. *Natl. Pap.* **2015**, *43*, 437–456. [CrossRef]
14. Kassenova, N. China's Silk Road and Kazakhstan's Bright Path: Linking Dreams of Prosperity. *Asian Policy* **2017**, *24*, 110–116. [CrossRef]
15. Jackson, S.L.; Dear, D. Resource extraction and national anxieties: China's economic presence in Mongolia. *Eurasian Geogr. Econ.* **2016**, *57*, 343–373. [CrossRef]
16. Ganbold, M.; Ali, S.H. The peril and promise of resource nationalism: A case analysis of Mongolia's mining development. *Resour. Policy* **2017**, *53*, 1–11. [CrossRef]
17. Reeves, M. Infrastructural Hope: Anticipating 'Independent Roads' and Territorial Integrity in Southern Kyrgyzstan. *Ethnos* **2017**, *82*, 711–737. [CrossRef]
18. Doolot, A.; Heathershaw, J. State as resource, mediator and performer: Understanding the local and global politics of gold mining in Kyrgyzstan. *Cent. Asian Surv.* **2015**, *34*, 93–109. [CrossRef]

19. Radnitz, S. *Weapons of the Wealthy: Predatory Regimes and Elite-Led Protests in Central Asia*, 1st ed.; Cornell University Press: New York, NY, USA, 2010.
20. McGlinchey, E. Central Asian Protest Movements: Social Forces or State Resources. In *The Politics of Transition in Central Asia and the Caucasus: Enduring Legacies and Emerging Challenges*, 1st ed.; Wooden, A.E., Stefes, C.H., Eds.; Routledge: New York, NY, USA, 2009; pp. 124–138.
21. Byambajav, D. The River Movements' Struggle in Mongolia. *Soc. Mov. Stud.* **2015**, *14*, 92–97. [CrossRef]
22. Wooden, A.E. Another way of saying enough: Environmental concern and popular mobilization in Kyrgyzstan. *Post-Sov. Aff.* **2013**, *29*, 314–353. [CrossRef]
23. Wooden, A.E. Images of Harm, Imagining Justice. In *ExtrACTION: Impacts, Engagements, and Alternative Futures*, 1st ed.; Jalbert, K., Willow, A., Casagrande, D., Paladino, S., Eds.; Routledge: New York, NY, USA, 2017; pp. 169–183.
24. Howard, K.W.; Howard, K.K. The new "Silk Road Economic Belt" as a threat to the sustainable management of Central Asia's transboundary water resources. *Environ. Earth Sci.* **2016**, *75*, 976–988. [CrossRef]
25. Escobar, A. Culture Sits in Places: Reflections on Globalism and Subaltern Strategies of Localization. *Pol. Geogr.* **2001**, *20*, 139–174. [CrossRef]
26. Urkidi, L. The Defence of Community in the Anti-Mining Movement of Guatemala. *J. Agrar. Chang.* **2011**, *11*, 556–580. [CrossRef]
27. Waters, K. The Fight for Community Justice against Big Oil in the Caspian Region: The Case of Berezovka, Kazakhstan. In *Environmental Justice and Sustainability in the Former Soviet Union*, 1st ed.; Agyeman, J., Ogneva-Himmelberger, Y., Eds.; MIT Press: Cambridge, MA, USA, 2009; pp. 153–189.
28. Strucker, D. Environmental Injustices, Unsustainable Livelihoods, and Conflict: Natural Capital Inaccessibility and Loss among Rural Households in Tajikistan. In *Environmental Justice and Sustainability in the Former Soviet Union*, 1st ed.; Agyeman, J., Ogneva-Himmelberger, Y., Eds.; MIT Press: Cambridge, MA, USA, 2009; pp. 237–275.
29. Kolstø, P. *Political Construction Sites: Nation-Building in Russia and the Post-Soviet States*, 1st ed.; Westview Press: Boulder, CO, USA, 2000.
30. Roy, O. *The New Central Asia: The Creation of Nations*, 1st ed.; I. B. Tauris: London, UK, 2000.
31. Suyarkulova, M. Reluctant Sovereigns? Central Asian States' Path to Independence. In *Sovereignty after Empire Comparing the Middle East and Central Asia*, 1st ed.; Cummings, S.N., Hinnesbusch, R., Eds.; Edinburgh University Press: Edinburgh, UK, 2011; pp. 127–157.
32. Olcott, M.B. Central Asia's catapult to independence. *Foreign Aff.* **1991**, *71*, 108–130. [CrossRef]
33. Isaacs, R.; Polese, A. Between "imagined" and "real" nation-building: Identities and nationhood in post-Soviet Central Asia. *Natl. Pap.* **2015**, *43*, 371–382. [CrossRef]
34. Akiner, S. The Struggle for Identity. In *After Empire: The Emerging Geopolitics of Central Asia*, 1st ed.; Snyder, J., Ed.; National Defense University Press: Washington, DC, USA, 2015; pp. 3–36.
35. Akbarzadeh, S. Nation-Building in Uzbekistan. *Cent. Asian Surv.* **1996**, *15*, 23–32. [CrossRef]
36. Akbarzadeh, S. National Identity and Political Legitimacy in Turkmenistan. *Natl. Pap.* **1999**, *27*, 271–290. [CrossRef]
37. Anceschi, L. Turkmenistan and the virtual politics of Eurasian energy: The case of the TAPI pipeline project. *Cent. Asian Surv.* **2017**, *36*, 409–429. [CrossRef]
38. Rasanayagam, J. Informal Economy, Informal State: The Case of Uzbekistan. *Int. J. Sociol. Soc. Policy* **2011**, *31*, 681–696. [CrossRef]
39. Engvall, J. The State as Investment Market: An Analytical Framework for Interpreting Politics and Bureaucracy in Kyrgyzstan. Ph.D. Thesis, Uppsala University, Uppsala, Sweden, 2011.
40. Satybaldieva, E. Political Capital, Everyday Politics and Moral Obligations: Understanding the Political Strategies of Various Elites and the Poor in Kyrgyzstan. *Eur.-Asia Stud.* **2015**, *67*, 370–387. [CrossRef]
41. Asia Times Kumtor Report Raises Corruption Concerns, 2014. Available online: http://www.atimes.com/atimes/Central_Asia/CEN-01-100914.html (accessed on 18 November 2017).
42. Transitions (TOL). In Kyrgyzstan, a Kumtor Shakedown or Setup? 2013. Available online: http://www.tol.org/client/article/23923-in-kyrgyzstan-a-kumtor-shakedown-or-setup.html (accessed on 18 November 2017).
43. Cooley, A.; Sharman, J.C. Blurring the line between licit and illicit: Transnational corruption networks in Central Asia and beyond. *Cent. Asian Surv.* **2015**, *34*, 11–28. [CrossRef]

44. Cutler, R.M. Kazakhstan's 'Resource Nationalism': Its Sources and Motives. Central Asian Economic Paper, No. 2. 2012, pp. 1–6. Available online: https://app.box.com/s/zs1iu7rg5kw4zwycs24c (accessed on 31 March 2018).
45. Humphreys, D. *Transatlantic Mining Corporations in the Age of Resource Nationalism*, 1st ed.; Transatlantic Academy: Washington, DC, USA, 2012.
46. Domjan, P.; Stone, M. A comparative study of resource nationalism in Russia and Kazakhstan 2004–2008. *Eur.-Asia Stud.* **2010**, *62*, 35–62. [CrossRef]
47. Robertson, G.B. *The Politics of Protest in Hybrid Regimes: Managing Dissent in Post-Communist Russia*, 1st ed.; Cambridge University Press: Cambridge, UK, 2011.
48. Institute for War & Peace Reporting (IWPR) Kyrgyzstan: Protests on Demand: Marginalised Groups Serve as Rent-a-Mob Demonstrators Whenever the Country is Convulsed by Unrest, 2010. Available online: http://www.iwpr.net/report-news/kyrgyzstan-protests-demand (accessed on 3 December 2017).
49. Satybaldieva, E. A mob for hire? Unpacking older women's political activism in Kyrgyzstan. *Cent. Asian Surv.* **2018**, *1*, 1–18. [CrossRef]
50. Guha, R. *Environmentalism: A Global History*; Longman: New York, NY, USA, 2000.
51. Martínez-Alier, J. Mining Conflicts, Environmental Justice, and Valuation. *J. Hazard. Mater.* **2001**, *86*, 153–170. [CrossRef]
52. Dalton, R.; Van Sickle, A.; Weldon, S. The individual–institutional nexus of protest behaviour. *Br. J. Political Sci.* **2010**, *40*, 51–73. [CrossRef]
53. Gurr, T.R. *Why Men Rebel, Boulder*; Paradigm Publishers: Boulder, CO, USA, 1970.
54. Kerkvliet, B.J. Everyday politics in peasant societies (and ours). *J. Peasant Stud.* **2009**, *36*, 227–243. [CrossRef]
55. Scott, J.C. *Weapons of the Weak: Everyday Forms of Peasant Resistance*; Yale University Press: New Haven, CT, USA, 1985.
56. Guha, R.; Martínez-Alier, J. *Varieties of Environmentalism. Essays North and South*; Earthscan: London, UK, 1997.
57. Schlosberg, D. *Defining Environmental Justice: Theories, Movements, and Nature*; Oxford University Press: Oxford, UK, 2009.
58. Buttel, F.H. New Directions in Environmental Sociology. *Annu. Rev. Sociol.* **1987**, *13*, 465–488. [CrossRef]
59. Tilly, C. *Contention and Democracy in Europe, 1650–2000*; Cambridge University Press: Cambridge, UK, 2004.
60. Lubell, M.; Vedlitz, A.; Zahran, S.; Alston, L.T. Collective Action, Environmental Activism, and Air Quality Policy. *Political Res. Q.* **2006**, *59*, 149–160. [CrossRef]
61. Smith, N. Geography, difference and the politics of scale. In *Postmodernism and the Social Sciences*, 1st ed.; Doherty, J., Graham, E., Malek, M., Eds.; Macmillan: London, UK, 1992; pp. 57–79.
62. Swyngedouw, E. Neither Global nor Local: Glocalization and the Politics of Scale. In *Spaces of Globalization*, 1st ed.; Cox, K., Ed.; Guilford Press: New York, NY, USA, 1997; pp. 137–166.
63. Brenner, N. The Limits to Scale? Methodological reflections on Scalar Structuration. *Prog. Hum. Geogr.* **2001**, *25*, 591–614. [CrossRef]
64. Vela-Almeida, D.; Kolinjivadi, V.; Kosoy, N. The building of mining discourses and the politics of scale in Ecuador. *World Dev.* **2018**, *103*, 188–198. [CrossRef]
65. Leitner, H.; Sheppard, E.; Sziarto, K.M. The spatialities of Contentious Politics. *Trans. Inst. Br. Geogr.* **2008**, *33*, 157–172. [CrossRef]
66. Robertson, R. *Social Theory and Global Culture*; Sage: London, UK, 1992.
67. Suopajärvi, L.; Ejdemo, T.; Klyuchnikova, E.; Korchak, E.; Nygaard, V.; Poelzer, G.A. Social impacts of the "glocal" mining business: Case studies from Northern Europe. *Miner. Econ.* **2017**, *30*, 31–39. [CrossRef]
68. Bebbington, A. NGOs and uneven development: Geographies of development intervention. *Prog. Hum. Geogr.* **2004**, *28*, 725–745. [CrossRef]
69. Triscritti, F. Mining, development and corporate-community conflicts in Peru. *Community Dev. J.* **2013**, *48*, 437–450. [CrossRef]
70. Barton, J.; Román, Á.; Fløysand, A. Resource extraction and local justice in Chile: conflicts over the commodification of spaces and the sustainable development of places. In *New Political Spaces in Latin American Natural Resource Governance*, 1st ed.; Haarstad, H., Ed.; Palgrave Macmillan: New York, NY, USA, 2012; pp. 107–128.

71. Bebbington, A. Capitals and capabilities: A framework for analyzing peasant viability, rural livelihoods and poverty. *World Dev.* **1999**, *27*, 2021–2044. [CrossRef]
72. Kurtz, H.E. Scale frames and counter-scale frames: Constructing the problem of environmental injustice. *Political Geogr.* **2003**, *22*, 887–916. [CrossRef]
73. Urkidi, L.; Walter, M. Dimensions of environmental justice in anti-gold mining movements in Latin America. *Geoforum* **2011**, *42*, 683–695. [CrossRef]
74. Keck, M.; Sikkink, K. *Activist beyond Borders. Advocacy Networks in International Politics*; Cornell University Press: Ithaca, NY, USA, 1998.
75. Aydin, C.I.; Ozkaynak, B.; Rodríguez-Labajos, B.; Yenilmez, T. Network effects in environmental justice struggles: An investigation of conflicts between mining companies and civil society organizations from a network perspective. *PLoS ONE* **2017**, *12*, e0180494. [CrossRef] [PubMed]
76. Bebbington, A.; Bebbington, D.H.; Bury, J.; Lingan, J.; Muñoz, J.P.; Scurrah, M. Mining and social movements: Struggles over livelihood and rural territorial development in the Andes. *World Dev.* **2008**, *36*, 2888–2905. [CrossRef]
77. Gudynas, E. Buen Vivir: Today's tomorrow. *Development* **2011**, *54*, 441–447. [CrossRef]
78. Urkidi, L. A Glocal Environmental Movement against Gold Mining: Pascua-Lama in Chile. *Ecol. Econ.* **2010**, *70*, 219–227. [CrossRef]
79. Mix, T.L. Rally the people: Building local-environmental justice grassroots coalitions and enhancing social capital. *Soc. Inq.* **2011**, *81*, 174–194. [CrossRef]
80. Schlosberg, D. Networks and Mobile Arrangements: Organisational Innovation in the US Environmental Justice Movement. *Environ. Politics* **1999**, *8*, 122–148. [CrossRef]
81. Asian Development Bank. Civil Society Briefs: The Kyrgyz Republic. Available online: https://www.adb.org/sites/default/files/publication/29443/csb-kgz.pdf (accessed on 29 March 2018).
82. Eurasianet. Kyrgyzstan: Could Kumtor Shakedown Backfire on Bishkek? Available online: http://www.eurasianet.org/node/63603 (accessed on 17 November 2017).
83. Yuldashev, F.; Sahin, B. The political economy of mineral resource use: The case of Kyrgyzstan. *Resour. Policy* **2016**, *49*, 266–272. [CrossRef]
84. Gullette, D. Kumtor: The Litmus Test for Kyrgyzstan's Mining Sector. *Zentralasien Analysen* **2014**, *76*, 2–5.
85. Hynes, T.P.; Harrison, J.; Bonitenko, E. The International Scientific Commissions' Assessment of the Impact of the Cyanide Spill at Barskaun, Kyrgyz Republic, 1998. Available online: https://s3.amazonaws.com/cg-raw/cg/final_report_of_the_international_commission_on_th_1998_cyanide_spill.pdf (accessed on 31 March 2018).
86. Institute for War & Peace Reporting. Toxic Spill Dispute Revived in Kyrgyzstan. Available online: https://iwpr.net/global-voices/toxic-spill-dispute-revived-kyrgyzstan (accessed on 16 December 2017).
87. Radio Free Europe/Radio Liberty. Kyrgyzstan: A Commercial Tragedy Revisited. Available online: https://www.rferl.org/a/1060387.html (accessed on 31 December 2017).
88. Toralieva, G. Environmental Reporting in Kyrgyzstan. *Probl. Post-Communism* **2011**, *58*, 58–66. [CrossRef]
89. Hill, A.F.; Minbaeva, C.K.; Wilson, A.M.; Satylkanov, R. Hydrologic Controls and Water Vulnerabilities in the Naryn River Basin, Kyrgyzstan: A Socio-Hydro Case Study of Water Stressors in Central Asia. *Water* **2017**, *9*, 325–341. [CrossRef]
90. Honkonen, T. Challenges of Mining Policy and Regulation in Central Asia: The Case of the Kyrgyz Republic. *J. Energy Nat. Resour. Law* **2013**, *31*, 5–32. [CrossRef]
91. Radio Free Europe/Radio Liberty. Many Kyrgyz Fail to Find A Glittering Future in Gold. Available online: https://www.rferl.org/a/why_no_gold_rush_in_kyrgyzstan/24339470.html (accessed on 20 March 2018).
92. Furstenberg, S. Consolidating global governance in nondemocratic countries: Critical reflections on the Extractive Industries Transparency Initiative (EITI) in Kyrgyzstan. *Extr. Ind. Soc.* **2015**, *2*, 462–471. [CrossRef]
93. GlacierHub. Protests over Weakening Glacier Protections in Central Asia. Available online: http://glacierhub.org/2017/11/21/protests-weakening-glacier-protections-central-asia/ (accessed on 11 January 2018).
94. Al Jazeera. Conflict continues at Kyrgyzstan's Massive Gold Mine. Available online: http://www.aljazeera.com/indepth/features/2016/01/conflict-continues-kyrgyzstan-massive-gold-160128071445334.html (accessed on 12 January 2018).
95. The Diplomat. Kyrgyzstan's Controversial Gold Mine. Available online: https://thediplomat.com/2015/02/kyrgystans-controversial-gold-mine/ (accessed on 11 January 2018).

96. Robbins, P. *Political Ecology*; Blackwell: Oxford, UK, 2004.
97. Carnegie Endowment for International Peace. What Was the True Cause of the Kyrgyz Demonstrations? Available online: http://carnegieendowment.org/2012/10/31/what-was-true-cause-of-kyrgyz-demonstrations-pub-49853 (accessed on 17 November 2017).
98. The Telegraph. Kyrgyzstan to Decide Fate of Canadian-Owned Gold Mine. Available online: https://www.telegraph.co.uk/news/worldnews/asia/kyrgyzstan/9355758/Kyrgyzstan-to-decide-fate-of-Canadian-owned-gold-mine.html (accessed on 2 March 2018).
99. Radio Free Europe/Radio Liberty. Hundreds Rally in Jalal-Abad for Release of Opposition Leaders. Available online: https://www.rferl.org/a/hundreds-rally-kyrgyzstan-jalal-abad-freedom-of-detained-opposition-leaders/24729905.html (accessed on 19 December 2017).
100. Radio Free Europe/Radio Liberty. Kyrgyz Opposition Protesters Demand Gold Mine's Nationalization. Available online: https://www.rferl.org/a/kyrgyzstan-kumtor/24966987.html (accessed on 18 December 2017).
101. Ikkevold. Was Kyrgyz Protest Really about Gold Mine or Face Time? Available online: http://www.ikkevold.no/2012/10/was-kyrgyz-protest-really-about-gold-mine-or-face-time/ (accessed on 7 December 2017).
102. Radio Free Europe/Radio Liberty. Questioning the Environmental Cost of Kyrgyzstan's Kumtor Gold Mine. Available online: https://www.rferl.org/a/kyrgyzstan-kumtor-gold-mine-environmental-/25003485.html (accessed on 16 November 2017).
103. Radio Free Europe/Radio Liberty. Kyrgyz Protesters Again Demand Nationalization of Major Gold Mine. Available online: https://www.rferl.org/a/kyrgyzstan-kumtor-mine-protest/25029473.html (accessed on 18 December 2017).
104. The Jamestown Foundation. Kumtor Gold Mine Protests and Unrest Rock Kyrgyzstan. Available online: https://jamestown.org/program/kumtor-gold-mine-protests-and-unrest-rock-kyrgyzstan/ (accessed on 2 March 2018).
105. Stratfor Worldview. Kyrgyzstan: Larger Issues behind the Kumtor Protests. Available online: https://worldview.stratfor.com/article/kyrgyzstan-larger-issues-behind-kumtor-protests (accessed on 2 March 2018).
106. Murzakulova, A.; Schoeberlein, J. The invention of legitimacy: Struggles in Kyrgyzstan to craft an effective nation-state ideology. *Eur.-Asia Stud.* **2009**, *61*, 1229–1248. [CrossRef]
107. CIA World Factbook. Kyrgyzstan. Available online: https://www.cia.gov/library/publications/the-world-factbook/geos/kg.html (accessed on 2 March 2018).
108. Kumtor. Centerra Gold Reports of the Status of Negotiations with the Kyrgyz Republic Government on the Restructuring of the Kumtor Project. Available online: https://www.kumtor.kg/en/centerra-gold-reports-on-the-status-of-negotiations-with-the-kyrgyz-republic-government-on-the-restructuring-of-the-kumtor-project/ (accessed on 9 January 2018).
109. Reuters. Canada's Centerra Settles Mine Dispute with Kyrgyzstan. Available online: https://www.reuters.com/article/centerra-gold-kyrgyzstan/canadas-centerra-settles-mine-dispute-with-kyrgyzstan-idUSL4N1LS454 (accessed on 5 January 2017).
110. The New York Times. Political Activism Begins to Take Hold in Kyrgyzstan. Available online: http://www.nytimes.com/2005/12/12/world/asia/political-activism-begins-to-take-hold-in-kyrgyzstan.html (accessed on 3 January 2018).
111. Radio Free Europe/Radio Liberty. Kyrgyz Villagers Block Road to Canadian-Operated Mine. Available online: https://www.rferl.org/a/1076331.html (accessed on 31 December 2017).
112. Reuters. Kyrgyz Protesters End Blockade of Gold Mine Road. Available online: https://uk.reuters.com/article/gold-kyrgyzstan/kyrgyz-protesters-end-blockade-of-gold-mine-road-idUKL5E8H48AG20120604 (accessed on 4 January 2017).
113. Radio Free Europe/Radio Liberty. Rally in Jalal-Abad Demands Release of Kyrgyz Opposition Leaders. Available online: https://www.rferl.org/a/rally-jalalabad-demand-release-kyrgyz-opposition-leaders/24732242.html (accessed on 17 December 2017).
114. Radio Free Europe/Radio Liberty. Kyrgyz Protesters Demand Release of Opposition Leaders. Available online: https://www.rferl.org/a/kyrgyzstan-ata-jurt-trial-protests/24925095.html (accessed on 17 December 2017).
115. Institute for War & Peace Reporting. Protests Put Pressure on Kyrgyz Authorities. Available online: http://www.refworld.org/docid/51b596914.html (accessed on 14 January 2018).

116. Radio Free Europe/Radio Liberty. Protesters Temporarily Block Road in Kyrgyz Mine Dispute. Available online: https://www.rferl.org/a/25130755.html (accessed on 17 December 2017).
117. K-News. In the Village of Saruu There Is a Kurultai with the Participation of More than 200 People. Available online: http://knews.kg/2013/10/v-sele-saruu-prohodit-kurultay-s-uchastiem-bolee-200-chelovek/ (accessed on 6 January 2018).
118. AKIpress. Saruu Activists to Protest against Dismissal of Claim Regarding Environmental Damage Caused by Kumtor (Corrected). Available online: https://akipress.com/news:538001 (accessed on 4 January 2018).
119. Hurtado, M.; Lungo, I. *Aproximaciones al Movimiento Ambiental en Centroamérica*; FLACSO: Ciudad de Guatemala, Guatemala, 2007.
120. Reuters. Mongolia Neo-Nazis Announce a Change of Tack—Pollution Control. Available online: http://news.trust.org//item/20130702082829-t2gpl? (accessed on 5 March 2018).

land

MDPI

Article

Groundwater Irrigation Management and the Existing Challenges from the Farmers' Perspective in Central Iran

Forough Jafary * and Chris Bradley

School of Geography, Earth and Environmental Sciences, University of Birmingham, Birmingham B15 2TT, UK; C.Bradley@bham.ac.uk

* Correspondence: foroughjafary@yahoo.com; Tel.: +44-7538377869

Received: 24 October 2017; Accepted: 17 January 2018; Published: 23 January 2018

Abstract: The sustainable management of natural resources, and particularly groundwater, presents a major challenge in arid regions to ensure security of water supply and support agricultural production. In many cases, the role of smallholder farmers is often neglected when managing irrigated water and land processes. However, management decisions have a major impact on farmers' livelihoods, and it is essential: first, to recognise the crucial role of regional and local social, political and economic systems; and second, to integrate farmers' perspectives in the governance and management of local groundwater practices. This is particularly important as the ways in which arid region farmers use land and water have wider implications for land degradation and salinization. This paper uses a community-based approach to identify and examine the social, economic and cultural dimensions to groundwater irrigation systems from the perspective of local farmers in central Iran. The paper utilises interviews with local farmers and water agencies in Iran to reflect on their respective roles within the irrigation system and in developing management plans for the sustainable use of groundwater. Through social research, we investigate the reasons why farmers might reject government irrigation management schemes and outline how local problems with land degradation and salinization and reduced water availability have arisen as a result of changing management policies. In conclusion, we identify future challenges and consider appropriate future management strategies.

Keywords: drip irrigation; groundwater; common-pool resource; water rights; local farming

---

1. Introduction: Groundwater Management Crisis in Iran

Groundwater is the principal source of water used for domestic and agricultural purposes in Iran. However, there are major challenges in managing groundwater sustainability given anticipated increases in the demand for food and potable water, which are leading to increasingly high rates of groundwater abstraction and contributing to widespread reductions in groundwater levels. In Iran, aquifers have been over-abstracted due to changing crop patterns, inefficient irrigation systems and high water wastage associated with traditional Iranian irrigation practices [1]. Currently, one of the major failures in groundwater resource management by national government is the inefficient agricultural water use, as the efficiency of irrigation water delivery is thought to be between 33 and 37 percent [2]. Other failures include uncompleted water projects, problems with water delivery and drainage, as well as inadequate resources for future water development schemes.

The consequences of inappropriate agricultural water management include environmental degradation, salinization, land subsidence and wetland desiccation. Soil salinization is widespread in Iran with an estimated economic loss of >$1 billion (U.S.) [3]. In arid regions, water scarcity has led to increased abstraction of brackish groundwater for agriculture, which can lead to salinization and accelerated land degradation and desertification [4]. As a result, it has been suggested that groundwater conditions in Iran are in a critical state [5–7]. This highlights the need for changes in domestic and agricultural water use practices, and specifically the adoption of more efficient irrigation technologies, which are adapted to local conditions and are economically viable for farmers. However, there have been problems with introducing new groundwater management practices, which have sought to increase irrigation efficiency (e.g., using micro-irrigation methods) and reduce groundwater abstraction rates. Hence, the impact of policies on irrigation modernization is uncertain; particularly the degree to which the introduction of new technologies might help to reduce rates of water use and consumption [8]. Although changing irrigation systems, from furrow to micro-irrigation (i.e., drip method), has reduced water use in some cases [9,10], other studies suggest it has been associated with increased water and energy consumption [11,12].

It is also important that ecosystem sustainability should be given more attention when managing water resources: particularly, the long-term consequences of changing irrigation policies on environmental sustainability. For example, at present, irrigated waters are prevented from recharging the common pool groundwater body due to: (i) the high aridity and evaporation rates; (ii) high rates of water consumption due to intensive cultivation practices; and (iii) the depth of the groundwater table. This raises questions about the current sustainability of groundwater aquifers. In response to these problems, the Iranian government is seeking to introduce new irrigation systems to reduce groundwater abstraction rates and improve water delivery efficiencies for irrigated agriculture, but this requires engagement with local farmers to maximise rates of adoption.

This paper investigates the challenges in introducing new irrigation technologies and management plans for local farming systems through top-down national policies. One potential outcome is that the projects are rejected by local farmers, which compromises any future improvement in irrigation delivery efficiency that was anticipated by the original policy. Hence, in developing such projects, policy makers should consider the socio-economic, biophysical and cultural dimensions of managing natural resources, particularly strategic common-pool resources such as groundwater aquifers. A crucial factor here is the flexibility of technologies to accommodate local irrigation traditions and water allocation rules as evidence suggests that technological solutions are not widely adopted if they are not socially acceptable [13]. Hence, any new technology, or management plan, must address the different dimensions of traditional irrigation and land management systems if it is to minimise social, economic and environmental impacts. In this empirical study, we discuss the main changes in this irrigation transformation, which are: first, the conversion from collective action around tube-wells, which occurred as farmers transferred from the historic Qanat irrigation system, to a new collective ownership/action for drip irrigation promoted by government (Qanats are underground water transmission canals that convey groundwater from upland areas to the lower plains by gravity) (Figure A1); second, the adoption of drip irrigation systems with associated changes in water right allotments (reduction), which create many challenges for farmers.

Water Crisis and Efficiency Management: Reconciling Different Viewpoints

In Iran, policy makers have largely attributed 'the water crisis' to climate change and particularly to the increased frequency of drought, which as a 'natural disaster' is outside their control and represents only a temporary problem [14]. However, in discussing Iran's water crisis, Foltz [15] suggested that many Iranian researchers believe that the water crisis is only partially attributed to drought events; and some water experts believe that about half of the problem is due to mismanagement of water resources.

The degradation of groundwater (both quantity and quality) thus reflects a combination of excessive groundwater abstraction (and/or drought, which reduces groundwater recharge), inappropriate groundwater governance and management plans and poor land use management. The lack of agreement between water experts and policy makers on the causes of the water crisis has inhibited the development of comprehensive water management plans. Consequently different solutions have been advanced to improve irrigation efficiency, which is defined differently by local communities and government decision-makers.

Recent work has challenged the concept of efficiency in irrigation practices, and there is a belief that what is usually known as improved irrigation efficiency is a misleading concept [16]. Under the concept of irrigation efficiency, the volume of water saved is considered an 'efficiency gain' (i.e., saved water to be used in environmental enhancement and non-farming activities). This is a paradox as the water saved is not allowed to return to the drainage basin (i.e., to recharge the groundwater body), and hence, true rates of water consumption are unchanged [16]. In an Iranian context, some of the proposed government schemes to improve irrigation efficiency strongly contrast with what local users perceive to be irrigation efficiency, and hence, there is a lack of cooperation between local water users and authorities.

In this paper, we illustrate how perspectives on water values, on the water crisis and on the rationale for groundwater protection and sustainable water resource management differ between local communities and the groundwater regulator, the Iranian Agricultural Jihad Agency. It is important to recognize the different viewpoints and to identify areas of common understanding. In Iran, improved efficiency for farmers' means increasing their accessibility to specific water rights to maintain or increase agricultural production levels. Farmers are likely to reject new schemes that threaten these rights, for example, the installation of drip irrigation systems to improve efficiency and manage groundwater sustainably. Furthermore, socio-economic conditions, livelihoods and technological constraints, all have a local context and must be understood separately as they can greatly affect water management and efficiency [17]. Conversely, government irrigation management schemes seek to protect groundwater bodies from over-abstraction and to ensure continued water availability to support projected water demands. Thus, the main difficulty is how efficiency should be measured by the water regulator and how permission for abstraction should be granted to avoid groundwater depletion and to satisfy the objectives of both farmers and government agencies. Currently, the Iranian Ministry of Energy (MoE) issues new permits for applicants on the basis of a hydrogeological assessment of the aquifer [18]. Legally new abstractions are allowed as part of a business plan, but under the traditional local system, water rights are associated with individuals through their historical allocation. The historic water distribution system (the Qanat) is based on cycles of specific numbers of days and nights (24 h) [19] (pp. 148–152), and this system of water allocation and distribution is still largely practiced by local farmers. The water cycle is divided into different time-slots (water right delivery might be received during the day or night). The water hour right for each person is the basis of assigning the entitlement, although the eligibility criteria for individual applicants is not necessarily transparent [18].

In this situation, controversy and conflict between environmental issues and farming practices can easily arise, usually resulting in conflict between agencies and farmers. In Iran, the government has sought to protect groundwater resources, while farmers' reliance on groundwater abstraction through pumped tube-wells hinders enhanced groundwater protection. The operation and maintenance of the traditional Qanat irrigation systems in Iran are established based on collective action and cooperation among farmers. This encourages farmers to share costs and responsibilities and is the main driving force for farmers to act collectively in maintaining the operation of currently-used pumped tube-wells. The traditional norms, rules and social capital around irrigation systems, e.g., water rights and allocation, have also been transferred from the Qanat system to the modern system of tube-wells to enable successful farming practices to continue. Generally, tube-well owners fund the construction, maintenance and operation of their wells, but not any external costs resulting from extensive groundwater abstraction on the environment or aquatic ecosystems [20]. Similar processes occur in developing countries where economic conditions act as a barrier for improving environmental protection, resulting in the general depletion of natural resources with inevitable environmental consequences [17].

In Iran, local farmers are confronted with a number of irrigation challenges that contribute to their unwillingness to join, or collectively participate, in government schemes to increase irrigation efficiency and manage scarce groundwater resources sustainably. Groundwater over-abstraction is potentially catastrophic as there is insufficient regulation (i.e., the issuing of new water rights/entitlement for applicants) and monitoring of groundwater abstraction rates (most tube-wells lack metering) to ensure the sustainable use of groundwater resources. At the same time, research on farmers' behavioural approaches to groundwater abstraction in arid regions is limited and detailed investigation of irrigators' reflections on concepts of efficiency and productivity are needed. Here, for the first time, we identify the factors that influence local farmers' adoption of drip irrigation within the government "Tooba Scheme" and investigate local definitions of irrigation efficiency. As part of the 'Tooba' scheme to install drip irrigation systems, farmers have to build small reservoirs on their land and consolidate scattered land holdings. The establishment of a reservoir is followed by acceptance of a drip irrigation system, which requires farmers' agreement to consolidate their fields if they are to be eligible for a government loan. For small-scale farmers, construction and maintenance of new irrigation systems is expensive, requiring new collective ownership of the infrastructure to share their cost.

Global and national water problems require local solutions, and local knowledge can enhance local capacity by utilising that knowledge to embrace alternative solutions [21]. Hence, local knowledge and community-managed systems have a crucial role in water management, particularly in Iran, given the importance of traditional and current irrigation management practices. Before discussing this, we first review water management practices and proposed government management schemes to provide a context for the discussion.

2. Materials and Methods

2.1. Case Study Characteristics: Kashan

Kashan city, and the surrounding rural villages (population ~400,000), is the largest city in Northern Isfahan province in central Iran (Figure 1). Given its location in Iran's Central Desert, this region has a dry climate and groundwater is the only source of water for agricultural, urban and industrial uses. At present, agriculture utilises 86 percent of the annual total of abstracted groundwater, while domestic and industrial activities use 7.5 and 1.7 percent, respectively. The remaining 4.8 percent is used to maintain green spaces.

There are two different climates within the region: First is an alpine climate with winter snow cover, where irrigation waters are supplied by springs, ephemeral rivers and Qanat systems. These alpine areas experience infrequent conditions of water scarcity, and farmers are able to cultivate a range of products including fruit trees. The second climate is associated with the plains adjacent to the central Kavir desert: a region of arid climate with a low annual precipitation (140 mm/year) and high temperatures (see Table 1). Here, the soil is sandy and the only water source is groundwater abstraction from deep wells, as most of the Qanats are dry. The main villages in Kashan are associated with areas of cultivation and include: Abu-Zeid Abad, Ali Abad, Rijen and Fakhreh. Farmers in these villages were interviewed and were the main participants in group discussions for this study (Table 2). The main criterion governing selection of the study area and participants was that the study should be in an arid region of Iran where water management practices reflected the rich indigenous knowledge of local irrigators. A further constraint was access to local communities with long experience in groundwater irrigation practices who were willing to cooperate in the study. The participants were mainly male farmers who practiced small-scale farming in Kashan (women are not generally involved in farming) (Table 3).

Traditional crops in Kashan include fruit trees (pomegranate, pistachio and apricot), cereals (wheat, barley, maize, millet and chickpea), vines, melons, cucumber and vegetables. Recent agricultural productivity has been threatened by reduced water availability, and farmers suggest that climate change, particularly a prolonged drought since 2010, has affected the diversity of crop cultivation patterns. In common with other arid and semi-arid regions in Iran, farmers in Kashan are experiencing critical declines in groundwater levels and increased water scarcity [1,22,23] to the extent that the future availability of groundwater is questionable. The majority of farmers have installed individual or collective tube-wells, and farmers have a vital role in the irrigation management process: both in maintaining irrigation practices and agricultural productivity. Farmers ensure the adoption of locally-based irrigation solutions and agree to collective management strategies, including selecting appropriate crop patterns given the conditions of water scarcity. However, the role of farmers has been largely neglected within the management-decision process in Iran.

The increasing frequency of conditions of water scarcity provides a strong incentive for farmers to identify and adopt coping mechanisms and adaptation strategies [24]. Local farmers in Kashan have sought to sustain their livelihoods despite this threat by: (i) extending their sources of income (e.g., husbandry, weaving carpets, working in factories); (ii) keeping livestock to be sold during periods of financial crisis; (iii) taking collective action to reduce labour costs; (iv) sharing the costs of purchasing land and drilling and maintaining tube-wells; (v) renting harvested lands to other farmers for livestock grazing in exchange for reducing labour costs for land clearance; (vi) relying on their social network for food or cash during times of financial crisis; and (vii) engaging in an informal market for water and land rental, as well as dairy products.

These local adaptive management strategies are alternative solutions used by farmers, which are neither governed, nor imposed by government. Where farmers feel that government management plans are poorly conceived, or if there are implications for future profitability, then farmers will follow their own management strategies and reject the range of different schemes proposed by the Iranian government to improve irrigation management efficiency. In the following section, we describe existing irrigation management schemes before examining the reasons that farmers gave for rejecting these schemes.

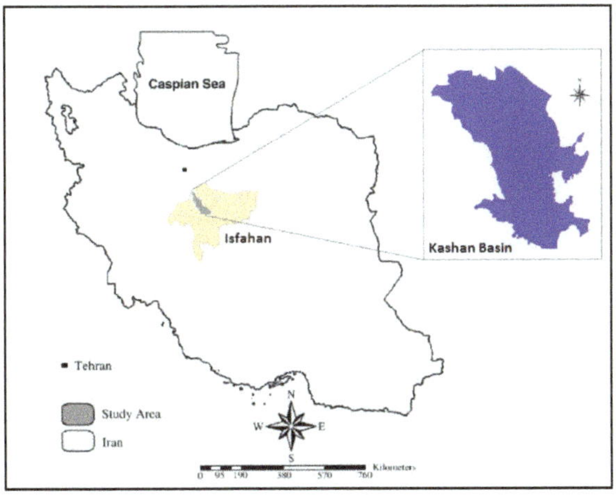

Figure 1. Kashan City in Isfahan province in Central Iran.

Table 1. Characteristics of farming systems in Kashan.

Agriculture Features	Farmlands Size	Crop Pattern	Water Right	Access/Ownership	Irrigation System	Water Uses m^3/ha	Irrigated Land
Kashan City Small-scale Large-scale	1470 km^2 1–5 ha >5 ha	Wheat, barley, alfalfa, pistachio	Is based on each land's size	Private or shared ownership of tube-wells	Furrow, flood, drip system	Total extraction 238 million cubic metre	22,500 ha (Kashan) 18 million ha in Iran

Table 2. An example of surveyed local farmers in Kashan region.

Farmer-Village / Farm Features	Farmland Size	Crop Pattern	Water Right	Access/Ownership	Irrigation System	Water Uses m3/ha	Irrigated Land (ha)
Taghavi, R. Rijen	2 ha	Barley, alfalfa	3 h water right every 5 days	Shared tube-well (5 small-scale farmer)	Furrow, flood	-	2
Saboohi, A. Ali-Abad	5 ha (integrated land)	Wheat, barley, beetroot	10 h water every 8 days	Shared tube-well	Furrow	-	5
Razavi, H. Rijen	32 ha (integrated land)	Pistachio	35 h every 5 days	Private land-purchased water rights	Furrow and drip system	-	32
Golnari, A. Ali-Abad	4 ha	Wheat, barley, pistachio	2 h water every 5 days	Shared tube-well	Furrow and flood	-	4
Nouri, T. Fakhreh	10 ha (Integrated land)	Pistachio, wheat, barely, alfalfa	24 h every 5 days	Shared tube-well	Furrow and drip	-	9
Ganji, A. Rijen	1 ha	Alfalfa, barley	2 h	Small-scale farmer	Flood	-	1

Table 3. Data collection in Kashan Villages (2011–2013) [1].

Methods \ Participants	Local Farmers Ali-Abad	Local Farmers Fakhreh	Local Farmers Rijen	Local Farmers Kashan
Deep Interviews	4	2	3	1
Semi-structured Interviews	3 (total 12 farmers)	2 (6 farmers)	1 (4 farmers)	1 (3 farmers)
Group Discussion	3 (total 38 farmers)	2 (25 farmers)	1 (8 farmers)	

[1] In four different villages and in the Kashan City, separate sessions of deep and semi-structured interviews, as well as group discussions were conducted. In Ali-Abad, 3 sessions of semi-structured interviews with a total of 12 farmers were carried out. In addition, 3 group discussions were organized, which in total, 38 farmers attended. The number of sessions and total participant farmers are summarised in Table 3.

2.1.1. Tooba' Drip Irrigation Scheme

In Kashan, the Iranian government encourages smallholder farmers to participate in large-scale (>10 ha) irrigation schemes, termed 'Tooba', comprising drip irrigation systems, which seek to improve irrigation delivery efficiency. The micro-drip irrigation system (in the 'Tooba' scheme) provides water and fertilizers through a network of narrow plastic tubes, and water is delivered directly to the root zone, or soil surface, to supply plant water requirements. In Iran, the area under micro-irrigation system is ~400,000 ha [25]. This method can prevent groundwater over-exploitation and reduce water wastage [26,27]; however, in arid regions, its implementation presents many challenges, which require detailed environmental monitoring of the irrigation system and engagement with local farmers. For example, Rodriguez Diaz et al. [11] evaluated water transmission from surface irrigation to a pressurized drip system in Guadalquivir (Spain) and reported a 40 percent reduction in water diversion, but with a 200 percent increase in cost. Given the expense, farmers with large land holdings and better access to water resources are in a better position than subsistence farmers to invest in the expensive irrigation equipment required [28]. In addition, studies using agro-economic models to evaluate pressurised systems to improve irrigation efficiency noted increases in water consumption rates, which is known as the rebound effect (e.g., [8]). This effect occurs where modern irrigation systems lead to increases in irrigated land areas, and the saved water is used to increase the extent of agricultural cultivation in situations where there are no limitations on water rights, leading to an increase in overall water consumption [29,30].

2.1.2. Sub-Surface Piped Irrigation System

Recent applications of sub-surface piped water technology for irrigation have included the use of 30 cm diameter polyethylene irrigation pipes installed at depths of ~1 m. These were introduced by the Iranian government in 2012 to reduce water transmission losses and improve irrigation efficiency. Different water distribution systems (such as canals or pipes) are usually tested by farmers, and their reflections on the practical utility of these systems can help local agencies to improve the system or to introduce different methods or devices.

The majority of studies promoting drip irrigation systems have had an experimental focus, whilst information on farmers' perspectives on these systems, and on collective irrigation management practices, has been limited [27,31]. There has also been little critical evaluation of farmers' perceptions of Iranian government schemes: there has been no local evaluation of the 'Tooba' irrigation scheme, nor have participatory and community-based management approaches to irrigation management been assessed.

2.2. Methods

This paper considers the perspectives of different groups of farmers who participated in a qualitative research study including a combination of group discussion, semi-structured interviews and a participatory irrigation simulation exercise (to be described in a subsequent paper). Four periods of field-work were undertaken over three years (2011–2013), each year for a period of between 1 and 2 months. Interviews and group discussions were conducted with experienced farmers and local water management authorities in the region (Tables 3 and 4). The purpose of the interviews were to: provide general background understanding of the irrigation system in Kashan; identify the main problems regarding irrigation efficiency, sustainable groundwater management, and the implementation of irrigation rules and management practices, as well as existing cultures regarding irrigation from the perspective of local farmers. Semi-structured interviews were designed on the basis of key statements extracted from preliminary 'deep interviews'. The former were conducted with 25 individuals, in farmers' houses or on their land (see Appendix B). Further interviews were conducted with officials in relevant local agencies, including the Regional Water Authorities of Kashan, with the responsibility of managing and maintaining the irrigation infrastructure (water canals, reservoirs and large-scale groundwater schemes) and delivering water to farmers. Officials in the Agricultural Jihad Agency were also interviewed: this body oversees the development and management of on-farm irrigation schemes. In total, 4 in-depth and semi-structured interviews were conducted with different management bodies in two sub-district regions of Kashan and Aran-Bidgol (Table 4).

Group discussions were invaluable in encouraging farmers to engage in debate, particularly prior to, and after, implementation of government irrigation schemes. A total of 6 group discussions was conducted to explore challenges in irrigation management and identify farmers' perspectives towards government schemes. These interviews were translated and transcribed into English for analysis. The transcripts were coded using NVivo, to enable the information to be classified to identify themes for further discussion.

Table 4. Key governmental departments interviewed (2011 and 2013).

Roles/Interviewees — Official Agencies	Official Role	Number of Interviewees and Types of Interviews
Agricultural Jihad Agency	Chief of rural cooperative	1 semi-structured interview
Water and sewage agency (in two sub-districts of Kashan and Aran-Bidgol)	The head of water agencies	1 in-depth interview and 1 semi-structured interview
Academic experts	University lecturer and PhD students	1 in-depth interview with academics 2 group discussions with PhD students

3. Results

3.1. Challenges for Local Stakeholders with Groundwater Irrigation Management in Kashan

In this section, we summarise the interviews with local farmers and water agencies on their role(s) in Kashan's irrigation systems and outline their perspectives on challenges regarding the sustainable use of groundwater resources. The interviews identified many challenges with current groundwater management and irrigation practices in Kashan, due to a combination of harsh climatic conditions, the long-term overexploitation of groundwater resources, decreasing water quality and the lack of effective controlling and monitoring mechanisms for groundwater abstraction. The introduction of pump irrigation technology has contributed to an increasing inequity in access water supplies between large-scale and small-scale farmers, and the latter have increasingly relied on their historic water allocation and collective management of tube-wells to sustain their farming practises. As a result, farmers and water agencies both consider that the sustainability of groundwater resources in Kashan is threatened. While some of the groundwater management problems can be attributed to national policy, others have arisen through a combination of water scarcity [14], the ineffectiveness (in some areas) of local institutions and as a result of farmers' local decision-making. Thus, a range of factors contribute to the challenges confronting farmers in Kashan and, importantly, analysis of the empirical outcomes of the interviews with local farmers show that perspectives on water values, on the rationale for groundwater protection, the definition of irrigation efficiency and on the sustainability of water resources differ between local communities, farmers and official authorities. Hence, there is a need to tackle different viewpoints and derive commonly-agreed statements of the problem prior to identifying appropriate irrigation management plans.

The empirical findings also suggest that farmers in Kashan face particular challenges that prevent them from taking collective action and agreeing on joint management decisions. This is the process whereby government devolves certain actions to small-scale farmers to encourage land consolidation and installation of drip irrigation systems and also transfers costs to farmers as part of a transition in water governance structures from state-led to community-based management. There is also an unhelpful focus on government schemes promoting technological solutions without considering socio-ecological conditions (i.e., importing technologies that may not be adaptable to local needs). Other factors include land consolidation agreements, as well as changing crop patterns, which might be counter to traditional social and cultural practices in a particular region (see the Discussion).

The interviews with farmers also revealed that historically, Iranian water rights were adjusted to provide a fair and equitable distribution of water to small and scattered land holdings. These small and scattered farmlands are increasingly economically disadvantageous for farmers, and inhibit farmers' participation in 'Tooba' irrigation schemes. However, while farmers are aware of the advantages of land consolidation, which can reduce water wastage and facilitate machinery use, in practice, land integration is difficult for farmers to achieve (see Section 4).

Regarding different perceptions on the efficiency and sustainability of groundwater use among farmers and state agencies, farmers in Ali-Abad (2013) defined the improved irrigation efficiency as:

> "Based on each farmers' water right (refer to the Qanat allocation system), a farmer needs to receive his water at time and if the total water reaches his plot increased by saving (water) from seepage through the transmission canals, then it means efficiency has improved so he can cultivate more lands ... and his final production level and revenue is higher".

The statement indicates the crucial importance of water right and allocation mechanisms amongst farmers, as this provides fair access and distribution of irrigated water. This also indicates that farmers recognize that significant quantities of water are lost during water transmission, when using poorly-maintained water canals. The main measurement for improved efficiency for local farmers seems to be improved accessibility to their allocated water rights, in order to maintain or increase their production level. Farmers also stated that cement-lined, or sub-surface, tube water canals can provide access to more water, and hence are more advantageous, as they can considerably improve water delivery efficiency. This can significantly affect farmers' decision-making behaviour in selecting suitable technology. When farmers evaluated the policy on reducing water rights to one-third of the current amount under the 'Tooba Scheme', they suggested initial estimates of groundwater abstraction rates need to be re-evaluated because they were incompatible with current needs. As one farmer in Fakhreh village (2013) stated:

> "Farmers cannot cope, government has to evaluate any new policy first with farmers and then set the rules, we cultivate based on our water right and if the policy interferes with this we cannot accept it".

In response to the question of whether the new water allocation rule had been re-assessed recently, the agricultural water agent responded:

> "No, we agree that there is an implementation problem for this scheme to be accepted by farmers, but on the other hand the agency has to deal with groundwater over-abstraction".

Many farmers stated that this state-oriented policy of water right re-allocation can potentially disturb the historic water allocation mechanism. The implementation of more restricted rules on water rights can potentially add extra uncertainty and increase the inequity between farmers in their access to water.

These empirical findings provide a deeper understanding of existing challenges when scarce groundwater resources are the only water source available to support domestic and irrigation practices for local communities. It is important, however, to understand the historical background to current irrigation practices, and particularly their irrigation context. The following section uses information from the interviews with farmers to assess the different factors influencing whether farmers might accept or reject the government irrigation schemes.

3.1.1. Introducing Pumped Technology and Associated Difficulties

Pumped-well technology in Iran was initially thought to have been successful introduced as it was widely adopted by farmers, although over time, it has adversely affected groundwater resources. This raises questions over the sustainability of pump technology, which has been described as 'the source of health, irrigation, power and control' [32]. Farmers initially valued the flexibility of pumped well systems, and the majority of farmers in Iran adopted these systems when they became available, leading to increased agricultural production. In Kashan, farmers indicated that when the government introduced pumped-well technology, they were encouraged to drill tube-wells and were provided with incentives including a cheap tariff on groundwater, subsidized electricity, long-term loans, free manure and aerial applications of pesticide on their farms to encourage uptake.

The introduction of irrigation wells created short-term economic benefits for many farmers, but has led to a number of subsequent problems. One major consequence is that following the introduction of pump technology, groundwater has become the main water source for agriculture: it offers a more resilient water source in drought-prone regions such as Kashan and can sustain local livelihoods and incomes. This illustrates how "groundwater development can improve socio-economic status of poor to a greater extent than traditional surface water" because it needs less initial investment to dig a new well, and it can be drilled inside the farmland [33]. However, as the technology was widely applied and was managed inappropriately, there were only short-term benefits. Extensive groundwater exploitation has resulted in salinization and degradation of soil and water quality. Interviews with representatives of the local water authorities indicated that the main reasons of soil salinization in Kashan included: the local geology, climate (high evapotranspiration, wind-borne salinity), saline groundwater intrusion, traditional irrigation methods and the usage of saline water in agriculture. As the situation has deteriorated over time, farmers' choice of crops has become increasingly limited to salt-tolerant crops, such as pistachio, barley and wheat, which has reduced agricultural incomes.

Farmers in Kashan expressed their dissatisfaction with increased farming costs and indicated that they had become more vulnerable as pumping costs have increased and groundwater abstraction rates have been controlled by the state. Irrigation costs have also increased while government subsidies have been removed. This has contributed to an increasingly uneven pattern of water distribution and water accessibility, which presents significant social and economic disadvantages for farmers. Inequity in access to water resources is one of the main social consequences that have arisen through pump technology: farmers in Kashan suggested that richer farmers have sufficient capital to purchase more land and water rights, and hence, their production levels and revenues are higher than poor farmers with limited access to water. Higher revenues also mean that rich farmers can easily pay the increasing labour and electricity costs and can secure loans to adopt modern irrigation facilities.

Although rich farmers work independently, in comparison to poor farmers, farming is not their main occupation. Small-scale farming systems include a large number of shared owners of wells who had previously shared water rights in a Qanat system. Where there are high numbers of shared owners, the benefits to the individual farmer is reduced with lower income and profit. Small-scale farmers, who cannot afford to purchase additional water rights above their historical allocation, also believe that richer farmers have secured better access to water sources through corruption in recent years. However, groundwater abstraction appears less susceptible to corruption in comparison to large-scale surface abstraction schemes.

3.1.2. Problems of Management Transfer

Current management problems in Iranian agriculture and irrigation management systems are associated with a change in the control and maintenance of water supply and distribution systems from individual communities to government authorities known as management transfer. Nationalisation of water and irrigation management practices and organisations commenced during a process of land reform in the 1960s, which has transferred irrigation management from a feudal basis to the state [34]. There have been several consequences for irrigation practices, particularly at a local level. Crucially, in contrast to the previous situation when the landlord was responsible for the distribution and maintenance of the water supply infrastructure, this responsibility has been transferred to the government, which enables the state to control and monitor farmers' abstraction rates. Farmers stated that in the traditional irrigation system, they worked collectively in activities such as irrigation, harvesting, land preparation and cultivation and in cleaning and maintaining Qanat systems and irrigation canals. Farmers tried to transfer these collective activities to the tube-well system management; however, collective management of water and irrigation systems largely ceased following the installation of private tube-wells by large-scale farmers. At present, farmers believe it is not their responsibility to protect and control groundwater abstraction rates from tube-wells, although when they used the Qanat systems, they were responsible for maintaining the water supply system.

In general, farmers have struggled to maintain their traditional water rights and allocation systems when adopting new irrigation technologies. One farmer in Abu-zeid Abad village in Kashan (2012) stated that:

> "Today it's the water that gives value to the land. The value of water is increased because its price has increased ... We stayed here because of available water".

Farmers also stated that collective ownership of tube-wells (by small-scale farmers) could ensure equitable access to groundwater resources and reduce pumping costs. However, government policies, such as drip irrigation systems, affect existing collective actions and water right regimes in ways that are particularly problematic for poorer farmers. Many small-scale farmers stated that this state-oriented policy disturbs their historical water allocation and the way they collectively manage irrigation practices. They considered it was neither suitable nor effective for them to use drip irrigation systems to support their existing farming patterns, and these farmers could not be persuaded to accept this method.

Small-scale farmers also outlined how a drip irrigation system was inappropriate for Kashan, as with their scattered land holdings and brackish groundwater, it was not feasible to use drip systems to irrigate pistachio trees. There were further problems with the labour-intensive maintenance requirements, high rates of water leakage, the cost of repair and worries about theft of key infrastructure components.

Consequently, the majority of farmers (both medium and small-scale) have tended to reject modern irrigation systems (i.e., drip irrigation methods) as this would require differing practices of collective ownership and action (see the Discussion). As a result, farmers would be unable to predict and accept future risks and would be unable to assume the responsibilities associated with modern irrigation technologies.

Another underlying factor in rejecting drip irrigation was provided by a farmer in Kashan who stated that:

> "This is why we cannot accept drip irrigation system, because then we have to establish a shared water reservoir; but each farmer wants to have his own authority on his 4–5 h of water within traditional water right allocation. The Jihad Agriculture (the agency) wants a farmer to manage his water in reservoir while it is not adapted to our practical water allocation mechanism".

When installing drip irrigation systems in the 'Tooba' Scheme, farmers have to construct small reservoirs to store their water allocation, which is then distributed through drip irrigation to their lands. This disturbs traditional water allocation mechanisms. It is also problematic given the scattered land ownership, and it is difficult to allocate water rights to farmers from the reservoir. Group discussions with farmers indicated that construction of a reservoir and the adoption of a drip irrigation system require farmers' agreement to consolidate their land holdings and become eligible for a government loan. For small-scale farmers, construction of the reservoir and/or drip systems and funding subsequent maintenance is expensive, requiring collective ownership of the infrastructure to share their cost equitably. Farmers pointed out that land consolidation requires extensive effort and agreement between farmers, which cannot be easily arranged.

As part of the contemporary water governance regime in Iran, the government has tried to promote a decentralized management approach to irrigation and to improve stakeholder engagement in irrigation management practices, but this appears to have been largely unsuccessful [35]. For example, in the new irrigation systems, the government has reduced financial support for farmers and has gradually transferred responsibility for irrigation management to farmers. However, adoption of this policy requires capacity building among small-scale farmers who rely on government financial support to implement and maintain their irrigation systems. The government policy of decentralization essentially transfers costs onto poorer farmers, with insufficient financial support and limited education and capacity building.

Farmers identified other factors that discourage their participation in large irrigation schemes, which mainly related to a lack of financial resources to invest in projects, the lack of agreements and cultural differences, inefficiency and poor coordination by rural institutions and unsuitable irrigation projects for farmers (with respect to profitability). However, poverty is the main factor affecting farmers' capacity to adapt to conditions of water shortage [36]. Other factors influencing the successful introduction of new irrigation technologies in Kashan are described below.

4. Discussion

4.1. Factors Influencing Farmers' Adoption of Government Irrigation Schemes

The main government schemes discussed here are the 'Tooba' scheme and associated policies (including the installation of drip irrigation systems, reduced water rights and land integration) and changing cultivation patterns with increasing pistachio cultivation. Different factors influence the acceptance or rejection of government schemes including: biophysical adaptability, economic return, cultural and social adaptability, which are analysed here from empirical research and by reviewing the literature. If a new government irrigation scheme includes all these factors, it is likely that local farmers will accept the project and participate in its adoption.

4.1.1. Biophysical Adaptation

As stated previously, the combination of changing water supply systems and mechanisms for water abstraction has led to significant changes in the Iranian irrigation and water distribution infrastructure. Government policy over the last 30 years has mainly focused on improving irrigation efficiency by introducing different irrigation technologies and encouraging farmers to adopt devices to control abstraction. Empirical evidence (e.g., [37]) suggests that farmers have failed to accept many of these policies, controlling mechanisms, or technologies, as in most cases, they are inappropriate for the local circumstances. The introduction of new technology (e.g., drip irrigation methods) requires social and material adaptability, and the most successful technologies are those that are flexible given the existing social, biophysical and cultural situation [34]. Bijker et al. ([38], p.13), argue that technological choices often require an iterative process of negotiation between members of each group to shape and confirm the best technological option. Implementation of each technological solution may bring different advantages or problems for local users, and it is important that the technology is fully evaluated and can be adapted to different ecological, social and cultural conditions ([34], p.28).

The climatic and biophysical characteristics of a region will also affect the adoption of a particular irrigation technology. In Kashan, farmers outlined their main reasons for rejecting proposals to reduce water abstraction rates to one-third of their current amount: farmers suggested that given the harsh regional climate, it was difficult to use less irrigation water. Currently, farmers are irrigating continually (for almost 24 h a day), and yet, they are still unable to achieve a good yield, so they question how they might increase production with much less water. For example, when discussing the adaptability of drip irrigation systems to the regional climate, one farmer in Ali-Abad village (2013) said:

> "This region is a desert land and it must be irrigated with flood method and consumes lots of water because the region is hot, and the soil is clay, so within 10 days we should irrigate again".

The other main limitation to adopting the drip irrigation system is that it is only suitable for wells with low salinity groundwater (i.e., with an electrical conductivity (EC) <3000 $\mu S\ cm^{-1}$). Where water has a higher salinity, drip systems typically become blocked with sediment, and the groundwater in most villages in the Kashan plain is above the salinity threshold.

Another farmer in Ali-Abad village (2013) stated that the permitted volumes of water abstraction and the degree to which farmers' must reduce their abstraction hours mirrors the practice in other arid countries and had not been assessed in the context of local water needs and climate. Farmers also suggested that water rights should be gradually reduced by government, as a sudden reduction would present financial problems.

It is evident that although farmers have struggled to maintain traditional water right systems, which could potentially ensure equitable access to groundwater resources, government policies (such as encouraging drip irrigation systems) affect existing water rights in ways that are problematic for poorer farmers. Farmers explained how drip irrigation systems were inappropriate as given their scattered land holdings, they could not use drip systems to irrigate pistachio trees. Farmers also explained that sub-surface piped water irrigation schemes could be more acceptable as the water outlets matched the time slots of the traditional water allocation mechanism. As a result, the water was not exposed to sunlight, and evaporation losses were much less in comparison to open cement-lined canals.

4.1.2. Economic Advantages/Returns

Economic factors act as both incentives (in terms of providing higher agricultural revenue) and as barriers (financial constraints) for farmers' acceptance and adoption of a particular irrigation scheme or technology. Under conditions of water scarcity and reduced farm income [39], any new system that could improve agricultural production by increasing water availability for farmers is an advantage that would help farmers accept technologies such as sub-surface piped irrigation systems. On the other hand, financial constraints can affect a farmer's choice of a particular irrigation scheme, and farmers' financial status should be considered when introducing government schemes. The government is required to provide sufficient financial support or loans for farmers, based on an estimate of the project's cost. However, in Kashan, the area of cultivated land has fallen due to increasing water scarcity, and farming costs have increased (due to the use of electric pumps and the removal of subsidies). This has prevented Iranian farmers from investing in schemes to improve the irrigation infrastructure.

Empirical findings from the interviews indicated that farmers were fully aware of the importance of sustainable abstraction and irrigation practices. However, as a result of their small and fragmented land ownership, farmers lacked the financial resources to address the problem.

One local farmer in Ali-Abad village (2013) indicated:

> "Land integration has many advantages, when we did it the government supported us, my lands which is around 1 hectare were scattered in 5 to 7 different locations, but now they are all gathered into one piece".

Agricultural productivity varies according to the type of farmer (particularly their farm size), the farm location and access to water supplies. Wealthier farmers, who could afford to purchase more water rights, were usually interested in consolidating any fragmented land holdings and increasing the size of their farm to enable adoption of modernized irrigation facilities to increase crop yields and hence income. Poorer farmers, with small- or medium-scale land holdings, had more difficulty in accepting and adopting new irrigation technologies. Farmers were also confronted with economic uncertainty in agricultural production as the state provides insufficient support for capital investment and there is a significant barrier for farmers in adopting new technologies, which might present an economic risk.

The other economic factor influencing acceptance of a particular scheme is crop value, which is the most important element relating to agricultural revenues. As farmers became more educated and wealthier, they moved from low value crops to higher value crops and cash crops as water availability increased with expansion of private tube-wells. This has compelled farmers to use more efficient technologies and develop national or regional markets for their products [20]. However, this is only possible for wealthier farmers: medium or small-scale farmers in Iran have retained traditional crop cultivation practices. The Iranian government has sought to persuade poorer farmers to cultivate crops requiring less water to conserve groundwater resources (although this scheme also requires land integration by farmers).

In this regard, and in response to the water shortage and salinity problem, one farmer in Kashan (2013) added: "We decided to change our cultivation to crops that use less water for irrigation, we came to this conclusion that the best way to irrigate with saline and scarce water is to plant pistachio in Kashan. The quality of groundwater is not good for cultivation of crops such as cucumber, tomato, fruit trees or vegetables; they will not grow and under extreme salinity those dry out quickly".

Regarding the government plan to encourage a change in cultivation patterns to pistachio trees, one major consequence would be the change to local farmers whose livelihood depends on short-term crops such as wheat and barley. It was found that small-scale farmers adopt an economic strategy based on the profitability of individual cultivated crops. The strategic behaviour of one farmer in choosing an appropriate crop pattern, in Rijen village (2013), was described as:

> "Our strategy is to see which crop is profitable for us in a short term, if it has economic return then we plant it otherwise we don't follow governmental crop pattern. Because we have to pay for everything so it must be profitable for ourselves. Officials just express their request blindly but they do not stick to their promises to support us".

If they are to implement the new irrigation systems provided by the government, farmers (particularly the poorer farmers) would require more financial support. However, farmers stated that as their production does not contribute to the national economy, the government has reduced financial support and has instead allocated the budget to large-scale industrial cultivation.

Some farmers expressed concern that by increasingly removing agriculture from this area and introducing industrial activities, they would no longer have a viable livelihood, particularly as they had no experience of working in industry.

The government is also seeking to reduce water rights to one-third of their current volume. This presents significant economic disadvantages for farmers, which is the main barrier to accepting this policy. One farmer indicated that the government would not allow farmers any choice and the scheme would have significant economic disadvantages for small-scale farmers. Hence, small-scale farmers would not adopt this policy as their reduction in water rights would result in a significant reduction in farm income. This would leave them with no choice other than abandoning farming and migrating to a city in search for work. However, whilst protecting groundwater from over-exploitation, significant quantities of water will be used in proposed industrial developments, which overshadows the gain for farmers. In situations where there is insufficient control and regulations over aquifer water levels (in terms of issuing new abstraction permits and the monitoring of groundwater abstraction rates) and where some stakeholders have better access to groundwater resource, any re-allocation of water rights may aggravate competition over water and contribute to environmental degradation.

4.1.3. Social and Cultural Adaptability

The historic irrigation rules and social relations around the use of traditional irrigation systems particularly Qanats, in different parts of Iran have developed over centuries and rely upon scientific and participatory management [38]. Thus, these rules have been integrated in each community and cemented through social capital and adaptive rules, which are very difficult to change through government intervention. Individual technologies can greatly affect the social relations between farmers, which make them reluctant to accept technological innovations. The importance of social norms (in the context of this study, water rights and allocation rules) is a key factor in influencing communities in acceptance of new technology or practices [40]. In general, society is reluctant to change traditionally recognized relationships, and new technologies may bring different social interactions to a society [34]. After introducing tube-well technology, the availability of water that provides profit has become the main criteria when valuing water and land.

Introducing pump-well technology has presented many challenges and difficulties for farmers who have adopted it. However, this technology can increase their access to water resources, and farmers have transferred the traditional rules of Qanat to tube-wells. Ultimately, any new technology that does not bring value for farmers, or does not improve their access to water (improve efficiency), or conform with their water allocation mechanisms is unlikely to be adopted. Kamash [34] argues that technology is interconnected with the history, culture and norms in each society, and hence, it is essential to consider social and cultural factors when adopting new technologies. Issues such as water rights and allocation mechanisms, as well as cultivation patterns and landholding systems are historical circumstances that are rooted in farmers' local culture and social norms, which create a sense of social capital between them to cooperate and trust each other and local agencies. Changes in any of these conditions will have consequences.

However, some of the decisions may be made on the basis of misconceptions as suggested by a manager in the Agricultural Jihad Agency when indicating why farmers might reject the drip system:

> "In flood system (under Qanat), because farmers used to see 5–10 cm water on top of soil now they don't accept drip method, they say that tree fails. Maybe under drip system, long rooted trees fail (because of insufficient water seepage through soil layers), it is based on farmers' experience but scientifically drip system is approved, and for their new crops they can adopt this system".

The other crucial factor that has been identified in Kashan is that the technologies and policy schemes are not localised [17] and the rules are not adaptable to a particular climate or to the socio-economic conditions of the region. For example, when questioned about changing cultivation patterns, one farmer in Ali-Abad (2013) suggested that farmers would regret changing the cultivation pattern to mono-cropping within a few years. He stated that:

> "I am not against pistachio cultivation but I'm saying that farmers get regret in 10 years. Now farmers say that cultivating tree is easier but if all cultivate pistachio, farmers would not bother to plant any wheat. Because the main livelihood is bread, if we do not have anything we can just eat bread but pistachio is an economic crop and cannot feed us. Under that situation, government may import more wheat... I think government is trying to make this region economic or industrial . . . "

The change in cultivation patterns threatens traditional livelihoods, which emphasis the integration of agriculture and husbandry activities. Traditionally agriculture and animal husbandry are interlinked [38], creating a resilient system for vulnerable farmers, and any minor impacts have a significant impact on other activities. One farmer explained that: "for example, if someone wants to have husbandry he must have alfalfa or barley production beside it". Farmers indicated that husbandry has reduced because the price of fodder has increased (with an unstable market for barley) and it is not profitable to continue husbandry. The majority of farmers found it unprofitable if they do not keep farming and husbandry together, and it seems that husbandry can help traditional farmers pay their agricultural costs. Thus, retaining the existing diverse cultivation pattern is a more strategic and resilient option for farmers rather than changing to industrial crops such as pistachio. Crop selection by farmers is thus undertaken on the basis of water availability and climate, as well as cultural traditions and economic need [41]. Changing cultivation patterns are one strategy that farmers use when one crop has a low yield; however, the crop they choose must be adapted to saline water and periodic water deficit. Government attempts to introduce new crops into the region are based on observations of a few large-scale farmers' experience in pilot studies, but these crop patterns usually differ from the majority of farmers' circumstances, as they practice small-scale farming and have significant financial constraints.

5. Conclusions

A lack of critical evaluation of management projects and governance systems in Iran has resulted in ineffective management strategies and outcomes, which are designed in a top-down manner. There is also a lack of analytical studies on the role of local farmers' behaviour in managing land and groundwater resources in arid regions of the world, and also in the specific characteristics of groundwater as an invisible common-pool resource for irrigation purposes. Government projects seeking to address water scarcity problems have been mainly technologically-oriented and lack sensitivity to the local social, cultural and biophysical context. This paper has reviewed the main challenges in groundwater irrigation management and practices in Kashan, Iran, from the perspective of local farmers. The main challenges confronting farmers include: management transfer, inappropriate technological solutions and incompatibility of proposed irrigation schemes with the biophysical, social-cultural and economic status of the region.

In the local irrigation system in Kashan, the rules used for irrigation management follow traditional mechanisms of water allocation and distribution, in which equity and timing play a major role in water delivery and division. Farmers conform to these rules and water right allocation because of the legacy of the Qanat system, and they believe that their irrigation practices satisfy their own perceptions of efficiency. This empirical study shows that farmers in Kashan collectively cooperate in regulating water distribution, in which each farmer receives his exact water right on time and in an equitable manner. For example, historical water rights were adjusted to provide a fair distribution of water to small and scattered pieces of lands, which is now problematic given the difficulties of land consolidation to implement 'Tooba' drip irrigation scheme. From the simulation mapping exercise and deep interviews, farmers collectively indicated that the agreements for land integration for installation of the 'Tooba' scheme require extensive effort between farmers, which is not possible within a short period of time. Farmers also consider groundwater resources to be threatened, but believe that their historical water allocation and irrigation rules provided each farmer with their own water rights, thus sustaining farming practices. If government approaches do not consider these mechanisms and collective rules, they will not be successful in managing groundwater resources.

The results indicate the importance of articulating local knowledge and identifying local solutions when making management decisions. The common-agreed policies and management options should be developed under suitable institutional settings, which empower farmers by engaging them in management processes as key stakeholders. By considering local irrigation culture and social economic factors when shaping water management regimes and identifying obstacles inhibiting farmers' participation in government schemes, it is possible to develop a more integrated approach to irrigation management. This highlights the need for more local adaptive management solutions in response to reduced governmental financial support. These solutions will become more necessary given climatic uncertainties (e.g., increased temperature and reduced precipitation) and the socio-political situation.

Regarding improved irrigation efficiency, groundwater over-abstraction and the occurrence of externalities are the consequence of dysfunctional basin management, and farmers are not solely responsible. Farmers are the main stakeholders who suffer from the impacts of water scarcity and are the main group who will be affected by water reallocation. In this basin, better efficiency in irrigation systems does not necessarily benefit farmers significantly. Therefore, it is essential for further empirical studies to investigate the local impacts of micro-irrigation systems' application; otherwise, land expansion (following land consolidation), new crop patterns and re-allocation of water rights may adversely impact water use and consumption rate. Interviews with farmers suggested that the irrigation technologies used to improve efficiency need to match the definitions used by farmers. Farmers want to increase water accessibility to their lands and to maintain agricultural production, and this is particularly the case for small-scale farmers. There are also challenges related to governance issues, which have resulted in a reduction in water rights and also shifting agricultural costs and responsibilities onto farmers. This has only raised dissatisfaction and destroyed trust and cooperation with local governmental agencies. The lack of capacity building programmes to empower farmers' decision-making abilities has also increased the social and financial constraints facing poorer farmers. Transferring management to communities will only be successful with increased participation of resource users in management-decision processes, but it will be impossible unless the government establishes a new relationship with the owners of the resource. The policy of decentralization seems to shift agricultural costs onto poorer farmers, as well as the cost of installation and maintenance of the recently introduced technologies, without sufficient financial support, education and establishment of capacity building among local farmers through the local institutions.

From this argument, it can be concluded that the government is not the final arbitrator for groundwater irrigation management and sustainability. Managing groundwater resources requires collaborative efforts between local farmer communities and responsive governmental organizations. While the role of government's financial supports in developing countries is crucial, farmers also need to have their own self-regulated strategies and alternative management solutions. The role of government should be eliminated and most responsibilities be transferred to the local institutions and farmers' communities. However, it is essential that the government still provide financial support and infrastructures for poor systems, as well as trying to establish effective institutions to engage farmers in management decisions and empower local capacity to undertake parts of governmental roles and responsibilities

This paper has also reflected on some of the main factors influencing farmers' decisions to select a particular irrigation scheme or technology. These include the adaptability of the irrigation management solution to biophysical, economic, social and cultural conditions within a local irrigation context. To introduce new irrigation schemes or new technologies, the system needs to be tested and evaluated through pilot studies, and after confirming the project's suitability, it could be implemented at a wider scale. The adaptability of the irrigation technology to the arid climate, the traditional culture and social norms of water allocation mechanisms must be carefully considered, in order to work with, rather than against farmers' prevailing values and behaviours. An in-depth understanding of these challenges and the main factors in the implementation of a successful irrigation management scheme in the region can facilitate introducing a more suitable irrigation technology and improve collective agreement for the adoption of governmental policy schemes.

According to the study's findings, there is the potential for farmers at a local level in Kashan to participate in irrigation management projects. Farmers are willing to improve collective action, and they accept strategies that are adaptable to their irrigation practices to improve irrigation efficiency. The study suggests the crucial role of local informal meetings as the main support for integrating locals' knowledge and perspectives into management decisions, which promotes social capital and empowers farmers' decision-making abilities. Under an appropriate framework approved and encouraged by political will and communities' capabilities for engaging in management decisions, the implementation of wider participatory irrigation management would be possible. Given the high uncertainty and low controllability of groundwater as a 'common-pool resource', there is a strong argument for more community-based and adaptive approaches to water and irrigation management. It is necessary to articulate local knowledge and local solutions in management-decisions in the region: by considering the role of local irrigation culture and social economic factors in shaping current water management regimes.

Acknowledgments: Forough Jafary is grateful to her supervisors in the U.K. and her professor at Tehran University (Iran). Special thanks to all participant farmers in Kashan city who deeply assisted in conducting this study. We are also grateful for the helpful comments from the reviewers and the guest editor on this paper.

Author Contributions: The work is extracted from Forough Jafary's doctorate thesis; this includes empirical field works and literature review studies. Chris Bradley contributed significantly in editing and clarification of the context, as well as the transparency and relevance of the literature and the arguments.

Conflicts of Interest: The authors declare no conflict of interest.

Appendix A

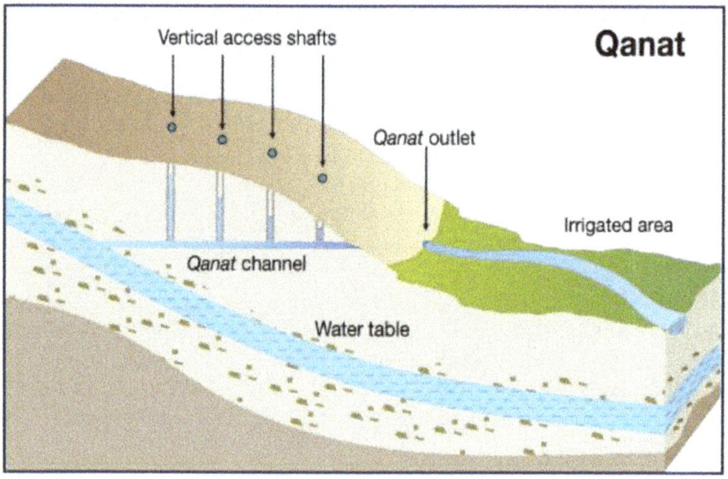

Figure A1. Source: The International Centre on Qanats (ICQHS), Yazd, Iran.

Appendix B

Interview Guide- Sample Questions: Local Farmers in Kashan and associated Villages
Personal Information:
Serial no:
Sector:
Rural:
Date of Interview:
Interviewee: Gender: Age: Mobile Number:
Name of Farm:
Location of Farm:
Farm establishment date:
General Questions on Occupations, Ownership Status, Cultivation Pattern, Irrigation Methods:
What are your responsibilities/roles in agricultural activities?
Are you the main land owner? How large is your farmland area?
If you are sharing your land how many farmers you are working with?

What is your cultivation pattern in different seasons?

What is your irrigation method?

Questions from local Farmers on Water Value and Traditional Irrigation Practices:

How modern allocation schemes affected farmers' traditional perspectives and behaviours towards agricultural water consumption?

How inappropriate governance in the past affected farmers attitudes towards new governmental water scheme projects?

What are farmer's perspectives towards reducing agricultural water consumption?

How different governance structures affect individuals' incentives and capabilities to cope with collective-action problems involved in system operation and maintenance?

Social and Cultural Questions: Kashan

What is your current agricultural water condition? Do you see any reasons to reduce/increase your water abstraction?

How modern system of water allocation has affected your agricultural activities and your income?

What is the main factor for rejecting new irrigation scheme?

Which factors would increase/affect your trust in agricultural water schemes?

Irrigation and agricultural Issues:

Would you apply 'Tooba scheme' for your farmland? Which factors affect your decision to accept or reject it?

Do you know farmers who have applied 'Tooba schemes' for their garden? How would they evaluate it?

Do you think of changing your cultivation pattern to halophyte crops? (Less water consumption plants?) What do you think about this method?

Questions relate to access to groundwater and social inequity

1. Are there any informal arrangements organised by small-scale farmers for accessing to groundwater? Why do they need this, and what is their approach?

2. Do you think access to groundwater is equal between small-scale and large-scale farmers? How land integration can improve farmers' access to water and improve irrigation efficiency?

3. How much land ownership status (small or large scale, assignees or tenant) has affected social inequity in terms of accessing groundwater and irrigation distribution systems?

4. How farmers have secured their access to water resources? Which obstacles exist against their water security?

Interview Guide: Governmental Organizations in Kashan

General Questions: Water and Climate Changes in the Region

Which organizations are involved in addressing key issues and problems related to impacts of Climate change/water resources problems?

What are the policy or strategy documents to guide their work?

What are the strengths and weaknesses of the institutions?

How people have access to information on current and future water problems?

What livelihood or economic sectors are the most vulnerable to climate change impacts? (i.e., to the water shortage problems).

References

1. Madani, K. Water management in Iran: What is causing the looming crisis? *J. Environ. Stud. Sci.* **2014**, *4*, 315–328. [CrossRef]
2. Panahi, D. Water crisis in Iran. *Entekhab (Morning Daily)* **2000**, *2*, 333–335.
3. Cheraghi, S.A.M. Institutional and scientific profiles of organizations working on saline agriculture in Iran. In Proceedings of the International Seminar on Prospects of Saline Agriculture in the GCC Countries, Dubai, UAE, 18–20 March 2001; pp. 399–412.

4. Zehtabian, G.R.; Amiraslani, F.; Khosravi, H. The reapplication of MEDALUS methodology in Kashan, desertified region, Iran. In Proceedings of the 14th International Soil Conservation Organization Conference, Water Management and Soil Conservation in Semi-arid Environments, Marrakech, Morocco, 14–19 May 2006.
5. Hojjati, M.H.; Boustani, F. *An Assessment of Groundwater Crisis in Iran, Case Study: Fars Province*; World Academy of Science, Engineering and Technology: Fars, Iran, 2010; Volume 70, pp. 476–480.
6. Izady, A.; Davary, K.; Alizadeh, A.; Ghahraman, B.; Sadeghi, M.; Moghaddamnia, A. Application of "Panel-data" modeling to predict groundwater levels in the Neishaboor Plain, Iran. *Hydrogeol. J.* **2012**, *20*, 435–447. [CrossRef]
7. Soltani, G.; Saboohi, M. Economic and social impacts of groundwater overdraft: The case of Iran. In Proceedings of the 15th Economic Research Forum (ERF) Annual Conference, Tehran, Iran, 21 March 2009.
8. Berbel, J.; Gutiérrez-Martín, C.; Rodríguez-Díaz, J.A.; Camacho, E.; Montesinos, P. Literature review on rebound effect of water saving measures and analysis of a Spanish case study. *Water Resour. Manag.* **2015**, *29*, 663–678. [CrossRef]
9. García-Mollá, M.; Sanchis-Ibor, C.; Ortega-Reig, M.V.; Avellá-Reus, L. Irrigation associations coping with drought: The case of four irrigation districts in eastern Spain. In *Drought in Arid and Semi-Arid Regions*; Springer: Dordrecht, The Netherlands, 2013.
10. Jackson, T.M.; Khan, S.; Hafeez, M. A comparative analysis of water application and energy consumption at the irrigated field level. *Agric. Water Manag.* **2010**, *97*, 1477–1485. [CrossRef]
11. Rodríguez Díaz, J.A.; Pérez Urrestarazu, L.; Camacho Poyato, E.; Montesinos, P. Modernizing water distribution networks—Lessons from the Bembézar MD irrigation district, Spain. *Outlook Agric.* **2012**, *41*, 229–236. [CrossRef]
12. Gómez, C.M.; Gutierrez, C. Enhancing irrigation efficiency but increasing water use: The Jevons' Paradox. In Proceedings of the 2011 International Congress on European Association of Agricultural Economists (EAAE 2011), Zurich, Switzerland, 30 August–2 September 2011.
13. Pannell, D.J.; Marshall, G.R.; Barr, N.; Curtis, A.; Vanclay, F.; Wilkinson, R. Understanding and promoting adoption of conservation practices by rural landholders. *Anim. Prod. Sci.* **2006**, *46*, 1407–1424. [CrossRef]
14. Madani, K. Iran's water crisis: Inducers, challenges and counter-measures. In Proceedings of the ERSA 45th Congress of the European Regional Science Association, Vrije University, Amsterdam, The Netherlands, 23–27 August 2005.
15. Foltz, R.C. Iran's water crisis: Cultural, political, and ethical dimensions. *J. Agric. Environ. Ethics* **2002**, *15*, 357–380. [CrossRef]
16. Lankford, B. *A New 'Commons'? The Rising Interest in Resources Efficiency*; GWF Discussion Paper 1405; Global Water Forum: Canberra, Australia, 2014.
17. Lankford, B. Localising irrigation efficiency. *Irrig. Drain.* **2006**, *55*, 345–362. [CrossRef]
18. Jaghdani, T.J.; Brümmer, B. Determinants of water purchases by pistachio producers in an informal groundwater market: A case study from Iran. *Water Policy* **2016**, *18*, 599–618. [CrossRef]
19. Bonine, M.E. Qanats, field systems, and morphology: Rectangularity on the Iranian Plateau. In *Qanat, Kariz and Khattara*; School of Oriental and African Studies: London, UK, 1989; pp. 35–57.
20. Rogers, P.; Silva, R.D.; Bhatia, R. Water is an economic good: How to use prices to promote equity, efficiency, and sustainability. *Water Policy* **2002**, *4*, 1–17. [CrossRef]
21. Von Korff, Y.; d'Aquino, P.; Daniell, K.A.; Bijlsma, R. Designing participation processes for water management and beyond. *Ecol. Soc.* **2010**, *15*, 1. [CrossRef]
22. Forootan, E.; Rietbroek, R.; Kusche, J.; Sharifi, M.A.; Awange, J.; Schmidt, M.; Omondi, P.; Famiglietti, J. Separation of large scale water storage patterns over Iran using GRACE, altimetry and hydrological data. *Remote Sens. Environ.* **2014**, *140*, 580–595.
23. Joodaki, G.; Wahr, J.; Swenson, S. Estimating the human contribution to groundwater depletion in the Middle East, from GRACE data, land surface models, and well observations. *Water Resour. Res.* **2014**, *50*, 2679–2692. [CrossRef]
24. Williams, B.K. Adaptive management of natural resources framework and issues. *J. Environ. Manag.* **2011**, *92*, 1346–1353. [CrossRef] [PubMed]
25. Dehghanisanij, H.; Akbari, M. Micro-irrigation in Iran—Current status and future needs. In Proceedings of the 8th International Micro Irrigation Congress, Tehran, Iran, 21 October 2011.

26. Lamaddalena, N.; Sagardoy, J.A. *Performance Analysis of on-Demand Pressurized Irrigation Systems*; Food & Agriculture Organization: Rome, Italy, 2000.
27. Benouniche, M.; Kuper, M.; Hammani, A.; Boesveld, H. Making the user visible: Analysing irrigation practices and farmers' logic to explain actual drip irrigation performance. *Irrig. Sci.* **2014**, *32*, 405–420. [CrossRef]
28. Albrecht, D.E. The adaptations of farmers in an era of declining groundwater supplies. *South. Rural Sociol.* **1990**, *7*, 46–62.
29. Berbel, J.; Mateos, L. Does investment in irrigation technology necessarily generate rebound effects? A simulation analysis based on an agro-economic model. *Agric. Syst.* **2014**, *128*, 25–34. [CrossRef]
30. Graveline, N.; Majone, B.; Van Duinen, R.; Ansink, E. Hydro-economic modeling of water scarcity under global change: An application to the Gállego river basin (Spain). *Reg. Environ. Chang.* **2013**, *14*, 119–132. [CrossRef]
31. Van der Kooij, S.; Zwarteveen, M.; Boesveld, H.; Kuper, M. The efficiency of drip irrigation unpacked. *Agric. Water Manag.* **2013**, *123*, 103–110. [CrossRef]
32. Kamash, Z. Irrigation technology, society and environment in the Roman Near East. *J. Arid Environ.* **2012**, *86*, 65–74. [CrossRef]
33. Roy, A.D.; Shah, T. Socio-ecology of groundwater irrigation in India. In *Intensive Use of Groundwater: Challenges and Opportunities*; Balkema Publishers: Lisse, The Netherlands, 2003.
34. Lambton, A.K.S. *The Persian Land Reform 1962–1966*; Clarendon Press: Oxford, UK, 1969.
35. Zand-razavi, S. The Factors Influencing People Cooperation in Sustaining Natural Renewable Resources. Ph.D. Thesis, Alameh Tabatabaee University, Tehran, Iran, 2004.
36. Balali, M.R. Towards Reflexive Land and Water Management in Iran Linking Technology, Governance and Culture. Ph.D. Thesis, Wageningen University, Wageningen, The Netherland, 2009.
37. Morris, C. Negotiating the boundary between state-led and farmer approaches to knowing nature: An analysis of UK agri-environment schemes. *Geoforum* **2006**, *37*, 113–127. [CrossRef]
38. Bijker, W.E.; Hughes, T.P.; Pinch, T. (Eds.) *The Social Construction of Technological Systems: New Directions in the Sociology and History of Technology*; MIT Press: Cambridge, UK, 1989.
39. Forouzani, M.; Karami, E. Agricultural water poverty index and sustainability. *Agron. Sustain. Dev.* **2010**, *13*, 415–432. [CrossRef]
40. Minato, W.; Curtis, A.; Allan, C. Social norms and natural resource management in a changing rural community. *J. Environ. Policy Plan.* **2010**, *12*, 381–403. [CrossRef]
41. Molle, F.; Mamanpoush, A.; Miranzadeh, M. *Robbing Yadullah's Water to Irrigate Said's Garden: Hydrology and Water Rights in A Village of Central Iran*; IWMI Research Report 80; International Water Management Institute: Colombo, Sri Lanka, 2004.

land

MDPI

Article

Vegetation in Drylands: Effects on Wind Flow and Aeolian Sediment Transport

Jerome R. Mayaud [1,*] and Nicholas P. Webb [2]

1 School of Geography and the Environment, Oxford University Centre for the Environment, South Parks Road, Oxford OX1 3QY, UK
2 USDA-ARS Jornada Experimental Range, Las Cruces, NM 88003, USA; nwebb@nmsu.edu
* Correspondence: jerome.mayaud@gmail.com

Received: 19 July 2017; Accepted: 14 September 2017; Published: 18 September 2017

Abstract: Drylands are characterised by patchy vegetation, erodible surfaces and erosive aeolian processes. Empirical and modelling studies have shown that vegetation elements provide drag on the overlying airflow, thus affecting wind velocity profiles and altering erosive dynamics on desert surfaces. However, these dynamics are significantly complicated by a variety of factors, including turbulence, and vegetation porosity and pliability effects. This has resulted in some uncertainty about the effect of vegetation on sediment transport in drylands. Here, we review recent progress in our understanding of the effects of dryland vegetation on wind flow and aeolian sediment transport processes. In particular, wind transport models have played a key role in simplifying aeolian processes in partly vegetated landscapes, but a number of key uncertainties and challenges remain. We identify potential future avenues for research that would help to elucidate the roles of vegetation distribution, geometry and scale in shaping the entrainment, transport and redistribution of wind-blown material at multiple scales. Gaps in our collective knowledge must be addressed through a combination of rigorous field, wind tunnel and modelling experiments.

Keywords: drylands; wind erosion modelling; drag partition; aerodynamic roughness; remote sensing; computational fluid dynamics; cellular automata

1. Introduction: The Drylands Context

Dryland environments, which constitute about 47% of the global land area [1], are extreme in their nature, typified by non-equilibrium conditions in climate, vegetation and geomorphology [2,3]. The strong interannual variability in precipitation characteristic of drylands [4,5] often results in ephemeral vegetative cover [6] or distinctive spatial patterning [7]. In turn, the limited nature of vegetation cover makes it an important control on the rate and extent to which geomorphological (and notably aeolian) processes operate.

Aeolian processes are increasingly viewed as major abiotic drivers that can have important consequences for landscape evolution [8,9], biogeochemical cycling [10,11], regional climate [12–14], land degradation [15–17] and public health [18,19]. Wind regime and sediment supply are known to largely control the formation of bare-sand dunes (e.g., [20–23]), but the complicating effects of vegetation, which form part of a wider, multi-scaled interplay of biophysical and anthropogenic factors, are less well understood [24–26].

Many of the potential effects of climate and land use change over the 21st century are likely to be mediated through aeolian activity in dryland landscapes [27,28]. Pressures on dryland resources are increasing as the population of the global arid zone, which is located mostly in the developing world, surpasses 2 billion people [29,30]. These pressures, combined with increasingly arid conditions owing to climatic change [13,31–33], are driving rapid, large-scale vegetation shifts that alter rates and patterns of soil erosion and modify ecosystem functioning and structure [6,34]. Land degradation, a form of regime shift whereby the environment changes potentially irreversibly to a new state [35], is likely to have a major impact on food security and environmental quality [6,36]. Changing vegetation patterns may lead to large-scale reactivation of dune fields and desert encroachment in many of the world's most vulnerable regions [8,9,33].

Given the key role that vegetation plays in modulating abiotic processes in drylands, this paper reviews recent studies to provide a holistic view of the effect of vegetation on wind flow and aeolian sediment transport. The paper examines: (i) the effect of vegetation on wind flow, both in terms of the wind velocity profile and on drag, drag partition and the distribution of surface shear stress; (ii) sediment transport dynamics on vegetated surfaces, collating information from wind tunnel, field, and modelling studies; and (iii) potential future avenues for research in these fields.

2. The Effect of Vegetation on Wind Flow

Semi-arid vegetation is often patchy and dynamic through time and space, primarily as a result of resource (water and nutrients) limitation [37–42]. In turn, the presence of patchy plant configurations significantly complicates wind flow near the land surface. The effect of non-erodible elements (e.g., vegetation) on wind flow can be examined both in terms of the wind velocity profile, which provides information at single points in space, and in terms of the partition of drag over the land surface at larger spatial scales.

2.1. *Effects on the Wind Velocity Profile*

At the scale of individual roughness elements, dryland vegetation modulates the erodibility of the surface and the erosivity of the wind through three primary mechanisms [43]. First, vegetation can directly shelter the soil from the wind by covering a fraction of the surface and initiating a wake region in its lee [44–46]; Second, vegetation directly affects wind velocity profiles by acting as a form of roughness that results in the growth of a boundary layer downwind (e.g., [47–52]); Finally, vegetation acts to trap windborne particles, thus reducing flux and providing loci for sediment deposition [27,53–55]. These mechanisms are explored in further detail below.

2.1.1. Plant Wake Dynamics

Single roughness elements such as plants shed turbulent eddies in a way that causes the flow in the wake to be separated from the surrounding flow [43]. For smooth, solid objects the interaction with the flow is relatively predictable [56], but characterising flow around live plants that are porous, pliable and of diverse geometry, is significantly more complex.

Using wind tunnel data, Judd et al. [44] built on the work of Wolfe and Nickling [43] to classify zones of flow around single porous roughness elements (see Figure 1). As wind approaches an obstacle (A), the air in the layer below the top of the element slows down and diverges upwind, with some air continuing through the porous obstacle to create a region of slower bleed flow in its immediate lee (C). As air is forced around the element and the flow lines are compressed, a region of accelerated wind develops above it and to its side (B). A low-velocity zone forms in the sheltered area of the obstacle (D), where wind velocity drastically decreases or even reverses in direction to form a recirculating eddy. This zone has variously been conceptualised as triangular in shape [57] or as a 'corridor' of asymptotic recovery [58]. Above and downwind of the low-velocity zone, a turbulent mixing zone grows downward from a thin layer at the top of the obstacle, eventually intersecting the ground surface (E). Ultimately the mixing zone merges into an equilibration zone (F) as the airflow recovers to equate to the upwind profile.

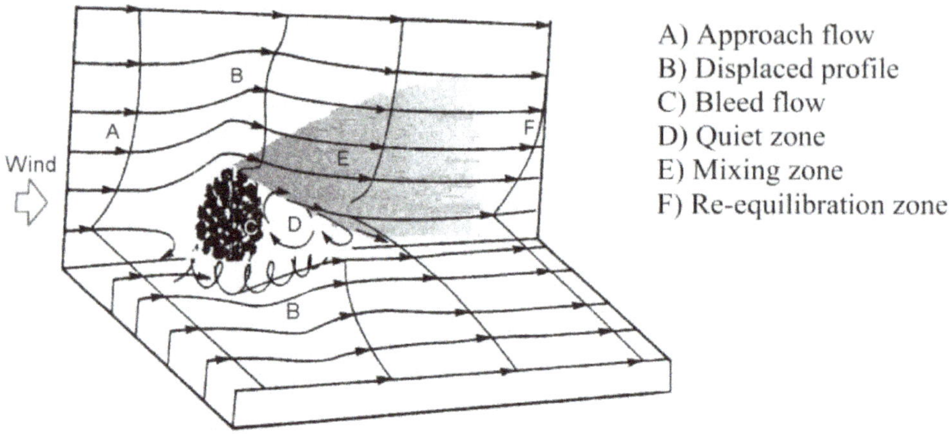

Figure 1. Flow visualisation around a single vegetation element, showing the development of six different flow regimes (from [46] (p. 1457), adapted from [43,44]).

The model of Judd et al. [44] for individual plants has subsequently been supported by wind tunnel data (e.g., [59–62]) and modelling experiments [63]. However, few studies have sought to examine the turbulent wake structures generated by dryland plants in the field. Leenders et al. [46], who observed distinct zones of flow acceleration and deceleration around single Sahelian shrubs, showed that turbulence intensity was higher in low-velocity zones than at other locations. They concluded that the net effect of a shrub is to reduce wind velocity and trap sand particles near the soil surface. Mayaud et al. [64] investigated differences in high-frequency wind flow around a single shrub and grass clump, and showed that the bluffer nature of the grass induced more intense wind slowdown in the lee than the shrub. This supports previous findings [65,66] that low-porosity elements induce turbulence in their lee, but allow flow to recover to upstream conditions sooner than more porous elements.

Trees affect wind flow differently to grasses and shrubs, owing to the presence of a trunk and canopy in the tree case [46,52,64,67]. Gross [67] and Dupont et al. [52] demonstrated using numerical models that wind flow accelerates over and around a single tree (Figure 2), and Leenders et al. [46] and Mayaud et al. [64] observed in the field that wind speed increases locally around the base of the trunk. A tree's elevated canopy produces a 'bottom gap' effect, whereby wind flowing near the top half of the tree moves over its crown [61] and streamline compression between the ground and underside of the crown results in wind speeding up [52,68].

Further downwind, the frontal impact of the tree crown produces a wake region similar to that of a shrub, although the greater size of a tree means that it reduces wind speed more efficiently than a shrub [46]. The length and strength of the reverse flow in the wake of a tree appears to increase with wind speed, and the location of the minimum wind speed to shift to higher levels as trunk height increases [67]. Tree leaves may also be important for altering local airflow patterns [69], by inducing updrafts and downdrafts in the upstream region that cause turbulent, upwelling recirculation zones to form [61].

Whilst single plants are present in drylands, vegetation often grows in clumps and patches that can have varying effects on wind flow. The interactions of individual plant wakes with their neighbours can be broadly described by three different flow regimes ([43] Figure 3). As roughness element density increases, the flow regime changes from one where the elements act individually on the flow (isolated roughness flow) to a regime where the airflow skims across the top of the elements (skimming flow). In dryland environments, vegetation is often spaced far enough apart that the wake shed by an individual plant can fully develop or is only slightly obscured by a downwind plant (wake-interference flow). Lee and Soliman [70] classified these regimes by a ratio of inter-row spacing length to element height. However, this relationship often breaks down in dryland systems because natural vegetative surface roughness changes are often irregular, thus affecting wind momentum extraction and surface sheltering in differing ways [71].

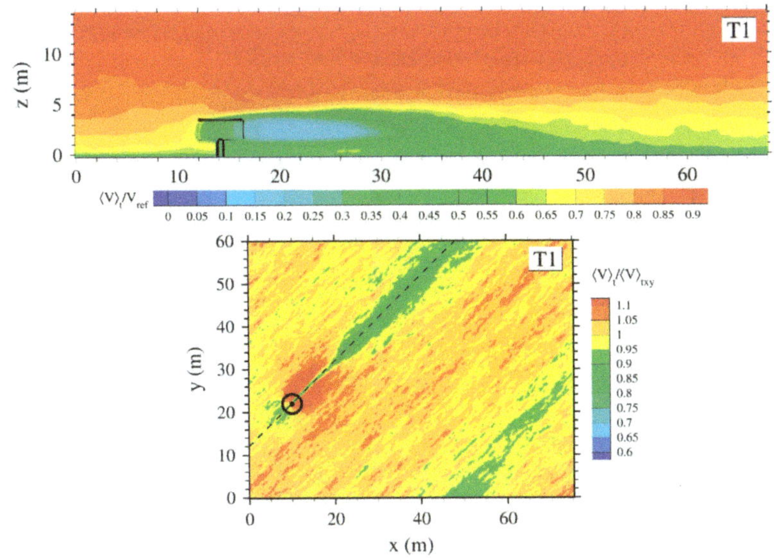

Figure 2. Flow visualisation downwind of a single tree, shown as: (top panel) the normalised time-averaged wind velocity (V_t/V_{ref}) simulated by Dupont et al.'s [52] model, shown in vertical cross-section. Solid black lines delimit the tree; (bottom panel) the normalised time-averaged wind velocity (V_t/V_{txy}) simulated by the model at 0.1 m height, shown in horizontal cross-section. The solid black circle represent the tree crowns (adapted from [52] (p. 176)).

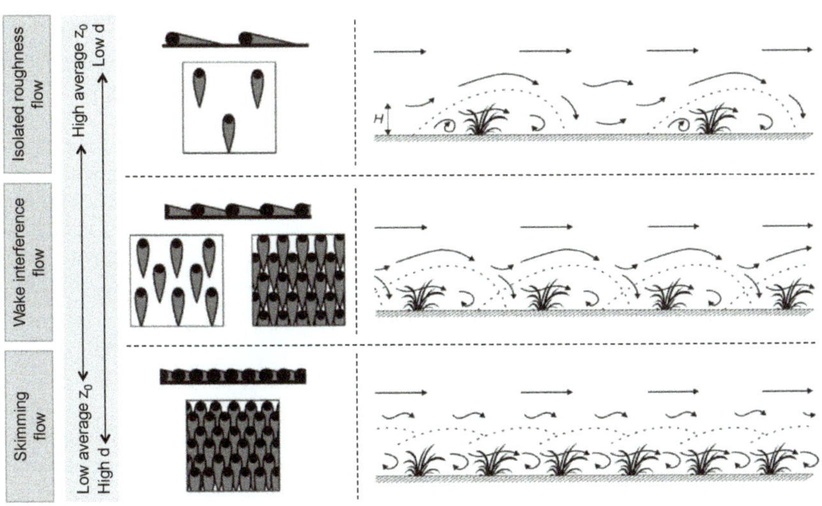

Figure 3. Flow regimes and associated theoretical wake development, shown in schematic plan and side view. Shaded areas are wake regions. The effect of different flow regimes on average z_0 (aerodynamic roughness) and d (displacement height) per plant unit is shown (adapted from [43] (p. 57), and [60] (p. 66)).

2.1.2. Vegetation as a Form of Roughness

Vegetation can directly affect wind velocity profiles on desert surfaces by acting as a form of roughness, extracting momentum from near-surface wind, which results in the growth of a boundary layer downwind [47,48,72]. Boundary layer growth differs depending on the type of flow regime (Figure 3). When isolated roughness elements populate the surface (<16% cover [43]), each plant sheds turbulent eddies by diverting wind flow around and above each plant (see Section 2.1.1). This increases drag, thus raising shear stress and the aerodynamic roughness (z_0), and potentially enhancing erosion locally [73]. An arch vortex with a reverse surface flow direction can develop directly downwind of individual roughness elements, upwind of a flow stagnation region where the outer flow reattaches to the ground [59,74].

In wake-interference flow (ca. 16–40% cover [43]), the increased drag and shear stress resulting from the presence of multiple elements may only be partly absorbed by the plants themselves. This results in stress being transferred to the inter-canopy surface, and potentially greater sediment transport [52,75]. In the case of skimming flow (>~40% cover [43]), the increased drag from the vegetation acts to displace z_0 upwards (establishing a zero-plane displacement height, d), which simultaneously extracts momentum from the surface wind and increases wind shear stress above the canopy (e.g., [47–52,76,77]). The absorption of additional stresses by the plants therefore decreases the erosion potential at the surface, despite the additional above-canopy shear stress induced by the increased surface roughness. That is, while increased roughness density may result in greater total shear stress, momentum partitioning between the plant canopy and the soil surface results in smaller surface shear stresses for entraining soil particles.

Skimming flow experiments have been mainly conducted in wind tunnels (e.g., [60,78,79]) and a few field and modelling studies of forest edges (e.g., [80–83]) and backward-facing steps (e.g., [84,85]). However, these configurations often have more extensive low-velocity zones and delayed reattachment points compared with vegetated cases due to their almost parallel-to-wall streamlines. More field research and empirical observations are needed to better understand skimming flow dynamics over full-scale vegetation patches in drylands.

2.1.3. Trapping of Windborne Sediment

Finally, vegetation acts to trap windborne particles, thus reducing the total horizontal and vertical sediment fluxes and providing loci for sediment deposition [27,86]. The prevalence of this effect can differ depending on vegetation height [55]: in the case of tall vegetation (e.g., trees and shrubs), saltation transport takes place primarily below canopy level in the open space between individual plants. In contrast, for short vegetation (e.g., grasses), sand transport takes place primarily within and above the plant canopy, so that grasses may trap aeolian sediment more readily than shrubs. Trees affect sediment trapping differently to grasses and shrubs, owing to their trunk and elevated crown [46,52,61,64,67].

The deposition of aeolian sediment in the presence of vegetation occurs via three principal processes: gravitational settling, inertial impaction and Brownian diffusion [87]. The interception of saltating particles and suspended sediment by vegetation may be considered a part of the impaction process, which in addition to vegetation height is influenced by vegetation porosity and the spatial distribution of roughness over the land surface [88]. Zhang and Shao [89] formulated a parameterisation of dry deposition for rough land surfaces that resolves the aerodynamic resistance of the surface, gravitational resistance, and surface collection resistance as a function of the component depositional processes. However, considerable uncertainty remains in how to effectively measure and model sediment deposition processes for areas with heterogeneous vegetation. New approaches to resolve the partition of wind momentum fluxes over heterogeneous land surfaces (see Section 3.2.1) may provide opportunities to reduce the uncertainty in sediment deposition schemes.

2.1.4. Effects of Plant Porosity, Pliability and Configuration

Theoretical calculations [57,58] and experimental measurements [46,54,62,63,79,90] suggest that protective wakes downwind of individual vegetation elements extend to approximately 7–10 h (where h is the height of the element). However, this can vary significantly depending on plant porosity, pliability and configuration.

Much of the research on the effects of plant porosity and pliability on wind flow has been conducted on sand fences and windbreaks, and findings from these are of some relevance to plants [46,91]. Taylor [92] found that drag coefficients decreased exponentially as the porosity of a two-dimensional barrier increased, but Grant and Nickling [91] demonstrated that a peak in drag coefficient occurred at an intermediate optical porosity (~20%) for three-dimensional trees. It has been shown that elements with a porosity less than 20% act to enhance wind speed recovery in their lee [68,93–95], so elements with intermediate porosity strike a compromise between shear stress reduction and the downwind distance for which it is effective [96,97]. However, even porosity values of 50% can cause wake effects to extend to at least 50 h downwind of a windbreak [98]. The sediment trapping mechanism of plants (see Section 2.1.3) also depends on plant porosity [53,54,96], with wind tunnel (e.g., [97]) and field (e.g., [91]) experiments showing that the trapping effect of vegetation is maximised at intermediate values of porosity (20–40%). Using wire mesh structures, Gillies et al. [86,99] showed that porous elements have a greater potential for modulating sediment flux than solid elements in large spatial arrays.

The pliability of plants affects their aerodynamic behaviour. Some pliable plant stems alter their form to become more streamlined in higher winds, thus extracting momentum less effectively as wind speed increases [53]. This results in a decreased sheltering effect at increasing wind speeds, due to a narrowing of the lee-side wake and suppression of horseshoe vortices [74,78]. Therefore, the rigid, non-porous roughness elements commonly employed in many wind tunnel studies to simulate vegetation arrays may provide inadequate approximations to live plants [74]. Indeed, Gillies et al. [54] showed that flow field responses between solid bluff body forms and elements of the same form and size that are covered with a porous outer layer surrounding a solid inner core are not equivalent. Moreover, scaling issues linked to boundary layer depths mean that full-scale elements are rarely used in wind tunnel experiments. This can limit the applicability of findings from wind tunnels. A few studies (e.g., [74,78,100]) have recently used live canopies in wind tunnels, but live plants are far more complicated to control in terms of appearance and behaviour than artificial imitations [78].

The configuration, or geometric arrangement, of windbreaks has also been shown to affect their aerodynamic behaviour. Using a numerical model, Liu et al. [72] showed that alternate and regular arrays of tree windbreaks provided the most protection, whereas a patchy arrangement, where trees were clumped together in space, resulted in the highest erosive forces. Liu et al. [72] hypothesised that this resulted from reduced diameter influences, whereby vegetation clumping decreases the total wake spread, thus reducing the protective effects of wake interference and allowing wind to flow through unimpeded. Wind tunnel studies [62,95] suggest that multiple-row barriers reduce wind velocities more efficiently than single-row barriers, at least in the immediate lee. This effect is likely due to the additional turbulence generated by a sequence of fences compared to an isolated barrier [101]. Wu et al. [62] also showed that using multiple vegetation species can provide optimal sheltering effects. There is little research on the impact of vegetation patch configuration at the meso scale, with the exception of some scaled-down wind tunnel experiments [60,78,79] and modelling studies of forest edges [81–83]. Wind tunnel and modelling experiments have also been used to understand flow behaviour around backward-facing steps (e.g., [84,85]), although these configurations often have more extensive low-velocity zones and delayed reattachment points compared to vegetated cases, due to their almost parallel-to-wall streamlines.

Collating information from windbreak and vegetation studies allows the main zones of flow change to be identified and partially quantified (see Figure 4). However, whilst windbreak studies can be useful proxies for vegetation elements, the three-dimensional nature of live plants (as opposed to the two-dimensional problem represented by a fence) means that findings may not be completely transferable [46,54,64,77]. The recovery length downwind of a vegetation patch or fence has been shown to be significantly longer than in the case of single vegetation elements [46,54,62,64,77,81,83,91,98,102]. The difference in velocity recovery length has been attributed to a lack of flow moving laterally around a patch [77] or a fence where the obstacle width is assumed to be infinite [63,86,99,103], in contrast to isolated vegetation elements where faster-moving airflow mixes with slower-moving airflow in the wake via counter-rotating vortices [59].

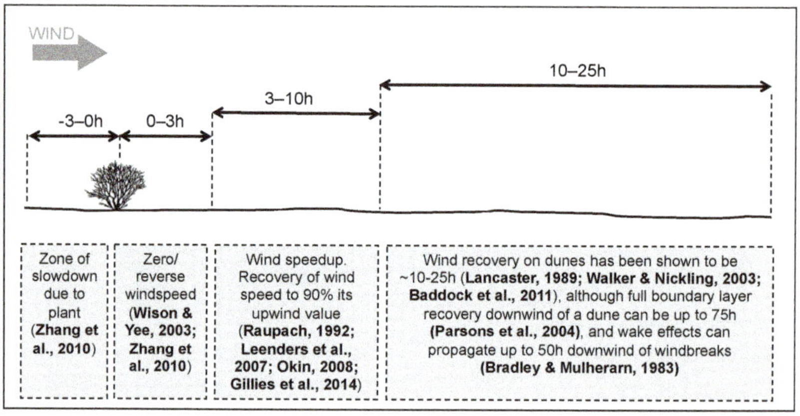

Figure 4. Main zones of possible speedup/slowdown around a roughness element, as collated from various vegetation and windbreak studies.

Furthermore, it is unclear whether windbreaks affect flow turbulence in the same way as live vegetation. This is particularly problematic in the field, where the dynamic complexity of the flow structures cannot be controlled in the same way as in modelled or wind tunnel environments [104]. This issue is pertinent given the evidence that turbulence is an important driving force behind sediment entrainment and transport in aeolian environments [105–114].

2.2. Effects on Drag

Alongside the effects on velocity profiles, it is useful to consider more complex scenarios where multiple vegetation elements affect drag across a surface. Shear stress (or drag) partitioning is traditionally used to quantify the proportion of the wind shear stress or shear velocity (u_*) that is acting on a soil surface, compared to the proportion acting on vegetation elements (e.g., [48,96,115]). The theoretical basis for partitioning comes from the work of Schlichting [116], who proposed that the total drag on a roughened surface (τ) can be expressed as:

$$\tau = \tau_S + \tau_R \quad (1)$$

where τ_S is the sum of the drag on the intervening surface and τ_R is the drag on all the individual roughness elements.

Marshall [117] proposed that partition should be mostly dependent on the dimensionless roughness density (λ), which characterizes the lateral roughness of the surface:

$$\lambda = \frac{nbh}{S} \quad (2)$$

where n is the number of roughness elements of width b and height h per unit surface area S (Figure 5). This definition implies that the absorption of momentum by elements is controlled to a large degree by the total frontal area of the elements [118,119]. Although the relationship between partition and λ was not fully supported by later field observations (e.g., [73]) and wind-tunnel experiments [48,115,120], the parameter was widely employed, particularly in the context of two principal partitioning models by Marticorena and Bergametti [118] and Raupach [57,121]. Both models have distinct approaches but are based on the 'shear stress ratio' (SSR) of shear stress required for erosion on a bare surface to the total shear stress threshold of the surface including roughness, relying on λ as a fundamental descriptor of surface roughness. King et al. [71] provide a comprehensive analysis of the Raupach [121] and Marticorena and Bergametti [118] drag partition schemes through a comparison with field and wind tunnel measurements. We detail benefits and limitations of the approaches in Section 3.2.

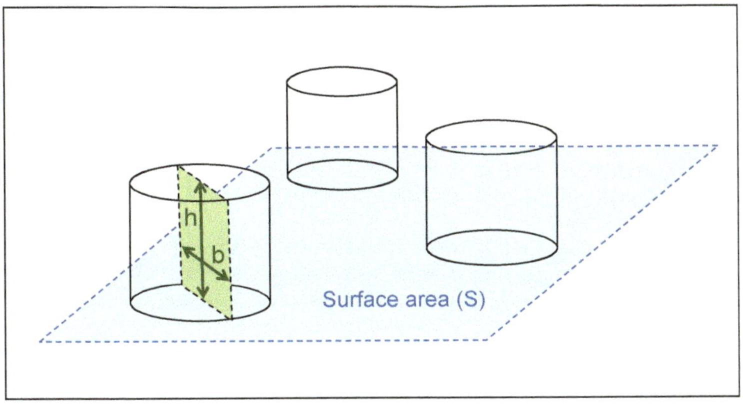

Figure 5. Illustration of the parameters used to calculate roughness density (λ) in roughness studies: h is element height, b is element base/width, n is the number of elements, and S is the surface area over which the calculation is made.

Shao and Yang [122,123] demonstrated that, in addition to τ_S and τ_R, it is sometimes useful to consider 'skin drag' (τ_C, i.e., the drag due to momentum transfer to roughness element surfaces). For ground surfaces with sufficiently large λ, Equation (1) becomes increasingly inadequate, so Shao and Yang [123] included τ_C in a modified version of Raupach's [57] drag partitioning model. This allowed Shao and Yang [123] to mathematically explain how increasingly closed canopies lead to a decrease in z_0 values, as is commonly observed in cases of skimming flow (see Section 2.1.2).

3. Sediment Transport on Vegetated Surfaces

The relationship between wind stress (erosivity) and the susceptibility of a surface to erosion (erodibility) broadly determines the potential for sediment transport in deserts [124]. Most of the evidence for the interactions between dryland vegetation and wind erosion comes from wind tunnel, windbreak and field experiments. However, an increasing number of modelling studies are exploring these relationships further, and introducing important predictive capabilities.

3.1. Evidence from Wind Tunnel, Windbreak and Field Studies

Wind tunnel studies have proved useful for studying the relationship between vegetation and wind-blown sediment transport under controlled conditions. Experiments have been conducted using dead vegetation [125], live vegetation [60,78], or artificial vegetation [79,95,126,127]. Suter-Burri et al. [60] used coloured sand to visualize spatial patterns of sediment redistribution within grass tussocks, allowing them to relate different canopy densities to their theoretical flow regimes (Figure 6). Wedge-shaped wake deposits developed fully in the low-density case (representing isolated roughness flow), overlapped with adjacent downstream tussocks in the medium-density case (wake-interference flow), and deposited grains were evenly distributed around the tussocks in the high-density case (skimming flow). Suter-Burri et al. [60] observed that the fraction of the surface that was exposed to erosion was substantially smaller (44–78%) than the area not covered by vegetation, owing to the turbulent wake structures around the tussocks.

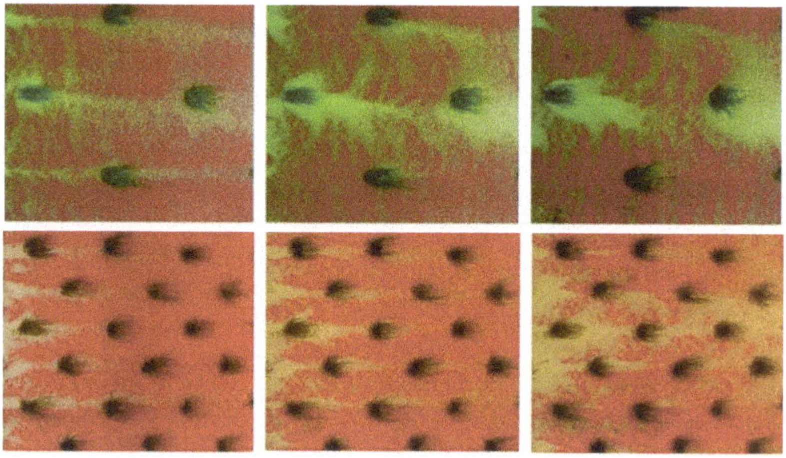

Figure 6. Image sequences of sediment deposition patterns in Suter-Burri et al.'s [60] wind tunnel experiments, on surfaces populated by low-density (**top row**) and medium-density (**bottom row**) canopies. White sand was blowing in from upstream, with wind direction from left to right (adapted from [60] (p. 69)).

Wind tunnel studies have also shown that vegetation can act to increase erosion in some circumstances. The experiments of Burri et al. [78] demonstrated that whilst sediment and dust flux was significantly reduced on high canopy-density surfaces compared to an unplanted configuration, low-density canopies (~3% cover) led to elevated shear stress on the bed resulting from flow acceleration, and thus higher sediment and dust fluxes. This region of elevated surface shear stress in the wake could result from counter-rotating vortices created by the roughness element [59], or from oscillating movements of grass blades on the sand surface [78]. In a 'regional' scale study (i.e., between neighbouring land units), Youssef et al. [79] used artificial 1:50 sized shrubs, laid out in different configurations in a wind tunnel, to measure the spatial distribution of sediment height after wind events. They found that shrubs bunched into patches resulted in far greater sediment fluxes than in the case of regular grids of single shrubs, likely because of a significant increase in turbulence between the patches. However, the length scales of saltation in wind tunnels has been shown to differ from those of natural saltation [128–130], so findings from wind tunnel studies should be treated with some caution.

Windbreak experiments (e.g., [87,88,131]) also provide some insights about the impact of vegetation on sediment transport. When oncoming particle-laden airflow approaches a windbreak, particles are filtered from the flow passing through the windbreak and deposited on the element, whereas in the air flowing over the obstacle, particle concentrations remain similar to upwind values [88]. Raupach and Lu [87] showed that wind speed reduction in the lee of a windbreak leads to lower particle concentrations, and that with increasing downwind distance, particles from the air flowing over the windbreak are progressively mixed downwards into the sheltered region. Using a high-speed camera system, Zhang et al. [131] observed that sand transport rates were reduced by 37% directly in front of a fence due to momentum loss of particles colliding with the fence surface, and by 80% up to the downstream location of 3 h. Similarly, the kinetic energy of saltating particles was reduced by more than 60% in the windward region, and up to 90% at 3 h.

In the field, decoupling the influences of plants on sediment transport from those induced by variations in the wind or surface conditions is very difficult at the plant or patch scale [74,79]. Several field studies have nonetheless attempted to demonstrate a gradient of surface activity in response to varying vegetation cover. There is evidence that vegetation cover of 12–15% provides a threshold for sand transport [55,132–134], but the use of percentage cover is limiting because it provides no information about the spatially heterogeneous impact of vegetation on erosion thresholds. It also does not account for the fact that erosion can still occur at relatively high (up to 45%) vegetation cover if the wind velocity is high enough or the vegetation height is small [47,73]. A more nuanced consideration of the impacts of vegetation on the balance between erosivity and erodibility factors is therefore needed.

3.2. *Modelling Sediment Transport on Vegetated Surfaces*

Wind erosion models form a key part of our understanding of sediment transport dynamics on partly vegetated surfaces, and are crucial for assessing the potential vulnerability of dryland regions to soil degradation [11,27,33,135]. The mobility potential of vegetated dunes over long timescales has been successfully modelled using wind-based indices that combine precipitation, temperature and wind data [8,9]. Whilst this approach is useful for simulating large dunefield dynamics, it does not resolve sediment transport dynamics at the same scale as plant-flow interactions. A variety of finer-scale models are examined here, from earlier theoretical models based on shear stress partitioning theory [116,118,121], to more recent approaches emphasising vegetation gaps [28,58] and numerical models simulating patch-scale sediment movement [52,136].

3.2.1. Drag Partition Schemes

Early methods for modelling the impact of roughness elements on sediment mass flux focused on the establishment of empirical relations between fractional ground cover and sediment transport rates [47,137–139]. During the 1990s, more attention was given to the development of drag partitioning techniques that could establish the wind momentum flux at the soil surface in the presence of roughness elements. The schemes of Raupach [57,121] and Marticorena and Bergametti [118] were widely adopted on the basis of their good agreement with wind tunnel experiments using solid objects on non-complex surfaces (e.g., [51,71,140–143]). However, the drag partition schemes have also been found to be limited in application in several respects.

By relying on the roughness density (λ) and aerodynamic roughness (z_0) parameters, both the Raupach [57,121] and Marticorena and Bergametti [118] schemes suffer from the inherent requirement for a given surface to be treated homogenously. The schemes seek to establish the mean (or maximum) shear stress ratio in the absence of information about the heterogeneity of roughness on the land surface [87]. The spatial distribution of vegetation has been shown to affect horizontal sediment flux, even at low roughness densities [28,51,119,144–146]. The roughness distribution effects on aeolian sediment transport are also nuanced, in that the sediment mass flux may not always be significantly affected as long as vegetation is not overly clumped together [147]. The so-called 'Telephone Pole Problem' (Figure 7), where two roughness scenarios have the same roughness density but different spatial distributions, illustrates the limitations of using average vegetation density information to describe transport dynamics [58].

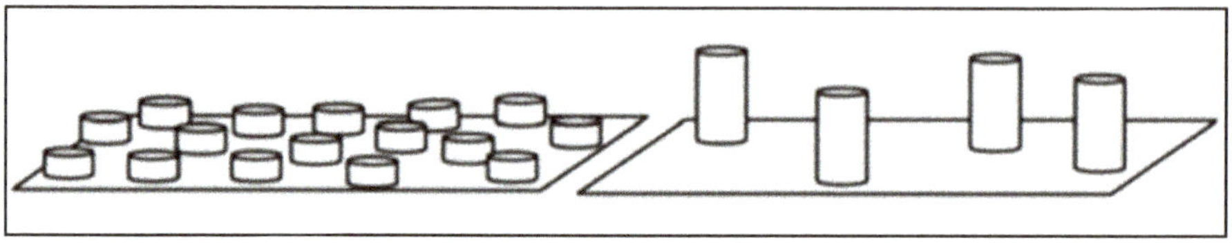

Figure 7. The Telephone Pole Problem. Both of these surfaces have the same roughness density, but on the left roughness density is partitioned into sixteen elements, whereas on the right, it is partitioned into four objects four times the height. Under the shear stress partitioning model of Raupach et al. [121], both surfaces experience the same horizontal flux, although intuitively this is not an accurate representation of physical reality (from [58] (p. 3)).

Other disadvantages of the λ parameter include scaling issues, whereby tall (>0.10 m) elements appear to influence sand transport in a way that cannot be accounted for based solely on knowledge of roughness density [141,142]. This deficiency is minimised in large homogeneous areas of randomly or regularly spaced vegetation of the same height, but becomes problematic in natural environments displaying structural anisotropy [27,58]. The λ parameter has also been shown to inadequately describe three-dimensional objects, due an intrinsic bias in using the frontal silhouette area [90,96]. Chappell and Webb [148] provided an in-depth analysis of the limitations of the drag partition scheme with respect to Raupach's [57] assumptions about how the sheltering afforded by roughness elements can be approximated by the λ parameter (denoted L_c in their text). Critically, interactions between sheltered areas are not captured by the drag partition scheme, but have demonstrably significant effects on sediment transport [148]. These interactions are also not captured by estimating the drag partition from approximations of z_0 [118].

Both λ and z_0 are difficult parameters to measure reliably in the field, especially at large scales and/or in the presence of large roughness elements (e.g., shrubs and trees). Rapid and repeatable methods have not been established to measure λ in the field, while the spatial variability and height of roughness elements often confounds measurements of z_0, which in response can have large spatial (including directional) and temporal variability [149,150].

Raupach and Lu [87] identified representation of the sub-grid scale heterogeneity in momentum transfer within the atmospheric boundary layer as a major challenge for reducing uncertainty in wind erosion and dust emission models. That challenge remains today, and despite considerable research into the parameterisation of drag partition schemes and their modification to account for the non-uniformity in surface shear stress in the presence of vegetation [52,74]. Brown et al. [119] demonstrated that roughness configuration may have a small effect on the average drag partition when basic descriptive statistics (e.g., mean, standard deviation) are used to represent variability in surface shear stress measurements, but parameters of the Raupach et al. [121] drag partition scheme are sensitive to roughness configuration. Webb et al. [146] used Brown et al.'s [119] data to further examine the effects of roughness configuration and the performance of the Raupach et al. [121] and Marticorena and Bergametti [118] drag partition schemes for estimating sediment mass flux. Both drag partition schemes were found to have difficulty in accurately, and realistically, representing sediment mass flux responses to changing free stream velocity (U_f), λ and z_0. This is due to uncertainties and potential inconsistencies in the parameter values and a lack of sensitivity of the schemes to roughness configuration when applied with invariant parameter values [146]. Drag partition schemes that explicitly represent the effects of roughness configuration on the distribution of surface shear stress are needed to improve the accuracy of wind erosion models [151].

3.2.2. Okin (2008) Model

An alternative wind erosion model presented by Okin [58] recognises the inherent irregularity of vegetation patterning in drylands, and emphasises the controlling influence that the spatial configuration of vegetation has on the distribution of surface shear stress. By moving away from homogenous mean values to describe the roughness density and drag partition, Okin's [58] approach resolves the heterogeneous wind shear velocity between roughness elements and over the soil surface. This notion is supported by field observations that flux is not ubiquitous across an entire landscape during a transport event [51]. Okin's [58] method implies that surface roughness moderates the wind erosivity, rather than the threshold shear velocity for soil entrainment.

The Okin model assumes that plants are porous objects [49,90], meaning that reverse flow in the lee of plants can be negligible and the surface shear velocities in the wake of plants can be greater than zero. Asymptotic recovery of shear stress has been observed downwind of porous fences (e.g., [98]) and nebkha vegetation [54,63], so the model is formulated probabilistically to simulate the distribution of surface shear stress as a function of the distance to the nearest upwind plant. The aerodynamic roughness length (z_0) is used to represent the soil roughness and establish u^* between roughness elements.

Okin's [58] model shows good agreement with aeolian sand flux data at the plant/patch scale (e.g., [28,152]). It simulates transport around solid objects well when the model-calculated shear stress ratio is compared with the experimental data of King et al. [71], which is noteworthy considering that the model omits return flow. A notable advantage of Okin's [58] scheme is its scale explicitness, which allows individual unvegetated gaps as well as entire landscapes to be simulated. The model can also be used to evaluate directional effects of wind flow on sediment transport for a given area.

Since the Okin model uses relatively simple spatial inputs, as opposed to the structural parameters of roughness elements needed for traditional shear stress partitioning models, data can be collected easily in the field using standardized measurement methods (e.g., [135,153]), and at the landscape scale using high resolution (<5 m) remote sensing techniques (e.g., [152,154,155]). As remote sensing approaches to estimating the precise spatial configuration of surface roughness are yet to be established at moderate spatial resolutions (e.g., 30–500 m), alternative data and models are needed to represent the heterogeneous effects of vegetation on aeolian sediment transport at regional to global scales.

3.2.3. Remote Sensing Approaches

Remote sensing data have been used to represent the spatial and temporal variability in vegetation within drag partition schemes, increasing wind erosion and dust model sensitivities to land cover dynamics (e.g., [156,157]). Normalised Difference Vegetation Index (NDVI) and Leaf Area Index (LAI) data are most commonly used to estimate the vegetation cover fraction and λ as input to drag partition schemes, and to estimate z_0 for the land surface [150]. However, while these approaches appear to improve modelled temporal responses of sediment transport, the data also introduce uncertainties that may compound those associated with selected drag partition parameter values [158]. As neither NDVI nor LAI adequately represent the effects of vegetation phenological changes on vegetation structure, which is central to the partitioning of wind momentum over the land surface [150], alternative approaches have been sought to obtain estimates of land surface aerodynamic properties at broad spatial scales.

Some attempts have been made to derive areal estimates of z_0 for desert surfaces, which can be implemented across several scales. Greeley et al. [159,160] established relations between z_0 obtained from wind velocity profile data and radar backscatter cross sections. Marticorena et al. [161] evaluated the application of a bidirectional reflectance model to estimate roughness properties, while Marticorena et al. [162] found a relation between radar backscatter coefficients in the C band and z_0 that suggested potential for remote sensing the roughness length at broad scales.

However, obtaining areal estimates of z_0 is not sufficient on its own to determine the drag partition and wind momentum transfer to exposed soil surfaces in vegetated landscapes. To address this need, Chappell and Webb [148] developed an approach to estimating the drag partition and surface shear stress for heterogeneous surfaces from shadow, which can be obtained globally at moderate spatial resolutions (e.g., 500 m) from albedo data. The approach built on Marshall's [117] wind tunnel study that underpinned Raupach's [57] drag partition theory, and simulations exploring the relation between surface illumination, shadow and z_0 conducted by Chappell and Heritage [163] and Chappell et al. [164]. Figure 8 shows an example of the relations established by Chappell and Webb [148] between normalised albedo (ω_{ns}) and scaled wind shear stress (u^*/U_f). Albedo (shadow) measurements scale linearly, can be collected in the field (e.g., using a net radiometer) or with airborne and satellite sensors, and are sensitive to vegetation spatial distribution. Therefore, the approach has potential to resolve the outstanding issue of representing sub-grid scale variability in the drag partition for aeolian sediment transport and land surface models [87].

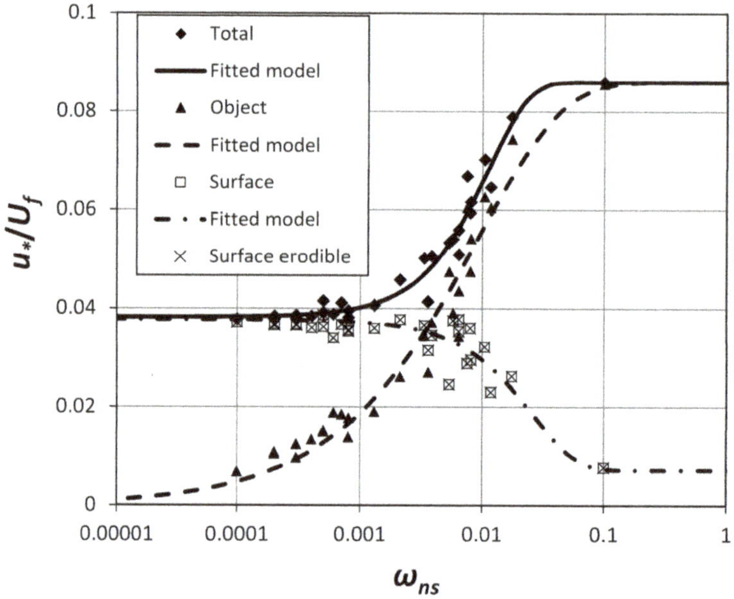

Figure 8. Chappell and Webb's [148] relation between normalised albedo (ω_{ns}, i.e., proportion of shadow) and the wind shear stress scaled by the velocity at a free stream height (u^*/U_f). Chappell and Webb [148] illuminated each of Marshall's [117] surfaces across a range of zenith angles to approximate the direct beam directional hemispherical albedo (or black sky albedo ω; viewed at nadir), then normalised ω by the reflectance of the surface, illuminated and viewed at nadir (ω_0), to remove any spectral influences, and used the inverse to reveal the shadow $\omega = (1 - \omega)/\omega_0$, and then rescaled ($\omega_{ns}$). The relation enables estimates of wind shear velocity at the soil surface to be produced from Moderate Resolution Imaging Spectroradiometer (MODIS) data globally, at a 500 m spatial resolution.

3.2.4. Computational Fluid Dynamics (CFD) Modelling

Computational fluid dynamics (CFD) modelling has been used to explicitly resolve turbulent wind flow and sediment flux around individual roughness elements to a high temporal and spatial resolution (e.g., [52,63,136,165]). CFD allows a more complete representation of flow structure than that obtained by field and wind tunnel experiments, so represents a promising approach for incorporating wind flow turbulence in vegetated contexts. Using CFD simulations, Turpin et al. [136] identified a peak in shear stress in the low-shear zone behind individual elements and showed that the "inversion point" at which an element switched from reducing shear stress to increasing it varied with coverage density and roughness height. Turpin et al. [136] could also test the effect of increasing the number and height of roughness elements on erosion patterns. Hesp and Smyth [63] used a CFD model to investigate the impact of nebkha width on wind flow recovery in the lee. By accurately quantifying turbulent wind structures around each simulated nebkha, Hesp and Smyth [63] could begin to link flow behaviour with resultant shadow dune formation.

Despite the high resolution and realistic flow structure representation that can be achieved using CFD, it is a computationally expensive method. To get around this issue, flow properties can be calculated using much coarser meshes, through large eddy simulations (LES) [52,165–167]. Dupont et al. [52] coupled a LES model with a physically based saltation model to resolve turbulent wind flow over heterogeneous vegetated landscapes composed of trees and shrubs. The erosion patterns simulated by Dupont et al.'s [52] model were qualitatively consistent with previous wind tunnel [59,78] and field [46,140] observations. Minimal dune models (e.g., [168,169]), which combine analytical descriptions of turbulent wind velocity with continuum saltation models, have also been used to successfully simulate transitions between barchan and parabolic dunes [169,170].

Nevertheless, the computational requirements of CFD and LES methods still limit their use for simulating long-term erosion at larger than the field scale. Such modelling approaches are also highly dependent on a large number of parameterisations and coefficients to describe, in high resolution, the physical processes they seek to simulate [171]. Given the significant uncertainties that exist in our understanding of the interactions between vegetation, wind flow and sediment transport, hard-coding the appropriate physics into CFD and LES models remains an ongoing challenge.

3.2.5. Cellular Automaton (CA) Modelling

Cellular automata (CA) are a class of numerical models consisting of regular, discrete grids of cells operated on by predefined rules, which capture full landscape-scale dynamics using only fundamental processes. The CA approach has been particularly productive in the context of dune-field patterning (e.g., [21–23,172]), and the spatial nature of CAs lends itself well to analysing ecogeomorphic systems [173]. The strength of CA models lies in their capacity to represent key processes underlying complex systems, using a far smaller number of parameters and assumptions than many detailed reductionist models [171].

In most CA models, local neighbourhood operations produce dynamic responses from basic rules centred on each grid cell. All cells hold a variety of attributes (in the drylands context, these could include sand surface height and vegetation characteristics such as plant type, height and porosity) that are altered by applying transition rules during each timestep. As model time progresses, the vegetation distribution can alter local wind flow characteristics, thus impacting sediment flux patterns over the surface. The grid formulation on which CAs are based allows wind velocity to effectively be 'mapped onto' the model domain as a function of vegetation height and porosity, fitting Okin's [58] conception of vegetated desert surfaces as a collection of gaps between roughness elements. The relative computational efficiency of CAs means that simulations can be run over larger domains (scale of ~10^4 m) and long timescales, which is particularly useful for landscape evolution modelling.

The CA approach not only allows sediment transport and wind flow behaviour to be modelled in isolation, but also enables explicit simulations of feedbacks between major components of dryland ecogeomorphic systems. For instance, surface changes resulting from wind erosion can be linked to subsequent vegetation growth/dieback within a coupled CA framework (e.g., [2,25,26,33,174]). This approach is rooted in the concept of 'landscape connectivity' [175,176], which considers dryland landscapes as a series of conduits for processes (e.g., fire, wind and water propagation) that link vegetation growth, aeolian processes and external forcing factors through various biotic and abiotic feedbacks [41,177,178].

The Discrete ECogeomorphic Aeolian Landscape model (DECAL) [2,25,174], later extended by Yan and Baas [179], provides an approach for simulating the dynamics of vegetation-dependent dunes such as parabolic dunes and nebkhas, by incorporating feedbacks between sedimentation balance and vegetation growth. DECAL introduces a length scale to resultant landforms that remains otherwise dimensionless in bare-sand CAs. Mayaud et al. [26] built on this approach by developing the coupled Vegetation and Sediment TrAnsport model (ViSTA). ViSTA allows plant-plant interactions to be specified and for realistic vegetation patterning to emerge in the domain, without the need for directly imposing factors such as seasonality in plant growth. By including a variety of ecogeomorphic feedbacks linking plant growth, wind dynamics and sediment transport, Mayaud et al. [26] successfully simulated the development of an equilibrium nebkha dunefield in the Skeleton Coast, Namibia (Figure 9).

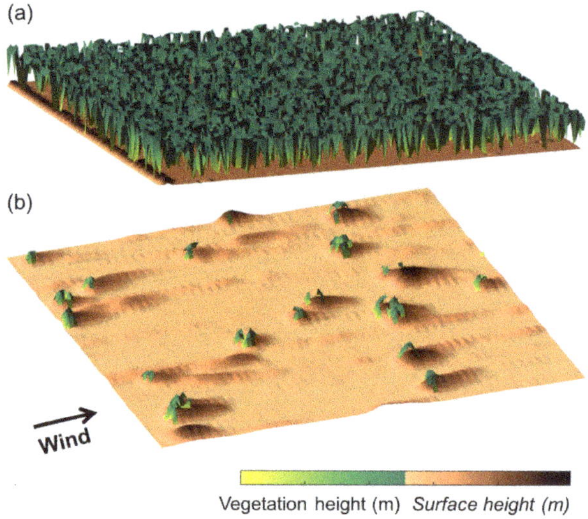

Figure 9. Three-dimensional representations of combined vegetation and surface morphology for a coupled Vegetation and Sediment TrAnsport model (ViSTA) simulation, with conditions similar to those observed in the Skeleton Coast National Park, Namibia. Figures show landscape at: (**a**) beginning of simulation, where the domain is artificially populated with shrubs; (**b**) after 25 years of arid conditions, where an equilibrium landscape of nebkha dunes emerges. Black arrow shows average wind direction (adapted from [26] (p. 22)).

4. Potential Future Avenues for Research

In general, there remains a need to collect more windflow data around a variety of dryland plant types in the field. Despite increasing evidence showing that two-dimensional wind fences do not approximate well to three-dimensional live plants, only a handful of studies have examined airflow dynamics in the wake of single rows of three-dimensional plants (e.g., [62]). These could display different behaviours to traditional wind fences, especially with regards to flow turbulence. Linked to this issue, the impact of roughness element width on wind flow recovery remains under-studied, perhaps because it intuitively does not have as great an impact on drag as element height or porosity. Recent numerical modelling of nebkhas [63] has revealed that nebkha width strongly controls wind flow recovery, so wind tunnel and field studies investigating element width would help to better parameterise wind flow in vegetated contexts.

Whilst it is clearly beneficial to conduct wind flow studies in field situations to account for natural processes and variability, it is not possible to control for vegetation parameters (height, width, porosity, pliability) and flow parameters (velocity, turbulence, intermittency) to the same degree as in wind tunnels. Wind tunnel experiments have proved valuable for understanding the impacts of a wide variety of parameters on flow behaviour, and should be extended to address some of the knowledge gaps highlighted here. More wind tunnel studies are needed that systematically investigate the effects of porosity and/or pliability of three-dimensional, live plants on wind flow. Porosity and pliability are relatively difficult to control for in live plants when attempting to keep height and width parameters constant. Some wind tunnel studies have begun to investigate porosity variability in high detail using porous mesh cubes and cylinders (e.g., [86]), although these remain crude approximations to live plants. Other studies have begun to account for plant pliability, for instance through measuring the temporally-averaged frontal area of flexible plants in wind tunnels (e.g., [100]).

Remote sensing data are proving increasingly useful for representing spatial and temporal variability in vegetation within modelling schemes [157]. Plant characteristics (e.g., height, width and porosity) can now be quantified rapidly and over relatively large scales thanks to the recent progress in image-based techniques (e.g., [155]) and high-resolution remote sensing [152,154,180]. However, ongoing challenges remain for identifying appropriate metrics of surface roughness distribution that could be used as inputs to schemes such as Okin's [58] over large (i.e., regional) areas, and for overcoming limitations of indices such as NDVI and LAI for representing vegetation (see Section 3.2.3). Direct approximation of the drag partition and surface shear stress from remote sensing is likely to alleviate these challenges, while reducing the uncertainty in wind erosion estimates [148].

Additionally, low-elevation surveys using aerial drones (unmanned aerial vehicles, UAVs) allow for vast vegetation transect measurements to be collected quickly and relatively inexpensively [181]. In turn, 'structure-from-motion' (SfM) photogrammetry can easily convert drone images of an element or landscape into a georeferenced digital elevation model (DEM) [182–184]. This technique can produce centimetre-scale horizontal and vertical precision that is comparable with airborne LiDAR [185] and terrestrial laser scanning [184], allowing landscape change before and after geomorphic events to be accurately quantified. However, issues arise with regards to the reliance of SfM's image-matching algorithms on image texture, which may result in low-texture images (e.g., bare sand surfaces) yielding poor point clouds [185].

Finally, an explicit link must be made between the impact of vegetation on wind flow, and the subsequent impact on sediment transport processes. Some wind tunnel studies (e.g., [60,78]) have used creative methods, such as coloured sand, for tracking erosion and deposition around roughness elements, but detailed exploration of flow field behaviour and the evolution of bedforms around plants is still required. For instance, high-resolution terrestrial laser scanning could be used to detect mm-scale changes in height around partially vegetated erodible surfaces [186], thus helping to identify erosional and depositional regions to a high level of accuracy. In terms of modelling vegetation/sediment interactions, Raupach and Lu [87] identify problems of scaling over heterogeneous surfaces, and raise questions over how sediment deposition processes in the presence of vegetation could be adequately measured. Whilst these issues remain pertinent, drag partition schemes that explicitly resolve the effects of roughness configuration on surface shear stress distribution could significantly reduce the uncertainty in sediment deposition schemes.

5. Conclusions

The effects of patchy vegetation on wind flow and sediment transport in drylands are multiple and varied. They depend on the vegetation type in question, as well as turbulence, porosity and pliability factors. The use of high-frequency anemometers in both field and wind tunnel contexts, as well as creative methodological approaches such as the use coloured sand, have helped to identify key aerodynamic zones of interest. Increasingly, sophisticated modelling techniques such as computational fluid dynamics (CFD) and cellular automaton (CA) modelling are being used to gain more holistic perspectives on the processes shaping partially vegetated surfaces. However, such models are often computationally expensive to run, and still suffer from an ongoing lack of adequate data describing flow dynamics at the plant/patch scale.

Moving forward, more empirically-derived flow and transport data collected in vegetated contexts must be used to parameterize, verify and validate wind flow/sediment transport models. Remote sensing methods are proving to be increasingly convenient, and in some cases relatively inexpensive, ways of rapidly measuring vegetation, wind flow and sediment characteristics over vast areas and at multiple scales. Remote sensing also offers opportunities to reduce the complexity, and increase the accuracy, of aeolian sediment transport modelling. Without these crucial data forming the basis for rigorous model parameterisation, significant uncertainty will remain surrounding the potential future responses of dryland landscapes to climate and land use change.

Acknowledgments: This research was funded by a UK Natural Environment Research Council Doctoral Training Grant (NE/L501530/1), a St Catherine's College (University of Oxford) Scholarship and a Hertford College (University of Oxford) Scholarship to Jerome Mayaud. We thank the Assistant Editor of Land (Sanford Su), the anonymous reviewers whose comments helped to improve this paper, and Richard Bailey and Giles Wiggs for their constructive feedback on a related version of this manuscript.

Author Contributions: Jerome Mayaud and Nicholas Webb both contributed to researching and writing the manuscript.

Conflicts of Interest: The authors declare no conflict of interest.

References

1. UNEP. *World Atlas of Desertification*, 2nd ed.; Edward Arnold: Sevenoaks, UK, 1998.
2. Baas, A.C.W.; Nield, J.M. Modelling vegetated dune landscapes. *Geophys. Res. Lett.* **2007**, *34*, L06405. [CrossRef]
3. Viles, H.A. Weathering systems. In *Arid Zone Geomorphology: Process, Form and Change in Drylands*, 3rd ed.; Thomas, D.S.G., Ed.; John Wiley & Sons Ltd.: Chichester, UK, 2011; pp. 85–101.
4. Thomas, D.S.G.; Shaw, P.A. *The Kalahari Environment*; Cambridge University Press: Cambridge, UK, 1991.
5. Goudie, A.S. *Great Warm Deserts of the World*; Cambridge University Press: Cambridge, UK, 2002.
6. Ravi, S.; Breshears, D.D.; Huxman, T.E.; D'Odorico, P. Land degradation in drylands: Interactions among hydrologic–aeolian erosion and vegetation dynamics. *Geomorphology* **2010**, *116*, 236–245. [CrossRef]
7. Kéfi, S.; Rietkerk, M.; Alados, C.L.; Pueyo, Y.; Papanastasis, V.P.; Elaich, A.; de Ruiter, P.C. Spatial vegetation patterns and imminent desertification in Mediterranean arid ecosystems. *Nature* **2007**, *449*, 213–217. [CrossRef] [PubMed]
8. Thomas, D.S.G.; Knight, M.; Wiggs, G.F.S. Remobilization of southern African desert dune systems by twenty-first century global warming. *Nature* **2005**, *435*, 1218–1221. [CrossRef] [PubMed]
9. Wang, X.; Yang, Y.; Dong, Z.; Zhang, C. Responses of dune activity and desertification in China to global warming in the twenty-first century. *Glob. Planet. Chang.* **2009**, *67*, 167–185. [CrossRef]
10. Reynolds, R.; Belnap, J.; Reheis, M.; Lamothe, P.; Luiszer, F. Aeolian dust in Colorado Plateau soils: Nutrient inputs and recent change in source. *Proc. Natl. Acad. Sci. USA* **2001**, *98*, 7123–7127. [CrossRef] [PubMed]
11. Ravi, S.; D'Odorico, P.; Breshears, D.D.; Field, J.P.; Goudie, A.S. Aeolian processes and the biosphere. *Rev. Geophys.* **2011**, *49*, 1–45. [CrossRef]
12. Goudie, A.S.; Middleton, N.J. *Desert Dust in the Global System*; Springer: Berlin, Germany, 2006.
13. Seager, R.; Ting, M.; Held, M.I.; Kushnir, Y.; Lu, J.; Vecchi, G.; Huang, H.-P.; Harnik, N.; Leetmaa, A.; Lau, N.-C.; et al. Model projections of an imminent transition to a more arid climate in southwestern North America. *Science* **2007**, *316*, 1181–1184. [CrossRef] [PubMed]
14. Bogle, R.; Redsteer, M.H.; Vogel, J. Field measurement and analysis of climatic factors affecting dune mobility near Grand Falls on the Navajo Nation, southwestern United States. *Geomorphology* **2015**, *228*, 41–51. [CrossRef]
15. Barbier, N.; Couteron, P.; Lejoly, J.; Deblauwe, V.; Lejeune, O. Self-organized vegetation patterning as a fingerprint of climate and human impact on semi-arid ecosystems. *J. Ecol.* **2006**, *94*, 537–547. [CrossRef]
16. D'Odorico, P.; Bhattachan, A.; Davis, K.F.; Ravi, S.; Runyan, C.W. Global desertification: Drivers and feedbacks. *Adv. Water Resour.* **2013**, *51*, 326–344. [CrossRef]
17. Middleton, N.J.; Sternberg, T. Climate hazards in drylands: A review. *Earth Sci. Rev.* **2013**, *126*, 48–57. [CrossRef]
18. Griffin, D.W.; Garrison, V.H.; Herman, J.R.; Shinn, E.A. African desert dust in the Caribbean atmosphere: Microbiology and public health. *Aerobiologia* **2001**, *17*, 203–213. [CrossRef]
19. Wiggs, G.F.S.; O'Hara, S.L.; Wegerdt, J.; Van Der Meer, J.; Small, I.; Hubbard, R. The dynamics and characteristics of aeolian dust in dryland Central Asia: Possible impacts on human exposure and respiratory health in the Aral Sea basin. *Geogr. J.* **2003**, *169*, 142–157. [CrossRef]
20. Wasson, R.J.; Hyde, R. Factors determining desert dune type. *Nature* **1983**, *304*, 337–339. [CrossRef]
21. Werner, B.T. Eolian dunes: Computer simulations and attractor interpretation. *Geology* **1995**, *23*, 1107–1110. [CrossRef]

22. Momiji, H.; Carretero-Gonzalez, R.; Bishop, S.R.; Warren, A. Simulation of the effect of wind speedup in the formation of transverse dune fields. *Earth Surf. Process. Landf.* **2000**, *25*, 905–918. [CrossRef]
23. Bishop, S.R.; Momiji, H.; Carretero-Gonzalez, R.; Warren, A. Modelling desert dune fields based on discrete dynamics. *Discret. Dyn. Nat. Soc.* **2002**, *7*, 7–17. [CrossRef]
24. Hesp, P. Foredunes and blowouts: Initiation, geomorphology and dynamics. *Geomorphology* **2002**, *48*, 245–268. [CrossRef]
25. Nield, J.M.; Baas, A.C.W. Investigating parabolic and nebkha dune formation using a cellular automaton modelling approach. *Earth Surf. Process. Landf.* **2008**, *33*, 724–740. [CrossRef]
26. Mayaud, J.R.; Bailey, R.M.; Wiggs, G.F.S. A coupled vegetation/sediment-transport model for dryland environments. *J. Geophys. Res. Earth Surf.* **2017**, *122*, 875–900. [CrossRef]
27. Okin, G.S.; Gillette, D.A.; Herrick, J.E. Multi-scale controls on and consequences of aeolian processes in landscape change in arid and semi-arid environments. *J. Arid Environ.* **2006**, *65*, 253–275. [CrossRef]
28. Li, J.; Okin, G.S.; Herrick, J.E.; Belnap, J.; Miller, M.E.; Vest, K.; Draut, A.E. Evaluation of a new model of aeolian transport in the presence of vegetation. *J. Geophys. Res. Earth Surf.* **2013**, *118*, 288–306. [CrossRef]
29. Huang, J.; Minnis, P.; Yan, H.; Yi, Y.; Chen, B.; Zhang, L.; Ayers, J. Dust aerosol effect on semi-arid climate over Northwest China detected from A-Train satellite measurements. *Atmos. Chem. Phys.* **2010**, *10*, 6863–6872. [CrossRef]
30. UNDP. *The Forgotten Billion: MDG Achievement in Drylands*; UNDP: New York, NY, USA, 2011.
31. Held, I.M.; Delworth, T.M.; Lu, J.; Findell, K.L.; Knutson, T.R. Simulation of Sahel Drought in the 20th and 21st centuries. *Proc. Natl. Acad. Sci. USA* **2005**, *102*, 17891–17896. [CrossRef] [PubMed]
32. IPCC. Summary for Policymakers. In *Climate Change 2013: The Physical Science Basis. Contribution of Working Group I to the Fifth Assessment Report of the Intergovernmental Panel on Climate Change*; Stocker, T.F., Qin, D., Plattner, G.K., Tignor, M., Allen, S.K., Eds.; Cambridge University Press: Cambridge, UK; New York, NY, USA, 2013.
33. Mayaud, J.R.; Bailey, R.M.; Wiggs, G.F.S. Modelled responses of the Kalahari Desert to 21st century climate and land use change. *Sci. Rep.* **2017**, *7*, 3887. [CrossRef] [PubMed]
34. Kinast, S.; Meron, E.; Yizhaq, H.; Ashkenazy, Y. Biogenic crust dynamics on sand dunes. *Phys. Rev. E* **2013**, *87*, 1–5. [CrossRef] [PubMed]
35. Bestelmeyer, B.T.; Okin, G.S.; Duniway, M.C.; Archer, S.R.; Sayre, N.F.; Williamson, J.C.; Herrick, J.E. Desertification, land use, and the transformation of global drylands. *Front. Ecol. Environ.* **2015**, *13*, 28–36. [CrossRef]
36. Webb, N.P.; Marshall, N.A.; Stringer, L.C.; Reed, M.S.; Chappell, A.; Herrick, J.E. Land degradation and climate change: Building climate resilience in agriculture. *Front. Ecol. Environ.* **2017**. [CrossRef]
37. Meron, E.; Gilad, E.; von Hardenberg, J.; Shachak, M.; Zarmi, Y. Vegetation patterns along a rainfall gradient. *Chaos Solitons Fractals* **2004**, *19*, 367–376. [CrossRef]
38. Borgogno, F.; D'Odorico, P.; Laio, F.; Ridolfi, L. Mathematical models of vegetation pattern formation in ecohydrology. *Rev. Geophys.* **2009**, *47*, RG1005. [CrossRef]
39. Wainwright, J. Desert Ecogeomorphology. In *Geomorphology of Desert Environments*, 2nd ed.; Parsons, A.J., Abrahams, A.D., Eds.; Springer: Berlin, Germany, 2009; pp. 21–66.
40. Bailey, R.M. Spatial and temporal signatures of fragility and threshold proximity in modelled semi-arid vegetation. *Proc. R. Soc. B* **2011**, *278*, 1064–1071. [CrossRef] [PubMed]
41. Stewart, J.; Parsons, A.J.; Wainwright, J.; Okin, G.S.; Bestelmeyer, B.; Fredrickson, E.L.; Schlesinger, W.H. Modelling emergent patterns of dynamic desert ecosystems. *Ecol. Monogr.* **2014**, *84*, 373–410. [CrossRef]
42. Getzin, S.; Wiegand, K.; Wiegand, T.; Yizhaq, H.; von Hardenberg, J.; Meron, E. Adopting a spatially explicit perspective to study the mysterious fairy circles of Namibia. *Ecography* **2014**. [CrossRef]
43. Wolfe, S.A.; Nickling, W.G. The protective role of sparse vegetation in wind erosion. *Prog. Phys. Geogr.* **1993**, *17*, 50–68. [CrossRef]
44. Judd, M.J.; Raupach, M.R.; Finnigan, J.J. A wind tunnel study of turbulent flow around single and multiple windbreaks, part I: Velocity fields. *Bound. Layer Meteorol.* **1996**, *80*, 127–165. [CrossRef]
45. Al-Awadhi, J.M.; Willetts, B.B. Sand transport and deposition within arrays of non-erodible cylindrical elements. *Earth Surf. Process. Landf.* **1999**, *24*, 423–435. [CrossRef]

46. Leenders, J.K.; van Boxel, J.H.; Sterk, G. The effect of single vegetation elements on wind velocity and sediment transport in the Sahelian zone of Burkina Faso. *Earth Surf. Process. Landf.* **2007**, *32*, 1454–1474. [CrossRef]
47. Wasson, R.J.; Nanninga, P.M. Estimating wind transport of sand on vegetated surfaces. *Earth Surf. Process. Landf.* **1986**, *11*, 505–514. [CrossRef]
48. Gillette, D.A.; Stockton, P.A. The effects of non-erodible particles on wind erosion of erodible surfaces. *J. Geophys. Res.* **1989**, *94*, 12885–12893. [CrossRef]
49. Gillies, J.A.; Nickling, W.G.; King, J. Drag coefficient and plant form-response to wind velocity in three plant species: Burning bush (*Euonymus alatus*), Colorado blue spruce (*Picea pungens glauca.*), and fountain grass (*Pennisetum setaceum*). *J. Geophys. Res.* **2002**, *107*, 4760. [CrossRef]
50. Crawley, D.M.; Nickling, W.G. Drag partition for regularly-arrayed rough surfaces. *Bound. Layer Meteorol.* **2003**, *107*, 445–468. [CrossRef]
51. Gillette, D.A.; Herrick, J.E.; Herbert, G.A. Wind characteristics of Mesquite Streets in the northern Chihuahuan Desert, New Mexico, USA. *Environ. Fluid Mech.* **2006**, *6*, 241–275. [CrossRef]
52. Dupont, S.; Bergametti, G.; Simoëns, S. Modeling aeolian erosion in presence of vegetation. *J. Geophys. Res. Earth Surf.* **2014**, *119*, 168–187. [CrossRef]
53. Gillies, J.A.; Lancaster, N.; Nickling, W.G.; Crawley, D. Field determination of drag forces and shear stress partitioning effects for a desert shrub (*Sarcobatus vermiculatus*, Greasewood). *J. Geophys. Res. Atmos.* **2000**, *105*, 24871–24880. [CrossRef]
54. Gillies, J.A.; Nield, J.M.; Nickling, W.G. Wind velocity and sediment transport recovery in the lee of a vegetated and denuded nebkha within a nebkha dune field. *Aeolian Res.* **2014**, *12*, 135–141. [CrossRef]
55. Davidson-Arnott, R.G.D.; Bauer, B.O.; Walker, I.J.; Hesp, P.A.; Ollerhead, J.; Chapman, C. High-frequency sediment transport responses on a vegetated foredune. *Earth Surf. Process. Landf.* **2012**, *37*, 1227–1241. [CrossRef]
56. Van Gardingen, P.; Grace, J. Plants and wind. *Adv. Bot. Res.* **1991**, *18*, 189–253.
57. Raupach, M.R. Drag and drag partition on rough surfaces. *Bound. Layer Meteorol.* **1992**, *60*, 375–395. [CrossRef]
58. Okin, G.S. A new model of wind erosion in the presence of vegetation. *J. Geophys. Res.* **2008**, *113*, F02S10. [CrossRef]
59. Sutton, S.L.F.; McKenna-Neuman, C. Sediment entrainment to the lee of roughness elements: Effects of vortical structures. *J. Geophys. Res.* **2008**, *113*, F02S09. [CrossRef]
60. Suter-Burri, K.; Gromke, C.; Leonard, K.C.; Graf, F. Spatial patterns of aeolian sediment deposition in vegetation canopies: Observations from wind tunnel experiments using colored sand. *Aeolian Res.* **2013**, *8*, 65–73. [CrossRef]
61. Lee, J.-P.; Lee, E.-J.; Lee, S.-J. Shelter effect of a fir tree with different porosities. *J. Mech. Sci. Technol.* **2014**, *28*, 565–572. [CrossRef]
62. Wu, X.; Zou, X.; Zhou, N.; Zhang, C.; Shi, S. Deceleration efficiencies of shrub windbreaks in a wind tunnel. *Aeolian Res.* **2015**, *16*, 11–23. [CrossRef]
63. Hesp, P.A.; Smyth, T.A.G. Nebkha flow dynamics and shadown dune formation. *Geomorphology* **2017**, *282*, 27–38. [CrossRef]
64. Mayaud, J.R.; Wiggs, G.F.S.; Bailey, R.M. Characterizing turbulent wind flow around dryland vegetation. *Earth Surf. Process. Landf.* **2016**, *41*, 1421–1436. [CrossRef]
65. Hagen, L.J.; Skidmore, E.L. Turbulent velocity fluctuations and vertical flow as affected by windbreak porosity. *Trans. Am. Soc. Agric. Eng.* **1971**, *14*, 634–637.
66. Perera, M.D. Shelter behind two-dimensional solid and porous fences. *J. Wind Eng. Ind. Aerodyn.* **1981**, *8*, 93–104. [CrossRef]
67. Gross, G. A numerical study of the air flow within and around a single tree. *Bound. Layer Meteorol.* **1987**, *40*, 311–327. [CrossRef]
68. Kim, H.B.; Lee, S.-J. The structure of turbulent shear flow around a two-dimensional porous fence having a bottom gap. *J. Fluids Struct.* **2002**, *16*, 317–329. [CrossRef]
69. Endalew, A.M.; Hertog, M.; Gebrehiwot, M.G.; Baelmans, M.; Ramon, H.; Nicolaï, B.M.; Verboven, P. Modelling airflow within model plant canopies using an integrated approach. *Comput. Electron. Agric.* **2009**, *66*, 9–24. [CrossRef]

70. Lee, B.E.; Soliman, B.F. An investigation of the forces on three-dimensional bluff bodies in rough wall turbulent boundary layer. *J. Fluids* **1977**, *99*, 503–509. [CrossRef]
71. King, J.; Nickling, W.G.; Gillies, J.A. Representation of vegetation and other non-erodible elements in aeolian shear stress partitioning models for predicting transport threshold. *J. Geophys. Res.* **2005**, *110*, F04015. [CrossRef]
72. Liu, S.; Wul, H.; Lytton, R.L.; Sharpe, P.J. Aerodynamic sheltering effects of vegetative arrays on wind erosion: A numerical approach. *J. Environ. Manag.* **1990**, *30*, 281–294. [CrossRef]
73. Ash, J.E.; Wasson, R.H. Vegetation and sand mobility in the Australian desert dunefield. *Z. Geomorphol.* **1983**, *45*, 7–25.
74. Walter, B.; Gromke, C.; Leonard, K.C.; Manes, C.; Lehning, M. Spatio-temporal surface shear-stress variability in live plant canopies and cube arrays. *Bound. Layer Meteorol.* **2012**, *143*, 337–356. [CrossRef]
75. Breshears, D.D.; Whicker, J.J.; Zou, C.B.; Field, J.P.; Allen, C.D. A conceptual framework for dryland aeolian sediment transport along the grassland–forest continuum: Effects of woody plant canopy cover and disturbance. *Geomorphology* **2009**, *105*, 28–38. [CrossRef]
76. Wiggs, G.F.S.; Livingstone, I.; Thomas, D.S.G.; Bullard, J.E. Effect of vegetation removal on airflow patterns and dune dynamics in the southwest Kalahari Desert. *Land Degrad. Rehabil.* **1994**, *5*, 13–24. [CrossRef]
77. Mayaud, J.R.; Wiggs, G.F.S.; Bailey, R.M. Dynamics of skimming flow in the wake of a vegetation patch. *Aeolian Res.* **2016**, *22*, 141–151. [CrossRef]
78. Burri, K.; Gromke, C.; Lehning, M.; Graf, F. Aeolian sediment transport over vegetation canopies: A wind tunnel study with live plants. *Aeolian Res.* **2011**, *3*, 205–213. [CrossRef]
79. Youssef, F.; Visser, S.M.; Karssenberg, D.; Erpul, G.; Cornelis, W.M.; Gabriels, D.; Poortinga, A. The effect of vegetation patterns on wind-blown mass transport at the regional scale: A wind tunnel experiment. *Geomorphology* **2012**, *159–160*, 178–188. [CrossRef]
80. Gash, J.H.C. Observations of turbulence downwind of a forest-heath interface. *Bound. Layer Meteorol.* **1986**, *36*, 227–237. [CrossRef]
81. Liu, J.; Chen, J.M.; Black, T.A.; Novak, M.D. E-ε modelling of turbulent air flow downwind of a model forest edge. *Bound. Layer Meteorol.* **1996**, *77*, 21–44. [CrossRef]
82. Belcher, S.; Jerram, N.; Hunt, J. Adjustment of a turbulent boundary layer to a "canopy" of roughness elements. *J. Fluid Mech.* **2003**, *488*, 369–398. [CrossRef]
83. Frank, C.; Ruck, B. Numerical study of the airflow over forest clearings. *Forestry* **2008**, *81*, 259–277. [CrossRef]
84. Le, H.; Moin, P.; Kim, J. Direct numerical simulation of turbulent flow over a backward-facing step. *J. Fluid Mech.* **1997**, *330*, 349–374. [CrossRef]
85. Wengle, H.; Huppertz, A.; Bärwolff, G.; Janke, G. Manipulated transitional backward-facing step flow: An experimental and direct numerical simulation investigation. *Eur. J. Mech.* **2001**, *20*, 25–46. [CrossRef]
86. Gillies, J.A.; Nickling, W.G.; Nickolich, G.; Etyemezian, V. A wind tunnel study of the aerodynamic and sand trapping properties of porous mesh 3-dimensional roughness elements. *Aeolian Res.* **2017**, *25*, 23–35. [CrossRef]
87. Raupach, M.R.; Lu, H. Representation of land-surface processes in aeolian transport models. *Environ. Model. Softw.* **2004**, *19*, 93–112. [CrossRef]
88. Raupach, M.R.; Woods, N.; Dorr, G.; Leys, J.F.; Cleugh, H.A. The entrapment of particles by windbreaks. *Atmos. Environ.* **2001**, *35*, 3373–3383. [CrossRef]
89. Zhang, J.; Shao, Y. A new parameterization of particle dry deposition over rough surfaces. *Atmos. Chem. Phys.* **2014**, *14*, 12429–12440. [CrossRef]
90. Minvielle, F.; Marticorena, B.; Gillette, D.A.; Lawson, R.E.; Thompson, R.; Bergametti, G. Relationship between the aerodynamic roughness length and the roughness density in cases of low roughness density. *Environ. Fluid Mech.* **2003**, *3*, 249–267. [CrossRef]
91. Grant, P.F.; Nickling, W.G. Direct field measurement of wind drag on vegetation for application to windbreak design and modeling. *Land Degrad. Dev.* **1998**, *9*, 57–66. [CrossRef]
92. Taylor, P.A. Turbulent wakes in the boundary layer. In *Flow and Transport in the Natural Environment: Advances and Applications*; Steffen, W.L., Denmead, O.T., Eds.; Springer: Berlin, Germany, 1988; pp. 270–292.
93. Wang, H.; Takle, E.S. Momentum budget and shelter mechanism of bounda-layer flow near a shelterbelt. *Bound. Layer Meteorol.* **1996**, *82*, 417–435. [CrossRef]
94. Vigiak, O.; Sterk, G.; Warren, A.; Hagen, L.J. Spatial modeling of wind speed around windbreaks. *Catena* **2003**, *52*, 273–288. [CrossRef]

95. Cornelis, W.M.; Gabriels, D. Optimal windbreak design for wind-erosion control. *J. Arid Environ.* **2005**, *61*, 315–332. [CrossRef]
96. Musick, H.B.; Trujillo, S.M.; Truman, C.R. Wind-tunnel modelling of the influence of vegetation structure on saltation threshold. *Earth Surf. Process. Landf.* **1996**, *21*, 589–605. [CrossRef]
97. Lee, S.-J.; Park, K.C.; Park, C.W. Wind tunnel observations about the shelter effect of porous fences on the sand particle movements. *Atmos. Environ.* **2002**, *36*, 1453–1463. [CrossRef]
98. Bradley, E.F.; Mulhearn, P.J. Development of velocity and shear-stress distributions in the wake of a porous shelter fence. *J. Wind Eng. Ind. Aerodyn.* **1983**, *15*, 145–156. [CrossRef]
99. Gillies, J.A.; Etyemezian, V.; Nikolich, G.; Glick, R.; Rowland, P.; Pesce, T.; Skinner, M. Effectiveness of an array of porous fences to reduce sand flux: Oceano Dunes, Oceano CA. *J. Wind Eng. Ind. Aerodyn.* **2017**, *168*, 247–259. [CrossRef]
100. Walter, B.; Gromke, C.; Leonard, K.; Clifton, A.; Lehning, M. Spatially resolved skin friction velocity measurements using Irwin sensors: A calibration and accuracy analysis. *J. Wind Eng. Ind. Aerodyn.* **2012**, *104–106*, 314–321. [CrossRef]
101. Wilson, J.D.; Yee, E. Calculation of winds disturbed by an array of fences. *Agric. For. Meteorol.* **2003**, *115*, 31–50. [CrossRef]
102. Hagen, L.J. *WEPS: Wind Erosion Prediction System, Technical Documentation*; Wind Erosion Research Unit: Manhattan, NY, USA, 1996.
103. McKenna-Neuman, C.; Bédard, O. A wind tunnel study of flow structure adjustment on deformable sand beds containing a surface-mounted obstacle. *J. Geophys. Res. Earth Surf.* **2015**, *120*, 1824–1840. [CrossRef]
104. Clifford, N.J.; French, J.R. Monitoring and modelling turbulent flows: Historical and contemporary perspectives. In *Turbulence: Perspectives on Flow and Sediment Transport*; Clifford, N.J., French, J.R., Hardisty, J., Eds.; John Wiley & Sons: New York, NY, USA, 1993; pp. 1–34.
105. Butterfield, G.R. Grain transport rates in steady and unsteady turbulent airflows. *Acta Mech.* **1991**, *1*, 97–122.
106. Sterk, G.; Jacobs, A.F.G.; van Boxel, J.H. The effect of turbulent flow structures on saltation sand transport in the atmospheric boundary layer. *Earth Surf. Process. Landf.* **1998**, *23*, 877–887. [CrossRef]
107. Namikas, S.L.; Bauer, B.O.; Sherman, D.J. Influence of averaging interval on shear velocity estimates for aeolian transport modelling. *Geomorphology* **2003**, *53*, 235–246. [CrossRef]
108. Schönfeldt, H.J.; von Löwis, S. Turbulence-driven saltation in the atmospheric surface layer. *Meteorol. Z.* **2003**, *12*, 257–268. [CrossRef]
109. Baas, A.C.W.; Sherman, D.J. Formation and behaviour of aeolian streamers. *J. Geophys. Res.* **2005**, *110*, F03011. [CrossRef]
110. Baas, A.C.W.; Sherman, D.J. Spatiotemporal variability of aeolian sand transport in a coastal dune environment. *J. Coast. Res.* **2005**, *22*, 1198–1205. [CrossRef]
111. Leenders, J.K.; van Boxel, J.H.; Sterk, G. Wind forces and related saltation transport. *Geomorphology* **2005**, *71*, 357–372. [CrossRef]
112. Weaver, C.M.; Wiggs, G.F.S. Field measurements of mean and turbulent airflow over a barchan sand dune. *Geomorphology* **2011**, *128*, 32–41. [CrossRef]
113. Wiggs, G.F.S.; Weaver, C.M. Turbulent flow structures and aeolian sediment transport over a barchan sand dune. *Geophys. Res. Lett.* **2012**, *39*, 1–7. [CrossRef]
114. Mayaud, J.R.; Bailey, R.M.; Wiggs, G.F.S. Modelling aeolian sand transport using a dynamic mass balancing approach. *Geomorphology* **2016**, *280*, 108–121. [CrossRef]
115. Musick, H.B.; Gillette, D.A. Field evaluation of relationships between a vegetation structural parameter and sheltering against wind erosion. *Land Degrad. Rehabil.* **1990**, *2*, 87–94. [CrossRef]
116. Schlichting, H. Experimentelle untersuchungen zum Rauhigkeitsproblem. *Ing.-Arch.* **1936**, *7*, 1–34. [CrossRef]
117. Marshall, J.K. Drag measurements in roughness arrays of varying densities and distribution. *Agric. Meteorol.* **1971**, *8*, 269–292. [CrossRef]
118. Marticorena, B.; Bergametti, G. Modeling the atmospheric dust cycle: 1. Design of a soil-derived dust emission scheme. *J. Geophys. Res.* **1995**, *100*, 16415. [CrossRef]
119. Brown, S.; Nickling, W.G.; Gillies, J.A. A wind tunnel examination of shear stress partitioning for an assortment of surface roughness distributions. *J. Geophys. Res.* **2008**, *113*, F02S06. [CrossRef]
120. Lyles, L.; Schrandt, R.L.; Schmeidler, N.F. How aerodynamic roughness elements control sand movement. *Trans. ASAE* **1974**, *17*, 134–139. [CrossRef]

121. Raupach, M.R.; Gillette, D.A.; Leys, J.F. The effect of roughness elements on wind erosion threshold. *J. Geophys. Res.* **1993**, *98*, 3023–3029. [CrossRef]
122. Shao, Y.; Yang, Y. A scheme for drag partition over rough surfaces. *Atmos. Environ.* **2005**, *39*, 7351–7361. [CrossRef]
123. Shao, Y.; Yang, Y. A theory for drag partition over rough surfaces. *J. Geophys. Res.* **2008**, *113*, F02S05. [CrossRef]
124. Lancaster, N. Development of linear dunes in the southwestern Kalahari, southern Africa. *J. Arid Environ.* **1988**, *14*, 233–244.
125. Molina-Aiz, F.D.; Valera, D.L.; Álvarez, A.J.; Madueño, A. A wind tunnel study of airflow through horticultural crops: Determination of the drag coefficient. *Biosyst. Eng.* **2006**, *93*, 447–457. [CrossRef]
126. Udo, K.; Takewaka, S. Experimental study of blown sand in a vegetated area. *J. Coast. Res.* **2007**, *23*, 1175–1182. [CrossRef]
127. Wuyts, K.; Verheyen, K.; De Schrijver, A.; Cornelis, W.M.; Gabriels, D. The impact of forest edge structure on longitudinal patterns of deposition, wind speed, and turbulence. *Atmos. Environ.* **2008**, *42*, 8651–8660. [CrossRef]
128. Sherman, D.; Farrell, E. Aerodynamic roughness lengths over movable beds: Comparison of wind tunnel and field data. *J. Geophys. Res. Earth Surf.* **2008**, *113*. [CrossRef]
129. Li, B.; Neuman, M.C. Boundary-layer turbulence characteristics during aeolian saltation. *Geophys. Res. Lett.* **2012**, *39*, L11402. [CrossRef]
130. Martin, R.L.; Kok, J.F. Linear scaling of wind-driven sand flux with shear stress. *Sci. Adv.* **2017**, 3. [CrossRef] [PubMed]
131. Zhang, N.; Kang, J.-H.; Lee, S.-J. Wind tunnel observation on the effect of a porous wind fence on shelter of saltating sand particles. *Geomorphology* **2010**, *120*, 224–232. [CrossRef]
132. Wiggs, G.F.S.; Thomas, D.S.G.; Bullard, J.E.; Livingstone, I. Dune mobility and vegetation cover in the southwest Kalahari Desert. *Earth Surf. Process. Landf.* **1995**, *20*, 515–529. [CrossRef]
133. Lancaster, N.; Baas, A. Influence of vegetation cover on sand transport by wind: Field studies at Owens Lake, California. *Earth Surf. Process. Landf.* **1998**, *23*, 69–82. [CrossRef]
134. Munson, S.M.; Belnap, J.; Okin, G.S. Responses of wind erosion to climate-induced vegetation changes on the Colorado Plateau. *Proc. Natl. Acad. Sci. USA* **2011**, *108*, 3854–3859. [CrossRef] [PubMed]
135. Webb, N.P.; Herrick, J.E.; Duniway, M.C. Ecological site-based assessments of wind and water erosion: Informing accelerated soil erosion management in rangelands. *Ecol. Appl.* **2014**, *24*, 1405–1420. [CrossRef]
136. Turpin, C.; Badr, T.; Harion, J.-L. Numerical modeling of aeolian erosion over rough surfaces. *Earth Surf. Process. Landf.* **2010**, *35*, 1418–1429. [CrossRef]
137. Chepil, W.S. Utilization of crop residue for wind erosion control. *Sci. Agric.* **1944**, *24*, 307–319.
138. Siddoway, F.H.; Chepil, W.S.; Armbrust, D.V. Effect of kind, and placement of residue on wind erosion control. *Trans. ASAE* **1965**, *8*, 327–331. [CrossRef]
139. Lyles, L.; Allison, B.E. Wind erosion: The protective role of simulated standing stubble. *Trans. ASAE* **1976**, *19*, 61–64. [CrossRef]
140. King, J.; Nickling, W.G.; Gillies, J.A. Aeolian shear stress ratio measurements within mesquite-dominated landscapes of the Chihuahuan Desert, New Mexico, USA. *Geomorphology* **2006**, *82*, 229–244. [CrossRef]
141. Gillies, J.A.; Nickling, W.G.; King, J. Aeolian sediment transport through large patches of roughness in the atmospheric inertial sublayer. *J. Geophys. Res.* **2006**, *111*, F02006. [CrossRef]
142. Gillies, J.A.; Nickling, W.G.; King, J. Shear stress partitioning in large patches of roughness in the atmospheric inertial sublayer. *Bound. Layer Meteorol.* **2007**, *122*, 367–396. [CrossRef]
143. Gillies, J.A.; Nickling, W.G.; King, J.; Lancaster, N. Modeling aeolian sediment transport thresholds on physically rough Martian surfaces: A shear stress partitioning approach. *Geomorphology* **2010**, *121*, 15–21. [CrossRef]
144. Okin, G.S.; Gillette, D.A. Distribution of vegetation in wind-dominated landscapes: Implications for wind erosion modeling and landscape processes. *J. Geophys. Res.* **2001**, *106*, 9673–9683. [CrossRef]
145. Lancaster, N.; Nickling, W.G.; Gillies, J.A. Sand transport by wind on complex surfaces: Field studies in the McMurdo Dry Valleys, Antarctica. *J. Geophys. Res.* **2010**, *115*, F03027. [CrossRef]
146. Webb, N.P.; Okin, G.S.; Brown, S. The effect of roughness elements on wind erosion: The importance of surface shear stress distribution. *J. Geophys. Res. Atmos.* **2014**, *119*, 6066–6084. [CrossRef]

147. Gillies, J.A.; Green, H.L.; McCarley-Holder, G.; Grimm, S.; Howard, C.; Barbieri, N.; Ono, D.; Schade, T. Using solid element roughness to control sand movement: Keeler Dunes, Keeler, California. *Aeolian Res.* **2015**, *18*, 35–46. [CrossRef]
148. Chappell, A.; Webb, N.P. Using albedo to reform wind erosion modelling, mapping and monitoring. *Aeolian Res.* **2016**, *23*, 63–78. [CrossRef]
149. Chen, Q.; Jia, L.; Hutjes, R.; Menenti, M. Estimation of aerodynamic roughness length over oasis in the Heihe River Basin by utilizing remote sensing and ground data. *Remote Sens.* **2015**, *7*, 3690–3709. [CrossRef]
150. Yu, M.; Wu, B.; Yan, N.; Xing, Q.; Zhu, W. A method for estimating the aerodynamic roughness length with NDVI and BRDF signatures using multi-temporal Proba-V data. *Remote Sens.* **2017**, *9*, 6. [CrossRef]
151. Shao, Y.; Nickling, W.; Bergametti, G.; Butler, H.; Chappell, A.; Findlater, P.; Gillies, J.; Ishizuka, M.; Klose, M.; Kok, J.F.; et al. A tribute to Michael R. Raupach for contributions to aeolian fluid dynamics. *Aeolian Res.* **2015**, *19*, 37–54. [CrossRef]
152. Sankey, J.B.; Law, D.J.; Breshears, D.D.; Munson, S.M.; Webb, R.H. Employing lidar to detail vegetation canopy architecture for prediction of aeolian transport. *Geophys. Res. Lett.* **2013**, *40*, 1724–1728. [CrossRef]
153. Herrick, J.E.; Van Zee, J.W.; McCord, S.E.; Courtright, E.M.; Karl, J.W.; Burkett, L.M. *Monitoring Manual for Grassland, Shrubland, and Savanna Ecosystems, Volume 1: Core Methods*, 2nd ed.; USDA-ARS Jornada Experimental Range: Las Cruces, NM, USA, 2017. Available online: http://www.landscapetoolbox.org/manuals/monitoring-manual/ (accessed on 1 January 2017).
154. Jupp, D.L.B.; Culvenor, D.S.; Lovell, J.L.; Newnham, G.J.; Strahler, A.H.; Woodcock, C.E. Estimating forest LAI profiles and structural parameters using a ground-based laser called 'Echidna'. *Tree Physiol.* **2008**, *29*, 171–181. [CrossRef] [PubMed]
155. Karl, J.W.; Duniway, M.C.; Schrader, T.S. A technique for estimating rangeland canopy-gap size distributions from high-resolution digital imagery. *Rangel. Ecol. Manag.* **2011**, *65*, 196–207. [CrossRef]
156. Shao, Y. *Physics and Modelling of Wind Erosion*; Kluwer Academic Publishers: London, UK, 2008.
157. Ito, A.; Kok, J.F. Do dust emissions from sparsely vegetated regions dominate atmospheric iron supply to the Southern Ocean? *J. Geophys. Res. Atmos.* **2017**, *122*, 3987–4002. [CrossRef]
158. Galloza, M.S.; Webb, N.P.; Bleiweiss, M.P.; Winters, C.; Herrick, J.E.; Ayers, E. Resolving dust emission responses to land cover change using an ecological land classification. *Aeolian Res.* **2017**, in press.
159. Greeley, R.; Gaddis, L.; Lancaster, N.; Dobrovolskies, A.; Iversen, J.; Rasmussen, K.; Saunders, S.; van Zyl, J.; Wall, S.; Zebker, H.; et al. Assessment of aerodynamic roughness via airborne radar observations. *Acta Mech.* **1991**, *2*, 77–88.
160. Greeley, R.; Blumberg, D.G.; McHone, J.F.; Dobrovolskis, A.; Iversen, J.D.; Lancaster, N.; Rasmussen, K.R.; Wall, S.D.; White, B.R. Applications of spaceborne radar laboratory data to the study of aeolian processes. *J. Geophys. Res.* **1997**, *102*, 10971–10983. [CrossRef]
161. Marticorena, B.; Chazette, P.; Bergametti, G.; Dulac, F.; Legrand, M. Mapping the aerodynamic roughness length of desert surfaces from the POLDER/ADEOS bidirectional reflectance product. *Int. J. Remote Sens.* **2004**, *25*, 603–626. [CrossRef]
162. Marticorena, B.; Kardous, M.; Bergametti, G.; Callot, Y.; Chazette, P.; Khatteli, H.; Le Hegarat-Mascle, S.; Maille, M.; Rajot, J.-L.; Vidal-Madjar, D.; et al. Surface and aerodynamic roughness in arid and semiarid areas and their relation to radar backscatter coefficient. *J. Geophys. Res.* **2006**, *111*, F03017. [CrossRef]
163. Chappell, A.; Heritage, G.L. Using illumination and shadow to model aerodynamic resistance and flow separation: An isotropic study. *Atmos. Environ.* **2007**, *41*, 5817–5830. [CrossRef]
164. Chappell, A.; Van Pelt, S.; Zobeck, T.; Dong, Z. Estimating aerodynamic resistance of rough surfaces using angular reflectance. *Remote Sens. Environ.* **2010**, *114*, 1462–1470. [CrossRef]
165. Dupont, S.; Bergametti, G.; Marticorena, B.; Simoëns, S. Modeling saltation intermittency. *J. Geophys. Res. Atmos.* **2013**, *118*, 1–20. [CrossRef]
166. Maruyama, T. Large eddy simulation of turbulent flow around a windbreak. *J. Wind Eng. Ind. Aerodyn.* **2008**, *96*, 1998–2006. [CrossRef]
167. Schröttle, J.; Dörnbrack, A. Turbulence structure in a diabatically heated forest canopy composed of fractal Pythagoras trees. *Theor. Comput. Fluid Dyn.* **2012**, *27*, 337–359. [CrossRef]
168. Schwämmle, V.; Herrmann, H.J. A model for barchan dunes including lateral shear stress. *Eur. Phys. J. E* **2005**, *16*, 57–65. [CrossRef] [PubMed]

169. Hermann, H.J.; Duran, O.; Parteli, E.J.R.; Schatz, V. Vegetation and induration as sand dunes stabilizators. *J. Coast. Res.* **2008**, *24*, 1357–1368. [CrossRef]
170. Durán, O.; Hermann, H.J. Vegetation against dune mobility. *Phys. Rev. Lett.* **2006**, *97*, 188001. [CrossRef] [PubMed]
171. Turnbull, L.; Hochstrasser, T.; Wieczorek, M.; Baas, A.; Wainwright, J.; Scarsoglio, S.; Tietjen, B.; Jeltsch, F.; Mueller, E.N. Approaches to modelling ecogeomorphic systems. In *Patterns of Land Degradation in Drylands: Understanding Self-Organised Ecogeomorphic Systems*; Mueller, E.N., Wainwright, J., Parsons, A.J., Turnbull, L., Eds.; Springer: Berlin, Germany, 2013.
172. Narteau, C.; Lajeunesse, E.; Métivier, F.; Rozier, O. Modelling of dune patterns by short range interactions. *River Coast. Estuar. Morphodyn.* **2006**, *2*, 1035–1046.
173. Baas, A.C.W. Complex systems in aeolian geomorphology. *Geomorphology* **2007**, *91*, 311–331. [CrossRef]
174. Nield, J.M.; Baas, A.C.W. The influence of different environmental and climatic conditions on vegetated aeolian dune landscape development and response. *Glob. Planet. Chang.* **2008**, *64*, 76–92. [CrossRef]
175. Peters, D.P.C.; Bestelmeyer, B.T.; Herrick, J.E.; Monger, H.C.; Fredrickson, E.; Havstad, K.M. Disentangling complex landscapes: New insights to forecasting arid and semiarid system dynamics. *BioScience* **2006**, *56*, 491–501. [CrossRef]
176. Okin, G.S.; Parsons, A.J.; Wainwright, J.; Herrick, J.E.; Bestelmeyer, B.T.; Peters, D.C.; Fredrickson, E.L. Do changes in connectivity explain desertification? *Bioscience* **2009**, *59*, 237–244. [CrossRef]
177. Turnbull, L.; Wainwright, J.; Brazier, R.E. A conceptual framework for understanding semi-arid land degradation: Ecohydrological interactions across multiple-space and time scales. *Ecohydrology* **2008**, *1*, 23–34. [CrossRef]
178. Turnbull, L.; Wilcox, B.; Belnap, J. Understanding the role of ecohydrological feedbacks in ecosystem state change in drylands. *Ecohydrology* **2012**, *183*, 174–183. [CrossRef]
179. Yan, N.; Baas, A.C.W. Environmental controls, morphodynamic processes, and ecogeomorphic interactions of barchan to parabolic dune transformations. *Geomorphology* **2017**, *278*, 209–237. [CrossRef]
180. Omasa, K.; Hosoi, F.; Konishi, A. 3D lidar imaging for detecting and understanding plant responses and canopy structure. *J. Exp. Bot.* **2007**, *58*, 881–898. [CrossRef] [PubMed]
181. Cruzan, M.B.; Weinstein, B.G.; Grasty, M.R.; Kohrn, B.F.; Hendrickson, E.C.; Arredondo, T.M.; Thompson, P.G. Small unmanned aerial vehicles (micro-UAVs, drones) in plant ecology. *Appl. Plant Sci.* **2016**, *4*. [CrossRef] [PubMed]
182. Fonstad, M.A.; Marcus, W.A. High resolution, basin extent observations and implications for understanding river form and process. *Earth Surf. Process. Landf.* **2010**, *35*, 680–698. [CrossRef]
183. James, M.R.; Robson, S. Straightforward reconstruction of 3D surfaces and topography with a camera: Accuracy and geoscience application. *J. Geophys. Res. Earth Surf.* **2012**, *117*, F03017. [CrossRef]
184. Westoby, M.J.; Brasington, J.; Glasser, N.F.; Hambrey, M.J.; Reynolds, J.M. "Structure-from-Motion" photogrammetry: A low-cost, effective tool for geoscience applications. *Geomorphology* **2012**, *179*, 300–314. [CrossRef]
185. Fonstad, M.A.; Dietrich, J.T.; Courville, B.C.; Jensen, J.L.; Carbonneau, P.E. Topographic structure from motion: A new development in photogrammetric measurement. *Earth Surf. Process. Landf.* **2013**, *38*, 421–430. [CrossRef]
186. Nield, J.M.; King, J.; Wiggs, G.F.S.; Leyland, J.; Bryant, R.G. Estimating aerodynamic roughness over complex surface terrain. *J. Geophys. Res. Atmos.* **2013**, *118*, 12948–12961. [CrossRef]

land

MDPI

Article

The Potential for Enhanced Water Decoupling in the Jordan Basin through Regional Agricultural Best Practice

Michael Gilmont [1,*], Lara Nassar [2], Steve Rayner [1], Nadav Tal [3], Erica Harper [2] and Hilmi S. Salem [4]

1 Institute for Science, Innovation and Society, University of Oxford, 64 Banbury Road, Oxford OX2 6PN, UK; steve.rayner@insis.ox.ac.uk
2 West Asia North Africa (WANA) Institute, Royal Scientific Society, P.O. Box 1438, Ahmad Al-Tarawneh Street 70, Jubeiha, Amman 11941, Jordan; lara.nassar@wana.jo (L.N.); erica.harper@wana.jo (E.H.)
3 EcoPeace Middle East, 90 Begin Rd, Tel Aviv 67138, Israel; nadav@ecopeaceme.org
4 (formerly of) Louise Building, Ras Al Tahouneh St, Al Bireh, Palestine; hilmisalem@yahoo.com
* Correspondence: mgilmont@gmail.com; Tel.: +44-1865278818

Received: 15 March 2018; Accepted: 9 May 2018; Published: 13 May 2018

Abstract: This paper examines the differences in agricultural water application per crop ton output in semi-arid jurisdictions in the Jordan Basin, focusing on Israel and Jordan, with some analysis relevant to Palestine. In order to understand differences in water application, it delivers a nationally averaged assessment of applied water application for 14 key regional crops, with most cases suggesting Israeli best practice in water application per unit crop. The paper draws on a secondary assessment of agricultural water intensity and primary data from farmer interviews to demonstrate differences in applied water productivity and agricultural context. The analysis suggests a conservative estimate that a difference of 168 Million Cubic Meters (MCM)/yr (33% of agriculture and 18% of national total) exists in terms of water application for a given crop production in Jordan when compared with Israel. The paper then proposes further work required to establish how differences in water application might translate into differences in agricultural water productivity, and thereby potential water savings that might enable growth of production within current agricultural allocations, allowing new future resources to be allocated to other economic and social needs. The paper also delivers a preliminary analysis of the political and institutional landscape for implementation, assessing the challenges of institutional silos and overlap that some policy stakeholders see as hindering cross-sectoral progress. The paper concludes by examining the limitations of the analysis, and it proposes future work to deepen the robustness of results and examines some of the challenges facing improved agricultural water productivity and changing farm behaviour in the region.

Keywords: Jordan River Basin; water productivity; Jordan; Israel; Palestine; agriculture; agricultural water intensity; decoupling; water security; institutional change

1. Introduction

Jordan, Israel, and Palestine are three of the most water-scarce jurisdictions in the world. All three have seen significant growth in recent decades in both population and economic activity, despite their limited water resources. These trends have been achieved through multiple mechanisms to circumvent their water limits, with the effect of 'decoupling' growth from national water needs (Gilmont, 2014) [1]. Israel in particular has demonstrated that four mechanisms, comprising diversification, food imports, agricultural water productivity, and non-natural water supplies (desalination and wastewater reuse), have enabled it to experience significant economic, population, and agricultural growth, while recently starting to redress decline in the environmental condition of freshwater resources. This paper focuses on agricultural water productivity as a mechanism for decoupling, and seeks to take a first step in understanding the potential savings that could be achieved in the region if agricultural best practice is targeted. For most examined crops, it is suggested that savings might accrue to Jordan if it moved towards Israeli levels of agricultural water productivity, although two key instances are highlighted for further research where the opposite might be true. This paper takes a first step towards understanding differences in water productivity, by looking at water application and yield. It deploys both a national secondary data assessment of agricultural water application, and farm-level primary interviews to understand farm-level detail on water application and yield on both sides of the Jordan River, including insights into the differences in the socio-economic context of agriculture. An initial assessment of the volumetric differences in water application between Israel and Jordan is made. The paper then briefly examines similarities and differences within basin jurisdictions in an agricultural context, before focusing on the institutional and political landscape of how improved water productivity might be deployed in Jordan, highlighting a number of policy barriers and some stakeholder-proposed solutions. The final section of the paper evaluates the methodology, and proposes future research to deepen both the understanding and robustness around the potential for enhanced agricultural decoupling through the assessment of differences in water consumption and the likely issues of return flows, as well as to contribute to the institutional challenges highlighted.

Context and Need

Israel is regarded as having one of the most water productive agricultural sectors in the world, owing to both water and crop technologies, cropping patterns, and crop choice (World Bank 2006) [2]. As a result of its limited available freshwater resources and growing population, over time, water has been allocated away from agriculture to other sectors, especially domestic uses (Teschner and Negev, 2013) [3]. This reallocation has occurred both steadily over time, and more dramatically during drought years, where agriculture has been cut faster and more severely than other sectors. Despite receiving less water following a peak in agricultural allocation in 1985, and with an increasing proportion of allocation comprised of treated wastewater, agricultural output tonnage has increased steadily (Figure 1).

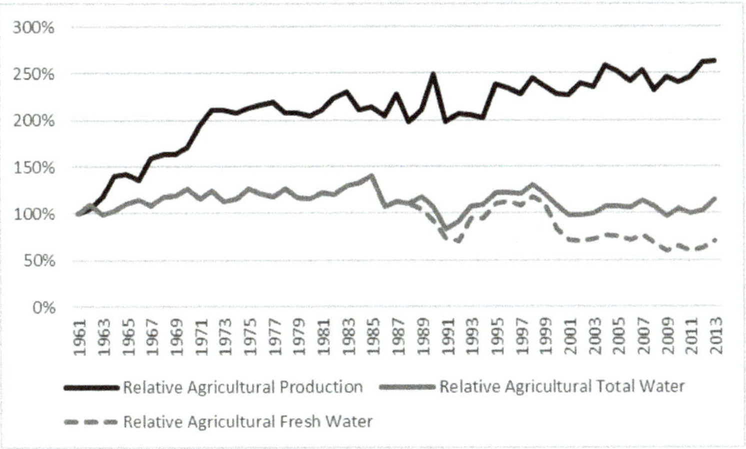

Figure 1. Agricultural production and relative water use in Israel, 1961–2013.

Jordan has not undertaken the same level of technical and managerial investment and innovation as its neighbour, but has seen considerable growth in agricultural outputs since the early 1990s, despite static or reduced water availability for agriculture (Figure 2). Despite these gains, Jordan's agricultural sector is regarded as having significant inefficiencies in water productivity (EcoPeace, 2013) [4]. Nationally, however, because of significant rain-fed production—especially of olives—agricultural production per unit water applied across the entire output appears to be higher in Jordan than in Israel. Based on 2016 Food and Agricultural Organization (FAO) food balance [5] and national water statistics, Jordan's total agricultural applied water productivity (agricultural water allocation per FAO recorded production tonne) was 194 m^3/t, while Israel's was 267 m^3/t. This data, however, excludes considerable illegal abstraction of agriculture water in Jordan, as well as conflating rain-fed and irrigated production. As the empirical work below will demonstrate, the aggregate numbers mask considerable differences in water application and yield for certain crops.

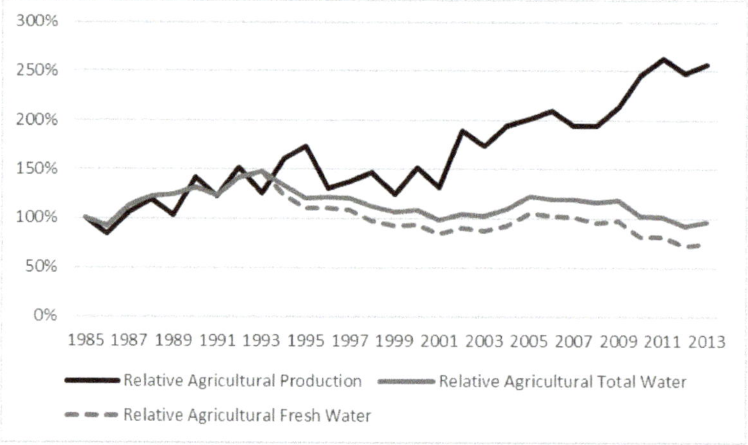

Figure 2. Agricultural production and relative water use in Jordan, 1985–2013.

Since 2012, Jordan has experienced additional pressure on its water resources due to its hosting of refugees from Syria. This has exacerbated its already extreme water scarcity, with available total supplies (across all sectors) dropping from 270 m^3/cap/yr in 1988 to 106 m^3/cap/yr in 2015 (based on Ministry of Water data and official population statistics). The increase in people within the territory required that the Ministry of Water and Irrigation rework their 2008–2022 water master plan 'Water for Life' (Ministry of Water and Irrigation (MWI) 2008) [6]. The resulting 2016–2025 plan (MWI, 2016) [7] accounts for increased demand due to Syrian refugees, and new supplies coming online from groundwater sources and the planned 'Red Sea Dead Sea' Conveyance. The proposed scheme, currently under tendering as of 2018, involves the production of desalinated water from the Red Sea, with brine discharged into the Dead Sea, helping to slow the decline in water level of the Dead Sea. The fresh desalinated water will be supplied directly to Jordan, and will form part of a water swap with Israel, and also through Israel to the Palestinians using natural water supplies from the Sea of Galilee (Lake Tiberias) and desalinated resources from the Mediterranean Sea (World Bank 2014) [8]. Increased domestic water use in Jordan, combined with new infrastructure capacity, will increase the availability of treated wastewater in agriculture (Seder and Abdel-Jabbar, 2011) [9], enabling reallocation of freshwater to the domestic and industrial sectors.

Despite these planned additional resources, there is still a significant deficit between anticipated demand by 2025 (Yorke, 2016) [10]. Figure 3 uses historic data from 1986 to 2015, combined with the authors' extrapolation of demand from present usage levels to meet with MWI's forecast demand by 2025, plotted with MWI's annual projection of available water. The MWI provides data for available supply and anticipated demand, but does not project where the shortfall will impact, or how new supplies will be distributed across sectors. As a result, there is no attempt by the 2025 plan authors to forecast how newly available water will be allocated across demands under conditions of supply deficit. In the absence of further data, our analysis here assumes a linear growth in demand from actual current to theoretical future demands. A significant deficit between available water and cross-sectoral demand exists in 2025, based on current resource projections. Given the gap between future supply and projected demand, this paper considers how a reduction in agricultural water demand, based on moving towards regional best practice in agricultural water productivity, could enable enhanced agricultural water decoupling, and thereby ameliorate Jordan's future water resource deficit.

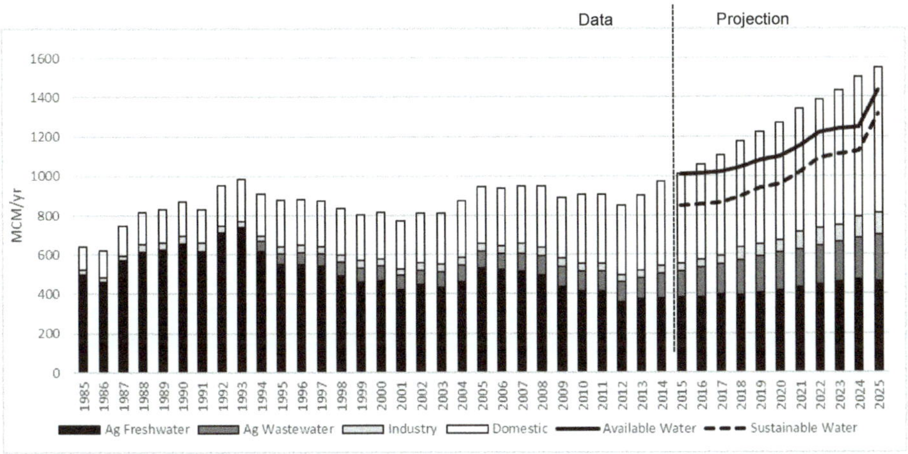

Figure 3. Projection of Jordan's water needs, based on Ministry of Water and Irrigation (MWI) 2016 demand and supply data in Million Cubic Meters (MCM)/yr. Adapted from Gilmont et al. (2017) [11].

This paper takes a first step towards establishing a potential reduction in agricultural water requirements in Jordan (or enabling growth within current water resource allocations), by examining differences in water application per unit crop output across the Jordan Basin. The analysis is focused on 14 crops that together comprise over 90% of production tonnage and water application in both Israel and Jordan, and 86% of Jordan's accounted water application as of 2010. Where data is available, the analysis also includes an assessment of agricultural water productivity within Palestine (considered in this paper as Palestinian-controlled/operated farms within the West Bank and Gaza Strip).

Two methods of assessments were pursued: The first to obtain an approximate aggregate national picture of relative water application per unit crop output; and the second to compare trends at the farm level. The two methods were also intended to verify the directions in differences of national level trends, and to provide an initial assessment of uncertainty in any trends being identified.

2. Methodology

The first methodology involved a comparison of agricultural yield per unit water application based on national average numbers, as assessed on a national level using aggregated country-specific yield data. Much work on crop water fluxes has been carried out to date at a regional scale, using large gridded, physically-based models (e.g., Hoff et al. 2010 [12]). It is recognised that this methodology ignores variations in crop water requirements in different regions of the countries, predominantly because of variations in topography, rainfall depth, soil conditions, and humidity. The methodology is adopted, however, as there is no long-term breakdown of annual production at the sub-national scale, and consequently no reliable means to weight national production by different hydro-climatic conditions. The first stage of the analysis involved deriving agricultural water productivity numbers from secondary sources and combining them with annual national tonnages.

For Israel, extension service data was assembled for 52 crops for all regions where the crops were grown, and yield per cubic meter of applied water was calculated for each region, using the most up-to-date value available from 2000 to 2015 (Ministry of Agriculture, 2016 [13]). The regional values were then averaged (using the arithmetic mean). As a result of the lack of regional crop distribution, no weighting could be applied to this calculation to capture regional bias in crop growth location. The average water intensity was then multiplied by reported national annual production data from the Israeli Central Bureau of Statistics (personal communication). Crops not accounted for in the calculations included ornamental flowers (water data unavailable) and feed crops (not irrigated). See Appendix A for water intensity calculations for most crop types.

For Jordan, agricultural water intensity was derived through a similar method, but using total water application and area data from International Resources Group (IRG) and Karablieh (2012) (Tables 23, 25, 27, and 29) [14], whose data is based on an analysis of cropping carried out in 2010, combined with official production tonnages. Fifty-five crops were available from this study in categories appropriate for reported national crop tonnages (See Appendix A for water intensity calculations). Olives, however, are produced under both irrigated and rain-fed conditions, and calculations on crop per unit water required isolating the irrigated production only. The IRG and Karablieh (2012) publication included its own assessment of rain-fed versus irrigated production proportions for 2010, and this proportional split was used to attribute an annual production proportion to irrigation. As with Israel, a national average water application and yield number was calculated, which, for Jordan, involved combining environmental diverse agricultural environments spanning the Jordan Valley and highlands. Again, this was necessary as annual tonnage reports do not attribute production to the different regions. No sector-wide secondary data on crop yield and water application was identified for Palestine.

In order to validate the national crop level numbers and provide an effective comparison between Israel and Jordan, especially given the impact of national averaging of the water numbers, the water application numbers were combined with historic agricultural tonnage. The results are shown in Figures 4 and 5. Figure 4, for Israel, shows significant under-estimation of agricultural water application, when derived from crop water needs multiplied by annual tonnage, by over 40%. A slight increase in efficiency is seen based on the simulations, with simulated agricultural water use growing faster than reported agricultural water use (gradient y = 6.83x vs. y = 3.4x).

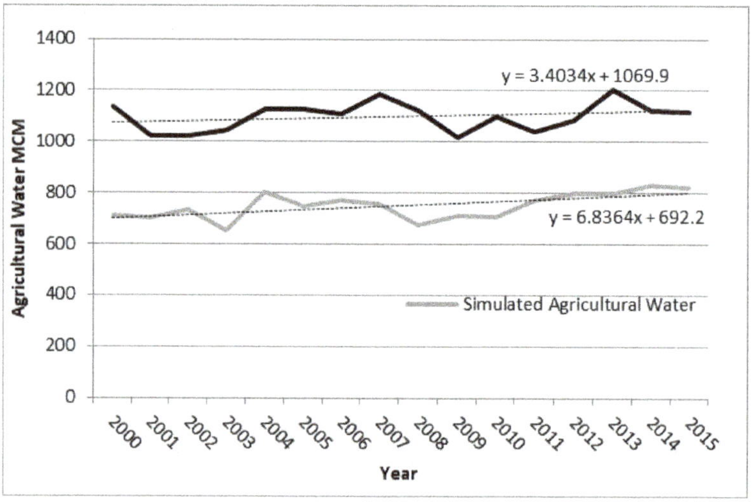

Figure 4. Plot of water data (Central Bureau of Statistics Agricultural Allocation) and simulated agricultural water (crop tonnage x water per ton) for the period 2000–2015 for Israel.

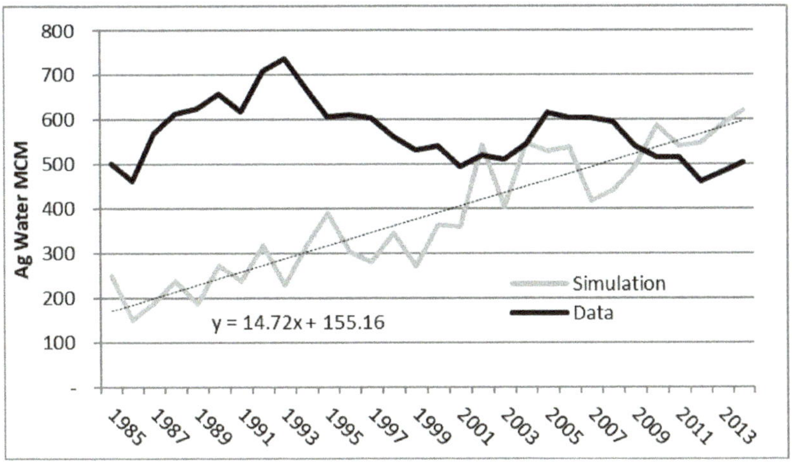

Figure 5. Plot of water data (Ministry of Water and Irrigation) and simulated agricultural water (crop tonnage x water per ton) for the period 1985–2014 for Jordan.

For Jordan, there is a strong efficiency gradient in the simulated value, indicated by a lower simulated water application historically (with static benchmarking, smaller historic yields would result in simulation of lower historic water use, whereas in reality, the historic water application would be higher for a given yield). Importantly, for the benchmark year of 2010 of the agricultural water data, there is a tendency towards convergence. For the year 2010, the simulation over-estimates agricultural water application by 72 MCM. This value is close to the 110 MCM of illegal groundwater abstraction that has been eliminated (Obeidat, 2016) [15]. Therefore, simulation over-estimation for the benchmark year should be expected given that reported tonnage includes all production, regardless of legality of water source.

The accountable discrepancy between simulated and official data in Jordan means that the average water application numbers are accepted for the purposes of this study. For Israel, there is a clear and unexplainable under-estimation of agricultural water application, which will in part be because of national averaging used in the methodology. Making allowances for crops not accounted for, (including cut flowers) and for water allocated to livestock and non-irrigated agricultural purposes, the agricultural water application numbers for Israel were inflated by 42% to ensure a highly conservative comparison between Israeli and Jordanian water application. This percentage was determined through a three-year moving-window average comparison between the simulated and official agricultural data. Across the entire 16-year time series, the discrepancy was 48% between the simulated and official agricultural allocation. However, focusing on the final five years of data (i.e., allowing for slight efficiency gradient), this moving average discrepancy was 42%. The simulated figures were thus inflated by 42%, resulting in the smallest sum of differences between simulated and official agricultural water over the 2009–2013 period. From the national simulations in Israel and Jordan, the basket of 14 crops (Table 1) was selected as the focus of deeper water analysis, and the identification of comparative trends across the region.

Table 1. Comparative water productivity and potential differences in water application across 14 Crops. Adapted from Gilmont et al. (2017) [11].

	Jordan Water Application MCM (2010)	Jordan Average Tonnage '000s (2009–2014)	Jordan Water/Crop Ton	Jordan Average Water Application 2009–2014 MCM/Year	Israel Water/Crop Ton @142% Inflation	% of Present Scenario	Water Application under New Scenario (MCM)	Difference in Water Application/Year for Jordanian Production Tonnages
Clover	84.0	190.1	376	71.41	Rain-fed	0%	0	71.4
Olives	167.1	110.5	1,627	179.75	1278	79%	141.2	38.6
Tomatoes	74.1	753.6	100	75.73	75	75%	56.3	19.4
Banana	26.6	42.4	609	25.81	449	74%	19.0	6.8
Apples	15.0	36.1	522	18.82	187	36%	6.7	12.1
Dates	24.6	10.7	2,187	23.44	1879	86%	20.1	3.3
Watermelon	14.8	117.6	96	11.33	81	84%	9.5	1.8
Grapes	14.6	34.7	491	17.02	426	87%	14.8	2.3
Wheat	3.5	21.6	160	3.46	57	35%	1.2	2.2
Onion, dry	4.2	27.0	265	7.14	185	70%	5.0	2.2
Citrus	35.1	108.7	293	31.89	227	77%	24.7	7.2
Eggplant	11.7	106.6	111	11.87	99	89%	10.6	1.3
Potato	25.2	159.8	144	23.00	189	132%	30.3	-
Cucumber	7.1	191.4	41	7.75	68	168%	13.0	-
Total	507.5	1910.7		508.4		352.5		**168.4**

A second assessment methodology aimed to verify the trends indicated by the methodology above. This involved gathering primary data from farm-level structured interviews carried out in the three study jurisdictions during late 2016. Farmer-based, semi-structured interviews have recently been used successfully by O'Keefe et al. (2016) [16] to understand farmer irrigation practices. Our interviews were carried out in summer 2016, comprising 19 respondents in Jordan, focusing on the North and Central Jordan Valley, along with farms near Amman; 12 farm interviews in Israel, focusing on the centre and north; and 27 in Palestine (comprising 22 interviews in the West Bank and 5 in the Gaza Strip). This farm-level analysis was compared with secondary data to confirm whether or not the trends seen in the secondary analysis were correct, as well as to identify variations in performance between individual farms. The farm interview questions elicited information of crop areas, yield per dunam (1 dunam = 1000 m^2, or 0.1 ha), and water application per dunam per year or growing season (then converted to annual numbers). Information on farm gate crop prices were also elicited, along with sources farmers used for agricultural production, irrigation methods used, and the role of agriculture in the farmers' economic lives.

A further series of semi-structured interviews with eight policy stakeholders and research scholars in Jordan was carried out in August 2017, reflecting on the initial empirical findings and exploring the institutional settings and opportunities for change along the lines identified through the study. This forms a final element of research on institutional barriers and opportunities for improved water productivity. All interviews were approved by the ethical committee of the School of Anthropology and Museum Ethnography Research Ethics Committee, University of Oxford, under references SAME_C1A_16_050 (for 2016 farmer interviews) and SAME_C1A_17_029 (for 2017 policy interviews), and carried out under the procedures laid out by the researchers and approved by the committee.

3. Results

This section reviews the results from the secondary and primary data, demonstrating the differences in water application per ton of production between Israel and Jordan. The farm-level primary analysis also illustrates the uncertainty associated with the direction of trends.

3.1. Secondary Agriultural Water Data Analysis

Table 1 shows the results of the secondary analysis for the 14 crops. For the 508 MCM (average 2009–2013) accounted for by the 14 crops selected in Jordan, agricultural water application may be reduced by up to 168 MCM/yr if Israeli trends in water application and crop yield were targeted. Key instances of lower water application appear in tomatoes, apples, clover, and olives. Potatoes and cucumber production in Jordan appear to be more water productive than in Israel, although in the case of potatoes, this is likely entirely due to the adjustment of the Israeli agricultural water figures discussed above.

Clover production is not irrigated in Israel as this is not considered an economically effective use of water. Feed crops are thus only grown in areas with sufficient rainfall, and a large majority of national needs are imported (Gilmont, 2014) [1]. Meanwhile, the data for Jordan suggests that the country devotes around 71 MCM/yr to irrigate clover for animal feed. One of the farmers interviewed indicated that much livestock feed irrigation is carried out by livestock farmers themselves to avoid purchasing feedstocks on the open market. While interview data suggests that some of this irrigation is carried out with treated wastewater, even this resource has an opportunity cost associated with it, especially in its potential use on tree crops. Review of assessments of agricultural land in Jordan suggests that sufficient rain-fed area theoretically exist nationally to support more rain-fed agriculture, including feed crops. Boller et al. (2005) [17] observed that *Trifolium Clypeatum* is grown near the Dead Sea in Israel in areas of less than 300 mm/yr rainfall. Al-Jaloudy (2006) [18] notes that 5.9% of Jordan's land area receives 200–300 mm of rainfall, the 'marginal zone' just above the threshold for aridity. Meanwhile, the total cultivated area in Jordan (combined rain-fed and irrigated lands) comprises only 3% of the national area [18]. This discrepancy suggests that further research is necessary into potential suitable rain-fed areas that could support feedstock, although it is likely that these areas would not be proximate to livestock farmers. Reducing or eliminating clover irrigation through either rain-fed production or assistance through enhanced import substitution would likely require structural support to assist farmers with purchasing feedstocks, and to ensure that supply and affordability does not place undue strain on livelihoods. Changes in currently irrigated feedcrops would, however, appear to be an important avenue to consider in enhanced agricultural water decoupling in Jordan. Savings of 71 MCM from such changes to rain-fed feedstock production would require significant regulatory power if these resources are to be mobilised outside the agriculture sector, as farmers would likely seek to grow other crops with their water and land resources. However, even this recapture of 'saved' water would assist decoupling in providing opportunities for support for the cultivation of crops that might enhance national food security or income in the agricultural sector.

3.2. Primary Interview Agricultural Water Data Analysis

The primary data collection included data on farm-level information on water application and yield. As per the secondary data, these were averaged across the sample, not to assemble a national average, but rather to identify a general direction of comparison between crops' water productivity in Israel and Jordan, as well as provide insights into behaviour in Palestine.

The results are shown in Table 2, while Figures 6 and 7 show the ranges (average, maxima, and minima) for yield and water application per hectare across the farm-level sample for the 14 crops investigated. Table 2 also includes extension service data for Israel for the same regions as the farm-level data, enabling comparison between farm-level results with recommended best practices.

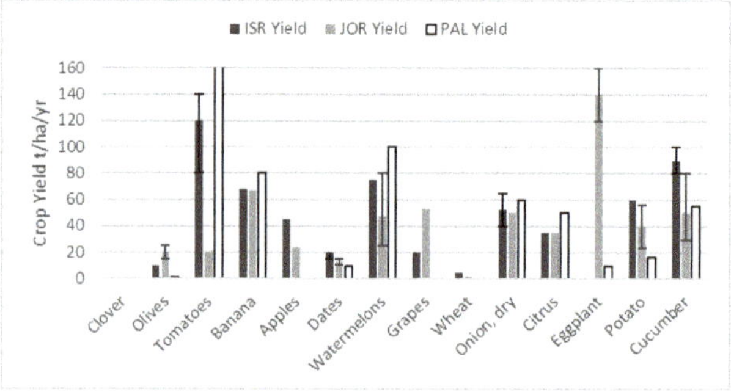

Figure 6. Crop yield per unit area comparison between Israeli, Jordanian, and Palestinian farmers by crop. Adapted from Gilmont et al. 2017 [11]. ISR—Israeli; JOR—Jordanian; PAL—Palestinian.

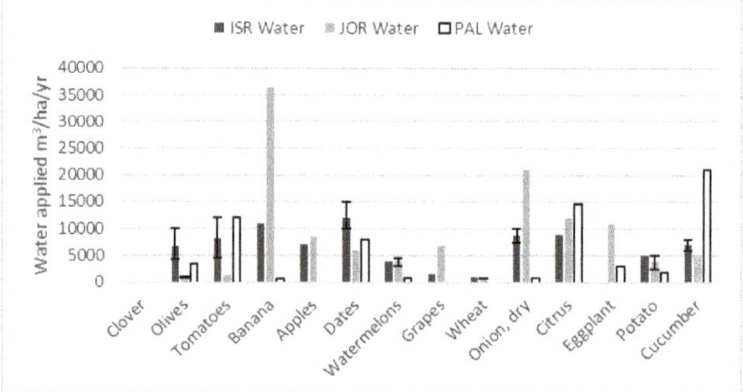

Figure 7. Water application per area comparison between Israeli, Jordanian, and Palestinian farmers by crop. Adapted from Gilmont et al. 2017 [11].

Table 2. Water application per tonne yield between secondary and farmer interview data, and between Israeli, Jordanian, and Palestinian cases.

	Israel Secondary Data as Calculated (m^3/t)	Extension Regional Data (m^3/t)	Farmer Interview (m^3/t)	Location	Farmer as% Secondary Data	Jordan Secondary Data (m^3/t)	Jordan Farmer (m^3/t)	Location	Jordan Farmer as% Secondary Data	Jordan Farmer as % Israel Farmer	Palestinian Farmer (m^3/t)	Location	Palestinian Farmer as % Israel Farmer
Clover	Rain-fed	-	-		-	375.58	No Data		-	-	No Data		-
Olives	900	355	675	Average	75%	1627.00	50.2	JV	3%	7%	3500	Jenin	519%
Tomatoes	52.6	38	38	Jordan Valley	72%	100.5	61.2	JV & North	61%	161%	43	Tubas	113%
Banana	316.1	204.7	217	Jordan Valley	69%	609.03	544	JV	89%	251%	8.75	Jericho	4%
Apples	131.5	153	156	North	119%	521.59	354	Highlands	68%	227%	ND		-
Dates	1323.5	1200	767	Jordan Valley	58%	2186.89	486	JV	22%	63%	889	Jericho	116%
Watermelons	57.1	45	150	Jordan Valley	263%	96.33	96	Average	100%	64%	6	Jericho	4%
Grapes	300	453	75	Jordan Valley	25%	491.06	130	Average	26%	173%	ND		-
Wheat	40	88	200	Jordan Valley	500%	160.04	350	Highlands	219%	175%	ND		-
Onion, dry	130	88	92	Jordan Valley	71%	264.77	422	JV	159%	459%	14	Tubas	15%
Citrus	160	161	203	Average	127%	293.45	342.9	JV	121%	169%	292	Gaza	144%
Eggplant	70		ND			111.31	143	JV	128%		700	Jericho	-
Potato	133.3	139	83	Centre	62%	143.9	125	Average	87%	151%	42	Tubas	51%
Cucumber	48		77	Jordan Valley	160%	40.51	113	JV	279%	147%	382	Jenin	496%

The water application values derived from farm interviews are generally lower than those suggested by the national average data. For Israel, this is likely in part because of the artificial adjustment of the numbers used in the calculation as discussed above. However, in the case of dates, potatoes, and grapes, farmer reported data is well below the uncorrected, non-averaged extension service data for the specific region in which interviews were carried out. The general direction of the evidence, however, suggests that Israeli farmers have lower application per unit yield when compared with neighbouring farmers. For Jordan, the farm-level results indicate that there has likely been some improvements in either water application and/or yield since the 2010 fieldwork of IRG and Karablieh (2012) [14]. Particular region disparity is noted for onions and bananas, which appear to have twice the water application in Jordan than in Israeli, for similar reported yield. It is possible that this result is skewed by only having one farm sample for each of these crops in Jordan, as a result of limited resources available for this study to date.

For tomatoes, banana, apples, wheat, onion, and citrus, Israel demonstrates, as with the secondary data, lower water applications than Jordan. For other crops, farm-level evidence suggests that Jordan may exhibit lower water application than Israel, especially in the case of olives and dates, where water application in Jordan is much lower. Follow-up questioning with farmers indicated that, while the fieldwork had isolated producers of irrigated crops, farmers were actually practicing rain-fed cultivation with supplementary irrigation. Yield data for dates is similar for Israel and Jordan (Figure 6), but water application is much lower in Jordan (Figure 7). In follow-up discussions with the Israeli agricultural extension service, it was confirmed that supplemental irrigation of rain-fed date and olive production in the North Jordan valley was possible, but according to their experience, was not profitable. Different species may play a role here, especially in the case of olives, where Jordanian experts suggested that much Israeli production was derived from more water-intensive species. Clover could not be tested at the farm level, as we were unable to access clover-producing farmers; the agricultural extension service, however, confirmed this as a rain-fed crop in Israel.

Analysis of Palestinian water productivity results finds significant potential for lower water application in olives, citrus, and cucumbers, when compared with Israel. For cucumbers, the farmer interview data suggests that water application is very high when compared with both Israel and Jordan. For citrus, while water application is high, reported yield is also higher than its neighbours, possibly because of limited land availability in the Gaza strip. By contrast, reported olive yield is low, while water application per hectare is comparable to the lowest water values reported for Israel, equating to a much higher water/crop ton. Other Palestinian crops appear to be far more water productive than their Israeli neighbours, although some results seem unusually low, especially for bananas, watermelons, and onions. The result for the banana crop does raise questions pertaining to the reliability of the data, as the single Palestinian banana farm surveyed was in an area receiving around 230 mm/yr rainfall, well below the depth required for rain-fed banana cultivation. The farmer also noted a significant decline in banana plantations in recent years due to the water requirements of the crop. It is suggested that it is likely that the reported yield and water data for bananas are not accurate. The results for olives, also when combined with supplemental irrigation practices, are also of questionable reliability.

While the interview results show some conflicting results with the secondary data, and are cause for concern over the reliability of the Palestinian data, there are some important implications to be drawn from the results. Importantly, for six crops, the direction of differences in water application are the same in both primary and second data between Israel and Jordan, thus there would appear to be relative confidence as to the secondary national trends. This represents 65 MCM/yr of secondary data reduced water application that appear reliable (out of the 97 MCM/yr non-clover related reduced application). Therefore, it is suggested that differences in water application due to non-clover related irrigated crops between Israel and Jordan may be between 65 and 97 MCM/yr, when scaled to Jordanian production tonnages, pending further in-depth research (discussed below).

3.3. *Primary Interview Data Analysis of Socio-Economic Farm Context*

This analysis draws on the interview data to understand some of the contextual differences in which farmers in the three jurisdictions operate, and the possible impact this may have on water productivity. The interviews included short-answer questions on farm size, education, income sources, perceptions on the future of farming, irrigation technology, and agricultural information.

Israeli farmers generally operate larger farm areas than Palestinian or Jordanian farmers (Figure 8). This potential provides an economy of scale in both operation and investment at the farm.

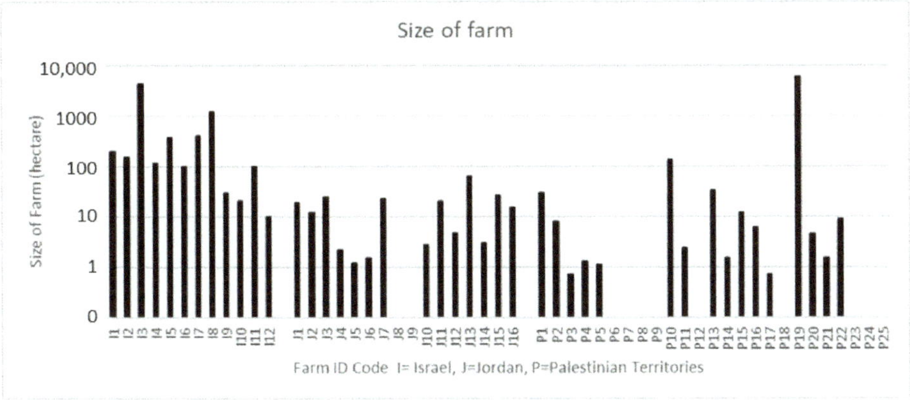

Figure 8. Indication of farm sizes across the three sets of samples (Israel, Jordan, and Palestine). Note that P19 reports a cooperative of multiple farms producing olives and olive oil. Gaps indicate no data given.

Nearly all Israeli farmers interviewed are dependent on agricultural activity for a majority of their income. For most Palestinian farmers, agricultural activities represent their main source of income. For Jordan, around 50% of the farmers interviewed had significant other incomes, including pensions from previous employment and property investments/trade. The presence of alternative income in the Jordanian case can be considered as potentially reducing the importance of increased agricultural returns, especially in the case of 'hobby farms' (Al-Rimawi, 2012) [19]. On the other hand, it is possible that alternative higher value income streams might enable larger investments than would be possible through agricultural returns alone. Investment decisions might also relate to the perceived sustainability of the agricultural industry, which is questioned by reliance on other sources of employment. While nearly all Israeli farmers believed that the next generation would enter farming (though possibly following earlier careers elsewhere), over half the Palestinian and Jordanian farmers surveyed perceived a lack of interest by the next generation, compounded by a lack of governmental promotion of agriculture. For Palestinian farmers, concerns over future water availability, yield reliability, land access, imported products, climate change, and limited donor support for the sector also contributed to the pessimistic outlook.

Farmer-reported adoption of irrigation technology did not produce notable differences between the three jurisdictions, with a mix of drip and sprinkler irrigation reported in nearly all cases. However, field evidence in Jordan suggests that some Jordanian 'drip' irrigation is, in the case of fruit trees, actually basin trickle irrigation, without the close application of water directly to the soil used by Israeli farmers. A difference in ground cover was also noted for citrus, with Israeli groves being mulched, while some Jordanian citrus groves were observed as having grass or scrub cover, likely consuming additional irrigation or rainfall moisture. For banana crops, the use of netting to change local canopy humidity is near-universal in Israel, but is only occasionally adopted in Jordan. Palestinian farmers raised an important observation about irrigation timing, highlighting that intermittent supply to many Palestinian users meant that farmers had to apply water to crops when it was available, rather than when the crops could optimally use it or needed it.

The role and use of government-sponsored agricultural extension support was also found to vary. Israeli farmers are well-supported by, and strong users of, the Ministry of Agriculture Extension service, which possesses field and research branches in each of the key agro-climatic regions. Palestinian farmers are also well served by the 300 extension service field staff, and farmers engaged with their advice. The Jordanian extension service, by contrast, suffers from lack of capacity, with a mere 54 field staff for the entire country. This limits the available advice and opportunities for engagement, and the depth of region/crop specific timely advice that can be provided. In the opinion of some farmers, certain extension agents lacked proactivity and indeed, some newly graduated extension officers lacked experience vis-a-vis the farmers they advise. Five farmers in the Jordanian sample reported actively avoiding advice from the extension service due to past poor experience.

The above analysis indicates that there are notable, although uncertain, differences in agricultural water productivity between Israeli, Palestinian, and Jordanian farmers, as well as socio-economic context. For Jordan, the potential difference in water productivity has the ability to make an important contribution to its future national water security. To this end, the next section discussed a further round of interviews that examined the potential and barriers to improve water productivity uptake in Jordan.

3.4. Interview-Based Analysis of Institutional and Policy Barriers and Opportunities to Improved Water Productivity

A further stage of this research involved discussing findings with policy practitioners and high-level stakeholders to elicit attitudes towards water productivity as a key instrument in Jordan's water future. Such research was not carried out in Palestine as initial data deficits obstructed a meaningful discussion to this end. Jordanian interviewees comprised individuals from within the water and agricultural bureaucracy, together with other academic and professional experts, and eight interviews were carried out during August 2017. The interviews covered a range of issues relating to future water challenges in Jordan, and the potential for new efforts in agricultural productivity to play a role in future resource management. The discussions also examined the institutional foundations of water and agricultural management with regard to improved water productivity.

Key water challenges that were highlighted included climate change and its potential role in reducing both rainfall and available water supply, as well as in increasing water demand by crops due to higher temperatures and CO_2 concentrations (noted by two interviewees). Domestic water allocations were also anticipated to increase in the coming years, as a result of both population growth and a stated government intention to increase per capita supply and supply reliability (noted by two interviewees). Three interviewees observed that the challenges are compounded by the Syrian refugee crisis and the increased pressure it has create in host communities. The refugee pressures especially have generated urgency to increase water supply capacity. Five interviewees highlighted that the resultant impetus to increase the use of treated wastewater in agriculture will also require improvements in wastewater quality to enable wider use of the resource without, as one interviewee stressed, a build-up of salt in the soil.

The interviews revealed a significant tension in perceptions over the role of improving agricultural water productivity as a contributing mechanism to assist in adjusting future water allocations. Three interviewees supported the potential of improved water productivity, and one provided an example of high-tech hydroponics near Mafraq in Northern Jordan, where the production of (predominantly export) tomatoes might rank as the most water productive in the region. This case demonstrates what can be achieved with high levels of investment and a guaranteed market, as well as the internal best practice learning that might occur across Jordan. Jordan's agricultural risk fund, while no longer providing explicit support for water-related investments, provides insurance for crops grown in appropriate regions, thereby encouraging improved sectoral resilience to environmental shocks. This indirect incentive to adhere to agricultural zoning should result in more appropriate use of water and greater returns per unit application. Furthermore, a Jordanian agricultural and water researcher observed that, with finite water available for agriculture, increased output in the future will necessitate improved productivity of water within agriculture. He noted further, however, that this needed to be part of a holistic cross-sectoral strategy, and would take 10–15 years to achieve, giving time and support for farmers to adapt.

Against the observed potential, a number of obstacles were raised. In particular, that poor agricultural performance is largely a result of ingrained behaviour by farmers. Indeed, an official within the water bureaucracy noted that force of habit and resistance to change is a key contributor to low water productivity. A forceful argument put forward by a Jordan Valley agricultural water representative highlights that current agricultural allocations to the Valley were around 60% of those mandated by policy. It would, therefore, be politically unacceptable to cut allocations further and force reduced water application in the Jordan Valley without impeding crop production. Therefore, it is argued that current performance indicates that tremendous efficiency is being achieved. By contrast, the Jordan highlands might have much greater improvements in productivity potential. This argument in the Jordan Valley, however, counters the empirical evidence presented above, that when compared with the environmentally similar context of Israel, it is possible to reduce water application per unit crop output. The argument also runs counter to the arguments by professionals discussed above, that improved productivity will have to feature in the future.

Existing regulatory arrangements and operational procedures were highlighted as a further obstacle to knowledge-led innovation. Water management and crop-choice decisions are embedded in complex webs of competing institutional commitments and customary expectations, echoing the findings of Lach et al. (2005) [20] and Rayner et al. (2005) [21]. Incorporating new knowledge is far from straightforward. Even existing rules and regulations may be hard to enforce in the face of entrenched agricultural practices. The case of bananas was cited as a particular example. Partly because of their water intensity, bananas are not zoned to be grown anywhere in Jordan. However, a senior official within the Jordanian hydrocracy argued that it is impossible, even undesirable, to completely stop their production because of the need to supply local markets and to maintain plant diversity within the Jordan Valley. The overall picture of water productivity and the policies that guide it presents a significant tension between those who argue that improved productivity is possible, and those who say Jordanian agriculture is already performing beyond its expected capabilities in water productivity.

The tension between attitudes towards potential to reduce agricultural water application and improved water productivity nests within a wider political challenge of structures laying out the rules and operations of water and agriculture in Jordan. This can be expressed as the institutional (rules) and organisational (actors) structures (as posited by the North, 1994 [22]) determining how resources are allocated. Raising tariffs for agricultural water might be a potential mechanism to incentivise investment and behavioural change at the farm level, however such a move would be politically challenging both to the organisational structure and its personnel, because of the social resistance that would be encountered. Socio-economic implications of change are also a barrier, with one interviewee highlighting that 10% of Jordan's prison population are indebted farmers. There is a strong aversion to increasing these impacts as a cost of changes in agriculture and practices. To overcome these barriers, representatives of the agricultural bureaucracy argued for solutions that are socially practical, in order to overcome resistance and enforce existing regulation.

A further barrier to improving on-farm water behaviour relates to coordination and responsibilities between relevant ministries and agencies. Five interviewees highlighted the need for improved coordination (a de-siloing) between water and agricultural policy, echoing findings by Al-Zu'bi (2016) [23] regarding climate change governance. In particular, it was noted that the Jordanian Ministry of Water and Irrigation's responsibilities for water end at the 'farm gate'; on-farm usage, regulation, and improvement of water productivity falls under the remit of the Ministry of Agriculture. Attention was drawn to an imbalance in attention between the water and agricultural sectors concerning water productivity, with much more attention in the former than in the latter. Coordination was seen to be further hindered by rivalries in policies and spheres of influence between the ministries of water and agriculture, with overlapping rules and regulations, and gaps in regulatory responsibilities. These findings echo previous work on Jordan's institutional water environment (Denny et al. 2008 [24], Yorke 2013 [25]), but offer a strong stakeholder-identified insight into the challenges for improved water and agricultural policy moving forward.

An example provided of the effects of siloed policy concerned case of new date trees, which have been encouraged by the Jordanian Ministry of Agriculture, and for which current water is available. Less consideration, however, had been given to the water requirements of the plantings as they matured over the next decade, implying a lack of sinuosity between water and agricultural policies.

Institutional challenges are compounded by deficits in operational and knowledge capacity. For the Jordan Valley, for instance, a more detailed understanding of specific agricultural and soil conditions for each farm might assist in delivering improved water productivity. However, this requires extension service and research investment. According to the extension service, while staff have deep in agricultural and environmental knowledge, the service suffers from a high turnover of staff, exacerbated by competition in salaries from the growing hi-tech agricultural sectors in the Gulf region. Dissemination of new knowledge and techniques is hindered by limited extension service budgets. The generation of that knowledge itself is hindered by donor dependence, bringing with it research that often matches donor or funder interests, rather than on-the-ground and policy needs. At the same time, as one agricultural scientist noted, there is failure on the part of science to speak the language of policy and to communicate ideas in a way that promotes policy salience. This perspective was shared by a representative of the Jordanian Ministry of Water and Irrigation who observed the need for greater engagement and understanding of conditions on the ground. Another suggested that Jordan had sufficient ideas and technical capacity, but lacked a deeper understanding, awareness, and action on how to achieve the required coordinated change at the organisational and institutional levels.

The perceived weaknesses of current institutional and organisational structure led to independent suggestions for change from four interviewees across the water and agricultural policy communities and water and agriculture research communities. First, that agricultural policies needed to be integrated into national economic planning with work on climate change and water. Second, that a high-level commission be formed with high-level representatives from the water, agriculture, trade, and labour sectors to plan a strategic cross-sectoral pathway, and to address institutional overlaps. Third, a related independent suggestion was a cabinet decision on a government package to determine the future balance of agricultural production and markets, and food imports and exports. Finally, an advisory committee with membership and/or input from policy and legal experts, as well as from applied practitioners and leading academics, drawing on global best practice cases of successful knowledge pooling.

Despite the challenges facing Jordan's water future, there is some professional evidence that changes in water application might be an avenue to improve the country's water situation through enhanced water resource decoupling in agriculture. Institutionally structural impediments must be overcome, with a clear demand for improved policy coordination. In addition to the immediate institutional challenges, there is also a longer-term need to consider water and agriculture's role within the broader political and social economy of Jordan.

4. Discussion and Conclusions

This paper has explored the potential for enhanced water productivity in Jordan and Palestine, based on regional best practice in water productivity from, in most cases, neighbouring Israel. Overall, the research found notable differences in water application per unit crop production in Israel and Jordan for at least six key crops. Scaled to Jordanian production tonnages, this difference amounts to 168 MCM/yr. 71 MCM (42%) of this is dependent on changing behaviour towards rain-fed clover cultivation, which, if socio-economic and regulatory conditions allowed it, could enable this volume of water to be devoted to higher value crops or even allocated to other economic opportunities outside agriculture. The remaining potential of reduced water application, based on the two methodologies used, is between 65 and 97 MCM/yr, (58% of reduction, or 13–20% of current agricultural water allocations). There is also evidence that Israel can learn from successful and profitable supplemental irrigation practices in Jordanian dates and olives. Trends for Palestine are less clear, with notable unreliability in the collected data, hindered by agricultural heterogeneity and geopolitical context. As discussed below, significant further work is needed to understand exactly if and how reduced water application can be translated into real savings that could be reallocated for either agricultural growth and/or other economic uses.

The methodology has a number of significant limitations, and should only be regarded as a first step in a deeper interrogation of regional agricultural trends. Key among these is the use of national averaging of water application. This was necessitated by an inability to disaggregate national crop production by sub-national areas. Only around half of the potential difference in water application was verified by farm-level data, and further work, including a larger number of farm samples, is needed. A larger farm sample size would capture the environmental diversity in the region and inter-annual variations in perennial crop water requirements related to variations in annual rainfall.

A deeper challenge relates to the idea of water saving. As recent FAO work highlights, improving water efficiency through reduced water application on farms does not necessarily result in water savings on a larger scale (Perry and Steduto, 2017) [26]. This is often because of a failure to account for return flows (direct runoff or infiltration), with the result that increased efficiency from the same abstraction results in more evapotranspiration and less return flow (ibid). Perry and Steduto (2017) suggest that disappointing environmental flow returns in Australia's Murray Darling Basin can be attributed to greater productive consumption of the same (or indeed less) actual abstraction, resulting in drastically reduced return flows. A similar picture is presented by Ward and Pulido-Velazquez (2008) in the Upper Rio Grande, in the Southern United States [27]. For Jordan, therefore, it is *essential* to understand what is happening to the water at the farm level, which, when compared with Israel, is apparently 'excess'. The presence of significant return flows or unaccounted water on farms will almost certainly result in increased water consumption if irrigation efficiency measures are implemented without simultaneous improved regulation (namely enforced reduction) of abstraction. Questions on runoff/return flows were included in farmer interviews, however no reports of any flows were made; in contrast, farmers were keen to stress that given that water and/or pumping energy has to be paid for, and given that allocations in the Jordan valley are below desired levels, every drop of water that farmers applied was used productively in their perception, or for maintaining healthy soils through salt flushing.

One potential future avenue of assessment in the area of water productivity and return flows is to use satellite data to compare evaporation and evapotranspiration rates with farmer-declared water application, plus local rainfall. This methodology has been used to uncover illegal water use in Jordanian agriculture (Al-Bakri, et al., 2016) [28], but might also have the potential to identify whether there are significant surface or sub-surface return flows, and indeed whether there are significant differences in non-productive evaporation between Israeli and Jordanian farm plots. A deeper interrogation of how the differences in water application relate to real differences in crop water use (evapotranspiration per yield) will be a crucial next step to further understand differences in agricultural water productivity and quantifying the real potential to produce savings at the system level. Only if real savings are possible in Jordan, can improved agricultural water productivity be an effective mechanism to enhance water resource decoupling.

If reduced water application can be translated into a real water saving, significant investment would likely be required to achieve these savings. However, similar efforts in Israel from 1995 for improvements in agricultural water productivity were economically assessed to provide the cheapest means of enhancing national water security (Arlosoroff, 1997 [29]). While infrastructure and regulation in Jordan may increase costs, agricultural investments might also be expected to have additional benefits to the rural economy and livelihoods. These multiple potential socio-economic gains beyond the immediate issue of enhanced water security should also be considered in any investment and policy decisions.

Politically, our research identified divergent views on the potential for changes in agricultural water application in Jordan, with strong arguments presented both ways, especially regarding the Jordan Valley. Policy progress, and thus the potential for improved coordination within and between ministries, appears to be hampered by siloed operations. Achieving this coordination, and forging future policy, however, is itself politically difficult. Our analysis of the policy conditions highlighted an awareness of the need for greater understanding and commitment to overcoming institutional barriers and coordination challenges. While some frustration was expressed that both research and donor work in Jordan was serving academic or donor interests, a genuine need for input in understanding and managing institutional challenges was apparent. Demand for applied research that meets stakeholder needs in this political arena would appear to exist, and should be given urgent attention as a key element for Jordan's future water, agricultural, and economic sustainability. Investment and attention in this area has clear salience to Jordan's pursuit of the 2015 Sustainable Development Goals.

Author Contributions: This research was conceived by Michael Gilmont, who also coordinated the work of the project research team, with conceptual development carried out in conjunction with the Principle Investigators Steve Rayner (UK), Nadav Tal (Israel), and Erica Harper (Jordan). Detailed project framing, fieldwork design, and fieldwork execution (including interviews) was carried out by Lara Nassar (Jordan), Nadav Tal (Israel), and Hilmi Salem (Palestine). Further development of the enquiry into Institutional analysis was carried out by Lara Nassar, Michael Gilmont, and Steve Rayner. This paper was drafted by the project team, under the leadership of Michael Gilmont.

Acknowledgments: This research was funded by British Council MENA Institutional Links Grant (STREAM Program) ID218186526 covering the core of this research into decoupling and water productivity, and by University of Oxford's John Fell Fund, grant number 162/047, for funding further work into institutional and organisational landscapes of uptake. The authors also acknowledge the support of the University of Oxford (including the Institute of Science, Innovation, and Society (InSIS), and the Environmental Change Institute (ECI)), and its respective country project partners, the West Asia North Africa (WANA) Institute and EcoPeace Middle East, in conducting this research. This paper is developed and extended from work included in a joint report produced by WANA, EcoPeace, and Oxford in July 2017.

Conflicts of Interest: The authors declare no conflict of interest. Aside specification by the British Council STREAM program of cooperation between U.K. researchers with research partner(s) in Israel and researcher partner(s) in Jordan, the funding sponsors have no role in the design of the study, in the collection, analyses or interpretation of data, in the writing of the manuscript, and in the decision to publish the results.

Appendix A. Water Intensity Calculations and Data

Table A1. Water Numbers for Israel.

Crop	Average Yield	Average Recommended Water Application	Average Water Intensity
Almonds	0.2	600.0	3750.0
Apricot	2.2	450.0	204.5
Avocado Etinger	2.3	1000.0	444.4
Avocado Has	1.6	1000.0	625.0
Avocado Average	1.9	1000.0	*534.7*
Bananas	5.7	1802.0	316.1
Beet	6.0	250.0	41.7
Cabbage and Brassicas	8.0	450.0	56.3
Carrot and turnips	7.5	550.0	73.3
Cauliflower	3.0	320.0	106.7
Celery	10,000 units (2t)	400.0	200.0
Chillies and Peppers	9.3	1246.0	133.4
Cotton fibres	0.2	470.0	2611.1
Cotton seeds	0.3	470.0	1615.1
Cucumbers	25.0	1200.0	48.0
Dates Barhi	1.7	1500.0	882.4
Dates Majul	0.9	1500.0	1764.7
Dates Average	12.6	835.0	*1323.5*

Table A1. *Cont.*

Crop	Average Yield	Average Recommended Water Application	Average Water Intensity
Other Vegetables	5.9	622.2	106.2
Dry onion	5.0	650.0	130.0
Eggplant	10.0	700.0	70.0
Figs Brazilan	1.5	850.0	566.7
Figs Hamadi	3.5	1700.0	485.7
Figs Average	7.5	675.0	*526.2*
Grapefruit white	8.0	800.0	100.0
Grapefruit red	7.0	900.0	128.6
Grapefruit average	7.5	850.0	*114.3*
Grapes for food	2.7	768.8	286.8
Grapes for wine	1.4	200.0	142.9
Ground nut	1.0	520.0	547.4
Lemons	5.0	750.0	150.0
Lettuce and chicory	8.0	180.0	22.5
Limes	3.0	750.0	250.0
Maize grains	1.8	550.0	305.6
Mangoes Maya	3.0	800.0	266.7
Mangoes Kit	4.5	900.0	200.0
Mango average	3.8	850.0	*233.3*
Melons	7.0	700.0	100.0
Nectarine middle	3.5	600.0	171.4
Nectarine late	4.3	700.0	164.7
Nectarine early	2.3	450.0	200.0
Nectarine average	3.3	583.3	*178.7*
Olives	1.3	1125.0	900.0
Other orchards	3.0	675.0	222.1
Other types of citrus	5.0	792.0	158.4
Pears kotsia	3.7	650.0	175.7
Pears kotsia valley	2.5	650.0	260.0
Pears average	3.1	650.0	*217.8*
Peaches middle	3.5	600.0	171.4
Peaches early	2.5	450.0	180.0
Peaches late	4.5	750.0	166.7
Peaches	3.5	600.0	*172.7*
Peas	0.6	125.0	227.3
Pecan	0.4	900.0	2571.4
Persimmon	5.0	900.0	180.0
Plums late	3.0	650.0	216.7
Plums early	2.0	450.0	225.0
Plums Average	2.5	550.0	*220.8*
Sorghum, grains	2.1	225.0	107.1
Strawberries	8.0	1100.0	137.5
Sweet potatoes	4.5	700.0	155.6
Tomato	19.0	1000.0	52.6
Oranges	5.0	800.0	160.0
Watermelon	7.0	400.0	57.1
Wheat for silage	1.0	40.0	40.0

Note: Derived from yield and recommended water application (Extension Service), as averaged from different regions where appropriate. Italic water intensity is an average of sub-species water intensity, not calculated by average yield and average application. Some water intensity numbers apply to multiple listed crop types (e.g., Other Citrus), so number of listed crops do not tally with crop numbers listed in text.

Table A2. Jordan Water Intensity Calculations.

Crop	2010 Yield (tonnes)	Area (dunam)	Water/Dunam (derived)	Total Water (m^3)	Water/Crop (m^3/t) (derived)
alfalfa	1	7	857.1	6000	12,000.0
almonds	2419	1260	870.6	1,097,000	453.5
apples	28,770	17,826	841.8	15,006,000	521.6
apricots	6796	8376	834.5	6,990,000	1028.5
bananas	43,753	18,527	1438.3	26,647,000	609.0
barley	10,659	12,623	258.2	3,259,000	305.8
beans broad	21,150	15,466	347.7	5,378,000	254.3
beans string	8218	5603	345.0	1,933,000	235.2
broom millet	14	11	545.5	6000	438.0
cabbage	20,317	6329	373.0	2,361,000	116.2
carrots	6770	1634	255.8	418,000	61.7
cauliflowers	54,734	18,025	395.1	7,121,000	130.1
chick peas	3935	1801	149.9	270,000	68.6
citrus	119,726	68,386	513.8	35,134,000	293.5
clover trifoil	223,591	65,771	1276.8	83,977,000	375.6
corn white	25,476	20,003	575.1	11,503,000	451.5
cow peas	650	456	339.9	155,000	238.5
cucumbers	176,179	20,047	356.0	7,137,000	40.5
date palms	11,241	17,079	1439.3	24,582,000	2186.9
eggplants	104,748	30,070	387.7	11,659,000	111.3

Table A2. *Cont.*

Crop	2010 Yield (tonnes)	Area (dunam)	Water/Dunam (derived)	Total Water (m^3)	Water/Crop (m^3/t) (derived)
figs	953	1051	882.0	927,000	972.9
garlic	486	149	684.6	102,000	210.1
grapes	29,683	21,660	672.9	14,576,000	491.1
guavas	2270	2155	781.9	1,685,000	742.3
jew's mallow	35,659	12,797	476.3	6,095,000	170.9
lentils	245	71	1436.6	102,000	416.0
lettuce	48,200	14,344	269.0	3,859,000	80.1
maize	29,006	14,963	728.4	10,899,000	375.7
marrows	69,655	11,620	244.9	2,846,000	40.9
melons	31,051	8112	846.9	6,870,000	221.2
okra	6814	7420	458.4	3,401,000	499.1
olives	171,672	249,729	527.8	1.32×10^8	2399.3
onions dry	15,765	8590	485.9	4,174,000	264.8
onions green	3512	836	793.1	663,000	188.8
other field crops	21	265	211.3	56,000	2654.0
other fruits	13,228	11,347	870.6	9,879,000	746.8
other vegetables	27,182	11,620	244.9	2,846,000	104.7
parsley	2543	980	259.2	254,000	99.9
peaches	23,153	15,982	841.0	13,441,000	580.5
pears	2141	3023	890.8	2,693,000	1258.1
peas	3512	1846	449.1	829,000	236.0
peppers hot	18,549	7823	452.1	3,537,000	190.7
peppers sweet	36,590	11,126	490.7	5,460,000	149.2
plums/prunes	2291	2863	836.5	2,395,000	1045.4
pomegranates	2146	2138	819.9	1,753,000	816.9
potatoes	174,931	57,969	434.2	25,173,000	143.9
radish	4161	1687	284.5	480,000	115.4
sesame	118	83	3831.3	318,000	2694.9
snake cucumber	20,987	3170	60.9	193,000	9.2
spinach	6368	2125	238.1	506,000	79.5
tomatoes	737,262	141,212	524.7	74,092,000	100.5
turnips	1516	390	276.9	108,000	71.2
vetch	540	78	6064.1	473,000	875.6
water melons	153,118	37,417	394.2	14,750,000	96.3
wheat	22,126	17,094	207.1	3,541,000	160.0

Note: Data in this table on area and water use are taken from International Resources Group (IRG) and Karablieh (2012) and calculations of area and water applied to crop during the year 2010, as cited in text. Data on annual yield taken from national statistical accounts for 2010. Calculations on water/dunam and water/crop are derived from this data.

References

1. Gilmont, M. Decoupling dependence on natural water: Reflexivity in the regulation and allocation of water in Israel. *Water Policy* **2014**, *16*, 79–101. [CrossRef]
2. World Bank. Water Resources Management in an arid environment: The case of Israel. In *World Bank Analytical and Advisory assistance (AAA) Program China: Address Water Scarcity*; Background Paper No. 3; East Asia and Pacific Region; World Bank: Washington, DC, USA, 2006.
3. Teschner, N.; Negev, M. The development of water infrastructures in Israel: Past, present and Future. In *Shared Borders, Shared Waters: Israeli-Palestinian and Colorado River Basin Water Challenges*; Megdal, S.B., Varadey, R.G., Eden, S., Eds.; CRC Press: Leiden, The Netherlands, 2013.
4. EcoPeace. Water for agriculture in Jordan: Strategies for reducing consumption in the agricultural sector. 2013. Available online: http://ecopeaceme.org/uploads/13622989291~%5E$%5E~Jordan_Agricultural_Policy_Paper_English.pdf (accessed on 28 June 2017).
5. FAO. Food balance tables 1961–2011. In *United Nations Food and Agricultural Organisation*; FAO: Rome, Italy, 2016.
6. Ministry of Water and Irrigation (MWI). Water For life. In *Jordan's Water Strategy 2008–2022*; MWI: Amman Jordan, 2008. Available online: http://www.mwi.gov.jo/sites/en-us/Documents/Jordan_Water_Strategy_English.pdf (accessed on 20 September 2016).
7. Ministry of Water and Irrigation (MWI). *National Water Strategy of Jordan 2016–2025*; MWI: Amman, Jordan, 2016.
8. World Bank. *Red Sea-Dead Sea Water Conveyance Study Program. Study of Alternatives*; Final Report; World Bank: Washington, DC, USA, 2014.
9. Seder, N.; Abdel-Jabbar, S. *Safe Use of Treated Wastewater in Agriculture Jordan Case Study*; Arab Countries Water Utilities Association (ACWUA): Amman, Jordan, 2011.

10. Yorke, V. Jordan's Shadow State and Water Management: Prospects for Water Security Will be Dependent on Politics and Regional Cooperation. In *Society-Water-Technology: A Critical Appraisal of Major Water Engineering Projects*; Hüttl, R.F., Bens, O., Bismuth, C., Hoechstetter, S., Eds.; Springer: Heidelberg, Germany, 2016.
11. Gilmont, M.; Rayner, S.; Harper, H.; Nassar, L.; Tal, N.; Simpson, M.; Salem, H.S. *Decoupling National Water Needs for National Water Supplies. Insights and potential for Countries in the Jordan Basin*; WANA Institute: Amman, Jordan, 2017.
12. Hoff, H.; Falkenmark, M.; Gerten, D.; Gordon, L.; Karlberg, L.; Rockstrom, J. Greening the global water system. *J. Hydrol.* **2010**, *384*, 177–186. [CrossRef]
13. Ministry of Agriculture. *Irrigation Coefficients, in Training and Professional Services Database*; Israel Ministry of Agriculture: Bet Dagan, Israel, 2016.
14. International Resources Group (IRG); Karablieh, E. *Institutional Support and Strengthening Program (ISSP) ISSP Water Valuation Study: Disaggregated Economic Value of Water in Industry and Irrigated Agriculture in Jordan*; USAID: Washington, DC, USA, 2012.
15. Obeidat, O. Ministry ends large-scale water theft in Jordan Valley. *Jordan Times*, 03 September 2016.
16. O'Keeffe, J.; Buytaert, W.; Mijic, A.; Brozovic, N.; Sinha, R. The use of semi-structured interviews for the characterization of farmer irrigation practices. *Hydrol. Earth Syst. Sci.* **2016**, *20*, 1911–1924.
17. Boller, B.; Willner, E.; Maggioni, L.; Lipman, E. *Report of a Working Group on Forages. Eighth Meeting, 10–12 April 2003*; International Plant Genetic Resources Institute: Linz, Austria; Rome, Italy, 2005.
18. Al-Jaloudy, M.A. *Country Pasture/Forage Resources Profiles: Jordan*; Suttie, J.M., Reynolds, S.G., Eds.; FAO: Rome, Italy, 2006.
19. Al-Rimawi, A.S. *Reorienting Agricultural Extension in Jordan. Expert Consultation workshop on Market Orientated Agricultural Advisory Services*; FAO/AARINENA: Amman, Jordan, 2012.
20. Lach, D.; Ingram, H.; Rayner, S. Maintaining the Status Quo: How Institutional Norms and Practices Create Conservative Water Organizations. *Univ. Tex. Law Rev.* **2005**, *83*, 2027–2053.
21. Rayner, S.; Ingram, H.; Lach, D. Weather Forecasts Are for Wimps: Why Water Resource Managers Do Not Use Climate Forecasts. *Clim. Chang.* **2005**, *69*, 197–277. [CrossRef]
22. North, D. Institutional Change: A Framework for Analysis. 1994. Available online: https://www.iei.liu.se/nek/730A22/filarkiv-2013/del-1a-andersson/1.516348/NorthInstchangeframework.pdf (accessed on 8 March 2014).
23. Al-Zu'bi, M. Jordan's climate change governance framework: From silos to an intersectoral approach. *Environ. Syst. Decis.* **2016**, *26*, 277–301. [CrossRef]
24. Denny, E.; Donnelly, K.; McKay, R.; Ponte, G.; Uetake, T. *Sustainable Water Strategies for Jordan*; International Economic Development Program, Gerald R. Ford School of Public Policy, University of Michigan: Ann Arbour, MI, USA, 2008.
25. Yorke, V. *Politics Matter: Jordan's Path to Water Security Lies through Political Reforms and Regional Cooperation*. NCCR Trade Regulation, Working Paper 2013/19. 2013. Available online: https://www.wti.org/media/filer_public/00/98/00989164-e76d-4887-be98-2acd3b193c59/working_paper_2013_19.pdf (accessed on 14 February 2018).
26. Perry, C.; Steduto, P. *Does Improved Irrigation Technology Save Water? A Review of the Evidence. Discussion Paper on Irrigation and Sustainable Water Resources Management in the Near East and North Africa*; FAO: Cairo, Egypt, 2017.
27. Ward, F.A.; Pulido-Velazquez, P. Water conservation in irrigation can increase water use. *Proc. Natl. Acad. Sci. USA* **2008**, *105*, 18215–28220. [CrossRef] [PubMed]
28. Al-Bakri, J.T.; Shawash, S.; Ghanim, A.; Abdelkhaleq, R. Geospatial techniques for Improved Water Management in Jordan. *Water* **2016**, *8*, 132. [CrossRef]
29. Arlosoroff, S. The public commission on the water sector reform (the general ideas underlying its recommendations). *Int. Water Irrig. Rev.* **1997**, *5*, 1–9.

land

MDPI

Review

Valuing Environmental Benefit Streams in the Dryland Ecosystems of Sub-Saharan Africa

Caroline King-Okumu [1,2]

[1] The Borders Institute (TBI), PO Box 13385-00100 Nairobi, Kenya; caroking@yahoo.com
[2] GeoData Institute, University of Southampton, Southampton SO17 1BJ, UK

Received: 14 October 2018; Accepted: 10 November 2018; Published: 19 November 2018

Abstract: Policy-makers and practitioners often struggle or fail to define and quantify the economic impacts that can be achieved through ecologically sustainable investments in dryland ecosystems. This paper reviews the current state of the art in the characterization and valuation of environmental benefits in drought-prone areas of Sub-Saharan Africa. Benefit streams from ecosystem services associated with the production of food, energy and water are characterized, as well as those from supporting and regulating hydrological systems. For each value type, valuation approaches and examples of their application in Sub-Saharan African contexts are presented. The review is drawn from a series of recent methodological discussions, working papers and field research reports focusing on the benefits of locally determined and ecosystem-based adaptations under dryland climate extremes. The focus is on the challenges faced by practitioners and researchers tasked with developing benefit-cost assessments for investments in the adaptive management and conservation of dryland ecosystems, particularly in marginalized dry and drought-prone areas of Sub-Saharan Africa. Recommendations could also interest a wider global community of dryland researchers and development practitioners.

Keywords: ecosystem services; economic valuation; drylands

1. Introduction

Policy-makers and practitioners often struggle or fail to define and quantify the impacts that development projects can have on dryland ecosystems. There is a lack of baseline assessments of benefit streams from dryland ecosystems against which to compare new management alternatives. This is because economic evaluation of the returns on the existing investments in adaptive management and conservation of the dryland ecosystems have been rare and limited. However, interest in the challenges associated with such assessments is building in Sub-Saharan Africa and in many other parts of the world [1–4].

In this paper, the current state of the art in the identification and valuation of selected key environmental benefit streams in the drought-prone dryland regions of Sub-Saharan Africa is reviewed. Insights on questions currently faced by researchers and practitioners developing benefit-cost calculations and economic impact assessments are drawn from a series of recent working papers and field research reports focusing on economic assessment of the benefits from ecosystem-based adaptations in the dry areas of East and West Africa. These are contextualised within the broader available published literature from across the region.

The selection and presentation of material responds to concerns expressed by practitioners tasked with presenting economic assessments of the anticipated returns on investments in dryland restoration, climate change adaptation, and disaster preparedness projects in the drylands of Sub-Saharan Africa. The same practitioners are often also tasked to design monitoring and evaluation systems that could verify these projections in the event that project proposals are funded.

The intention of the paper is that it should provide a useful and accessible reference source for project managers, practitioners and researchers developing benefit-cost assessments of small-scale adaptation and resilience-building interventions for environmentally sustainable management in the drylands, particularly in marginalized dry and drought-prone areas of Sub-Saharan Africa. This could be of interest to a wider global community of dryland researchers and practitioners as well. The primary question for the international scientific community is: what are the existing and potential benefits from adaptive management of the dryland ecosystems? [5–7]. This is particularly important where there are proposals to transform, replace or simply remove these under-recognized benefits through development initiatives. Since such proposals will introduce changes and tradeoffs among different benefit streams, there is an urgent need to identify and value the baseline values and potential for adaptive management.

Following a brief conceptual overview of the assessment challenge in the dryland ecosystems in Section 2, key ecosystem services are summarized. Methods used to generate this review are described in Section 3. Section 4 then presents the techniques and methods that are available to practitioners for the assessment and valuation of benefit streams that may be anticipated from adaptive management and conservation of the dryland ecosystems. The section is structured around four key benefit streams and includes case studies sourced from published assessments from Sub-Saharan Africa, both West and East. In addition to describing techniques for valuing the benefit streams, practical examples demonstrate how these can be connected to some of the many available approaches and scientific tools for quantification of the benefit streams.

The discussion, recommendations and conclusion in Sections 5 and 6 underline the opportunity and feasibility for practitioners to apply the available assessment methods. Synergies between the benefit streams that may not be quantified, but will merit additional attention following the valuation are also highlighted.

2. Overview of the Challenge to Assess the Benefit Streams and Management Options in Dryland Systems

2.1. Adaptive Management in the Dryland Ecosystems

Resource-dependent populations living in dryland ecosystems use their accumulated knowledge of environmental conditions and climatic patterns to adapt practices for managing shared or common pool resources [8–11]. These practices and adaptations enable communities to sustain their land and water resources for the long term while also drawing benefits from them in the short term. In cases where well-informed practices ensure that dryland ecosystems can retain and enhance their own endogenous production of goods and services, this reduces dependence on external cooperation and aid in the forms of disaster relief and social assistance [12–14]. Recognizing, maximizing and restoring ecosystem function, self-regulation and resilience can play an important role in ensuring that growth can be self-sustaining both for the ecosystem and for the current members of society.

The international community considers investments that it could make to boost adaptive resource management practices in relation to its current priorities and commitments for sustainable land management (SLM) [15], enabling community-based adaptations to climate change and variability [16], building resilience to drought by supporting adaptations that are considered to be ecosystem-based [17–19] or nature-based [20] (see example of common community-scale practices for adaptive management in Appendix A, Box A1). However, to establish the economic value of the endogenous productivity of the drylands and the business case for how it could best be improved requires baselines, data and models. These remain rudimentary and/or unavailable in many of the dryland production systems of Sub-Saharan Africa and elsewhere.

The economics of adaptation and returns on adaptation investments have received significant attention [1,3,21–27] (www.aboutvalues.net). But the economic value of returns on these investments remains challenging for practitioners to quantify and monitor systematically—especially under erratic, uncertain and drought-prone conditions that commonly occur in Sub-Saharan Africa [13,14,28–33].

There has not previously been a review that focuses on bringing together assessment methods and examples applied in both East and West Africa and presenting them in a way that could be convenient for practitioners from these regions.

2.2. Identifying the Major Benefit Streams from the Dryland Ecosystems

Economically valuable public goods and services provided by ecosystems in the drylands (as described in [34]) include water, energy and other essential provision services (Figure 1). These are supported and regulated by other services that sustain the system and its functions.

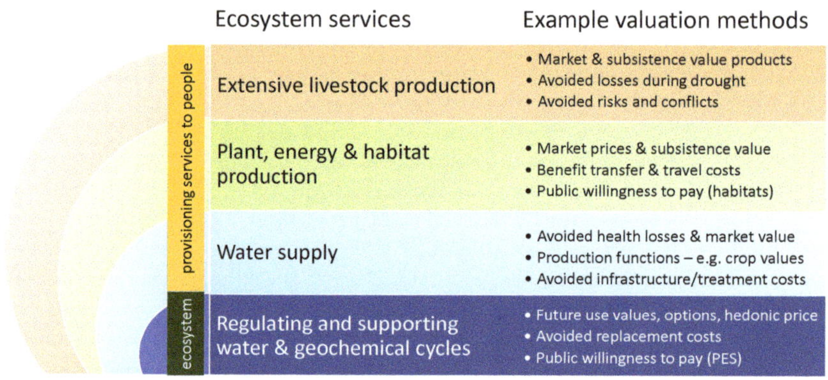

Figure 1. Goods and services in dryland ecosystems.

In dry and drought-prone environments, hydrology and climatic extremes can have a particularly strong influence over the functioning of the ecosystems and their productivity. When human populations are not able to manage these effects, natural hazards such as droughts can have disastrous effects on ecosystem services that support the well-being of dependent populations. In light of this, the analysis of benefit streams that is presented in this paper includes consideration of the economic value both of the provisioning services and also the essential underlying supporting and regulating services. This is a departure from the approach taken by many of the current available assessments which have not assigned a value to the supporting and regulating services for fear of double-counting the value of provisioning services (e.g., [35]).

Ecosystem accounting is an emerging field that aims to provide a consistent approach to analyzing environment–economy interactions [36–38]. But for most of the dry areas of Sub-Saharan Africa, this is still in its very early stages. Across Sub-Saharan Africa, assessing economic productivity is complicated by the nature of dryland products and services requiring valuation and the presence of various informal, undocumented economies. Often, national statistical systems fail to effectively record the productivity of goods and services from dryland systems that are highly relevant to livelihoods and the economy—including many of the products from extensive livestock raising, natural heat, energy sources, underground water storage and other environmental services.

Rather than insisting on monetizing all benefits and adding them together, a well-recognized alternative is to use a multi-criteria analysis. This weighs benefits expressed in diverse units that are not directly comparable either instead of or alongside the economic valuation of tradeoffs and can accommodate graduated values derived from subjective qualitative assessments, if necessary. In Botswana, multi-criteria decision analysis has been explored as a means to capture the breadth of benefits from communal grazing systems [39–41].

Economic valuation is helpful to weigh tradeoffs between some of the multiple effects caused by different investments—whereby one effect may cancel out the benefit of another. For example, a drip irrigation scheme may save water on a per hectare basis, as compared to a gravity-controlled system. But if the introduction of such systems causes a larger area to be irrigated—then the tradeoff between volume and extent of irrigation may mean that the adaptation does not achieve any overall water-saving at all. Economic assessment provides a means to place a comparable value on the effects on water versus food production. Other tradeoffs can be harder to weigh up—especially in multi-criteria analyses. For example, if a grazing system generates less income overall, but contributes in small ways to the well-being of a greater number of low-income households, opinions may differ concerning whether or not the tradeoff is worthwhile.

The scarcity or availability of goods and services is often driven by contextual processes operating at scales that are larger than the intervention areas of development projects within which economic assessments may be proposed. For example, pastures and wildlife habitats require landscape-level assessments, and water scarcity will also be affected by basin level processes and demands. Timescales for assessment of the value of benefits from investments in resource management also do not always match well to project and assessment timeframes. Where benefits are anticipated to accrue in the distant future, these anticipated benefits may be considered to have a net present value, but this is usually discounted because benefits that may arrive in the future are generally considered less attractive than benefits secured in the present [42].

2.3. Assigning Value

Research on the monetary valuation of natural resources dates back to the early 1960s but has progressively received wider attention through growing discussion of the capital and service values provided by nature and ecosystems [43]. An inherent challenge in any attempt to place an economic value on public goods and services is that different people will make different judgements about what to value, and how much it is worth. Local resource users' views of what should be valued can differ considerably from those of national or international agencies and partners. This is important because an economic assessment of returns on investments should focus on returns to society as a whole (which is different from a financial assessment of returns on investments to private individuals) [42].

Valuation methods for the benefit streams from dryland ecosystems can include market- and non-market-based approaches [3,44–46]. Ten broad types of valuation methods have previously been identified for the valuation of ecosystem services globally, but few of the cases identified were located in Sub-Saharan Africa [46] (Table 1). The 10 methods include use of available market values for ecosystem goods and services, or the costs that would need to paid to obtain them. Where there are no markets to buy the ecosystem goods and services, willingness to pay for access to them may suggest that such markets or payment systems could be created.

Table 1. Classification of ecosystem service valuation methods [46].

Type		Name of Method	Acronym
Market-based		Direct Market Pricing	DMP
		Payment for Ecosystem Services	PES *
		Factor Income/Production Function	FI/PF
Cost-based		Avoided Cost	AC
		Mitigation and Restoration Cost	MC/RC
		Replacement Cost	RC
Willingness to pay (WTP)	Revealed preference	Hedonic Pricing	HP
		Travel Cost	TC
	Stated preference	Contingent Valuation	CV
		Group Valuation	GV

* A PES can be seen as an expressed WTP through the market.

Provisioning services are most often valued through direct market pricing methods. However, there are also other alternatives where market prices do not fully reflect their value (see direct use values in [44]) (Figure 1). For habitat services, for example, methods such as factor income and contingent valuation are often used as an alternative to direct market pricing if these are not considered adequate [46]. Valuations of regulating services often use avoided cost and replacement cost methods in addition to direct market prices, where available (see description of indirect use values in [44]).

3. Materials and Methods

To identify some of the key benefit streams in the dryland systems of Sub-Saharan Africa and methods that can and have been applied to assess their economic value, this review draws on insights and examples from a series of recent working papers and field research reports focusing on economic assessment of the benefits from ecosystem-based adaptations in the dry areas of East and West Africa [47–52]. A series of meetings were held with planners and NGO staff implementing, monitoring and evaluating investments in shared resource management systems in a range of dry areas of Kenya, Senegal and Mali, as described in the respective project reports.

Findings were shared and discussed with practitioners tasked with the challenge to present economic assessments of the anticipated returns on proposed future investments in dryland restoration, climate change adaptation and disaster preparedness projects in the drylands of sub-Saharan Africa [53,54]. Practitioners highlighted the challenges to prepare a business case for such investments and referred to available examples of project proposals that have previously been approved by financing institutions including the Green Climate Fund (see: https://www.greenclimate.fund/what-we-do/projects-programmes). Additional relevant assessment challenges were identified through a broader overview of the nature of benefits that project managers and practitioners would wish to capture in their assessments as compiled for the World Overview of Conservation Approaches and Technologies (WOCAT) (see: https://www.wocat.net/en/).

Having identified the practical challenges, and methods available in context at selected locations in Sub-Saharan Africa, further reviews of the available published literature across a wider geographical area of the region's drylands were made using keyword searches in SCOPUS online database. The keyword searches were filtered to select materials referring directly to cases in Sub-Saharan Africa. These could be used and further developed by practitioners and researchers who are working or studying there.

4. Assessing and Valuing Benefit Streams

4.1. Livestock Production in the Drylands

The production of livestock is the most widespread and traditional economic activity in the more arid regions of Sub-Saharan Africa [55,56]. Therefore, it is one of the most critical benefit streams for the societies that live in these regions. In addition to cows, sheep, goats, camels, etc., livestock production in the drylands may also include the keeping of poultry, rabbits, and bee-keeping for honey [35,57]. Livestock production can include production of live animals for sale, meat, milk, dung, skins and other products. Sometimes livestock also produce other services, such as draught power from donkeys, horses and camels, pollination from bees—or others. Increasingly, wildlife services are also attracting attention in the drylands of Sub-Saharan Africa.

Valuation of livestock and their services is usually dependent on market prices. Often, local or national statistics will be available on some—but generally not all—aspects of livestock production in the drylands of Sub-Saharan Africa (e.g., as in [58] for Senegal). Milk has a particular value in the diets and cultural traditions of pastoral communities of both East and West Africa [59,60], and is a growth industry [61–65]. Where livestock products such as milk are consumed for subsistence, statistics may not be available, but the volumes of production can still often be estimated, based on knowledge of herd size and lactation patterns (as demonstrated in Mopti, Mali, by [52], in Niger by [66],and in Kenya by [67,68]).

Resource economists use bioeconomic models to assess productivity in the rangeland ecosystems and the potential impacts of altered water and fodder availability [69–73]. Such models have been explored in drought-prone Sub-Saharan environments in the Senegalese Ferlo (see Box 1) [74], Niger [66], Ethiopia [75,76] and Kenya [77]. They require information on the relationships between herd size, lactation and reproduction rates, as well as animal nutrition and health (for further discussion of these in the context of the arid lands of Kenya, see: [67,78]).

Grass grown for fodder is one of the best documented successful investments to increase and maintain income and reduce vulnerable households' expenditures in the most drought-prone regions of Somalia, Ethiopia and Kenya [14]. Not only does improved forage availability directly affect the availability of milk and improve nutrition levels in vulnerable households (as described above), but it does so quickly—which can be important for project stakeholders. Because it grows rapidly, grass can give a positive return within a short space of time, and can continue to provide an economically valuable crop twice per year [79]. In light of this, it is attractive to households and agencies seeking to secure immediate livelihood improvements under drought conditions. A recent study has identified the economic value of animal feeding during droughts in terms of avoided emergency food relief supplies to drought-affected households [80]. Furthermore, in pastoral and agro-pastoral areas where most households keep livestock, fodder availability reduces animal malnutrition that can otherwise cause the loss of these assets during droughts.

A recent global assessment [4] has identified an estimated return on investment in grassland restoration of 35:1 over a twenty year period across the drylands and other climatological zones [81]. However, the bioeconomic models that are available to capture economic benefits from rangeland livestock production are only as good as the underlying datasets and field observations that they use. Where there is institutional support in place, they can be complemented by long-term field monitoring and measurements, as has been done to study the effects of enclosures in Kenya [82]. These can be verified using rapid participatory methods described above (e.g., [83,84]). Where support for these methods is not available, economic assessments can be based on input values drawn from the best available literature.

In extensive rangeland systems, physical inputs may be few, but institutional investments are still needed to coordinate communal land and water management systems [85]. Often these necessary investments and transaction costs are overlooked in assessments of the economics of pastoral livestock production. Also, in extensive systems, during the dry seasons and the droughts, transhumant livestock migrations occur. These are challenging to model effectively. The costs of livestock production increase during these times due to water shortages and pressures on grazing resources, conflicts and others. Climatic factors can aggravate increased risks of diseases and threaten the value of livestock production. A theoretical model of the economic connection between communal grazing, disease transmission risk, risk perceptions, and antimicrobial use has been developed and tested in Kenya [86]. The analysis supported by this model suggested that factors influencing subjective perception of disease risk and antimicrobial uses included disease history, and types of grazing systems—with greater levels of concern regarding disease prevalent in communal grazing systems, as compared to private grazing areas.

With intensification of the commercialization of livestock production, production models become more complex, and farmers no longer depend only on pastures to nourish their animals. Increasingly, they manufacture or buy supplements and they also often pay for water, labor and medical attention for their animals. Increasingly, livestock water productivity has gained attention in Sub-Saharan Africa [87,88]. The global average water footprint for beef produced from grazing systems is 243 m^3/ton for grey water and 465 m^3/ton for freshwater, whereas for industrial beef systems this average increases to 712 and 683 respectively [89]. Livestock raised in the extensive grazing systems across much of

Sub-Saharan Africa often consume less water than those raised in other grazing systems because in the extensive systems they do not necessarily have access to water every day.

Some products from livestock, such as leather and wool, require significant amounts of processing to add value before they can be marketed as textiles, garments and furnishings. This is mediated by market access, and other conditions. A key methodological question for the valuation of rangeland productivity concerns the addition of value through value chains, and the extent to which the economic assessment is able to consider the dispersal of benefits and positive feedbacks along the value chain [90]. It has been argued that economic multipliers can be applied to represent the value of cash circulating through the regional economies in the drought-prone regions of Kenya and Ethiopia [91,92].

Box 1. Case Study: Valuing rangeland production in the Senegalese Ferlo using a bioeconomic model (based on Hein et al. [74]).

The anticipated effects of climate change in Senegal on the economic value of rangeland production were assessed by Hein et al. [74]. The assessment used an ecological-economic modelling approach. This enabled simulation of the effects of climatic changes and management responses via modifications of the stocking rates. The model was based on local data collected over a 10-year period (1981–1990) at the Widou Thiengoly research station in the western part of the Ferlo on vegetation production under different grazing densities and rainfall conditions. The model is a dynamic systems model, running with time increments of 1 year. The model calculates the annual income for the pastoralists as a function of annual rainfall and the long-term stocking rate maintained by the pastoralists. The model is spatially homogeneous, and changes in livestock routes or other adaptation strategies not involving changes in stocking densities are not considered.

For Senegalese pastoralists, the main source of income is the sale of animals for meat, with milk production coming in second place. The role of milk production and agriculture has decreased in the last decades with the increased focus on livestock herding for meat production. For reasons of simplicity, in the model it is assumed that income is only derived from the sale of animals. Income depends upon the amount of surplus livestock that can be sold annually on the market, as well as on the livestock price.

In years of drought, when there is not enough grass production to feed the livestock number corresponding to the long-term stocking rate, it is assumed that the pastoralists maintain the amount of livestock that can be fed, and that they sell the surplus on the market. However, the price will be low, as there will be a high supply, and low demand on the local markets. If a dry year is followed by a wet year, the pastoralists will purchase livestock in the market in order to stock up to the long-term stocking rate. If rainfall decreases, the optimal stocking density is reduced, and so are the maximum profits the pastoralists can obtain. Pastoralists can adapt to climate change by destocking, which would increase the societal profits gained per livestock unit as well as the overall profits compared to the current grazing strategy.

As the pastoralists maintain an open-access individual profit maximization strategy, the optimal rate of destocking is not expected to occur. But reductions in stocking rates are anticipated to result from predicted climatic changes due to reduced availability of fodder. The model does not allow analysis of consequent changes in the local labor markets (this would require a general circulation model). However, Hein et al. [74] concluded that climatic changes leading to a reduction in livestock numbers would result in reduced employment opportunities in the Ferlo and create further pressure on local resources.

4.2. Energy, Plants and Habitat Production

In the dryland systems, where sufficient water is available, solar energy is rapidly converted to net primary production, providing valuable plant products and habitats. Plant production can include wild or cultivated plants for use by humans (e.g., as food, medicines, and other non-timber forest products) or for the habitat and amenities that they provide collectively in situ (e.g., as described in [93]). These are in addition to supporting animal production (see previous section). Often, plant and energy production can be valued using market prices. However, due to value-adding processes for energy, food and other products, production functions may reveal higher values than the available market prices for the raw material.

Valuation of wild plant products is often challenging due to lack of statistical information. In some cases, annual inventories of the value of the non-wood products from the forests and also the revenues gained from permits and taxes, etc., may also be available (e.g., [94] for Senegal). It can be possible to obtain information on agricultural crop production and prices through the national statistical systems (e.g., [58] for Senegal, or [95] for Kenya]. Where information about plant products and their value is not complete, or not available at a suitable scale, rapid appraisal techniques are frequently used to obtain relevant information, when required (see e.g., Box 2).There are also many modeling tools available and applied for understanding the production of crops in the sub-Saharan agroecosystems (see e.g., [96] in West Africa, [97–99] in East Africa).

Across sub-Saharan Africa, the amounts of energy consumed from different sources including wood-fuels for cooking and others are calculated in the national emissions inventories based on population numbers and assumptions concerning the locally available fuel sources. For example, in Kenya, it is estimated that 50% of households depend on charcoal and one person uses about 1.9 kg of charcoal a day [100]. The majority of producers do not have tree-replanting schemes. They also use wet wood in earth mould kilns to convert wood to charcoal. This produces a low conversion of solar energy and woody biomass to charcoal. Increasing the conversion efficiency would increase fuel value achieved per volumetric unit of wood, as well as reducing pollution and habitat degradation where wood is over-harvested (see [101]). As yet, direct use of solar energy without conversion to biomass is still relatively limited in Sub-Saharan Africa. The use of fossil fuels from within the region is also not yet widespread—although often such fuels are imported.

At present, charcoal-burning is often practiced as an income-generating strategy and can be one of the few available sources of income for households in Sub-Saharan dryland ecosystems either alongside livestock-raising or as a substitute for households who do not have access to livestock. Charcoal can be sold and transported to supply urban households with fuel. In Kenya, use of charcoal in urban areas has risen by 64% in two decades [102]. The charcoal industry is estimated to employ 500,000 people and generate more than US$427 million that benefits grassroots communities. Yet it remains illegal and is not captured in national statistics or fiscal systems.

Where communities produce crops for subsistence uses, available statistics may be incomplete and provide an inadequate indication of their economic value. Market gardening production—which can generate numerous harvests throughout the year may also not be well-captured in national statistical systems. Irrigated market gardening can generate significant contributions to household income in marginal dry areas and is a popular investment choice for development projects in many parts of Sub-Saharan Africa (see e.g., an economic assessment in the Volta Basin [103]).

Where plants have been cultivated, there will be input costs including labor and others that must be weighed against the output value of the crop. If the inputs are subsidized, the economic costs to society will be higher than the financial costs paid by individual farmers. There may also be opportunity costs to society associated with inputs such as water or conversions of land-use—even if these are not considered to have a high market value. Examples include irrigation schemes constructed in the East African pastoral rangelands [104–106]. Also in northern Nigeria [107–110], the Inner Niger delta in Mali [111–114] and parts of Mauritania [115,116]. A recent study in the arid lands of Kenya [48] demonstrated that irrigated agriculture could generate a value of US$0–4 per unit of water applied, whereas extensive production could generate US$13–22 and other domestic uses up to US$90.

Output values for wood, charcoal and crops are usually identified in relation to market prices. Where there are no markets, benefit transfer is often proposed as a valuation method. Examples of cases where benefit transfer is frequently applied include medicinal plants, such as incense, that may have a high value in some markets, even if this value is not yet accessible to communities in many dryland areas. Some medicinal and aromatic plants require processing to extract high-value products, such as essential oils and dyes. This raises questions about real and hypothetical value chains, similar to those identified for livestock products. In addition, some of the use values of plants e.g., for medicines may not yet be fully understood and could require years of future research to uncover. It is very difficult to assess the cost to society if these plants are not conserved due to habitat loss and extinctions.

Plants, trees and natural fibers are used to generate agreeable habitats for humans, including shaded areas, cooler, well-ventilated buildings, fans, etc. They create cleaner air and reduce urban heat island effects. Bamboo, raffia and other fibres provide lightweight, portable and biodegradable furniture, ornaments and accessories that are aesthetically pleasing. Many relevant practices are well-known and widely available in vernacular building styles and traditional handicrafts. More are also continually evolving.

Beyond individual plant products, the value of plant assemblages and habitats for wildlife tend to be assessed based on wildlife tourism revenues, according to the travel cost method [42]. Sometimes, conservationists are willing to pay also through other arrangements—e.g., in eastern Africa, Maasai land owners whose grazing lands fall within the ecologically important Kitengela wildlife dispersal area in Kenya have entered into land leasing agreements with conservation groups who lease land from households in return for commitments to maintain the land unfenced and open to both wildlife and livestock [117–119].

Willingness to pay among the Nairobi public for securing the Kitengela wildlife dispersal area has been estimated at US$1.2 million which could provide a longer-term source of financing [120]. In the Simanjiro district of Tanzania, a similar scheme has been established where tour companies make an annual payment of US$4500 to communities to prevent agricultural encroachment onto a 23,000-acre wildlife corridor [121].

Box 2. Case study: valuing energy, plant and habitat production in West African agroforestry using participatory rapid appraisal of social returns (based on Weston et al. [101]).

The anticipated effects of agroforestry and improved stoves on the economic value of energy and plant provisioning in the district of Talensi in the semi-arid Upper East Region in Ghana have been assessed by Weston et al. [101]. The assessment used focus groups and a household survey to generate a social return on investment analysis (after [122,123]), which identifies proxy financial values for non-market as well as market benefits.

This enabled valuation of a large number of benefits that were generated by the agroforestry interventions, including their asset value as wood resources that could be used or sold in difficult times. But also, until then, other benefits that they would provide in situ, including effects on the production of crops, livestock and wild foods. Crop production is boosted by the presence of trees and their effects on soil quality. Livestock production is increased by the availability of shade and fodder from the trees. Wild foods are either produced directly by the trees (fruits) or by the habitat that they provide e.g., for rabbits and partridges. The presence of the trees also creates other beneficial effects that were considered valuable by the human population, including shade, reduced windspeeds, airborne dust, aesthetic and cultural effects.

Weston et al. [101] used this information to calculate social returns over three time periods: immediately after close of project activities; 4 years after project; and 10 years after the project using a discount rate of 8.9% p.a., which was Ghana's predicted inflation rate for 2012–2017. The values for each social impact were added to further values associated with the presence of the project (such as increased credit, optimism and community solidarity), aggregated into a single total value and divided by the total cost of project inputs to arrive at a social return on investment ratio for each of three time periods.

4.3. Water Availability and Supply

Access to water is usually highly valued by society in water-stressed environments. This value provides for a range of uses by different groups and individuals—including for household needs, watering of livestock, market gardening and firefighting—even if sometimes these are free of charge [124].

Water supply often has a market value—paid to service-providers or vendors. Nevertheless, the costs that society pays to secure water supplies may often be much higher than the unit prices for the water on the local markets —which may sometimes be low or non-existent due to market inefficiencies, state support or subsidies. Market prices for water supply can also rise sharply in times of drought and scarcity. Because of this complication in drought-prone dryland areas, it is particularly problematic to rely on the use of market values to assess the benefits of improved water supplies.

Valuation of water supply in the drylands can include the avoided costs of disease [125–128] and replacement expenses such as travel costs to obtain water supplies [129,130] or infrastructure, pumping and treatment costs. The input costs to society to secure water availability are very context-dependent and may be higher in remote or hard-to-reach areas than they are in densely populated areas. Furthermore, in addition to infrastructure, the costs of gaining water supplies must include investments in human capacities. These costs are often difficult to fully quantify or generalize. They are also context-specific and influenced by many other factors.

'Production-function' approaches to the valuation of water supplies focus on estimating how much value is added to the economy per unit of water supplied [109,131]. This can be relatively easily estimated for agricultural products where there is a market and the relationship of water inputs to outputs is reasonably well understood [132,133]. Production-functions for water can give values that are higher than the market prices for water due to value-adding processes. However, there are usually other inputs to production that may also need to be accounted for, in addition to the units of water.

There is often a hierarchy of different production values for water according to its various uses. Frequently, the most readily accessible value for water is the value of plant production that it can support (see previous section) (e.g., as in [109,134]). Normally, the value to society of a cubic metre of water for basic human needs should be greater than the value of a cubic metre of water to be used for irrigation.

The actual economic value of the productive uses of water will be sensitive to supply and demand factors affecting prices that are more difficult to understand and predict than the physical production aspects [57]. Often, households will use a combination of multiple water sources to support their needs and economic activities, and will change these configurations under different conditions of water scarcity [109,135]. As a result of this, measuring water's contribution to all sectors, and to the total value chain of a good or service, or its total economic benefit can also be complex [37,132,136–138].

Sometimes there is an opportunity cost for water supply if this may reduce its availability for other possible uses—e.g., where water is extracted in the upstream part of a basin and so becomes unavailable to other users downstream (see Box 3) [57]. Taking a catchment level approach to assessing the value of water availability can be particularly important to ensure that water is available for all vulnerable households' basic needs during droughts. Using infrastructure, water can be transferred across the system or catchment to be made available where it is most needed for particular uses at specific points in time.

Box 3. Case study: valuing water supply in a Kenyan rangeland using a water evaluation and planning model (based on Mutiga et al. [139] and Silvestri et al. [35]).

The anticipated effects of upstream water extraction in the Ewaso Ng'iro North Catchment, Kenya, on the economic value of ecosystem service provision across the catchment as a whole were assessed by Silvestri et al. [35] using market valuation methods to identify the values of provisioning services in the irrigated farmlands and rangelands. The study drew on exploratory work by Mutiga et al. [139], using a water evaluation and planning model (WEAP) ([140] see: www.weap21.org).

The economic value assigned to upstream irrigation extractions did not take into account the opportunity costs and externalities caused by the extraction of water from the system, which reduced its availability downstream during a period of severe drought. Losses and damage to livestock and livelihoods in the downstream rangelands were also not taken into consideration. In the valuation of services to downstream areas, livestock production was assigned a value that may have been underestimated. Furthermore, human domestic uses of water were not valued at all—neither upstream nor downstream. Based on this assessment, continued increases in upstream water use for irrigation could be justified in light of their apparently major contribution to the economy.

Subsequent studies suggested that reduced access to water supply for humans and livestock in downstream areas during droughts caused unsustainable losses of lives, assets and livelihoods, which in turn seriously affected the growth of the national economy [141]. Furthermore, the value of downstream livestock production and domestic water uses has increasingly gained in recognition [48,142,143]. These observations suggest that the rationale for continued upstream extractions should be re-evaluated in light of increased understanding of the resulting externalities downstream.

Although in these studies the WEAP was demonstrated to be a useful tool for modelling the physical aspects water allocation across the catchment, there is a critical need for careful selection of the assumptions and valuation methods to be assigned to the different water-use alternatives represented through the model.

4.4. Supporting and Regulating Services for the Functioning of the Dryland Ecosystems

Critical supporting and regulating processes for the functioning of the dryland ecosystems can include regulation of water flows, flood defense, groundwater recharge, water purification, storage, and soil conservation, including for carbon sequestration (see Box 4). The value of the supporting services can be captured in terms of the avoidance of damage (e.g., for flood-prevention) and replacement costs to reproduce these services if they were not provided by nature (e.g., costs of infrastructure for pumping and treating water) [44,144,145].

Various studies have explored public willingness to pay for ecosystem services that can be achieved through soil and water management, groundwater recharge and revegetation to sustain ecosystems (e.g., by regulating stream-flow and reducing flood risks) or to increase downstream water quality. For example, in South Africa, Turpie et al. [146] assessed public demand and willingness to pay for groundwater recharge, flood attenuation and flow regulation using a geographical information system. Due to water scarcity, they assumed that groundwater recharge was fully demanded and that therefore the entire public would be willing to pay for it, but that demand for flood attenuation should be adjusted by a demand factor applied at a secondary catchment scale. If urban areas or mines occurred within 100 m of rivers then the service was considered fully demanded, if irrigated agriculture was present, the service was considered 50% demanded, otherwise it was not considered to be demanded. Maps of households and their estimated willingness to pay were used to generate estimates of the economic value of the services to society.

For over a decade, researchers have been exploring scope for payments for watershed services in Kenya's Tana River catchment to avert downstream water shortages and treatment costs. Regulating runoff and preventing soil erosion upstream is expected to increase the quality of water downstream but farmers will not maintain these practices without sufficient economic incentives [147]. On the other hand, construction of dams upstream will reduce the availability of drinking water downstream [148,149]. Recently, a business case has been presented for a water fund through which a company that supplies water to Nairobi by diverting water from the Tana river can pay for upstream soil conservation [150–152]. The water fund agreed to provide payments to upstream tea and coffee farmers to regulate water flows, runoff and erosion by maintaining the upstream parts of the catchment. No payments will be made to the downstream users of the catchment whose supply is reduced but the Kenyan government is investing in a seawater desalination plant to boost downstream urban water supplies.

Other aspects of the economics of soil erosion control can involve on-site productivity and carbon sequestration [153]. International markets for payments for carbon sequestration services are increasingly becoming established through global initiatives to mitigate climate change. Such schemes have been explored in various parts of Africa. Some additional relevant valuation approaches include option values which have been used for valuation of rights to future services e.g., as they would for energy exploration. These have a market value and can be traded. No examples of trading in rights to water regulation and storage have been identified in Sub-Saharan Africa during the preparation of this review, but trading in water rights has been introduced in water scarce areas of North America and Australia [154]. Access to water affects the value of land in many parts of the world, including Sub-Saharan Africa [155]. A relevant valuation approach, known as hedonic pricing, focuses on capturing the difference in market value of properties with or without access to the service [156,157]. Such an approach has been applied in other parts of the world, e.g., to assess willingness to pay to avoid erosion and sedimentation of lakes and improve water quality [158,159] or flood risk [160]. Recently, this approach has been used to identify the value of reduced flood risks and other improvements to living conditions in African cities [161].

Where water storage in dry areas enables populations to reduce vulnerability to drought, this can be valued in terms of the avoided costs e.g., avoided reductions in livestock productivity and/or avoided mortalities [28,141]. However, sometimes the lack of proven counterfactual evidence reduces the credibility of claims of damages avoided. To assess the value of increases in water storage and availability—e.g., through increased groundwater recharge, the anticipated future use of the water can have a value.

Assessing the economics of groundwater management effects can require determination of an optimal extraction path for an aquifer over time. A spatial externality in which excessive pumping at one location on the aquifer lowers the water table at another location may also require attention [162–164]. Practical challenges for economic assessment of this and other effects on groundwater availability under different land management conditions include the problem that groundwater conditions and extraction rates are subject to strategic private management and information constraints [165,166].

Box 4. Case study: assessing water regulation and carbon sequestration using InVEST model (based on: Sahle et al [167,168] and others cited).

The anticipated effects on ecosystem service provision to be achieved through introducing agroforestry in the Wabe River catchment of the Gurage Mountain chain landscape, Ethiopia, over the period 2017–2030 have been modelled by Sahle et al [167,168] using the InVEST model [169]. Under a future land-use scenario with increased agroforestry, improvements could include increased food production (102%), water yields (17%) and carbon sequestration (19%), and reduced sediment export (21%) and soil loss (18%). Varying the extent of agroforestry activities would generate altered estimates, enabling assessment of tradeoffs among the different options. Quantified estimates of water yield and carbon sequestration generated using InVEST can be assigned economic values using market prices or other methods to facilitate the assessment of tradeoffs.

Water yield in InVEST is defined as the amount of water that runs off the landscape (precipitation minus storage and evapotranspiration losses) [170]. Limitations of InVEST include an annual timeframe, and also that surface and base flow partitioning and inter-annual water delivery timing are not captured so it cannot assess effects on groundwater recharge and all other estimates are insensitive to these critical processes in the dryland ecosystems [169]. This means that the model may be of limited use in assessing the effects of investments other than revegetation.

InVEST has proved useful for obtaining broad estimates of the effects on water yield based on limited input information where such information is not available to support other modelling tools—e.g., in the Nile Basin [171]. The InVEST model has also been applied in other areas of Sub-Saharan Africa, including Rwanda [172], Ghana, and Cote D'Ivoire [173].

Other hydrological modelling approaches that have been explored to generate assessments of effects of vegetation on hydrology, including flow regulation and storage that are more sensitive to variations in land surface conditions include the soil water assessment tool (SWAT) [145,151,174], among others (see e.g., [144]).

5. Discussion

Even though economic evaluation of ecosystem service benefit streams from the dryland ecosystems have been relatively rare and limited, a significant body of relevant examples is available in Sub-Saharan Africa, as identified in this review. These include assessments of critical benefit streams from ecosystem service values associated with water availability, energy, food, and the underlying ecosystem support functions. Increasingly, assessments are taking into consideration the value of water and energy resources—both as a supporting and regulating features in the drylands as well as essential commodities that contribute directly to human well-being.

Examples of assessments applied across the region demonstrate that it has already been possible for practitioners to assess the extent and economic value of some these benefit streams on a piecemeal basis, even if comprehensive long-term assessments have not yet been attempted. A range of adaptive management practices that affect these benefit streams are also identified. It is important to acknowledge that the selection of benefit streams considered in this review was not exhaustive and there can be many other benefit streams worthy of attention.

For the benefit streams considered, the compilation of examples that is presented in this review is also not exhaustive. Yet, the clear affirmation that the selected benefit streams can and have been valued in both West and East Africa should encourage practitioners that further such assessments are feasible. Collectively, the precedents and examples drawn together in this review demonstrate the potential for more systematic assessment of the current level of benefits from the dryland systems and of the scope for investments in further improving management practices.

This should encourage practitioners to continue to identify and value the baseline values of benefit streams from the existing management systems in the drylands and the potential for adaptive management. This will create a stronger business case for investment—both for donors and for local communities to reinvest.

Where effects on benefit streams are predicted at the outset of projects, it is important to recognize that not all investments succeed as planned (personal communication: Obadiah Mungai, 26 May 2018). Some investments can fail or may not secure the benefits that are anticipated. Therefore, careful monitoring ex-post assessment following the investment process is important. To build necessary institutions and capacities for these processes in order to support adaptive management may require additional investments in human capacities.

The diversity and layering of multiple benefits that dryland ecosystems provide can pose a practical challenge for assessment. Not just one but often a range of different types of goods and services are generated and can be affected by investments in local resource management in the drylands. For example, a water-harvesting investment can simultaneously generate all of the types of benefit streams described. So can an investment in agroforestry.

In most cases, some very important aspects of the benefits streams and adaptations will escape economic evaluation. The real costs of illnesses in terms of missed opportunities and conflicts in the communities are difficult to identify but the benefits from avoiding these problems are likely to be enormous. Where these are secured, the local institutions can also secure further benefits, such as increased confidence and capacities, and other changes in perceptions and social processes.

Synergies between benefits can introduce further complexity and often escape economic assessment. This problematizes the additive summation of benefits since, with synergies, the final sum-total of the benefits will be more than the combined value of the individual benefit streams. For example, rangeland restoration may improve the availability of fodder, milk and tree growth, while also enabling lactating livestock to stay closer to settlements. The end result of this will be of greater benefit to household than the sum of the value of the milk production and the reduced journey times for herding and wood-collecting because the households having access to free supplies of milk and fuel for cooking will have better diets, lower expenditures, and more time to engage in productive activities. These may enable households to multiply the benefits that they obtain.

There may also be cost-savings to the humanitarian and international development communities, national governments and society more broadly due to reduced vulnerability to droughts and floods and avoidance of the associated losses and damages.

6. Recommendations and Conclusions

The review has identified some key benefits from local investments in dryland ecosystems that are worthy of additional consideration because they are essential to the livelihoods of the local populations—such as the value of livestock production, water resources, energy in the forests, and others. We have also identified available methods to fill the strategic gaps in the available statistical systems. It is recommended that decisionmakers should choose the tools, systems and calculations that will best capture the value of their investments and the benefits that are most critical to society.

This paper has demonstrated that economic assessment of some of the returns on investments in dryland ecosystems is possible for project managers and researchers to undertake even at the early stages of the implementation of projects. Practical insights presented in the paper demonstrate that such assessments can at least include one or two benefits that will appear during the first year. In the exploration of the assessment process, it is likely that other benefits will be identified that will accumulate over the longer term and hence may be more challenging to assess. Wherever, possible, mainstreaming of systems to monitor and assess critical benefit streams into national statistical systems is strongly recommended.

As decision-makers and practitioners increase their use of the available assessment methods to value the benefit streams that are generated in dryland ecosystems, this will improve the availability of baseline information. Improving the baselines will facilitate the assessment of potential impacts from investment alternatives. The wider global community of dryland researchers and development practitioners could accelerate the current state of the art in the assessment of benefits from investments in dryland ecosystems by building these institutional capacities to generate such assessments as well as through continued support for physical land and water management improvements both upstream and downstream in dryland ecosystems.

Funding: The preparation of this manuscript received no external funding.

Acknowledgments: In-kind support received from the following sources should be acknowledged: IUCN, KEFRI, WOCAT, IIED and the ADA Consortium, GIZ ELD Initiative, the University of Southampton and the Oxford Deserts Conference. None of the above bears any responsibility for the final content of the manuscript, including any errors or oversights that it may contain. The author also wishes to thank two anonymous reviewers and the editorial office for outstanding Editorial Assistance and support.

Conflicts of Interest: The author declares no conflict of interest

Appendix A

Box A1. Examples of community-scale practices and investments for adaptive management in dryland ecosystems.

To maximize services from livestock production:

- Increase fodder availability or introducing supplementary feeding;
- Improve watering facilities and transhumance routes and patterns in extensive systems enabling access to water and pasture during dry seasons and droughts;
- Improve veterinary facilities, vaccination and disease-reduction programmes.

To maximize services from plant production:

- Enclosure of vegetated areas to prevent over-grazing or extraction by humans;
- Water-harvesting and irrigation systems to prevent crop failure during droughts and enable more frequent harvests e.g., for market gardening;
- Improve storage facilities to prevent post-harvest losses due to floods and fires.

To maximize services from energy production:

- Improve conversion efficiencies of stoves and burners;
- Increase use of solar-powered systems for pumping and treatment of water;
- Establish mini-grids for distribution of renewable energy supplies to households.

To maximize services from water supply:

- Improve surface water pans, reservoirs and other water-harvesting structures;
- Deepen shallow wells and digging boreholes to accelerate access to groundwater reserves;
- Install and improve water supply and treatment networks.

To maximize supporting and regulating services:

- Soil and water management to prevent erosion and increase infiltration and groundwater storage and boost carbon sequestration;
- Revegetation—e.g., afforestation or construction of wetlands to improve soil quality, settle contaminants, increase infiltration and boost carbon sequestration;
- Institutional strengthening to manage the above and other practices, as needed.

References

1. Chambwera, M.; Heal, G.; Dubeux, C.; Hallegatte, S.; Leclerc, L.; Markandya, A.; McCarl, B.A.; Mechler, R.; Neumann, E.J. Chapter 17: Economics of Adaptation. In *Climate Change 2014: Impacts, Adaptation, and Vulnerability. Part A: Global and Sectoral Aspects. Contribution of Working Group II to the Fifth Assessment Report of the Intergovernmental Panel on Climate Change*; Field, C.B., Barros, V.R., Dokken, D.J., Mach, K.J., Mastrandrea, M.D., Bilir, T.E., Chatterjee, M., Ebi, K.L., Estrada, Y.O., Genova, R.C., et al., Eds.; Cambridge University Press: Cambridge, UK; New York, NY, USA, 2014; p. 33.
2. Economics of Land Degradation Initiative (ELD); United Nations Environment Programme (UNEP). *The Economics of Land Degradation in Africa: Benefits of Action Outweigh the Costs*; Economics of Land Degradation Initiative: Bonn, Germany, 2015; p. 156.
3. Emerton, L. *Valuing the Benefits, Costs and Impacts of Ecosystem-Based Adaptation Measures—A Sourcebook of Methods for Decision-Making*; Deutsche Gesellschaft für Internationale Zusammenarbeit (GIZ) GmbH: Bonn, Germany, 2017; p. 288.
4. IPBES Secretariat. *Chapters of the Thematic Assessment Report on Land Degradation and Restoration of the Intergovernmental Science-Policy Platform on Biodiversity and Ecosystem Services*; IPBES Secretariat: Bonn, Germany, 2018.
5. United Nations Convention to Combat Desertification (UNCCCD). *Global Land Outlook*; UNCCD: Bonn, Germany, 2017; p. 340.
6. Davies, J. *The Land in Drylands: Thriving in Uncertainty through Diversity*; United Nations Convention to Combat Desertification (UNCCD): Bonn, Germany, 2017; p. 18.
7. Cohen, S. *African Drylands Commodity Atlas*; Secretariat of the United Nations Convention to Combat Desertification, the Common Fund for Commodities Intergovernmental Group on Grains of the United Nations Food and Agriculture Organization, International Crops Research Institute for the Semi-Arid Tropics: Bonn, Germany, 2009; p. 82.
8. O'Keefe, L.; Howes, M. A Select Annotated Bibliography: Indigenous Technical Knowledge in Development. *IDS Bull.* **1979**, *10*, 51–58. [CrossRef]
9. Brokensha, D.W.; Warren, D.M.; Werner, O. (Eds.) *Indigenous Knowledge Systems and Development*; University Press of America: Lanham, MD, USA, 1980; p. 466.
10. Warren, D.M. *Using Indigenous Knowledge in Agricultural Development*; World Bank: Washington, DC, USA, 1991; p. 58.
11. Villamor, G.B.; Badmos, B.K. Grazing game: A learning tool for adaptive management in response to climate variability in semiarid areas of Ghana. *Ecol. Soc.* **2016**, *21*, 39. [CrossRef]
12. *Return on Investment for Emergency Preparedness Study*; The Boston Consulting Group, World Food Programme (WFP)/The United Nations International Children's Emergency Fund (UNICEF): Boston, Germany, 2015; p. 35.
13. Bond, C.A.; Strong, A.; Burger, N.; Weilant, S.; Saya, U.; Chandra, A. *Resilience Dividend Valuation Model Framework Development and Initial Case Studies*; RAND Corporation: Santa Monica, CA, USA, 2017; p. 178.
14. Venton, C.C. *Economics of Resilience to Drought*; United States Agency for International Development (USAID): Washington, DC, USA, 2018; p. 43.
15. King, C. Sustainable Management of Marginal Drylands (SUMAMAD): Sharing Success Stories from Indigenous, Adaptive and Innovative Approaches. In *The Future of Drylands*; Lee, C., Schaaf, T., Eds.; Springer: Dordrecht, The Netherlands, 2008; pp. 747–761.
16. Reid, H. Ecosystem- and community-based adaptation: Learning from community-based natural resource management. *Clim. Dev.* **2016**, *8*, 4–9. [CrossRef]
17. Bertram, M.; Barrow, E.; Blackwood, K.; Rizvi, A.R.; Reid, H.; Scheliha-Dawid, S.V. *Making Ecosystem-Based Adaptation Effective: A Framework for Defining Qualification Criteria and Quality Standard*; International Institute for Environment and Development (IIED): London, UK; International Union for Conservation of Nature (IUCN): Gland, Switzerland, 2017; p. 14.
18. Vignola, R.; Locatelli, B.; Martinez, C.; Imbach, P. Ecosystem-based adaptation to climate change: What role for policy-makers, society and scientists? *Mitig. Adapt. Strat. Glob. Chang.* **2009**, *14*, 691–696. [CrossRef]

19. Doswald, N.; Munroe, R.; Roe, D.; Giuliani, A.; Castelli, I.; Stephens, J.; Möller, I.; Spencer, T.; Vira, B.; Reid, H. Effectiveness of ecosystem-based approaches for adaptation: Review of the evidence-base. *Clim. Dev.* **2014**, *6*, 185–201. [CrossRef]
20. WWDR. *The United Nations World Water Development Report 2018: Nature-Based Solutions for Water*; United Nations World Water Assessment Programme/UN-Water/UNESCO: Paris, France, 2018; p. 154.
21. Trærup, S.; Stephan, J. Technologies for adaptation to climate change. Examples from the agricultural and water sectors in Lebanon. *Clim. Chang.* **2014**, *131*, 435–449. [CrossRef]
22. *Economic Evaluation of Climate Change Adaptation Projects Approaches for the Agricultural Sector and Beyond*; World Bank: Washington, DC, USA, 2010; p. 66.
23. *Guidelines for the Economic Analysis of Projects*; Asian Development Bank (ADB): Manilla, Philippines, 2017; p. 170.
24. Sartori, D.; Catalano, G.; Genco, M.; Pancotti, C.; Sirtori, E.; Vignetti, S.; Bo, C.D. *Guide to Cost-Benefit Analysis of Investment Projects Economic Appraisal Tool for Cohesion Policy 2014–2020*; European Commission: Brussels, Belgium, 2015; p. 364.
25. Watkiss, P.; Hunt, A.; Blyth, W.; Dyszynski, J. The use of new economic decision support tools for adaptation assessment: A review of methods and applications, towards guidance on applicability. *Clim. Chang.* **2015**, *132*, 401–416. [CrossRef]
26. *The Economic Advantage: Assessing the Value of Climate Change Actions in Agriculture*; International Fund for Agricultural Development (IFAD): Rome, Italy; International Center for Tropical Agriculture (CIAT): Cali, CO, USA; CGIAR Research Program on Climate Change, Agriculture and Food Security (CCAFS): Wageningen, The Netherlands, 2016; p. 8.
27. Vermeulen, S.; Richards, M.; Pinto, A.D.; Ferrarese, D.; Läderach, P.; Lan, L.; Luckert, M.; Mazzoli, E.; Plant, L.; Rinaldi, R.; et al. *The Economic Advantage: Assessing the Value of Climate Change Actions in Agriculture: Economic Evidence from Research by IFAD, CGIAR and Partners*; IFAD: Rome, Italy; CGIAR: Monpellier, France, 2016.
28. Venton, C.C.; Fitzgibbon, C.; Shitarek, T.; Coulter, L.; Dooley, O. *The Economics of Early Response and Disaster Resilience: Lessons from Kenya and Ethiopia*; Department for International Development (DFID): London, UK, 2012; p. 84.
29. Vardakoulias, O.; Nicholles, N. *Managing Uncertainty: An Economic Evaluation of Community-Based Adaptation in Dakoro, Niger*; New Economics Foundation, CARE: London, UK, 2014; p. 53.
30. Siedenburg, J. *Community-Based Cost Benefit Analysis (CBCBA). Findings from DFID Kenya's Arid Lands Support Programme*; Landell Mills: London, UK, 2016; p. 62.
31. Bond, C.A.; Strong, A.; Burger, N.; Weilant, A.S. *Guide to the Resilience Dividend Model*; RAND Corporation: Santa Monica, CA, USA, 2017; p. 43.
32. Shreve, C.M.; Kelman, I. Does mitigation save? Reviewing cost-benefit analyses of disaster risk reduction. *Int. J. Disaster Risk Reduct.* **2014**, *10*, 213–235. [CrossRef]
33. Logar, I.; van den Bergh, J.C.J.M. Methods to Assess Costs of Drought Damages and Policies for Drought Mitigation and Adaptation: Review and Recommendations. *Water Resour. Manag.* **2013**, *27*, 1707–1720. [CrossRef]
34. Adeel, Z.; Safriel, U.; Niemeijer, D.; White, R. *Ecosystems and Human Well-Being: Desertification Synthesis: A Report of the Millennium Ecosystem Assessment*; World Resources Institute: Washington, DC, USA, 2005; p. 36.
35. Silvestri, S.; Zaibet, L.; Said, M.Y.; Kifugo, S.C. Valuing ecosystem services for conservation and development purposes: A case study from Kenya. *Environ. Sci. Policy* **2013**, *31*, 23–33. [CrossRef]
36. Peskin, H.M. A national accounting framework for environmental assets. *J. Environ. Econ. Manag.* **1976**, *2*, 255–262. [CrossRef]
37. United Nations (UN). *System of Environmental Economic Accounting for Water*; United Nations: New York, NY, USA, 2012; p. 216.
38. *System of Environmental-Economic Accounting for Energy (SEEA-Energy) Prepared under the Auspices of the United Nations Committee of Experts Environmental-Economic Accounting*; United Nations Statistics Division (UNSD), United Nations Department of Economic and Social Affairs (DESA): New York, NY, USA, 2015; p. 174.
39. Favretto, N.; Stringer, L.C.; Dougill, A.J.; Dallimer, M.; Perkins, J.S.; Reed, M.S.; Atlhopheng, J.R.; Mulale, K. Multi-Criteria Decision Analysis to identify dryland ecosystem service trade-offs under different rangeland land uses. *Ecosyst. Serv.* **2016**, *17*, 142–151. [CrossRef]

40. Favretto, N.; Luedeling, E.; Stringer, L.C.; Dougill, A.J. Valuing Ecosystem Services in Semi-arid Rangelands through Stochastic Simulation. *Land Degrad. Dev.* **2017**, *28*, 65–73. [CrossRef]
41. Reed, M.S.; Stringer, L.C.; Dougill, A.J.; Perkins, J.S.; Atlhopheng, J.R.; Mulale, K.; Favretto, N. Reorienting land degradation towards sustainable land management: Linking sustainable livelihoods with ecosystem services in rangeland systems. *J. Environ. Manag.* **2015**, *151*, 472–485. [CrossRef] [PubMed]
42. Hanley, N.; Barbier, E.B.; Barbier, E. *Pricing Nature: Cost-Benefit Analysis and Environmental Policy*; Edward Elgar: Cheltenham, UK, 2009.
43. Costanza, R.; d'Arge, R.; DeGroot, R.S.; Farber, S.; Grasso, M.; Hannon, B.; Limburg, K.; Naeem, S.; O'Neill, R.V.; Paruel, J.; et al. The value of the world's ecosystem service and natural capital. *Nature* **1997**, *387*, 253–260. [CrossRef]
44. Emerton, L. *Environmental Planning and Management in Eastern Africa.*; International Union for Conservation of Nature (IUCN)—The World Conservation Union, Eastern Africa Regional Office: Nairobi, Kenya, 1999.
45. *The Economics of Ecosystems and Biodiversity Ecological and Economic Foundations*; Kumar, P. (Ed.) Earthscan: London, UK; Washington, DC, USA, 2010; p. 133.
46. de Groot, R.; Brander, L.; van der Ploeg, S.; Costanza, R.; Bernard, F.; Braat, L.; Christie, M.; Crossman, N.; Ghermandi, A.; Hein, L.; et al. Global estimates of the value of ecosystems and their services in monetary units. *Ecosyst. Serv.* **2012**, *1*, 50–61. [CrossRef]
47. King-Okumu, C. *A Framework to Assess Returns on Investments in the Dryland Systems of Northern Kenya*; International Institute for Environment and Development (IIED): London, UK, 2015.
48. King-Okumu, C.; Wasonga, O.V.; Jarso, I.; Salah, Y.M.S. *Direct Use Values of Climate-Dependent Ecosystem Services in Isiolo County, Kenya*; International Institute for Environment and Development (IIED): London, UK, 2016.
49. King-Okumu, C. *Distilling the Value of Water Investments*; International Institute for Environment and Development (IIED): London, UK, 2016; p. 2.
50. King-Okumu, C.; Diop, D. *Economic Valuation of Benefits from Adaptation Investments: A Methodological Note for Assessment of Returns on Locally Determined Investments in Adaptation to Climate Extremes in the Senegalese Region of Kaffrine [Evaluation économique des bénéfices de l'adaptation: Note Méthodologique pour évaluer la valeur du retour sur les investissements localement déterminés pour s'adapter aux extrêmes climatiques dans la Région de Kaffrine au Ferlo du Sénégal]*; International Institute for Environment and Development (IIED): London, IK, 2017; p. 70.
51. King-Okumu, C.; Diop, D.; McPeak, J. Approche d'evaluation des retours sur investissements des projets d'adaptation au changement climatiqe a Kaffrine. *AGRIDAPES Revue de L'agriculture Durable a Faibles Apports Externes*, 1 February 2018; Spécial DFC. 43–47.
52. Coulibaly, B.; King-Okumu, C.; Keita, A. Analyse preliminaire des incidences socio-economiques des projets de renforcement de la resilience a Mopti. *AGRIDAPES Revue de L'agriculture Durable a Faibles Apports Externes*, 1 February 2018; Spécial DFC. 39–42.
53. The Government of Kenya (GoK). *Submission in the Area of Ecosystems, Interrelated Areas Such as Water Resources and Adaptation under the Nairobi Work Programme*; NEMA (National Environment Management Authority)/National Drought Management Authority (NDMA)/The Adaptation (ADA) Consortium: Nairobi, Kenya, 2017; p. 6.
54. *Adaptation Planning, Implementation and Evaluation Addressing Ecosystems and Areas Such as Water Resources—Synthesis Report by the Secretariat*; The United Nations Framework Convention on Climate Change (UNFCCC): Bonn, Germany, 2017; p. 33.
55. *World Atlas of Desertification*; Cherlet, M., Hutchinson, C., Reynolds, J., Hill, J., Sommer, S., von Maltitz, G., Eds.; Publication Office of the European Union, Joint Research Centre (JRC): Luxembourg, Germany, 2018; p. 248.
56. World Overview of Conservation Approaches and Technologies (WOCAT). *Sustainable Rangeland Management in Sub-Saharan Africa—Guidelines to Good Practice*; World Overview of Conservation Approaches and Technologies: Bern, Germany, 2018; forthcoming.
57. Turpie, J.; Ngaga, Y.; Karanja, F. *Catchment Ecosystems and Downstream Water: The Value of Water Resources in the Pangani Basin, Tanzania*; The World Conservation Union, Ecosystems and Livelihoods Group Asia, International Union for Conservation of Nature (IUCN): Vientiane, Laos, 2005; p. 108.

58. Fofana, I.; Tankari, M.; Traore, F. *Economic Accounts for Agriculture and Farm Income in Senegal*; International Food Policy Research Institute (IFPRI): Dakar, Senegal, 2017; p. 44.
59. Guilhem, D.; Boëtsch, G. Breastmilk or cow's milk? Production of the child's body by two liquid foster among the Fulani of Ferlo (Senegal) [Lait maternel ou lait de vache? La production du corps de l'enfant par deux liquides nourriciers chez les Peuls du Ferlo (Sénégal)]. *Anthropozoologica* **2017**, *52*, 91–101. [CrossRef]
60. Sougou, N.; Boëtsch, G. Diet and growth of young Fulani children in Widou Thiengoli (Ferlo, Senegal) [Alimentation et croissance des jeunes enfants Peuls à Widou Thiengoly (Ferlo—Sénégal)]. *Bulletins et Memoires de la Societe d'Anthropologie de Paris* **2016**, *28*, 145–154.
61. Musinga, M.; Kimenye, D.; Kivolonzi, P. *The Camel Milk Industry in Kenya*; Resource Mobilization Centre/SNV: Nanyuki, Kenya, 2008; p. 101.
62. Gaye, M. *Tiviski Dairy: Africa's First Camel Milk Dairy Improves Livelihoods for Semi-Nomadic Herders in Mauritania*; United Nations Development Programme (UNDP): New York, NY, USA, 2008.
63. Anderson, D.M.; Elliott, H.; Kochore, H.H.; Lochery, E. Camel herders, middlewomen, and urban milk bars: The commodification of camel milk in Kenya. *J. East. Afr. Stud.* **2012**, *6*, 383–404. [CrossRef]
64. Parisse, M. Developing local dairy production: The laiterie du Berger, Senegal. *Field Actions Sci. Rep. (FACTS Rep.)* **2012**, *6*.
65. *Technical Report: Mapping Camel-Milk Corridors in Wajir County*; Mercy Corps (MC): Nairobi, Kenya, 2017; p. 37.
66. Moussa, B.; Nkonya, E.; Meyer, S.; Kato, E.; Johnson, T.; Hawkins, J. Economics of Land Degradation and Improvement in Niger. In *Economics of Land Degradation and Improvement—A Global Assessment for Sustainable Development*; Nkonya, E., Mirzabaev, A., von Braun, J., Eds.; Springer: Basingstoke, UK, 2015; pp. 499–539.
67. Behnke, R.; Muthami, D. *The Contribution of Livestock to the Kenyan Economy*; IGAD Livestock Policy Initiative: Addis Ababa, Ethiopia, 2011; p. 62.
68. Elhadi, Y.A.M.; Wasonga, O.V. *Economic and Nutritional Contribution of Camel Milk in Northern Kenya: A Field Study in Isiolo County*; International Institute for Environment and Development (IIED): London, UK, 2015; p. 40.
69. Karp, L.; Pope, A. Range Management under Uncertainty. *Am. J. Agric. Econ.* **1984**, *66*, 437–446. [CrossRef]
70. Huffaker, R.; Cooper, K. Plant Succession as a Natural Range Restoration Factor in Private Livestock Enterprises. *Am. J. Agric. Econ.* **1995**, *77*, 901–913. [CrossRef]
71. Perrings, C.; Walker, B. Conservation in the Optimal Use of Rangelands. *Ecol. Econ.* **2004**, *49*, 119–128. [CrossRef]
72. Janssen, M.A.; Anderies, J.M.; Walker, B.H. Robust Strategies for Managing Rangelands with Multiple Stable Attractors. *J. Environ. Econ. Manag.* **2004**, *47*, 140–162. [CrossRef]
73. Finnoff, D.; Strong, A.; Tschirhart, J. A Bioeconomic Model of Cattle Stocking on Rangeland Threatened by Invasive Plants and Nitrogen Deposition. *Am. J. Agric. Econ.* **2008**, *90*, 1074–1090. [CrossRef]
74. Hein, L.; Metzger, M.J.; Leemans, R. The local impacts of climate change in the Ferlo, Western Sahel. *Clim. Chang.* **2009**, *93*, 465–483. [CrossRef]
75. Baumgärtner, J.; Gilioli, G.; Tikubet, G.; Gutierrez, A.P. Eco-social analysis of an East African agro-pastoral system: Management of tsetse and bovine trypanosomiasis. *Ecol. Econ.* **2008**, *65*, 125–135. [CrossRef]
76. Gutierrez, A.P.; Gilioli, G.; Baumgärtner, J. Ecosocial consequences and policy implications of disease management in East African agropastoral systems. *Proc. Natl. Acad. Sci. USA* **2009**, *106*, 13136–13141. [CrossRef] [PubMed]
77. Mulinge, W.; Gicheru, P.; Murithi, F.; Maingi, P.; Kihiu, E.; Kirui, O.K.; Mirzabaev, A. Economics of land degradation and improvement in Kenya. In *Economics of Land Degradation and Improvement—A Global Assessment for Sustainable Development*; International Food Policy Research Institute: Washington, DC, USA, 2015; pp. 471–498.
78. Herlocker, D.J.; Shaaban, S.B.; Wilkes, S. *Range Management Handbook of Kenya*; Ministry of Agriculture, Livestock Development and Marketing: Nairobi, Kenya, 1993.
79. Onduru, D.D.; Muchena, F.N. *Cost-Benefit Analysis of Land Management Options in the Upper Tana, Kenya*; ISRIC-World Soil Information: Wageningen, The Netherlands, 2011; p. 136.
80. FAO. Boosting Emergency Drought Response in Kenya. Available online: http://www.fao.org/resilience/news-events/detail/en/c/1027789/ (accessed on 21 May 2018).
81. De Groot, R.S.; Blignaut, J.; Van Der Ploeg, S.; Aronson, J.; Elmqvist, T.; Farley, J. Benefits of Investing in Ecosystem Restoration. *Conserv. Boil.* **2013**, *27*, 1286–1293. [CrossRef] [PubMed]

82. Mureithi, S.M.; Verdoodt, A.; Njoka, J.T.; Gachene, C.K.K.; Van Ranst, E. Benefits Derived from Rehabilitating a Degraded Semi-Arid Rangeland in Communal Enclosures, Kenya. *Land Degrad. Dev.* **2016**, *27*, 1853–1862. [CrossRef]
83. *Portrait de la Famille de Moussa Ka, village de Maodo Peulh*; Projet Décentralisation des fonds climat (DFC)/Innovation Environnement Dévéloppement (IED): Dakar, Senegal, 2017; p. 38.
84. King-Okumu, C. *Adaptation to Climate Change: Economic Value and Return on Investments*; International Institute for Environment and Development (IIED)/New Economics Foundation (NEF): London, UK, 2017; p. 22.
85. Tari, D.; King-Okumu, C.; Jarso, I. *Strengthening Local Customary Institutions: A Case Study in Isiolo County, Northern Kenya*; Ada Consortiumj: Nairobi, Kenya, 2015; p. 52.
86. Ahmed, H.; Call, D.R.; Quinlan, R.J.; Yoder, J.K. Relationships between livestock grazing practices, disease risk, and antimicrobial use among East African Agropastoralists. *Environ. Dev. Econ.* **2018**, *23*, 80–97. [CrossRef]
87. Peden, D.; Taddesse, G.; Haileslassie, A. Livestock water productivity: Implications for sub-Saharan Africa. *Rangel. J.* **2009**, *31*, 187–193. [CrossRef]
88. Amede, T.; Tarawali, S.; Peden, D. Improving water productivity in crop-livestock systems of drought-prone regions: Editorial comment. *Exp. Agric.* **2011**, *47*, 1–5. [CrossRef]
89. Mekonnen, M.M.; Hoekstra, A.Y. A global assessment of the water footprint of farm animal products. *Ecosystems* **2012**, *15*, 401–415. [CrossRef]
90. King-Okumu, C.; Wasonga, O.V.; Yimer, E. *Pastoralism Pays: New Evidence from the Horn of Africa*; International Institute for Environment and Development (IIED): London, UK, 2015; p. 4.
91. Venton, C. *Economics of Resilience to Drought—Ethiopia Analysis*; United States Agency for International Development (USAID): Washington, DC, USA, 2018; p. 47.
92. Venton, C. *Economics of Resilience to Drought—Kenya Analysis*; United States Agency for International Development (USAID): Washington, DC, USA, 2018; p. 43.
93. Ba, C.O.; Bishop, J.; Deme, M.; Diadhiou, H.D.; Dieng, A.B.; Diop, O.; Garzon, P.A.; Gueye, B.; Kebe, M.; Ly, O.K.; et al. *Évaluation Économique des Ressources Sauvages au Sénégal—Évaluation Préliminaire des Produits Forestiers non Ligneux, de la Chasse et de la Pêche Continentale*; International Union for Conservation of Nature (IUCN): Gland, Switzerland; Cambridge, UK, 2006; p. 79.
94. *Inspection Régionale des Eaux et Forêts de Kaffrine Rapport Annuel 2014*; Inspection Régionale des Eaux et Forêts (IREF): Kaffrine, Senegal, 2014; p. 68.
95. Kenya National Bureau of Statistics (KNBS). *Economic Survey, 2018*; Kenya National Bureau of Statistics: Nairobi, Kenya, 2018; p. 340.
96. Belem, M.; Saqalli, M. Development of an integrated generic model for multi-scale assessment of the impacts of agro-ecosystems on major ecosystem services in West Africa. *J. Environ. Manag.* **2017**, *202*, 117–125. [CrossRef] [PubMed]
97. Holden, S.; Shiferaw, B. Land degradation, drought and food security in a less-favoured area in the Ethiopian highlands: A bio-economic model with market imperfections. *Agric. Econ.* **2004**, *30*, 31–49. [CrossRef]
98. Holden, S.; Shiferaw, B.; Pender, J. Policy analysis for sustainable land management and food security in Ethiopia: A bioeconomic model with market imperfections. *Res. Rep. Int. Food Policy Res. Inst.* **2005**, *140*, 1–76.
99. Berger, T.; Troost, C.; Wossen, T.; Latynskiy, E.; Tesfaye, K.; Gbegbelegbe, S. Can smallholder farmers adapt to climate variability, and how effective are policy interventions? Agent-based simulation results for Ethiopia. *Agric. Econ.* **2017**, *48*, 693–706. [CrossRef]
100. Njenga, M. *Banning Charcoal Isn't the Way to Go. Kenya Should Make It Sustainable*; International Council for Research in Agroforestry (ICRAF): Nairobi, Kenya, 2018.
101. Weston, P.; Hong, R.; Kaboré, C.; Kull, C.A. Farmer-Managed Natural Regeneration Enhances Rural Livelihoods in Dryland West Africa. *Environ. Manag.* **2015**, *55*, 1402–1417. [CrossRef] [PubMed]
102. Njenga, M.; Karanja, N.; Munster, C.; Iiyama, M.; Neufeldt, H.; Kithinji, J.; Jamnadass, R. Charcoal production and strategies to enhance its sustainability in Kenya. *Dev. Pract.* **2013**, *23*, 359–371. [CrossRef]
103. Somda, J.; Naba, M.; Onadja, A. *Economic Performance of Climate Change Adaptation Activities in the Volta Basin*; International Union for Conservation of Nature (IUCN) Central and Western Africa Office: Ouagadougou, Burkina Faso, 2013; p. 40.

104. Behnke, R.; Kerven, C. *Climate Resilience, Productivity and Equity in the Drylands Counting the Costs: Replacing Pastoralism with Irrigated Agriculture in the Awash Valley, North-Eastern Ethiopia*; International Institute for Environment and Development (IIED): London, UK, 2013; p. 49.
105. Niemi, E.; Manyindo, J. *Economic Importance of Goods and Services Derived from Dryland Ecosystems in the IGAD Region: Case Studies*; International Union for Conservation of Nature (IUCN): Nairobi, Kenya, 2010; p. 48.
106. Aboud, A.A.; Kisoyan, P.K.; Said, M.Y.; Notenbaert, A.; de Leeuw, J.; Gitau, J.W.; Manzano, P.; Davies, J.M.; Roba, G.M.; Omondi, S.O.; et al. *Natural Resource Management and Biodiversity Conservation in the Drylands of Eastern and Central Africa*; Association for Strengthening Agricultural Research in Eastern and Central Africa (ASARECA): Entebbe, Uganda, 2012; p. 91.
107. Barbier, E.B. Upstream dams and downstream water allocation: The case of the Hadejia-Jama'are floodplain, northern Nigeria. *Water Resour. Res.* **2003**, *39*, WES11–WES19. [CrossRef]
108. Barbier, E.B. Wetlands as natural assets [Les zones humides en tant que biens naturels]. *Hydrol. Sci. J.* **2011**, *56*, 1360–1373. [CrossRef]
109. Acharya, G.; Barbier, E.B. Valuing groundwater recharge through agricultural production in the Hadejia-Nguru wetlands in northern Nigeria. *Agric. Econ.* **2000**, *22*, 247–259. [CrossRef]
110. Acharya, G.; Barbier, E. Using domestic water analysis to value groundwater recharge in the Hadejia-Jama' are floodplain, Northern Nigeria. *Am. J. Agric. Econ.* **2002**, *84*, 415–426. [CrossRef]
111. Aich, V.; Koné, B.; Hattermann, F.F.; Paton, E.N. Time series analysis of floods across the Niger River Basin. *Water* **2016**, *8*, 165. [CrossRef]
112. Wymenga, E.; Zwarts, L.; Kone, B. *Water Sharing in the Upper Niger Basin*; Wetlands International, Bamako/Sévaré Mali & Altenburg & Wymenga Ecological Consultants: Feanwâlden, The Netherlands, 2012; p. 63.
113. *The Niger, a Lifeline. Effective Water Management in the Upper Niger Basin*; Zwarts, L.P., Beukering, V., Kone, B., Wymenga, E., Eds.; RIZA: Lelystad, The Netherlands; Wetlands International: Sévaré, Mali; Institute for Environmental Studies (IVM): Amsterdam, The Netherlands; A&W Ecological Consultants: Veenwouden, The Netherlands, 2005; p. 169.
114. Morand, P.; Kodio, A.; Andrew, N.; Sinaba, F.; Lemoalle, J.; Béné, C. Vulnerability and adaptation of African rural populations to hydro-climate change: Experience from fishing communities in the Inner Niger Delta (Mali). *Clim. Chang.* **2012**, *115*, 463–483. [CrossRef]
115. Hamerlynk, O.; Duvail, S. Ecosystem restoration and livelihoods in the Senegal River Delta, Mauritania. In *Linking Conservation and Poverty Reduction: Landscapes, People and Power*; Fisher, R.J., Maginnis, S., Jackson, W.J., Barrow, E., Jeanrenaud, S., Eds.; Earthscan: London, UK, 2008; pp. 68–77.
116. Shine, T.; Dunford, B. What value for pastoral livelihoods? An economic valuation of development alternatives for ephemeral wetlands in eastern Mauritania. *Pastor. Res. Policy Pract.* **2016**, *6*, 18. [CrossRef]
117. Gichohi, W.H. Direct payments as a mechanism for conserving important wildlife corridor links between Nairobi National Park and its wider ecosystem: The wildlife conservation lease programme. In Proceedings of the Vth World Parks Congress, Durban, South Africa, 8–17 September 2003.
118. Silvestri, S.; Osano, P.; de Leeuw, J.; Herrero, M.; Ericksen, P.; Kariuki, J.; Njuki, J.; Bedelian, C.; Notenbaert, A. *Greening Livestock: Assessing the Potential for Payment for Environmental Services in Livestock Inclusive Agricultural Production Systems in Developing Countries*; International Livestock Research Institute (ILRI): Nairobi, Kenya, 2012; p. 55.
119. McGahey, D.; Davies, J.; Hagelberg, N.; Ouedraogo, R. *Pastoralism and the Green Economy—A Natural Nexus?* International Union for Conservation of Nature (IUCN): Gland, Switzerland; United Nations Environment Programme (UNEP): Nairobi, Kenya, 2014; p. 58.
120. Rodríguez, L.C.; Henson, D.; Herrero, M.; Nkedianye, D.; Reid, R. Private farmers' compensation and viability of protected areas: The case of Nairobi National Park and Kitengela dispersal corridor. *Int. J. Sustain. Dev. World Ecol.* **2012**, *19*, 34–43. [CrossRef]
121. Mwangi, H.M.; Lariu, P.; Julich, S.; McDonald, M.A.; Feger, K.-H. Characterizing the intensity and dynamics of land-use change in the Mara River Basin, East Africa. *Forests* **2017**, *9*, 8. [CrossRef]
122. Lingane, A.; Olsen, S. Guidelines for social return on investment. *Calif. Manag. Rev.* **2004**, *46*, 116–135. [CrossRef]
123. Nicholls, J.; Eilis, L.; Neitzert, E.; Goodspeed, T. *A Guide to Social Return on Investment*; The SROI Network: London, UK, 2012; p. 55.

124. Wasike, W.S.K.; Hanley, N. The pricing of domestic water services in developing countries: A contingent valuation application to Kenya. *Int. J. Water Resour. Dev.* **1998**, *14*, 41–54. [CrossRef]
125. *A Guide to the Social Return on Investment*; The SROI Network, Nef: London, UK, 2012; p. 110.
126. Metaferia. *Productive Safety Net Program (APL III) Final Report Consulting Services for Conducting Productive Safety Net Program 2nd 2011 Public Works Impact Assessment Credit#: 4666—ET*; Metaferia Consulting Engineers: Addis Ababa, Ethiopia, 2013; p. 128.
127. Hutton, G. *Benefits and Costs of the Water Sanitation and Hygiene Targets for the Post-2015 Development Agenda—Working Paper as of 26 January, 2015*; World Bank/Copenhagen Consensus Center: Copenhagen, Denmark, 2015; p. 32.
128. Luedeling, E.; Oord, A.L.; Kiteme, B.; Ogalleh, S.; Malesu, M.; Shepherd, K.D.; Leeuw, J.D. Fresh groundwater for Wajir—Ex-ante assessment of uncertain benefits for multiple stakeholders in a water supply project in Northern Kenya. *Front. Environ. Sci.* **2015**, *3*, 16. [CrossRef]
129. Ndirangu, W. *Catalysing Low Cost Green Technologies for Sustainable Water Service Delivery in Kenya—Feasibility Study Report*; Climate Technology Centre and Network (CTCN), The UN Environment Programme (UNEP), Technical University of Denmark (DTU) & Water Services Trust Fund (WSTF): Nairobi, Kenya, 2017; p. 147.
130. Mwaura, J.; Koske, J.; Kiprotich, B. Economic value of water harvesting for climate-smart adaptation in semi-arid Ijara Garissa, Kenya. *Environ. Syst. Res.* **2018**, *6*, 10. [CrossRef]
131. O'Farrell, P.J.; De Lange, W.J.; Le Maitre, D.C.; Reyers, B.; Blignaut, J.N.; Milton, S.J.; Atkinson, D.; Egoh, B.; Maherry, A.; Colvin, C.; et al. The possibilities and pitfalls presented by a pragmatic approach to ecosystem service valuation in an arid biodiversity hotspot. *J. Arid. Environ.* **2011**, *75*, 612–623. [CrossRef]
132. Hassan, R.M.; Okbrich, B. Comparative analysis of the economic efficiency of water use by plantation forestry and irrigation agriculture in the Crocodile catchment. *Agrekon* **1999**, *38*, 566–575. [CrossRef]
133. Lange, G.-M.; Mungatana, E.; Hassan, R.M. Water accounting for the Orange River Basin: An economic perspective on managing a transboundary resource. *Ecol. Econ.* **2007**, *6*, 660–670. [CrossRef]
134. Bouma, J.A.; Hegde, S.S.; Lasage, R. Assessing the returns to water harvesting: A meta-analysis. *Agric. Water Manag.* **2016**, *163*, 100–109. [CrossRef]
135. Rebelo, L.-M.; McCartney, M.P.; Finlayson, C.M. Wetlands of Sub-Saharan Africa: Distribution and contribution of agriculture to livelihoods. *Wetl. Ecol. Manag.* **2010**, *18*, 557–572. [CrossRef]
136. Nieuwoudt, W.L.; Backeberg, G.R.; Du Plessis, H.M. The value of water in the South African economy: Some implications. *Agrekon* **2004**, *43*, 162–183. [CrossRef]
137. Crafford, J.G.; Hassan, R.M.; King, N.A.; Damon, M.C.; de Wit, M.P.; Bekker, S.; Rapholo, B.M.; Olbrich, B.W. *An Analysis of the Social, Economic, and Environmental Direct and Indirect Costs and Benefits of Water Use in Irrigated Agriculture and Forestry*; Water Research Commission: Pretoria, South Africa, 2004.
138. Moolman, C.E.; Blignaut, J.N.; Van Eyden, R. Modelling the marginal revenue of water in selected agricultural commodities: A panel data approach. *Agrekon* **2006**, *45*, 78–88. [CrossRef]
139. Mutiga, J.K.; Mavengano, S.T.; Zhongbo, S.; Woldai, T.; Becht, R. Water Allocation as a Planning Tool to Minimise Water Use Conflicts in the Upper Ewaso Ng'iro North Basin, Kenya. *Water Resour. Manag.* **2010**, *24*, 3939–3959. [CrossRef]
140. Stockholm Environment Institute (SEI). *WEAP: Water Evaluation and Planning System, User Guide for WEAP21*; Stockholm Environment Institute: Boston, MA, USA, 2001.
141. *Kenya: Post-Disaster Needs Assessment (PDNA) for the 2008-2011 Drought*; Republic of Kenya with Technical Support from the European Union, United Nations, and World Bank and Financial Support from the European Union and the Grand Duchy of Luxembourg; Republic of Kenya: Nairobi, Kenya, 2012; p. 188.
142. Gies, L.; Agusdinata, D.B.; Merwade, V. Drought adaptation policy development and assessment in East Africa using hydrologic and system dynamics modeling. *Nat. Hazards* **2014**, *74*, 789–813. [CrossRef]
143. Agusdinata, D.B. Evaluating water infrastructure and agriculture practices for drought adaptations in East Africa: A combined hydrological and system dynamics approach. In Proceedings of the GHTC 2016—IEEE Global Humanitarian Technology Conference: Technology for the Benefit of Humanity, Conference Proceedings 7857362, Seattle, DC, USA, 13–16 October 2016; pp. 753–760.
144. Blignaut, J.; Mander, M.; Schulze, R.; Horan, M.; Dickens, C.; Pringle, C.; Mavundla, K.; Mahlangu, I.; Wilson, A.; McKenzie, M.; et al. Restoring and managing natural capital towards fostering economic development: Evidence from the Drakensberg, South Africa. *Ecol. Econ.* **2010**, *69*, 1313–1323. [CrossRef]

145. Ashagre, B.B.; Platts, P.J.; Njana, M.; Burgess, N.D.; Balmford, A.; Turner, R.K.; Schaafsma, M. Integrated modelling for economic valuation of the role of forests and woodlands in drinking water provision to two African cities. *Ecosyst. Serv.* **2018**, *32*, 50–61. [CrossRef]
146. Turpie, J.K.; Forsythe, K.J.; Knowles, A.; Blignaut, J.; Letley, G. Mapping and valuation of South Africa's ecosystem services: A local perspective. *Ecosyst. Serv.* **2017**, *27*, 179–192. [CrossRef]
147. Emerton, L. *Local-Level Investments in Natural Water Infrastructure: Economic Preferences, Constraints and Incentives for Sustainable Land Management among Farmers in the Upper Tana Basin, Kenya*; BC3 Report for the WISE UP Project; International Union for Conservation of Nature (IUCN): Leioa, Spain, 2017; p. 31.
148. *The Economics of Ecosystem Services of the Tana River Basin Assessment of the Impact of Large Infrastructural Interventions*; Beukering, P.V., de Moel, H., Eds.; Institute for Environmental Studies (IVM)/The Economics of Ecosystems and Biodiversity (TEEB): Amsterdam, The Netherlands, 2015; p. 163.
149. Emerton, L. *Case Studies in Wetland Valuation #6: May 2003 Tana River, Kenya: Integrating Downstream Values into Hydropower Planning*; International Union for Conservation of Nature (IUCN): Nairobi, Kenya, 2003; p. 6.
150. Apse, C.; Bryant, B.; Droogers, P.; Hunink, J.; Kihara, F.; Leisher, C.; Vogl, A.; Wolny, S. *Upper Tana-Nairobi Water Fund A Business Case*; Natural Capital Project and FutureWater; The Nature Conservancy: Washington, DC, USA, 2014; p. 36.
151. Vogl, A.L.; Bryant, B.P.; Hunink, J.E.; Wolny, S.; Apse, C.; Droogers, P. Valuing investments in sustainable land management in the Upper Tana River basin, Kenya. *J. Environ. Manag.* **2017**, *195*, 78–91. [CrossRef] [PubMed]
152. *Upper Tana-Nairobi Water Fund (UTNWF) Detailed Design Report*; International Fund for Agricultural Development (IFAD): Rome, Italy, 2017; p. 56.
153. Schmiedel, U.; Kruspe, M.; Kayser, L.; Oettlé, N. The Ecological and Financial Impact of Soil Erosion and its Control—A Case Study from the Semiarid Northern Cape Province, South Africa. *Land Degrad. Dev.* **2017**, *28*, 74–82. [CrossRef]
154. Yoo, J.; Simonit, S.; Connors, J.P.; Maliszewski, J.P.; Kinzig, A.P.; Perrings, C. The value of agricultural water rights in agricultural properties in the path of development. *Ecol. Econ.* **2013**, *91*, 57–68. [CrossRef]
155. Choumert, J.; Stage, J.; Uwera, C. Access to water as determinant of rental values: A housing hedonic analysis in Rwanda. *J. Hous. Econ.* **2015**, *26*, 48–54. [CrossRef]
156. Freeman, A.M., III. Hedonic prices, property values and measuring environmental benefits: A survey of the issues. *Scand. J. Econ.* **1979**, *81*, 154–173. [CrossRef]
157. Miyata, Y.; Abe, H. Measuring the Effects of a Flood Control Project: Hedonic Land Price Approach. *J. Environ. Manag.* **1994**, *42*, 389–401. [CrossRef]
158. Yoo, J.; Simonit, S.; Connors, J.P.; Kinzig, A.P.; Perrings, C. The valuation of off-site ecosystem service flows: Deforestation, erosion and the amenity value of lakes in Prescott, Arizona. *Ecol. Econ.* **2014**, *97*, 74–83. [CrossRef]
159. Poor, P.J.; Pessagno, K.L.; Paul, R.W. Exploring the hedonic value of ambient water quality: A local watershed-based study. *Ecol. Econ.* **2007**, *60*, 797–806. [CrossRef]
160. Daniel, V.E.; Florax, R.J.G.M.; Rietveld, P. Flooding risk and housing values: An economic assessment of environmental hazard. *Ecol. Econ.* **2009**, *69*, 355–365. [CrossRef]
161. Gulyani, S.; Talukdar, D.; Bassett, E.M. A sharing economy? Unpacking demand and living conditions in the urban housing market in Kenya. *World Dev.* **2018**, *109*, 57–72. [CrossRef]
162. Brozovic, N.; Sunding, D.L.; Zilberman, D. On the Spatial Nature of the Groundwater Pumping Externality. *Resour. Energy Econ.* **2010**, *32*, 154–164. [CrossRef]
163. Epanchin-Niell, R.; Kuwayama, Y.; Walls, M. Spatial-dynamic complexities of climate challange for rural areas: Integrating resource and regional economic insights. *Am. J. Agric. Econ.* **2017**, *99*, 447–463. [CrossRef]
164. Speir, C.; Han, J.; Brozovic, N. Spatial Dynamic Optimization of Groundwater Use with Ecological Standards for Instream Flow. *Water Econ. Policy* **2016**, *2*. [CrossRef]
165. Saak, A.E.; Peterson, J.M. Groundwater Use under Incomplete Information. *J. Environ. Econ. Manag.* **2007**, *54*, 214–228. [CrossRef]
166. Athanassoglou, S.; Sheriff, G.; Siegfried, T.; Huh, W.T. Optimal Mechanisms for Heterogeneous Multi-Cell Aquifers. *Environ. Resour. Econ.* **2012**, *52*, 265–291. [CrossRef]

167. Sahle, M.; Saito, O.; Fürst, C.; Demissew, S.; Yeshitela, K. Future land use management effects on ecosystem services under different scenarios in the Wabe River catchment of Gurage Mountain chain landscape, Ethiopia. *Sustain. Sci.* **2018**, 1–16. [CrossRef]
168. Sahle, M.; Saito, O.; Fürst, C.; Yeshitela, K. Quantifying and mapping of water-related ecosystem services for enhancing the security of the food-water-energy nexus in tropical data-sparse catchment. *Sci. Total Environ.* **2019**, *646*, 573–586. [CrossRef] [PubMed]
169. Sharp, R.; Tallis, H.T.; Ricketts, T.; Guerry, A.D.; Wood, S.A.; Chaplin-Kramer, R.; Nelson, E.; Ennaanay, D.; Wolny, S.; Olwero, N.; et al. *VEST + VERSION + User's Guide*; The Natural Capital Project; Stanford University, University of Minnesota, The Nature Conservancy and World Wildlife Fund: Stanford, CA, USA, 2015.
170. Tallis, H.T.; Ricketts, T.; Guerry, A.D.; Wood, S.A.; Sharp, R.; Nelson, E.; Ennaanay, D.; Wolny, S.; Olwero, N.; Vigerstol, K.; et al. *VEST 2.2.1 User's Guide*; The Natural Capital Project; Stanford University: Stanford, CA, USA, 2011.
171. Belete, M.; Deng, J.; Zhou, M.; Wang, K.; You, S.; Hong, Y.; Weston, M. A new approach to modelingwater balance in Nile River Basin, Africa. *Sustainability* **2018**, *10*, 810. [CrossRef]
172. Rukundo, E.; Liu, S.; Dong, Y.; Rutebukaa, E.; Asamoah, E.F.; Xu, J.; Wu, X. Spatio-temporal dynamics of critical ecosystem services in response to agricultural expansion in Rwanda, East Africa. *Ecol. Indic.* **2018**, *89*, 696–705. [CrossRef]
173. Leh, M.D.K.; Matlock, M.D.; Cummings, E.C.; Nalley, L.L. Quantifying and mapping multiple ecosystem services change in West Africa. *Agric. Ecosyst. Environ.* **2013**, *165*, 6–18. [CrossRef]
174. Duku, C.; Rathjens, H.; Zwart, S.J.; Hein, L. Towards ecosystem accounting: A comprehensive approach to modelling multiple hydrological ecosystem services. *Hydrol. Earth Syst. Sci. 2015 19*, 4377–4396.

land

MDPI

Article

Pastoralism and Land Tenure Transformation in Sub-Saharan Africa: Conflicting Policies and Priorities in Ngamiland, Botswana

Lenyeletse V. Basupi *, Claire H. Quinn and Andrew J. Dougill

Sustainability Research Institute, School of Earth and Environment, University of Leeds, Leeds LS2 9JT, UK; c.h.quinn@leeds.ac.uk (C.H.Q.); a.j.dougill@leeds.ac.uk (A.J.D.)

* Correspondence: eelvb@leeds.ac.uk or vbasupi@gmail.com

Received: 2 November 2017; Accepted: 27 November 2017; Published: 11 December 2017

Abstract: In dryland Africa, access to land and water resources are central to pastoral livelihood activities. Policy intervention in these regions represents the outcome of concerted post-independence processes in which countries have committed to land tenure transformation as a policy objective. This was meant to create private, liberal property rights to replace communal customary tenure systems which were considered to be a constraint to development. Despite these efforts, decades of scientific research indicate that countries are still struggling to meet environmental sustainability objectives. Land degradation where it existed has not been halted and traditional pastoral livelihoods have been disrupted. The overall evidence base for policymaking remains weak as deficiencies in data or information on which management decisions were based led to poor policy performance. In a bid to strengthen understanding in this area, this study has a dual aim: 1. Using a systematic review of the literature, we examine the impact of land tenure transformation in pastoral areas in sub-Saharan Africa; 2. We analyse user-perspectives on land tenure transformation and pastoralists' rights in Ngamiland, Botswana, so as to draw out the salient issues that must be addressed in order to reconcile pastoral tenure conflicts and land management in sub-Saharan Africa. Results from meta-analysis and case study show that land tenure transformation policies across pastoral areas are subject to similar challenges and consequences. Protecting pastoral land rights requires deliberate policy interventions that recognise pastoralism as a productive and efficient use of resources. Policymakers need to overcome anti-pastoral prejudice and focus on Sustainable Land Management goals. This entails establishing negotiated and flexible tenure frameworks that strengthen pastoralists' participation in decision-making arenas by working with pastoral communities on the basis of understanding their livelihood system.

Keywords: communal rangelands; property rights; environmental impacts; policy implementation; drylands

1. Introduction

In drylands, access to grazing and water resources are central to rural pastoral livelihoods activities. In these environments, pastoralism is one example of a resource use system that is highly dynamic and uncertain. For rural community dwellers, communal lands and their resources are the mainstay of most economic activities and rural livelihoods, including farming, hunting and the day to day gathering of natural resources such as veld products [1,2]. However, concerns over the demise of traditional pastoral resource use systems due to rangeland degradation, impacts of climate change, impacts of land tenure policies, expansion in commercial agricultural activities and conservation areas continue to occupy the central agenda in pastoralism literature [3–5]. Many pastoral communities are faced with challenges of shifts in land tenure as their communal rights are considered by development practitioners as a constraint that hinders development with a need to be modernised [6]. Moreover, climate change, population growth and land use policies that focus on sedentarisation of pastoral communities continue to cause accelerated pressure on natural resources leading to rangeland resource degradation, wildlife declines and pastoralists vulnerability [7,8]. As drylands are characterised by low and spatiotemporally variable precipitation [9], sustainable land and livestock management are dependent on adaptive mobility and pastoralists' flexibility to make use of the highly variable rangeland resources [10]. Historically, pastoralists have been able to follow rainfall or specific pasture resources through space and time in order to meet the needs of their animals and prevent rangeland degradation caused by the concentration of animals in smaller territories [11,12].

In sub-Saharan Africa (SSA), competition over land has intensified over the last few decades due to urbanisation, agricultural intensification, conservation initiatives and privatisation of communal lands through rangeland policies that have sought to create private, liberal property rights to replace communal customary systems [13]. The two opposing views in this debate are focussed on either supporting tenure reform through the registration of land to individuals, state or strengthening customary tenure. The proponents of tenure reform have received support since Hardin's argument that communal tenure arrangements fail to regulate irrational behaviour, leading to overexploitation of communal resources [14]. Hardin's thesis also provided the rationale for World Bank programs calling for privatisation of communal grazing lands so as to commercialise the livestock sector in developing countries [15]. De Soto's [16] support has been singled out with his theoretical argument stating that the conditions and terms of negotiation under which land is held under customary tenure only encourage low rates of productivity-enhancing investments. De Soto refers to land held under customary tenure as 'dead capital' because it cannot be used as collateral in a formal banking system.

However, these views have been widely contested as not representing customary land rights and management systems that were in place for African rural communities [17,18]. Ostrom [19] and others have argued against solutions that are imposed on users by external authorities, arguing that traditional group property regimes are able to self-organise, that local users are capable of designing and changing their own rules, implementing the agreed upon rules and most importantly draw on inherited skills to learn strong locally crafted rules as well as evolved norms of behaviour [19,20], especially reciprocity [21]. Ostrom further argues that undermining local resource users through privatisation or rangeland enclosure schemes increases the vulnerability of resources to degradation including the increased vulnerability of their users [19]. In SSA, land use policies have ignored the multi-purpose goals of traditional group property regimes as practiced in communal lands and emphasised rangeland enclosures, privatisation of communal grazing lands and commercialisation of the livestock sector, leading to weakening and marginalisation of traditional land and pastoral management regimes [11,22]. Mobility and flexibility have diminished as land ownership has become more rigid and fixed, with different land uses separated by fences and other administrative barriers [23]. In Kenya's Maasailand for example, researchers describe the impact of government enclosure policy in which rangeland development schemes have not only privatised the best land but have also led to overgrazing, violent conflicts and increased wealth inequalities [3,24].

Implementing property rights that are equitable and can enhance the sustainability of both pastoral livelihoods and resources has remained a challenge for public policy across Africa's drylands [25]. The performance of land tenure policies has had mixed results and issues of impacts and implementation of such policies for sustainability remains debated in the research literature. The overall evidence base for policymaking remains weak and insufficient as deficiencies in the data or information on which planning and management decisions are based often leads to poor performance of different policies. In a bid to strengthen understanding in this area, this study has a dual aim: 1. Using a systematic review of the literature, we examine the impact of land tenure transformation in pastoral areas in SSA; 2. We analyse user-perspectives on land tenure transformation and pastoralists' rights in Ngamiland, Botswana, to draw out the salient issues that must be addressed to reconcile pastoral tenure conflicts and land management in dryland areas.

2. Materials and Methods

2.1. Systematic Literature Review on Pastoralism and Land Tenure Discourse

Data for this study were collected from primary and secondary sources. Data from secondary sources were obtained through a systematic procedure (Figure 1). This was developed using keywords and other search options such as Boolean operations (AND, OR), using methods described by Waddington et al. [26]. Resources available through databases: Web of Science, SCOPUS, Springer Link, JSTOR, Google Scholar and other library resources were used. We used both published literature (Journal articles and books) and grey literature in the form of government policies and legislative documents, technical reports, land use and land management plans obtained from government departments and government printing and publishing agencies.

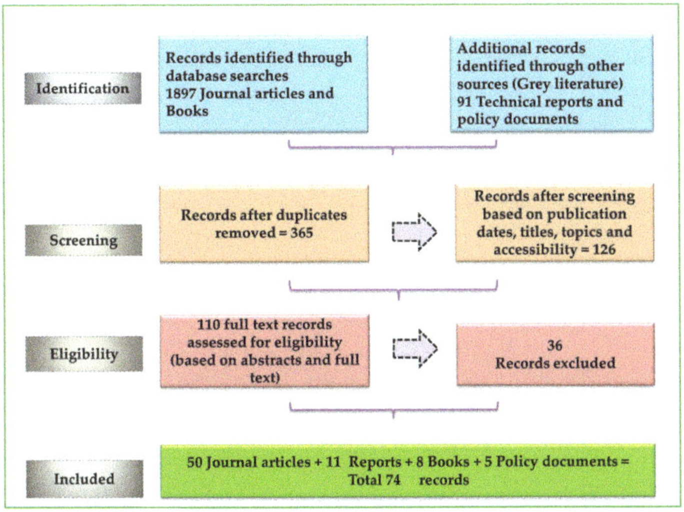

Figure 1. Flowchart of literature search, process and results.

2.1.1. Eligibility

For eligibility, we considered articles that examine land tenure, property rights and pastoralism in SSA or other developing country contexts. We considered empirical studies published between 1990 and 2017. Specifically, the eligibility criteria based on abstracts and full text were as follows;

- Articles that have the words: land tenure AND/OR communal land privatisation, pastoralism AND/OR property rights in abstracts.
- Publications based on outcomes of land tenure transformation in SSA.
- Publications that study a clearly defined policy intervention in drylands pastoral areas.
- Publications that provide adequate methodological information.
- Publications that asses the outcome at an appropriate level of analysis, specific case studies, e.g., district or village level.

2.1.2. Exclusion Criteria

- Publications whose methodologies were considered difficult to assess in a systematic manner.
- Nonrelated articles, e.g., commentary, simulations and modelling.
- Review papers.
- Articles whose main focus is in developed countries.

- Research articles (not policies) published prior to the year 1990.
- Publications that were produced in a language other than English.

2.1.3. Data Extraction

A data extraction template was developed and study variables and characteristics systematically collected were as follows (1) Document type e.g., Journal, Book, Policy; (2) Title of the study; (3) Research category/field; (4) Study geographical area e.g., country, region; (5) Keyword and meaning from the abstract e.g., pastoralism, land tenure, property rights, commons, communal land privatisation and (6) Date of publication.

2.2. Mixed Participatory Methods

Empirical data from six study villages in Ngamiland (Sehithwa, Toteng, Bodibeng, Bothatogo, Makakung and Semboyo) were used to understand the policy impacts of land tenure transformation on pastoralism and provide comparisons with issues emanating from the wider literature. A mixture of qualitative participatory rural appraisal methods were used to allow pastoral communities to share their experience and knowledge with regard to pastoralism, livelihoods and land tenure transformation. This included focus group discussions, key informant interviews and semi-structured interviews. A total of 6 focus group discussions, 26 key informant interviews and 97 semi-structured interviews were undertaken in the 6 villages. Purposive sampling and snowballing techniques [27] were used to identify key informants. Farmers' committees, village leadership and village development committees were used to solicit names of participants for focus groups. Key informants were selected based on their pastoral and local environmental knowledge. For semi-structured interviews, a structured sampling procedure was employed and participants were taken from a cross-section of the pastoral community and included both males and females (Table 1). The qualitative data were coded and analysed through iterative content analysis [27] in order to identify major themes. Structuring themes permitted the comparison of the responses with themes identified through the systematic literature review.

Table 1. Study villages and demography.

Village	Human Population	Total Numbers in Semi-Structured Interviews	Male (Semi-Structured Interviews)	Female (Semi-Structured Interviews)
Sehithwa	2748	28	18	10
Toteng	909	19	12	8
Bodibeng/Bothatogo	1333	28	11	17
Semboyo/Makakung	691	22	10	11
Total	5681	97	51	46

Data source: Central Statistics Office, Botswana, 2011 census report and authors' interview transcripts.

3. Results and Discussion

3.1. Pastoralism and Land Tenure Discourse in Sub-Saharan Africa

Most of the literature from the systematic review e.g., [22,28–31] identifies the complexity of land tenure and pastoralism in SSA. Many SSA governments have committed to land tenure transformation as a policy goal [22,32–34]. Most policy documents and official reports have depicted pastoralists as unable to manage communal resources in a rational way, thus providing a strong justification for privatising communal pastures and controlling pastoralists' movement, stocking rates and access [35,36]. Almost two third of the papers (63%) assessed the impacts on pastoral livelihoods and how the individual tenure transformation policies in pastoral areas had failed to meet their environmental objectives. The remaining papers investigated issues related to implementation, policy conflicts and priorities. The issue of land tenure in SSA has been mentioned as of significant importance for agricultural development and food security in documents such as the UN Millennium Project [37],

NEPAD's Comprehensive Africa Agriculture Report [38] and a significant number (37/50) of the journal articles point to the challenges of land tenure transformation in dryland Africa. Priority issues in land tenure policies and implementation in pastoral areas consisted of:

- A decline in land productivity—reduction in communal managed grazing, constriction in livestock mobility (53% of the papers).
- Loss of critical common property management regimes—reduced coping mechanisms (49%).
- Implementation challenges—actors' priorities and inadequate resources to support policy implementation (39%).
- Increased inequality and social conflicts, breakdown of social networks and safety nets (33%).
- Limited coping mechanisms and adaptive capacity by pastoralist remaining in communal areas (29%).
- Lack of accountability by local level institutions and authorities (26%).
- Low level of economic development and deficiencies in markets (23%).
- Wildlife management areas and traditional use of rangelands in conflict (19%).

With respect to pastoralism, most of the papers (77%) highlight land tenure security and land expropriation as key problems in pastoral land development. Expansion of competing land-uses, land tenure transformation, individualisation and enclosures have reduced the net availability of rangeland resources, often with significant consequences for pastoral livelihood and the environment. For example, in Kenya, the group ranch concept is said to be in its fourth decade, but there is general consensus among scholars and researchers, including policymakers, that the policy has failed to meet its objective (of commercializing production, improving pastoral wellbeing, improving environmental management) and has also jeopardised the socio-economic welfare of the Massai community [23,39,40]. The group ranch concept, a world bank-sponsored Kenya livestock development project, allowed for the setting aside of certain areas of land to be collectively owned by a group of people legally registered as members of a particular ranch for collective management [40]. Analysts point out that the positive aspects of the concept were overshadowed by problem such as continued trespassing of ranch boundaries, loss of land to elites' members, refusal to control stock numbers and that no real transformation to a market-oriented livestock production system was made [23,40,41]. Inequitable access to land and socio-political factors were identified in 33% of the articles analysed. Galaty [41] found that the group ranch subdivisions in Kenya have benefited elites and outside investors, undermining the traditional livelihoods of poorer Massai pastoralists.

In Ethiopia, according to Tache [42], the practice of reserving some pastures for drought was widely practiced by Borana, Guji and Gabra Oromo communities long before the arrival of externally/donor founded land tenure and pastoral development projects. Tache argues that these reserved areas were not fenced, but word of mouth was enough to restrict access. Over the years' pastoralists in Ethiopian drylands have experienced a shrinkage in available dry season grazing, a reduction in communally managed grazing reserves and growing individualisation of land use rights through privatisation. Similarly, in Sudan, Babiker [43] found that the process of land resource individualisation has severely fragmented the Central Sudan rangelands as land is expropriated for large-scale commercial farming and wildlife conservation.

While the formalisation of the commonage under the Transformation of Certain Rural Areas Act of 1998 (TRANCRAA) in South Africa's rural Namaqualand has increased tenure security for individual plot holders, in respect to de Soto's hypothesised benefits of formalisation and privatisation, tenure security for users of the commons, especially pastoralists, has decreased [44]. Formalisation has led to privatisation, increased fencing, reduced communal rangelands and closed corridors so undermining local grazing patterns [45].

Lack of accountability and conflicting policy priorities and objectives were also cited as an implementation challenge in 26% of the papers. Often economic development objectives were prioritised over environmental concerns or pastoralists wellbeing. Another deficiency discussed was the low level of economic development and deficiencies in markets which make it almost impossible to achieve environmental sustainability objectives (23%).

In summary, the meta-analysis shows that in SSA, land tenure transformation policies were based on western, classical rangeland ecological models [22,46], economic theories [47], rangeland degradation narratives and tragedy of the commons theory [14,22], rather than the socio-ecological realities of drylands' rangelands dynamics. Consequently, traditional grazing territories have been shrinking while pastoralists dependent on rangeland resources and ecosystem services were displaced and exposed to incremental risks; poverty, livestock diseases and a breakdown of social networks and safety nets as well as a decline in rangeland productivity. This compression has suppressed the flexibility and spatial extent necessary for pastoralism in dryland environments. Evidence from the review suggests that the perceived benefits of tenure transformation has acted as a justification for the concentration of land in the hands of a few individuals, exacerbating insecurity of land tenure for the rural poor.

3.2. *Reviewing Botswana's Rangeland Policy*

Botswana is a semi-arid country whose rural population depends largely on livestock production. Botswana's rural people are mostly rural village dwellers and their pastoral activities assume the form of transhumance under a three-tier settlement system, whereby rural village dwellers commute between villages, land areas and temporary, encampments known as cattle posts, where livestock are kept [48]. Traditionally, communal rangelands have been managed by traditional institutions which allow for inter-territorial grazing between unfixed tribal boundaries so that animals can access forage and water even in times of stress, such as drought years [49]. Change in environmental conditions has always influenced pastoral livelihoods in Botswana [2]. Unfavourable ecological conditions and pastoralist's vulnerability have increased since the 1980s due to increased fragmentation of landscapes as a result of rangeland policies [50]. The literature on Botswana's rangeland transformation policies points to a situation where the design and implementation of rangeland policy were based on an insufficient or poor understanding of the problem [50–53]. Deficiencies in the data or information on which planning and management decisions were based means the policy assumptions were not supported by concrete scientific evidence.

Botswana registered its concern for rangeland degradation and what was termed 'unsustainable livestock keeping' in 1975 through the Tribal Grazing Land Policy (TGLP) [35]. Hardin's "The Tragedy of the Commons" [14] theory was widely used to blame communal grazing for land degradation. TGLP had three objectives: (1) to stop overgrazing and degradation of the range, (2) to promote greater equality and incomes in the rural areas and (3) to allow growth and commercialisation of the livestock industry on a sustainable basis [41]. Through this policy, the government hoped that pressure in communal lands would be alleviated through demarcation of ranches and allowing large herds owners to transfer their cattle to these ranches, thus leaving the communal lands for communal subsistence pastoralists [51]. According to the White Paper on the TGLP and feasibility reports by the Ministry of Agriculture, development had to start with granting exclusive rights and fencing of specific areas. Land Boards and Land Use Officers in the Ministry of Agriculture were given the responsibility of surveying the Tribal areas of Botswana (making up 71% of the total area of the country at that time) and zoning them into three categories: (1) commercial areas where exclusive rights would be granted to individuals and groups with a minimal rental payment, (2) Communal areas, where the land tenure system would remain the same but stock limitations would be imposed and (3) reserved areas which would not be allocated to anyone but rather set aside for the future, thus ensuring 'safeguards for the future generation and poor members of the population' [54].

The planning stage of TGLP focussed on economic gains and administrative initiatives [55]. In spite of complaints from local people at the consultation stage of the uncertainty of potential benefits [51], the policy was implemented without proper mapping to provide the necessary spatial baseline information on how much land would be available for the policy's different objectives.

In addition, there was no plan to monitor the progress of activities in the different zones stated above [56].

Lack of spatial information and a good monitoring plan made implementation in its original form difficult. The zoning process revealed that many parts of the country which were assumed to have been unutilised actually contained a substantial number of people [55]. Some areas were used by migratory pastoralists during drought years or wet seasons because they had natural water ponds [50,56]. Most of the smallholders, including hunter-gatherers, lost the rights to utilise the land. Communal land privatisation and creation of wildlife management areas illustrate how conflicts have arisen in the interface between customary and formal statutory tenure [51]. As zoning was done, it was found that in most districts there was too little land available to permit reserved areas to be set aside, so the reserved category was dropped and the establishment of commercial ranches became the major focus of attention on the part of land use planners and development personnel. This was in spite of the TGLP's original emphasis on ensuring 'safeguards for the future generation and poor members of the population' [54]. Large-scale cattle owners, especially those with boreholes, were allocated ranches and were encouraged to transfer their herds into ranches and leave the communal grazing land to subsistence farmers [35,51]. However, no legislation was put in place to enforce this as those allocated ranches continued to enjoy dual grazing rights by keeping their livestock in communal areas and ranches. This led to environmental threats through concentration of livestock in reduced areas [50]. Such problems are not unique to Botswana as 53% of the articles in the systematic review mentioned reduction in communal managed grazing, constriction in livestock mobility and decline in land productivity as of particular challenge in tenure transformation in pastoralists' areas.

In spite of difficulties in implementation of TGLP [57], Botswana continued with communal land privatisation in the subsequent National Policy on Agricultural Development (NPAD) issued in 1991 [36], prompting fears that the concentration of rural poor on the diminished communal lands may cause further social and environmental problems [51]. NPAD included a wide range of objectives for the development of the agricultural sector in Botswana. As regards to fencing and privatisation of communal lands, NPAD emphasized that TGLP would be intensified and expanded into all communal areas. Under NPAD, the ranches would not have a fixed size as originally stated in the first stage of TGLP (8 km $\times$ 8 km); the size of the ranch would depend on the number of cattle the applicant for a ranch owned, the availability of land and its carrying capacity, and most importantly individuals could apply to fence areas within the vicinity or around boreholes, regardless of their location in communal areas [36]. This policy implied a major land tenure change since the zones that were originally identified as communal lands (notably grazing lands around cattle post areas) in the earlier TGLP zoning process would gradually be privatised [36]. The policy recommended the fencing of a significant part of the communal areas as commercial leasehold ranches or privatised land and the consequences for pastoralists' tenure security, livestock mobility and flexibility would later prove to be significant.

3.3. Experiences of Land Tenure Transformation from Ngamiland District, Botswana

Ngamiland District is situated in North Western Botswana and includes the Okavango Delta. This makes it an important ecosystem characterised by both drylands and water bodies, abundant wildlife and enthralling traditional cultures. The district's major economic activity is substance pastoralism, with limited arable agriculture on drainage plains [48]. The district was selected for this analysis because previous studies by Basupi et al. [50,56] identified the region as characterised by land use competition, conflicts and environmental problems, some of which are attributed to animal health and land tenure transformation policies. The case study provides critical lessons on crucial issues such as the interplay between pastoralists' interests and larger national conservation goals.

In communal areas south of the Okavango Delta, blocks of commercial ranches were demarcated and allocated under both TGLP and NPAD policies [50]. Wildlife Management Areas (WMAs) emerged in the 1980s as a result of a national land zoning exercise following the introduction of the TGLP [48].

As such, land is broadly zoned into distinct uses consisting of communal areas, National Parks, Game Reserves, Conservation areas (operated as tourism concession areas), privatised ranches and Wetlands (Figure 2). However, land use allocation and overlaps of different land use types create a complicated system resulting in pressure and conflict among people, wildlife and livestock [50]. With an average annual rainfall of 350 mm, the area falls into the semi-arid or dryland transition zone. Surrounded by fences, the remaining communal area south of the Setata veterinary cordon fence is about 7500 km^2 in extent. The livestock numbers are high (at 203,269 cattle; 27,552 goats; 10,148 sheep; 5432 horses and 5644 donkeys (Department of Veterinary Services, Botswana, 2016 livestock statistics report)) which makes the regulation of grazing areas difficult. Although some members of the population are involved in arable farming, it is usually flooded recession agriculture in riverbeds and Lake Ngami floodplains because the soils are too poor elsewhere to support meaningful agricultural activity. As a result, pastoralism remains the main economic activity. The case study, therefore, offers an ideal setting for studying user-perspectives on land tenure transformation and pastoralists' rights. Owing to the district's unique ecological and socio-cultural variation, land tenure transformation and landscape fragmentations has resulted in a new generation of social, economic and environmental challenges; livestock diseases which are difficult to control in crowded areas, limited adaptive capacity and conflicting policies and actor priorities. Table 2 summarises the key issues/themes that emerged from the review of the wider literature and shows how these issues manifest in Ngamiland. Figure 2 shows the current land use zones and the case study area/villages while Figure 3 shows the livestock and wildlife spatial distribution patterns and density.

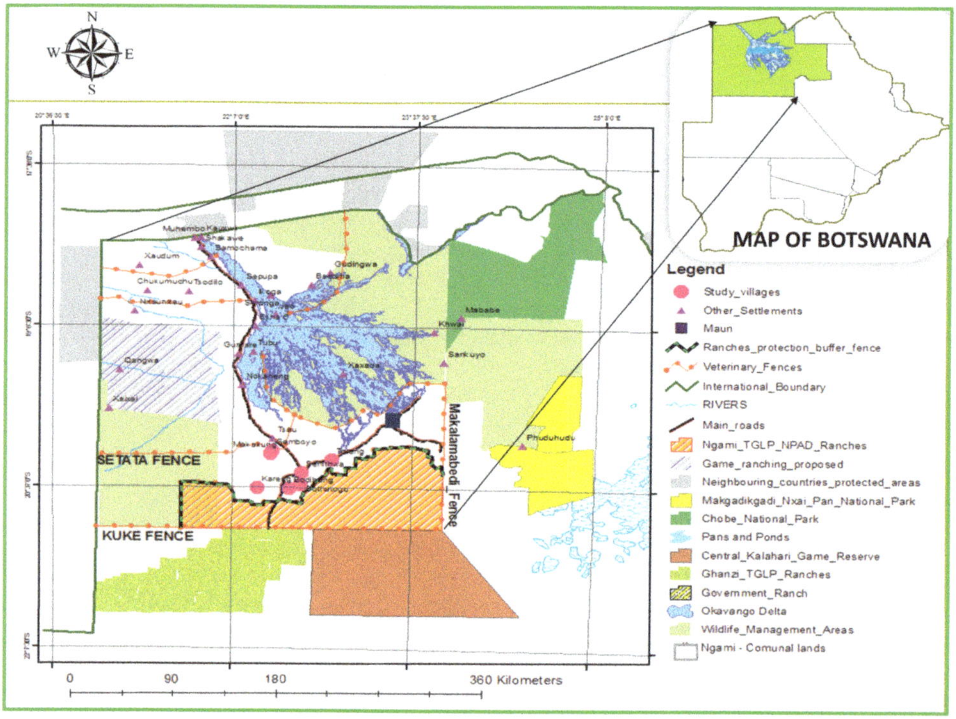

Figure 2. Map showing land use and tenure transformation in Ngamiland, Botswana; increased landscape fragmentation due to rangeland enclosures, privatisation and veterinary fences.

Table 2. Summary of key issues/themes that emerged from the systematic review and how the issues manifest in Ngamiland.

Key Issues That Emerged from the Wider Literature	Example from the Wider Literature in Sub-Saharan Africa	How Do These Issues Manifest in Ngamiland? Data from Semi-Structured Interviews (n = 97) & Except Interviews	Central Findings (Issues That Must Be Addressed in Order to Reconcile Pastoral Tenure Conflicts and Land Management in SSA)
A decline in land productivity—reduction in communal managed grazing, constriction in livestock mobility	Perkins [52]; Bassett [58]; Bennett et al. [29]; Basupi et al. [50]; Fernendez-Gimenez and Le Febre [59]; Benjaminsen et al. [28], Rohde et al. [22]	77 (79.6% of respondents pointed to a decline in land productivity). Figure 1: continued land scape fragmentation, pastoralists squeezed between fences, livestock biomass indicates cattle concentration in the reduced communal area and around water resources	Environmental impacts • Little consideration of environmental dynamics during policy planning and implementation • Conversion of rangelands to other uses, including game ranches • Current tenure systems causing significant land use pressures and conflicts • Bush encroachment and invasion of alien species observed in communal lands together with some privatised ranches-congestion around water points
Increased inequality and social conflicts, breakdown of social networks and safety nets	Peters [60]; Lesorogol [61]; Thebaund and Batterbury [62]. Benjaminsen et al. [28]; Perkins [52]	51 (53.1%) Complex ranch allocation processes exclude poor communal area pastoralists [56]	Social and Economic Impacts • Lack of resources means majority of pastoralists were seen to be unable to use land effectively which resulted in their marginalisation and vulnerability • Countries have yet to reap any significant economic benefits from the process of tenure transformation • Loss of pastoral identity—gains from land privatisation remain mostly theoretical • Inequalities in access to communal lands as elites have taken over and control access
Limited coping mechanisms and adaptive capacity by pastoralist remaining in communal areas	Goldman and Riosmena [63], Greiner and Mwaka [64], Headey et al. [65], Herrero et al. [66]	33 (34.4%), Remoteness, lower levels of literacy, limited infrastructural development, limited access to resources and productive lands, inadequate policy support and lack of alternatives limits pastoralists adaptive capacity	Social and economic impacts • Little support and insecure property rights means opportunities are limited • Lack of political will to address pastoralists problems at marginal areas • Inadequate institutional, human and financial resources to build capacity at the local level • Increased labour costs as labour is reallocated to more productive pursuits
Wildlife management areas and traditional use of rangelands in conflict	Mulale et al. [67], Homewood [68], Western et al. [7], Basupi et al. [56]	63 (65.6%) Fences have bisected ungulates migratory corridors and fragmented habitats, increasing conflicts with bigger game especially elephants as fences have diverted elephants from their traditional migratory corridors into cattle posts and arable lands	Environmental and social impacts • Loss and fragmentation of habitats-land cover and wildlife declines • Decreased diversity of vegetation patches, increased pastoralists vulnerability • Change in behaviour and movement of wild animals, e.g., elephants • Human-wildlife conflicts • Changing patterns of livestock diseases
Loss of critical common property management regimes-reduced coping mechanisms	Swallow and Bromley [69], Wily [70], Lesorogol [71]	65 (67.7%) Pastoralists have relinquished their control of the management of communal areas and they do not regard themselves as responsible	Implementation • Community consultation only focused on the related benefits but very little on the perceived impacts • Pastoralists participation in local decision making lacking • Pastoral production system has been disrupted by the land privatisation and tenure systems
Conflicting policy objectives, actors' priorities and inadequate resources to support policy implementation	Bennett et al. [72]; Toulmin [73]; Nelson and Agrawal [74]; Mulale et al. [67]. Bennett et al. [29], Adams [33]	31 (32.2%), Increasing demand for tourism land, Conversion of ranches to game farms, Figure 3. Encroachment of privatised areas in communal lands especially village grazing areas exacerbate land use competitions and conflicts	Implementation • Weak institutions making implementations problematic, lack of enforcement of legislation remains commonplace leading to increased conflicts and resource degradation e.g., dual grazing by ranchers • Challenges of developing and implementing policies for people whose livelihoods depend on access to the remaining communal rangelands • Lack of integrated planning and coordination between sectors • Failure to communicate effectively with affected pastoral communities • Inability to enforce rights to grazing resources and competing claims and conflicts

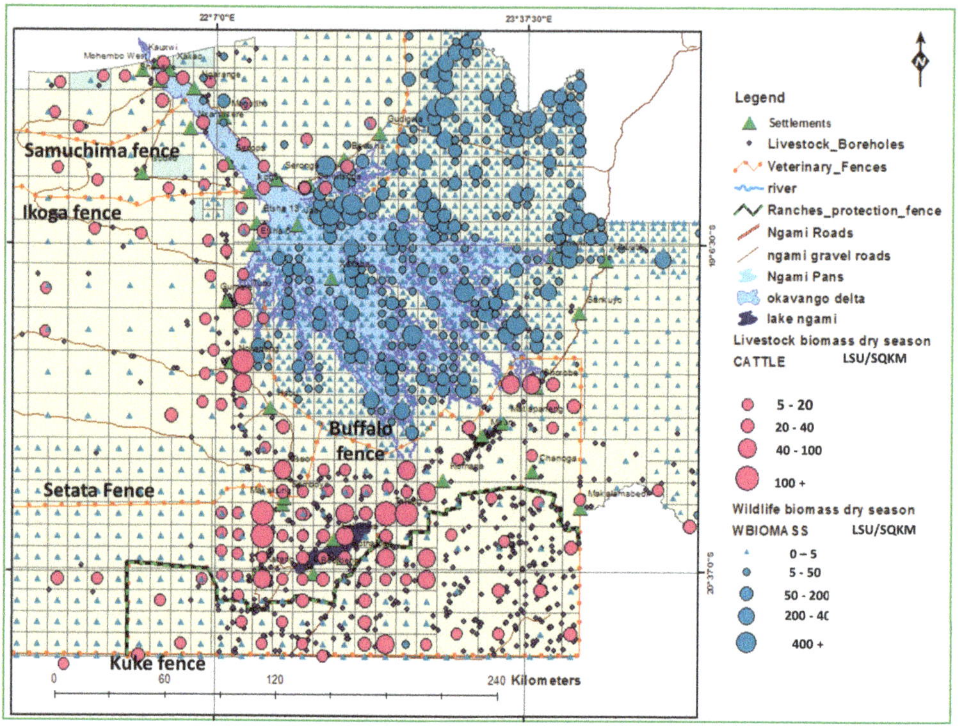

Figure 3. Dry season wildlife and livestock biomass in Ngamiland (based on wildlife aerial survey counts by the Department of Wildlife and National Parks)—livestock distribution pattern shows a high density in communal areas especially around water resources such as Lake Ngami.

3.3.1. Livestock Diseases, Pastoralists' Vulnerability and Limited Adaptive Capacity

Pastoralists' livelihoods are heavily dependent on the availability and access to natural resources [75]. In Ngamiland, respondents reflected on a stark decrease in availability of quality pastures, restricted access to traditional water resources, land use conflicts and livestock diseases. Pastoralists have in recent time's experienced continuous and severe livestock disease outbreaks especially Foot and Mouth Disease (FMD) [55], at a time when communal management institutions and pastoral landscapes are least structured to cope with such crises. In Ngamiland, veterinary disease control fences have formed an integral part of the land use system and have led to increasingly fragmented parcels of land. Such zonal methods of disease control aim to prevent and contain disease outbreaks. These fences have since resulted in several kilometres of fences that aim to separate livestock from wildlife (particularly buffalo as carriers of FMD). Despite increasingly fragmented and fenced landscapes, the period since 2007 has seen the frequency and duration of FMD outbreaks increasing significantly [56]. This terminated beef exports from Ngamiland since 2007. This reality shows that existing disease control practices are failing in the face of constrained livestock mobility, diminishing communal lands and human-wildlife conflicts.

During the interviews, the overwhelming majority of households (73%) stated that a lack of cattle sales is making their lives difficult. The new pastoral environment where a complex sectorial based institutions are used to manage livestock and land resources is such that pastoralists have relinquished their control of the management of communal areas and they do not regard themselves as responsible which in turn has created a liability gap in communal resource management and fight against livestock diseases. Pastoralists do not believe themselves to be responsible for range management or fight against diseases instead blame the government departments for problems in communal areas. While a combination of external factors can also be attributed to the increase in livestock diseases such as climate change [76,77], respondents argued that the increasing vulnerability is a result of weakened coping mechanisms especially decreased mobility resulting from rangeland enclosures and concentration of livestock in reduced lands.

Virtually all the households interviewed (89%) stated that the need for alternative livelihood was increasing. As such, some households have diversified their activities into fishing, flood recession cultivation, wage labour and small businesses in the form of petty trade. The increase in petty trade and diversion into waged labour, including migration to towns, can benefit some pastoralists' households but it also leaves general pastoralism and care of livestock exposed as critical labour is lost. With a reduction in labour availability, many households struggled to round up their cattle for vaccination and some interviewed households reported that they now preferred to leave their cattle to roam unattended. This has contributed to a decline in the quality of herding practices and also increasing environmental stress around water resources and villages.

Under these semi-arid conditions, temporal and spatial fluctuations and the scale of the assessment are critical in assessing the impacts associated with rangeland and tenure transformation policies. It is evident that there are some serious ecological and land use pressures between the ranches and veterinary fences that arise from the current land use practices. This was revealed by studies using participatory research methods and GIS mapping techniques [50]. Out of these studies emerged maps that show the concentration of livestock activities between the ranches and the veterinary fence and around Lake Ngami. In Botswana, rangeland degradation is evident and spreading [78]. However, there are widely contrasting views as to its severity [51]. Some studies emphases that large tracts of the Kalahari sandveld are severely degraded, with indicators of declining productivity such as soil erosion, loss of vegetation cover, and a declining groundwater table evident in communal areas [79]. Others emphasize the localised nature of range degradation around livestock water points and settlements [52]. A recent drought (2016) killed more than 16,000 cattle in Ngamiland (according to the Ministry of Agriculture) of which 14,000 were in communal areas between the ranches protection fence and the Setata veterinary fence. This occurrence raises the fear that the depleted grazing pastures in these communal areas will not cope with increasing grazing pressures. Respondents acknowledged facing increasing resource depletion and indicated diminishing communal grazing land as the main cause. The resilience of this semi-arid rangeland system is being damaged, and if continuously stressed could lead to increased declines in ecosystem services such as production of grazing pastures, ground water recharge and carbon sequestration.

3.3.2. Conflicting Policies, Priorities and Implementation Gaps

Some 39% of the reviewed articles mentioned issues related to policy implementation as an issue in land tenure policies in SSA. It is argued that structural defects in how policies are formulated, especially policies formulated through external influence with limited involvement of local level structures fail to reflect the complexity and necessary flexibility of customary tenure arrangements.

In Ngamiland, pastoralists' perceptions reflected on government policy and their priorities including interactions with government officials in the management of land resources and fight against livestock diseases. Land use conflicts including those between traditional pastoral land rights, human-wildlife conflicts, livestock predation and crop losses were attributed to conflicts in policies and priority of actors. Demarcation of ranches and the provision of veterinary cordon disease control fencing were said to have exacerbated conflicts, particularly when the fence alignments have bisected key pastureland, wildlife habitats and movements, rather than strengthened existing land uses. These conflicts manifest themselves in encroachment of land uses such as settlements into arable land, arable into communal grazing lands, commercial ranching into communal grazing lands and grazing into wildlife areas. Land use competitions are a problem in that where they are left to market forces, essential land use perhaps with lower economic rent run the risk of being out-competed and relegated to less suitable areas. Expansion of competing land-uses has reduced the net availability of rangeland resources. Wildlife conflicts especially elephants are viewed as a permanent threat to pastoralism as they compete for available water resources, and also destroy veterinary fences that separate livestock from diseases carrying animals such as buffaloes. Some respondents (65%) blamed the elephants threats on blockage of ungulates migratory corridors by fences, especially where fences have bisected ungulates migratory corridors and fragmented habitats. The concern for pastoralists is that while wild animals are protected by national and international laws including enjoying long-term security in wildlife management areas, game farms, national parks and game reserves, pastoralists do not enjoy such security.

Both TGLP and NPAD represent policies whose poor results could be attributed to structural defects that characterised their formulation and implementation. Numerous reviews and studies associated with the TGLP shows that after more than three decades the policy has not yet realised its objectives, especially of reducing pressure on communal grazing land or promoting equality and incomes in the rural areas [50,51,57], with some arguing that the policy has reduced environmental and societal resilience to environmental variability [80]. The idea that there was ample empty land that could be reserved for future use was misleading [55]. During the planning phase of TGLP, Potential conflicts in accessibility to grazing resources between ranch owners and communal land dwellers were identified and a regulation was imposed to protect villages with a 20-km buffer zone within which no ranch would be allocated. This was done to prevent the ranches from encroaching into village grazing areas so as to reduce land use pressure and conflicts [35]. This was reiterated by the NPAD and subsequent feasibility studies which stated that to safeguard the interest of the poor households, the village grazing area should cover a radius of 20 km. However, with an emphasis on rangeland enclosures and commercialisation and without the use of a proper spatial technique to monitor the expansion of the demarcation of ranches the buffer zone was difficult to enforce and some ranches are now less than 10 km from the villages. Figure 4 shows some 20 km wide buffer zones generated around each study village using ArcGIS spatial analyst tools (buffering or proximity analysis) and encroachment of ranches into these buffer zones. With no proper monitoring mechanism, commercial ranches continued to encroach onto village grazing areas in spite of the TGLP's original emphasis on ensuring 'safeguards for the poor members of the population'. The encroachment has restricted livestock mobility and increased livestock congestion creating a zone of conflict and pressures between villages and the ranches leading to rangeland degradation.

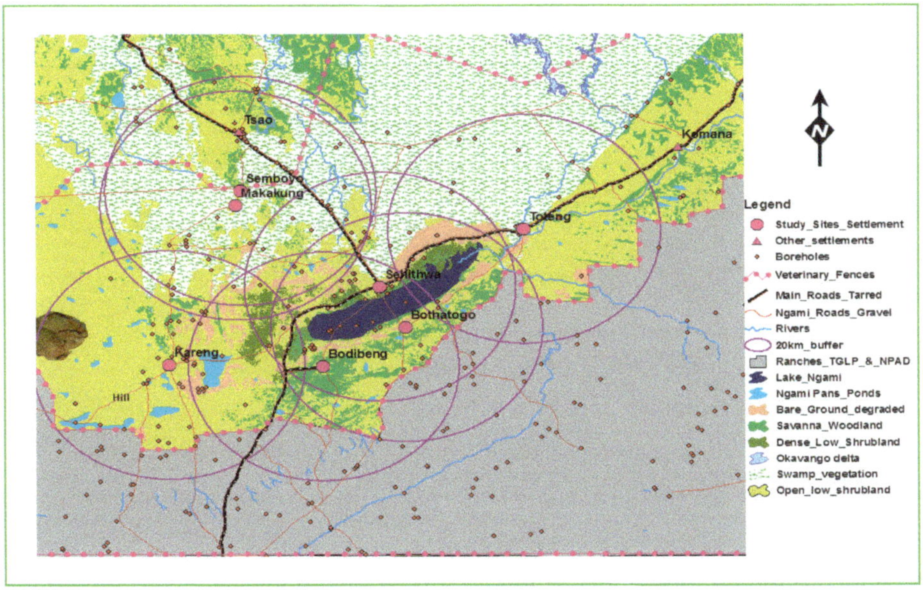

Figure 4. 20 km buffer zones around study villages and the encroachment of ranches in village grazing areas.

4. Conclusions and Recommendations

Our meta-analysis and case study both show that most land tenure transformation policies in pastoral areas have not yet yielded the intended benefits, with rangeland degradation remaining a significant problem. Most of these policies lacked a firm understanding of traditional sociological as well ecological basis of pastoral production systems. In most cases traditional management institutions have been disrupted and pastoralists are increasingly squeezed into smaller territories, so undermining livestock management systems of transhumance and coping mechanisms.

The paper contributes to the land tenure discourse by providing empirical evidence and comparative analysis to deepen our understanding of the challenges of land tenure transformation on pastoralism in SSA. Many of the issues identified in our case study relate to those identified in the systematic review. However, owing to its unique ecological and socio-cultural variation and abundant wildlife, the case study area presents new land management issues and challenges such as the uncontrollable livestock diseases especially FMD. Human-wildlife conflicts and the increasing demand for game ranches (including the conversion of some livestock ranches to game ranches) has introduced complexity in solving the land tenure problem. Control of diseases in the diminished and crowded communal lands has proved to be problematic despite efforts made through the creation of veterinary cordon fences and vaccination campaigns. As a result, markets are severely restricted putting pressure on communal pasture resources as there is no offtake. The complexities in resource management policies and stakeholder/actors interests and priorities also come to play.

Analysts agree that no one range management intervention can be recommended as a 'blanket solution' in pastoral areas, but rather adaptive management and flexible decision making through learning, stakeholder engagement and a bottom-up approach are necessary prerequisites for sustainability. Protecting pastoral land rights and migratory corridors requires deliberate policy interventions that recognises pastoralism as a productive and efficient use of resources. Land use planning should support and provide for economic mobility for pastoralists. The spatial relationship between local communities and the natural environment in which they make their living is often poorly understood and misrepresented in rangeland policies. Geospatial technologies such as Geographic Information Systems (GIS) have the potential to become an information technology enabling decision makers in pastoral areas to sustainably plan and monitor pastoral regions and pastoralists spatiotemporal land rights [50]. Spatiotemporal land rights in this case entails pastoralists seasonal movements, distance that pastoralists travel and areas covered. Moreover, policymakers and government land managers need to reorient relationship with pastoralists so as to overcome anti-pastoral prejudice and focus on Sustainable Land Management goals in communal areas by establishing participatory negotiating and flexible frameworks that strengthen local communities' participation in decision-making arenas by working with pastoral communities on the basis of understanding their livelihood system. There is a need for an appropriate communication programme in pastoral areas where key stakeholders including those that represent pastoralists will share information about pastoral system functionality including mutual understanding of strategic choices for conservation and sustainable use.

Acknowledgments: This research was supported by funding from the Botswana International University of Science and Technology (BIUST), Government of Botswana and the University of Leeds Sustainable Agriculture Fund Bursary. The research was conducted with the permission of The Ministry of Environment, Wildlife and Tourism in Botswana (Government research permit number EWT 8/36/4 XXX (73)). This research would not have been possible without the active participation, support and cooperation of Ngamiland pastoral communities in the villages of Toteng, Bothatogo, Bodibeng, Sehithwa, Makakung and Semboyo. Many thanks to the anonymous reviewers for their insightful comments and suggestions.

Author Contributions: Lenyeletse V. Basupi conceived, researched and wrote the paper; Claire H. Quinn and Andrew J. Dougill provided substantial contribution throughout the process through supervision, research framing, ideas and wordsmithing. All authors read and approved the final manuscript to be published.

Conflicts of Interest: The authors declare no competing interests.

References

1. Bennett, J.; Barrett, H. Rangeland as a Common Property Resource: Contrasting Insights from Communal Areas of Central Eastern Cape Province, South Africa. *Hum. Ecol.* **2007**, *35*, 97–112. [CrossRef]
2. Chanda, R.; Totolo, O.; Moleele, N.; Setshogo, M.; Mosweu, S. Prospects for subsistence livelihood and environmental sustainability along the Kalahari Transect: The case of Matsheng in Botswana's Kalahari rangelands. *J. Arid Environ.* **2003**, *54*, 425–445. [CrossRef]
3. Lesorogol, C.K. Land privatization and pastoralist well-being in Kenya. *Dev. Chang.* **2008**, *39*, 309–331. [CrossRef]
4. Lebert, T.; Rohde, R. Land reform and the new elite: Exclusion of the poor from communal land in Namaqualand, South Africa. *J. Arid Environ.* **2007**, *70*, 818–833. [CrossRef]
5. Thornton, P.K.; van de Steeg, J.; Notenbaert, N.; Herrero, M. The impacts of climate change on livestock and livestock systems in developing countries: A review of what we know and what we need to know. *Agric. Syst.* **2009**, *101*, 113–127. [CrossRef]
6. Elhadary, Y.A.E. Challenges facing land tenure system in relation to pastoral livelihood security in Gedarif State, Eastern Sudan. *J. Geogr. Reg. Plan.* **2010**, *3*, 208–218.
7. Western, D.; Groom, R.; Worden, J. The impact of subdivision and sedentarization of pastoral lands on wildlife in an African savanna ecosystem. *Biol. Conserv.* **2009**, *142*, 2538–2546. [CrossRef]
8. Meadows, M.E.; Hoffman, T.M. Land degradation and climate change in South Africa. *Geogr. J.* **2003**, *169*, 168–177. [CrossRef]
9. Kaptue, A.T.; Hanan, N.P.; Prihadko, C.; Ramirez, J.A. Spatial and temporal characteristics of rainfall in Africa: Summary statistics for temporal downscaling. *Water Resour. Res.* **2015**, *51*, 2668–2679. [CrossRef]
10. Turner, M.D. The New Pastoral Development Paradigm: Engaging the Realities of Property Institutions and Livestock Mobility in Dryland Africa. *Soc. Nat. Resour.* **2011**, *24*, 469–484. [CrossRef]
11. Oba, G. The sustainability of pastoral production in Africa. In *Pastoralism and Development in Africa: Dynamic Change at the Margins, Pathways to Sustainability;* Catley, A., Lind, J., Scoones, I., Eds.; Routlege: New York, NY, USA, 2013; pp. 29–36.

12. Adriansen, H.K.; Nielsen, T.T. Going where the grass is greener: On the study of pastoral mobility in Ferlo, Senegal. *Hum. Ecol.* **2002**, *30*, 215–226. [CrossRef]
13. Kisamba-Mugerwa, W.; Pender, J.; Kato, E. Impacts of individualization of land tenure on livestock and rangeland management in Southwestern Uganda. In Proceedings of the 11th Biennial Conference of the International Association for the Study of Common Property, Bali, Indonesia, 19–23 June 2006; IFPR Institute: Bali, Indonesia, 2006.
14. Hardin, G. The tragedy of the commons. *Science* **1968**, *162*, 1243–1248. [CrossRef] [PubMed]
15. Fratkin, E. Pastoralism: Governance and Development Issues. *Annu. Rev. Anthropol.* **1997**, *26*, 235–261. [CrossRef]
16. De Soto, H. *The Mystery of Capital: Why Capitalism Triumphs in the West and Fails Everwhere Else*; Buntum Press: London, UK, 2000.
17. Cousins, B.; Scoones, I. Contested paradigms of 'viability' in redistributive land reform: Perspectives from southern Africa. *J. Peasant Stud.* **2010**, *37*, 31–66. [CrossRef]
18. Leach, M.; Mearns, R.; Scoones, I. Environmental Entitlements: Dynamics and Institutions in Community-Based Natural Resource Management. *World Dev.* **1999**, *27*, 225–247. [CrossRef]
19. Ostrom, E. *Governing the Commons: The Evolution of Institutions for Collective Action*; Cambridge University Press: New York, NY, USA, 1990.
20. Ostrom, E. Coping with tragedies of the commons. *Annu. Rev. Political Sci.* **1999**, *2*, 493–535. [CrossRef]
21. Bendor, J. In good times and bad—Reciprocity in an uncertain world. *Am. J. Political. Sci.* **1987**, *31*, 531–558. [CrossRef]
22. Rohde, R.F.; Moleele, N.M.; Mphale, M.; Allsopp, N.; Chanda, R.; Hoffman, M.T.; Magole, L.; Young, E. Dynamics of grazing policy and practice: Environmental and social impacts in three communal areas of southern Africa. *Environ. Sci. Policy* **2006**, *9*, 302–316. [CrossRef]
23. Letai, J.; Lind, J. Squeezed from all Sides Changing resource tenure and pastoralist innovation on the Laikipia Plateau, Kenya. In *Pastoralism and Development in Africa: Dynamic Change at the Margins*; Routledge: New York, NY, USA, 2013; pp. 164–176.
24. Galaty, J.G. "The land is yours": Social and economic factors in the privatization, sub-division and sale of Maasai ranches. *Nomadic Peoples* **1992**, *30*, 26–40.
25. Mwangi, E. Property rights and governance of Africa's rangelands: A policy overview. *Nat. Resour. Forum* **2009**, *33*, 160–170. [CrossRef]
26. Waddington, H.; White, H.; Snilstrveit, B.; Hombrados, J.G.; Vojtkova, M.; Davies, P.; Bhavsar, A.; Eyers, J.; Koehlmoos, T.P.; Petticrew, M.; et al. How to do a good systematic review of effects in international development: A tool kit. *J. Dev. Eff.* **2012**, *4*, 359–387. [CrossRef]
27. Tongco, M.D.C. Purposive Sampling as a tool for Informant Selection: Research Methods. *Ethnobot. Res. Appl.* **2007**, *5*, 147–158. [CrossRef]
28. Benjaminsen, T.A.; Rhode, R.; Sjaastad, E.; Wisborg, P.; Lebert, T. Land Reform, Range Ecology, and Carrying Capacities in Namaqualand, South Africa. *Ann. Assoc. Am. Geogr.* **2006**, *96*, 524–540. [CrossRef]
29. Bennett, J.; Ainslie, A.; Davis, J. Fenced in: Common property struggles in the management of communal rangelands in central Eastern Cape Province, South Africa. *Land Use Policy* **2010**, *27*, 340–350. [CrossRef]
30. Beyene, F. Customary Tenure and Reciprocal Grazing Arrangements in Eastern Ethiopia. *Dev. Chang.* **2010**, *41*, 107–129. [CrossRef]
31. Coleman, E.A.; Mwangi, E. Conflict, Cooperation, and Institutional Change on the Commons. *Am. J. Political Sci.* **2015**, *59*, 855–865. [CrossRef]
32. Deininger, K.W. *Land Policies for Growth and Poverty Reduction*; World Bank Publications: Washington, DC, USA, 2003.
33. Adams, M. Reforming communal rangeland policy in southern Africa: Challenges, dilemmas and opportunities. *Afr. J. Range Forage Sci.* **2013**, *30*, 91–97. [CrossRef]
34. Benjaminsen, T.A.; Holden, S.; Lud, C.; Sjaastad, E. Formalisation of land rights: Some empirical evidence from Mali, Niger and South Africa. *Land Use Policy* **2009**, *26*, 28–35. [CrossRef]
35. Government Printers. *Republic of Botswana, National Policy on Tribal Grazing Land: Government Paper No. 2 of 1975*; Government Printers: Gaborone, Botswana, 1975.
36. *Republic of Botswana, National Policy on Agricultural Development (NPAD)*; Government Printers: Gaborone, Botswana, 1991.

37. United Nations. *The UN Millennium Project: A Practical Plan to Achieve the MDGs*; UN: London, UK, 2005.
38. *Comprehensive Africa Agriculture Development Programme (CAADP)*; The New Partnership for Africa's Development: Midrand, South Africa, 2003.
39. Mwangi, E.; Dohrn, S. Securing access to drylands resources for multiple users in Africa: A review of recent research. *Land Use Policy* **2008**, *25*, 240–248. [CrossRef]
40. Mwangi, E. Subdividing the commons: Distributional conflict in the transition from collective to individual property rights in Kenya's Maasailand. *World Dev.* **2007**, *35*, 815–834. [CrossRef]
41. Galaty, J.G. Land grabbing in the Eastern African rangelands. In *Pastoralism and Development in Africa: Dynamic Change at the Margins, Pathways to Sustainability*; Catley, A., Lind, J., Scoones, I., Eds.; Routlege: New York, NY, USA, 2013; pp. 143–153.
42. Tache, B. Rangeland enclosures in Southern Oromia, Ethiopia: An innovative response or the erosion of common property resources? In *Pastoralism and Development in Africa: Dynamic Change at the Margins, Pathways to Sustainability*; Catley, A., Lind, J., Scoones, I., Eds.; Routledge: New York, NY, USA, 2013; pp. 37–46.
43. Babiker, M. Mobile Pastoralism and Land Grabbing in Sudan. In *Pastoralism and Development in Africa: Dynamic Change at the Margins*; Routledge: New York, NY, USA, 2013; pp. 177–185.
44. Wisborg, P.; Rohde, R. Contested land tenure reform in South Africa: Experiences from Namaqualand. *Dev. S. Afr.* **2005**, *22*, 409–428. [CrossRef]
45. Benjaminsen, T.A.; Sjaastad, E. Where to draw the line: Mapping of land rights in a South African commons. *Polit. Geogr.* **2008**, *27*, 263–279. [CrossRef]
46. Klintenberg, P.; Verlinden, A. Water points and their influence on grazing resources in central northern Namibia. *Land Degrad. Dev.* **2008**, *19*, 1–20. [CrossRef]
47. Simbizi, M.C.D.; Bennett, R.M.; Zevenbergen, J. Land tenure security: Revisiting and refining the concept for Sub-Saharan Africa's rural poor. *Land Use Policy* **2014**, *36*, 231–238. [CrossRef]
48. Department of Lands (DoL). *Ngamiland Integrated Land Use Plan, Final Report*; Ministry of Lands and Housing: Gaborone, Botswana, 2009.
49. Makepe, P.M. The Evolution of Institutions and Rules Governing Communal Grazing Lands in Botswana. *East. Afr. Soc. Sci. Res. Rev.* **2006**, *22*, 39–61. [CrossRef]
50. Basupi, L.V.; Quinn, C.H.; Dougill, A.J. Using participatory mapping and a participatory geographic information system in pastoral land use investigation: Impacts of rangeland policy in Botswana. *Land Use Policy* **2017**, *64*, 363–373. [CrossRef]
51. White, R. *Livestock Development and Pastoral Production on Communal Rangeland in Botswana*; The Botswana Society: Gaborone, Botswana, 1993.
52. Perkins, J.S. Botswana: Fencing out the equity issue. Cattleposts and cattle ranching in the Kalahari Desert. *J. Arid Environ.* **1996**, *33*, 503–517. [CrossRef]
53. Peters, P.E. *Dividing the Commons: Politics, Policy, and Culture in Botswana*; University Press of Virginia: Charlottesville, VA, USA, 1994; p. 277.
54. APRU (Animal Production Research Unit). *An Integrated Programme of Beef Cattle and Range Research in Botswana: 1970–1976*; Animal Production Research Unit (APRU), Ministry of Agriculture: Gaborone, Botswana, 1976.
55. Childers, G.W. *Western Ngwaketse Remote Area Dwellers: A Land Use and Development Plan for Remote Area Settlements in Southern District*; Government Printer: Gaborone, Botswana, 1981.
56. Basupi, L.V.; Quinn, C.H.; Dougill, A.J. Historical perspectives on pastoralism and land tenure transformation in Ngamiland, Botswana: What are the policy and institutional lessons? *Pastoralism* **2017**, *7*, 24. [CrossRef]
57. Tsimako, B. *The Tribal Grazing Land Policy (TGLP) Ranches: Performance to Date*; Ministry of Agriculture: Gaborone, Botswana, 1991.
58. Bassett, T.J. Mobile pastoralism on the brink of land privatization in Northern Cote d'Ivoire. *Geoforum* **2009**, *40*, 756–766. [CrossRef]
59. Fernandez-Gimenez, M.E.; Le Febre, S. Mobility in pastoral systems: Dynamic flux or downward trend? *Int. J. Sustain. Dev. World Ecol.* **2006**, *13*, 341–362. [CrossRef]
60. Peters, P.E. Challenges in Land Tenure and Land Reform in Africa: Anthropological Contributions. *World Dev.* **2009**, *37*, 1317–1325. [CrossRef]

61. Lesorogol, C.K. Privatizing pastoral lands: Economic and normative outcomes in Kenya. *World Dev.* **2005**, *33*, 1959–1978. [CrossRef]
62. Thébaud, B.; Batterbury, S. Sahel pastoralists: Opportunism, struggle, conflict and negotiation. A case study from eastern Niger. *Glob. Environ. Chang.* **2001**, *11*, 69–78. [CrossRef]
63. Goldman, M.J.; Riosmena, F. Adaptive capacity in Tanzanian Maasailand: Changing strategies to cope with drought in fragmented landscapes. *Glob. Environ. Chang.* **2013**, *23*, 588–597. [CrossRef] [PubMed]
64. Greiner, C.; Mwaka, I. Agricultural change at the margins: Adaptation and intensification in a Kenyan dryland. *J. East. Afr. Stud.* **2016**, *10*, 130–149. [CrossRef]
65. Headey, D.; Taffesse, A.S.; You, L. Diversification and Development in Pastoralist Ethiopia. *World Dev.* **2014**, *56*, 200–213. [CrossRef]
66. Herrero, M.; Thornton, P.K.; Gerber, P.; Reid, R.S. Livestock, livelihoods and the environment: Understanding the trade-offs. *Curr. Opin. Environ. Sustain.* **2009**, *1*, 111–120. [CrossRef]
67. Mulale, K.; Chanda, R.; Perkins, J.S.; Magole, L.; Sebego, R.J.; Atlhopheng, J.R.; Mphinyane, W.; Reed, M.S. Formal institutions and their role in promoting sustainable land management in Boteti, Botswana. *Land Degrad. Dev.* **2014**, *25*, 80–91. [CrossRef]
68. Homewood, K.M. Policy, environment and development in African rangelands. *Environ. Sci. Policy* **2004**, *7*, 125–143. [CrossRef]
69. Swallow, B.M.; Bromley, D.W. Institutions, governance and incentives in common property regimes for African rangelands. *Environ. Resour. Econ.* **1995**, *6*, 99–118. [CrossRef]
70. Wily, L.A. 'The Law is to Blame': The Vulnerable Status of Common Property Rights in Sub-Saharan Africa. *Dev. Chang.* **2011**, *42*, 733–757. [CrossRef]
71. Lesorogol, C.K. Transforming Institutions among Pastoralists: Inequality and Land Privatization. *Am. Anthropol.* **2003**, *105*, 531–541. [CrossRef]
72. Bennett, J.; Ainslie, A.; Davis, J. Contested institutions? Traditional leaders and land access and control in communal areas of Eastern Cape Province, South Africa. *Land Use Policy* **2013**, *32*, 27–38. [CrossRef]
73. Toulmin, C. Securing land and property rights in sub-Saharan Africa: The role of local institutions. *Land Use Policy* **2009**, *26*, 10–19. [CrossRef]
74. Nelson, F.; Agrawal, A. Patronage or Participation? Community-based Natural Resource Management Reform in Sub-Saharan Africa. *Dev. Chang.* **2008**, *39*, 557–585. [CrossRef]
75. Schmidt, M.; Pearson, O. Pastoral livelihoods under pressure: Ecological, political and socioeconomic transitions in Afar (Ethiopia). *J. Arid Environ.* **2016**, *124*, 22–30. [CrossRef]
76. Bett, B.; Kiunga, P.; Gachohi, J.; Sindato, C.; Mbotha, D.; Robinson, T.; Lindahl, J.; Grace, D. Effects of climate change on the occurrence and distribution of livestock diseases. *Prev. Vet. Med.* **2017**, *137*, 119–129. [CrossRef] [PubMed]
77. Rojas-Downing, M.M.; Nejadhashemi, A.P.; Harrigan, T.; Woznicki, S.A. Climate change and livestock: Impacts, adaptation, and mitigation. *Clim. Risk Manag.* **2017**, *16*, 145–163. [CrossRef]
78. Dougill, A.J.; Akanyang, L.; Perkins, J.S.; Eckardt, F.D.; Stringer, L.C.; Favretto, N.; Atlhopheng, J.R.; Mulale, K. Land use, rangeland degradation and ecological changes in the southern Kalahari, Botswana. *Afr. J. Ecol.* **2016**, *54*, 59–67. [CrossRef]
79. Ringrose, S.; Chanda, R.; Nkambwe, M.; Sefe, F. Environmental change in the mid-Boteti area of north-central Botswana: Biophysical processes and human perceptions. *Environ. Manag.* **1996**, *20*, 397–410. [CrossRef]
80. Thomas, D.S.G.; Sporton, D.; Perkins, J. The environmental impact of livestock ranches in the Kalahari, Botswana: Natural resource use, ecological change and human response in a dynamic dryland system. *Land Degrad. Dev.* **2000**, *11*, 327–341. [CrossRef]

land

MDPI

Article

Identifying Hot Spots of Critical Forage Supply in Dryland Nomadic Pastoralist Areas: A Case Study for the Afar Region, Ethiopia

B. G. J. S. Sonneveld [1,*], C. F. A. van Wesenbeeck [1], M. A. Keyzer [1], F. Beyene [2], K. Georgis [3], F. Urbano [4], M. Meroni [4], O. Leo [4], M. K. Yimer [5] and M. Abdullatif [6]

1 Amsterdam Centre for World Food Studies, Vrije Universiteit Amsterdam, 1081 HV Amsterdam, The Netherlands; c.f.a.van.wesenbeeck@vu.nl (C.F.A.v.W.); m.a.keyzer@vu.nl (M.A.K.)
2 College of Agriculture and Environmental Sciences, Haramaya University, P.O. Box 138, Dire Dawa, Ethiopia; keneefbk@gmail.com
3 GEOSAS Consultancies, P.O. Box 787, Addis Ababa, Ethiopia; kidanegeorgis@yahoo.com
4 Sustainable Resources, Food Security Unit, Joint Research Centre, European Commission, Via Fermi 2749, 21027 Ispra, Italy; ferdinando.urbano@ext.jrc.ec.europa.eu (F.U.); michele.meroni@jrc.ec.europa.eu (M.M.); olivier.leo@jrc.ec.europa.eu (O.L.)
5 Water Resource and Irrigation Engineering department. Haramaya Institute of Technology, Haramaya University, P.O. Box 138, Dire Dawa, Ethiopia; hamernoah@gmail.com
6 Afar Pastoral and Agro-pastoral Research Institute, P.O. Box 16, Semara, Ethiopia; amame120@yahoo.com
* Correspondence: b.g.j.s.sonneveld@vu.nl

Received: 26 October 2017; Accepted: 13 November 2017; Published: 18 November 2017

Abstract: This study develops a methodology to identify hot spots of critical forage supply in nomadic pastoralist areas, using the Afar Region, Ethiopia, as a special case. It addresses two main problems. First, it makes a spatially explicit assessment of fodder supply and demand extracted from a data poor environment. Fodder supply is assessed by combining rainfall-based production functions and rule-based assessment for prevailing land use. Fodder demand is based on a data consistency check of livestock statistics concerning herd size, composition and geographical distribution. Second, individual herd movements have to be evaluated jointly in concurrent migration patterns to assess local pressures on fodder resources. We, therefore, apply a transition model that relates stock levels to seasonal migration routings for all Afar sub-clans jointly so as to localize the hot spots where feed demand exceeds forage supply. Critical areas come to the fore, especially, near fringes of Highlands and in the southern part of the Afar. A sensitivity test shows that 'Baseline' scenario is close to the 'Best' but under 'Worst', the Afar region would fall into despair. We conclude that the model is a useful tool to inform policy makers on critical areas in the Afar region.

Keywords: nomadic pastoralism; spatial migration model; Afar; livestock; fodder demand; fodder supply

1. Introduction

Drylands cover 40 percent of the world's land area and host around two billion people, 90 per cent of which lives in developing countries [1]. In these drylands some 30–40 million people practice nomadic pastoralism, an extensive grazing system that uses flexible migration patterns to follow rainy seasons in arid regimes. For a long time nomadic pastoralism was synonymous for overgrazing [2–4] and archaic production systems [5,6], that echoed the 'Tragedy of the Commons' [7]. Yet, the stance that nomadic pastoral systems are unsustainable came under serious criticism [8–13]) and was categorically rejected in 2009 when Elinor Ostrom was awarded the Nobel Prize in Economics for her lifetime scholarly work on the management of common pool resources. Ostrom's studies show that communities devise ways to govern the commons through organizing collective action to assure its survival for their needs and future generations. Many studies on nomadic pastoralism [14–16] refer to her paradigm on successful sharing of communal rangelands to spread risk. Indeed, nomadic[1] pastoralism is nowadays considered an epitome of sustainability [17–19] that unrelentingly depends on traditional institutions that have proven to be instrumental in the management of the shared natural resource base. Yet, the question is if these institutions are sufficiently resilient to cope with new challenges that are often influenced by external stress factors that go beyond their control. For example, population growth and appropriation of land for irrigation [20] has caused pressure to mount in the nomadic pastoralist societies, with increasing incidences of overgrazing and violent conflicts as most visible symptoms [21]. Moreover, development plans for massive expansion of biofuel plantations [22] on marginal drylands can be expected to further restrict land's accessibility to pastoralists [23]. This creates an enormous challenge for policy makers to come up with appropriate policy decision if poverty alleviation is to be achieved.

There are also good economic and environmental reasons for investing in pastoral development. First, pastoralism is an efficient land use system that can cope with the extreme prevailing climatic variations in arid environments [24,25]. Studies show that livestock grazing has a positive effect on development and sustainment of plant biodiversity [26,27]. Second, new opportunities for livestock production arise as global markets are rapidly expanding due to a growing and more affluent population in the urban areas, demanding more meat products [28]. Indeed, in the next two decades the livestock sector is projected to become the world's most important agricultural subsector in terms of added value and land use whereby the developing world is projected to be the major supplier of this growing market [29]. Hence, the urgent calls for more research to support the pastoral communities and explore the potential of livestock production under climate change conditions in dryland areas seem justified.

1.1. The Study Area

Afar Regional State[2] in North East Ethiopia in the horn of Africa (Figure 1) is a typical case in point. This semi-arid to arid region of 94,436 km^2, hosts 1.4 million people, divided over 112 sub-clans[3], 78 percent of which are involved in nomadic pastoralism, with herds that mainly consists of cattle, camels, sheep and goats [30]. Yet, accessibility to rangelands and watering points is increasingly hampered by expansion of sedentary agricultural settlements along the Awash River, implementation of large scale agricultural projects like the state-owned Tenaha sugar plantation and the increasing incidence of contested territorial claims by different ethnic groups from outside the region [31–33]. Tensions are sharpened further by regulatory legislation and taxation discouraging traditional trans-boundary movements to Djibouti and Eritrea [34]. Current plans for expansion of biofuel plantations of sugarcane and Jatropha shrubs, while opening new economic prospects for the Afar region, will also encroach further on the land available to pastoralists [35]. Indeed, with further restrictions on rangeland accessibility, the threats of land degradation, negative climate change effects and violent conflicts over scarce remaining land and water resources will wreak havoc on the pastoral societies in the Afar. Yet, there are also development opportunities for the Afar region. Studies show that potential meat production of the lowlands remain largely untapped [36], while meat and hides demand from neighboring countries is increasing rapidly [37–39]. These prospects for pastoralists should, however, be studied in relation to a sustainable development of the natural resources base with a strong geographical component to account for spatial and temporal variability of natural endowments and

---

1 According to the type of trekking patterns pastoralists are called nomadic (irregular movements) or transhumant (regular movements between fixed locations).

2 Ethiopia is administratively divided into regional states and chartered cities, zones, woredas (districts) and kebeles (wards).

3 The Afar as ethnic group occupy a territory that comprises the Afar Region of Ethiopia, northern Djibouti and southern point of Eritrea. The territorial and political unit in the Afar Region is the sub-clan which retains a relatively high degree of political, social, and economic independence.

land uses. Yet, studies in Ethiopia concentrated basically on sedentary agriculture in the Highlands [40]. In recent years also national planning strategies for the rangeland areas came to the fore [41–43].

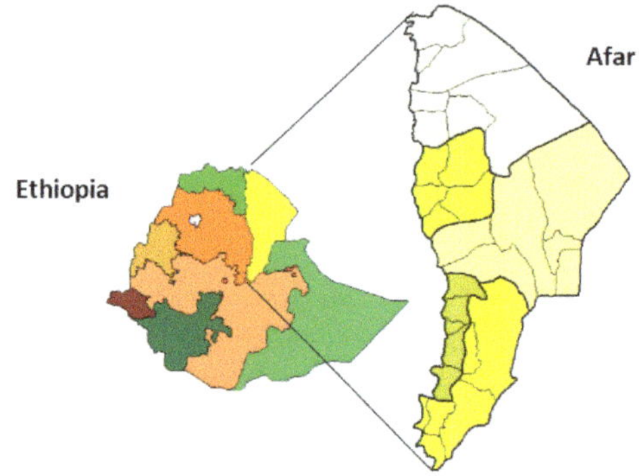

Figure 1. Ethiopia (scale 1:41,000,000) Afar State, zones and woredas (scale 1:9,200,000).

The policy document for land use and administration of the Afar Regional State (Land Administration and Use Proclamation No. 49, enacted in 2009) emphasizes sedentarization of pastoralists and the establishment of formal institutions to manage land use and administration. Meanwhile, the intention is to retain customary systems that do not contradict the formal land administration. Its effectiveness has been questioned by a recent study [44] carried out on the feasibility of land use policy in the region where environmental factors largely affect land resource uses while customary systems were efficient in enabling pastoralists to manage risks. In the policy document, the communal system has been viewed as dysfunctional and destructive. However, it acknowledges the contribution of the customary institutions only towards conflict management rather than land resource management [45]. The pastoral development policy of the country does not recognize pastoralists' environment that is uniquely different from the sedentary farming systems. The policy narrates voluntary settlement while it imposes it in practice. Sedentarization as a policy choice has been perceived as exposing pastoralists to greater ecological risk and vulnerability [46]. However, such a choice provides the government an option to provide potential pastoral land to large-scale foreign direct investment—a practice that has created hostile relationship between pastoralists and investors since the strategy hinders pastoral mobility and response to environmental risk [47].

1.2. This Study

This complex whole of threats and opportunities motivates the current study that conducts a spatially explicit analysis to identify hot spots where the hazard of overgrazing looms. For this, the study addresses two issues. First, the generally data poor environments of the drylands requires elaborate evaluations and careful assessments of sources on livestock distribution and fodder supply. Various estimates of livestock distribution from different sources are compared and evaluated for consistency in reporting. Concerning the spatial assessment of fodder supply we combine the Afar land use map indicating percentage of grass land by land use type with rainfall production functions that estimate palatable fodder production. Second, sites under risk can only be identified when movements of clans are followed simultaneously in their geographical and temporal dependence. In absence of information on migration routes, Sonneveld et al. [48] reported on stylized results by releasing boundary restrictions from woreda to zonal and state level. Yet, these attempts are far from realistic. Therefore, during the Period October-November 2015, key figures related to the 112 sub-clans were interviewed about migration routes and the share of migrating herd during the four prevailing seasons in the year. The survey was conducted under the auspices of the Afar Pastoral and Agro-pastoral Research Institute by an experienced rangeland management expert. Data on migration were harmonized for further analysis during a two-week workshop held in December 2015 in Amsterdam. A sub-clan map was produced that reflects the boundaries of the sub-clan territories. The model that we present follows over time and space migration routes of all clans and evaluates locally the pressure of livestock presence on the produced fodder. Hot spots are identified when demand exceeds supply.

This article is organized as follows. Section 2 presents the data and methodology that have been used in this study. Section 3 presents the results of the migration model for the four seasons. Section 4 performs a sensitivity analysis for assumed uncertainty of parameters. Section 5 concludes.

2. Material and Methods

This section presents the sub-clan map of the Afar (Section 2.1), the methodologies for spatial assessment of forage supply (Section 2.2) and spatial distribution of livestock in the Afar region (Section 2.3) and, finally, the migration model that is used for identification of sites at risk (Section 2.4).

2.1. The Base Map

Figure 2 presents the map of sub-clan areas. In total there are 112 sub-clans with areas varying from 20 to 4333 square kilometers. Larger areas are found in the dryer North Eastern part bordering Eritrea and Djibouti. To our knowledge this is the first sub-clan map that is operationalized for migration modelling.

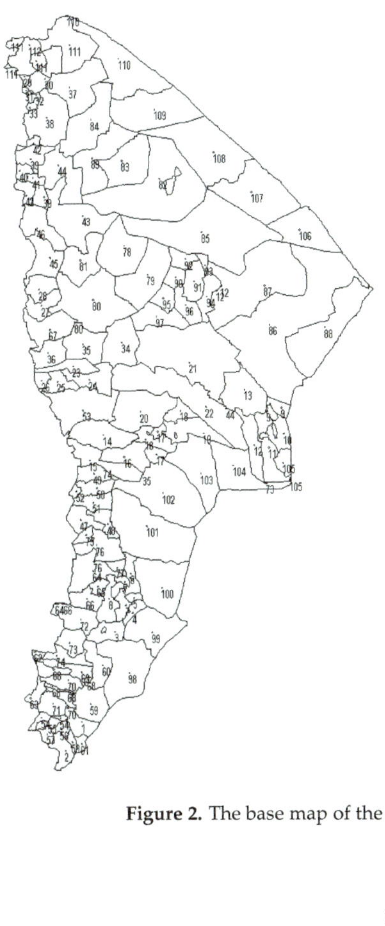

Code	Sub Clan	Code	Sub Clan	Code	Sub Clan
1	Harkemella	46	Alihammedo	91	Mandayta
2	Awash-Park	47	Adeali	92	Dahimela_Modaytu(92)
3	Tekiele	48	Ayti_Asosa	93	Muhoto
4	Mahesara	49	Wegeb	94	Dahimela_Modaytu
5	Ulukuto	50	Tamitmella	95	Assabkari Nassar
6	Seka(6)	51	Asandosara	96	Askaki Meli
7	Madimak_Mesara	52	Arebeta	97	Dahimela_mohoto
8	Gideboso	53	Arebta	98	Mandeta
9	Gelela	54	Deben	99	Deryto
10	Ouluato	55	Ayreso	100	Geliela
11	Syereo	56	Gahelu	101	Beadul
12	Kaxabura	57	Algeala	102	Herto
13	Asmalik	58	Weima(58)	103	Haysantokboosali
14	Hassen_mohammed	59	Deben(59)	104	Kutebula
15	Seka(15)	60	Weima(60)	105	Adahiso
16	Bagisase	61	Adayela	106	Wadijma
17	Harbis_Mella	62	Argoba	107	Halola
18	Adukum_Aslubak	63	Argoba/chano	108	Galealu
19	Asabakri	64	Modaytu	109	Humed
20	Molak_Mayabe	65	Abeamela	110	Taela
21	Maqosora	66	harmela	111	Barito
22	Dathamado	67	Seha_Muhammed	112	Udumu
23	Kikhenkeba	68	Asoba		
24	Hondogole_Maado	69	Adealli		
25	Nasarkagen	70	Fediha		
26	Harso	71	Ameassa		
27	Domohoyota	72	Sidehabura		
28	Haydermo	73	Gidbossa		
29	Bedal	74	Ayidahisso		
30	Diahimela	75	Deriham		
31	Ruhna	76	Huguto		
32	Welwelu	77	Meanddita		
33	Abamela	78	Kedeburra		
34	Aferera	79	Seha_mohammed		
35	Hadermo(35)	80	Abelelek_Adealli		
36	Daheimelea	81	Asabekerri		
37	Dahimela	82	Gedentiok Manndita		
38	Domohoyta(38)	83	Ayrobuldo		
39	Galasiso	84	Ferdan and Dahimella		
40	Hadermo(40)	85	Sheika(85)		
41	Maandeta	86	Afkak Madu(Afsek)		
42	Gardia	87	Arebtak Assabbekri		
43	Dermella	88	Melk Mela		
44	Dahi	89	Shekah Hulufi		
45	Dehimela	90	Hamedi _Demoyta		

Figure 2. The base map of the sub-clan areas (scale 1:3,700,000).

The sub-clan map was compiled using interviews with clan leaders, local authorities of Woredas and key figures from the Afar Pastoral and Agro-Pastoral Research Institute. When indicated by the interviewees sub-clan areas were delineated by one or more Kebele boundaries or by using clear landmarks. Experts of the Afar Pastoral and Agro-Pastoral Research Institute confirmed that the map was a fair assessment of sub-clan areas [49].

2.2. Fodder Supply

Estimation of spatial and temporal production of palatable biomass in the Afar is seriously hampered by lack of data as few direct observations on forage yields are available. Therefore, we decided to take the Woody Biomass land use map [50], depicting land use categories and percentage grassland (Figure 3), as baseline information for our assessment in the following step-wise approach. First, a set of rainfall dependent forage production functions designed for arid and semi-arid regimes in Africa (Table 1) is used to calculate the annual forage production for grassland cover. The spatial fodder estimates used annual rainfall maps derived from the Global Agro-Ecological Zones data set [51] covering the period 1991–2000. The 5 minute resolution of the rainfall maps is resampled using a bi-cubic filter to match the 200 $\times$ 200 m^2 grid of the land use map. The reported fodder estimates derived from the production functions are the average over the 10-year period. Second, forage production data by land use type (Table 2), provided by the Regional Werer Station of the Ethiopian Institute for Agricultural Research (EIAR) are applied to the remaining land use/cover categories (Table 3). Results of the assessments are presented in Table 4, with the first column referring to source of production function, second and third columns to total and per ha forage production, respectively, based on rain dependent production functions and the fourth column presenting the maximum number of Tropical Livestock Unit (TLU)[4] that could be fed by the corresponding forage assessment, assuming an annual intake of 2.3 ton per TLU; column five to seven give similar information as column two to four but with forage production adjusted for land use characteristics. Our estimations are within the bounds of forage production estimates in arid regions (Table 5) using reference data found in [52].

Table 1. Rainfall dependent forage production functions.

Area	Equation (Number of Observation; R^2)		Reference
Amboseli (Tanzania)	$Y = -367 + 3.8X$	(6; 0.99)	[53]
Kiboko (Kenya)	$Y = 262 + 4.41X$	(38; 0.78)	[54]
Serengeti (Tanzania)	$Y = 262 + 4.8X$	(7; 0.93)	[55]
Tsavo (Kenya)	$Y = 380 + 8.0X$	(89; 0.65)	[56]
Serengeti (Tanzania)	$Y= -1644 + 10.7X$	(12; 0.62)	[55]
Serengeti (Tanzania)	$Y = -185 + 6.6X$	(24; 0.90)	[57]
Athi (Kenya)	$Y = -251 + 1.2X + 0.01X^2$	(24; 0.95)	[58]
Serengeti (Tanzania)	$Y = -1052 + 8.6X$	(10; 0.56)	[55]
East Africa	$Y = -195.77 + 8.49X$	(32; 0.67)	[59]

Y = Biomass in kg DM ha^{-1}; X = rainfall in mm.

[4] Tropical livestock units allow to compare grazing demand of different species in common units.

Table 2. Biomass production by land use [60].

Land Use/Cover Types	Production (Herbage Yield, ton/ha)
Moderately cultivated	0.9
Open grassland	2.3
Open grassland shrub-bed	1.4
Dense shrub-land	0.3
Open shrub land (Open bush shrub land)	0.8
Open woodland	0.9
Wooded grassland shrub bed	1.5
Riparian wood/shrub land/bush-land	0.5

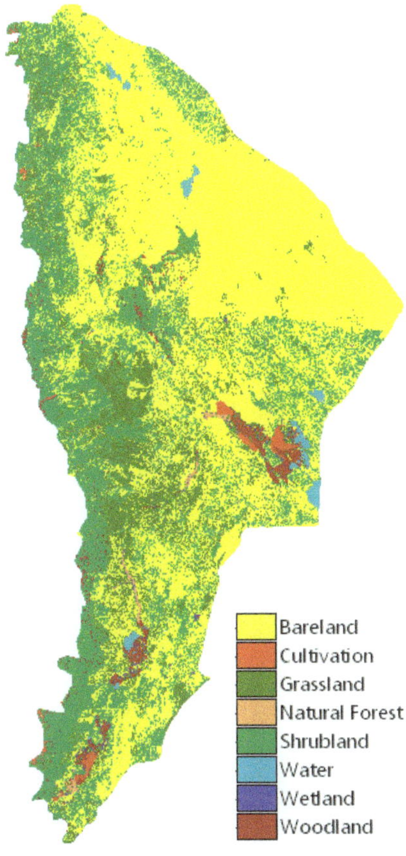

Figure 3. Afar Land Use Map (scale 1:5,500,000).

Table 3. Rule based procedure to map biomass assessment for land use categories in the Woody Biomass project.

Land Use Category	Rule Based Procedure to Assign EIAR/ Werer Biomass to LUC of the Woody Biomass	Herbage Yield ton/ha
Bareland; Exposed rock		0
Bareland; Exposed sand/soil		0
Cultivated Land; Irrigated	50% of Moderately cultivated	0.45
Cultivated Land; Rainfed; Cereal Land Cover System; lightly stocked	50% of Moderately cultivated	0.45
Cultivated Land; Rainfed; Cereal Land Cover System; moderately stocked	50% of Moderately cultivated	0.45
Cultivated Land; Rainfed; Cereal Land Cover System; unstocked (woody pl)	50% of Moderately cultivated	0.45
Forest; Montane coniferous; Open (20–50% crown cover)	Open woodland	0.9
Forest; Riparian; Dense (5–80% crown cover)	50 % Riparian wood/shrub land/bush-land	0.25
Forest; Riparian; Open (2–50% crown cover)	Riparian wood/shrub land/bush-land	0.5
Grassland; lightly stocked	Open grassland	2.3
Grassland; moderately stocked	50% Open grassland	1.2
Grassland; unstocked (woody plant)	25% Open grassland	0.6
Shrubland; Dense (>50% woody cover)	Dense shrub-land	0.3
Shrubland; Open (20–50% woody cover)	Open shrub land (Open bush shrub land)	0.8
Wetland; Open water	Riverside	0.5
Wetland; Perennial Swamp/Marsh	25% of Riverside	0.13
Wetland; Seasonal Swamp/Marsh	50% of Riverside	0.25
Woodland; Dense (>50% tree cover)	50% of Open woodland	0.45
Woodland; Open (20–50% tree cover)	Open woodland	0.9

Source: Sutcliffe, 2006 [50] and own computations.

Table 4. Forage production, yield and maximum Tropical Livestock Unit (TLU) estimates for Afar State using rain dependent production functions and rain dependent production functions adjusted for land cover characteristics.

Source Equation	Forage Production, Rain Dependent (1000 ton)	Forage Yield Rain Dependent (ton per ha)	Max TLU Rain Dependent (in 1000)	Forage Production adj. Land Charact. (1000 ton)	Forage Yield adj. Land Charact. (ton per ha)	Max TLU. adj. Land Charact. (in 1000)
[53]	5029	0.7	2624	3714	0.5	1615
[54]	7506	1.0	3916	5405	0.8	2350
[55]	8113	1.1	4233	5846	0.8	2542
[56]	13,384	1.9	6983	9652	1.3	4197
[55]	12,679	1.8	6615	9485	1.3	4124
[57]	9832	1.4	5130	7170	1.0	3117
[58]	12,010	1.7	6266	9413	1.3	4093
[55]	10,844	1.5	5658	8052	1.1	3501
[59]	12,751	1.8	6652	9290	1.3	4039
Average	10,239	1.4	5342	7559	1.0	3286

Table 5. Estimation of total forage production (t DM ha-1) under various rainfall regimes [52].

Rainfall (mm)	200	400	600	800	Source
West Africa	0.6	1.1	1.7	2.2	[61]
Zimbabwe (water holding cap.: 100 mm)	0.5	1.7	2.2	2.5	[62]
Zimbabwe (water holding cap.: 200 mm)	0.7	2.6	3.2	3.7	[62]
Kenya	1.1	2.3	3.6	-	[63]

2.3. Total Livestock Herds in the Afar Region

Like the fodder supply, estimating feed demand of livestock population in the Afar region is a challenging exercise due to prevailing data paucity. Especially in nomadic systems livestock numbers are inherently difficult to obtain due to herd mobility and limited resources for enumeration. Hence, assessments are often based on rough estimates of scaled surveys and different sources are bound to give conflicting results. Hence, data selection requires a careful evaluation of available sources and a wider range of possibilities should be part of the assessment. We, therefore, compare various sources listed in Table 6 to come to an informed assessment on prevailing number of livestock.

Specifically, we check for consistency of number of TLU at regional level, by Woreda and for the herd composition. For reporting on livestock assessments, we concentrate first on total number of TLU for the entire Afar. Figure 4 presents total TLU by source[5]. The [64] livestock data for cattle, camel, goats, sheep and equine presented for zone 1 and 3 were used to create scaling factors to bring the CSA_03 data [48] the year 2011 level for the entire Afar Region. This data set is presented in Figure 4 as TLU_CSA03_scale We observe that the scaled CSA03 data (TLU_CSA03_scale) has the highest number of TLU followed by four more or less equal numbers (TLU_CSA03, TLU_regA13, TLU_LDMPS, TLU_ESGPIP). TLU_bofed is somewhat lower with respect to the previous numbers, TLU_rega08 is very low. TLU_CSA03, TLU_rega08, TLU_regA13 are based on inventories at Woreda level for which an inventory is available; other assessments are by Zone or region-wide.

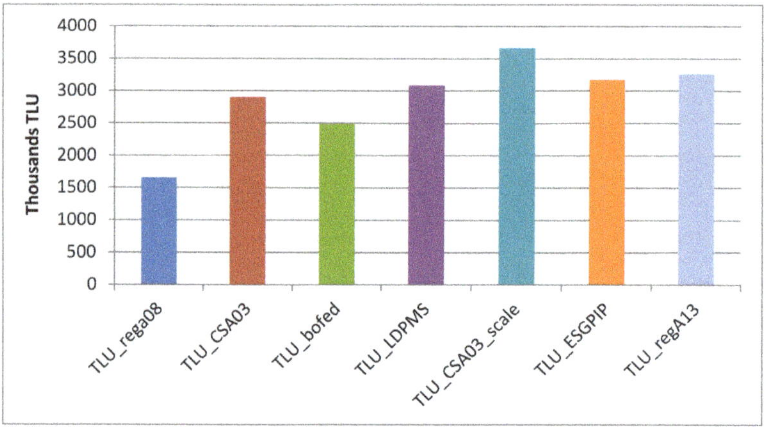

Figure 4. TLU in the Afar by data source.

Table 6. Sources of livestock numbers in the Afar.

Data	Source	Spatial coverage	Year
Gebremeskel (2012)	[60]	Zone 1, 3, 4	2006
CSA03	[65]	29 Woredas	2003
REGA_08	[66]	29 Woredas	2008–2009
CSA (2011)	[64]	Zone 1,3	2011
ESGPIP	[67]	Region-wide	2005
LDMPS	[68]	Region-wide	2006
BOFED	[69]	Region-wide	2009
REGA13	[70]	31 woredas	2013

5 A clear reference of TLU numbers in Gebremeskel 2012 was missing and is not further elaborated.

For an assessment that accounts for livestock distribution by Zone we analyse Figure 5, where total TLU is depicted by Zone and data source. Data indicate a more or less similar pattern for number of TLU by Zone. Zone 1 in most of the cases the highest followed by Zone 3 and 4 while zone 2 and 5 report lowest numbers. The TLU_rega08 data set clearly deviates from this pattern.

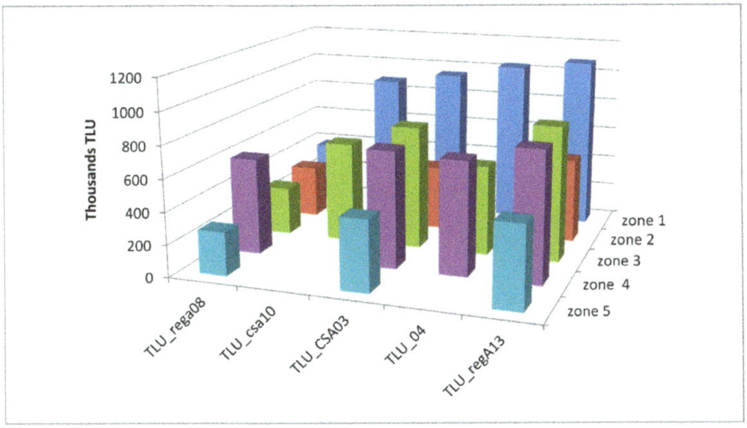

Figure 5. TLU by zone and data source.

Finally, we look at the distribution of livestock species for the three Woreda based inventories CSA03, REGA08 and REGA13. For comparison of the herd composition we correct for differences in total number of livestock by using relative numbers by Woreda. The scatterplots in Appendix show in CSA03 data against REGA08 (Figures A1–A5, panels A1, B1, C1, D1 and E1), CSA03 against REGA13 (Figures A1–A5, panels A2, B2, C2, D2 and E2) and REGA08 against REGA13 (Figures A1–A5, panels A3, B3, C3, D3 and E3). Figure A1, panels A1–A3 show the relative number of camels, Figure A2, panels B1–B3 for cattle, Figure A3, panels C1–C3 for equine, Figure A4, panels D1–D3 sheep for goats jointly and, finally Figure A5, panels E1–E3 for the TLU at the fifth row. We note a deviation from the expected 1:1 line in Figures A1–A5, panels A1, A3, B1, B3, C1, C3, D1, D3 and E1, E3, REGA08 data are involved. In Figures A1–A5, panels A2, B2, C2, D2 and E2the REGA13 and CSA03 data show a nice correlation indicating that livestock species composition is more or less the same over corresponding Woredas.

We conclude that REGA13 and CSA03 data confirm the distribution of livestock species and are in accordance with totals of TLU numbers of TLU_regA13, TLU_LDMPS, TLU_ESGPIP and TLU_bofed. Hence we decide to use the mean of TLU-regA13 and CSA03 data set which covers an average over a time period that is also used for the forage demand. CSA03 data were used earlier in a land degradation assessment [47]. Finally, using a constrained downscaling procedure the TLU by Woreda were distributed over corresponding[6] sub-clan areas, proportionally to the fodder availability (see Section 2.2) by sub-clan in that Woreda.

2.4. Following Herds: Seasonal Migration in the Afar Region

The migration model follows, simultaneously, movements of migrating herds of Afar sub-clans while accounting for the share of the herd that remains behind. The model's geographical dimension constitute of sub-clan territories while weekly times steps over four prevailing seasons cover the

[6] Sub-clan areas corresponding to Woreda's are identified by crossing both maps.

temporal dimension. In this study the model highlights the pressure on natural resources by aggregating herds' presence over the year that is confronted with estimated biomass available for grazing. This provides stakeholders with information on where local institutional constraints seem not able to regulate the use of natural resources sufficiently to prevent overuse. Below we give a formal introduction to the model.

Specification requirements of the migration model imposed on the geographical and temporal representation of migratory movements are: (1) discreteness, movements follow adjacent sub-clan map units and weekly time steps; (2) contiguity, movements cannot skip neighbors in space and time (no jumps); (3) time follows logical sequence, no movements to the past.

The model implements these requirements as follows. For an annual cycle of months indexed $t = 1, \ldots, T$, let subscripts $i = 1, \ldots, I$ and $r = 1, \ldots, R$ denote home and destination areas of sub-clans, respectively and variable x_{it} the stock of animals at *it* (home front; the area claimed by the sub-clan). For one year we describe livestock distribution and migratory movements as:

$$x_{r't'} = M_{r't',it}x_{it}$$

where M is a square transition matrix of dimension $IT \times RT$ with elements $m_{r't',it}$ representing fraction of total herd size (x_{it}) expressed in TLU that moves to $r't'$ with diagonal elements $1 - \sum_{it} m_{r't',it}$, one minus column sum of non-diagonal elements, accounting for remaining herd share at sub-clan area. Information on $m_{r't',it}$, the share, location and timing of migrating herd of a sub-clan is derived from interviews with key persons at sub-clan level and expert knowledge from research stations and local authorities. Sub-clans stay a number of weeks in a specific sub-clan territory and then move to the next, where they again reside for a number of weeks. Stress on the resource base is quantified by confronting the fodder demand with supply. As we do not avail of a comprehensive data base on water resources we assume that water is available at the moment that herds visit the sites.

Ideally, the model would have a time-space dimension that accommodates sequential movements in time over space that follow the herd in a real-time mode. However, this is unrealistic for several reasons. First, there is no detailed information available that would allow us to follow the individual herds. Second, as argued above, fodder supply assessments are hard to obtain and are available only at the annual level. Yet, given the scope of the model—to identify where seasonal hotspots appear, we are also confident that aggregations in time and space are at a sufficient fine resolution to reveal vital patterns in herd management. In terms of assumptions made, aggregation over time implies that contiguity requirements might be violated when herds migrate across multiple sub clan territories within one week.

3. Results

In this section we start with an annual supply-demand balance expressed in TLU for each sub-clan territory, under static conditions, that is, all herds stay at their place without migration (Section 3.1). Next, for illustrative purposes we show the temporal and spatial movements for individual clans (Section 3.2). Finally, we present the results of the full model for the supply-demand balance at annual scale and analyse the seasonal variation of herd movements (Section 3.3).

3.1. Supply-Demand Balance under Static Conditions

Livestock distribution expressed in TLU on sub-clan territory without migration is shown at the left side of Figure 6 (panel a). The highest concentrations are found on the western fringes of the Afar bordering the highlands. Following the west-east line we observe a more or less uniform density with some near empty pockets in the Centre while lowest densities are reported for the North East bordering Eritrea, at least partly explained by the ongoing hostilities between the countries. The right side of Figure 6 (panel b) shows annual food supply also expressed in number of TLU that can be fed. We observe that fodder supply mirrors livestock distribution to a certain extent. The Western areas report high fodder supply while low TLU densities in the Centre correspond to lower fodder supply. Figure 7, finally, shows fodder surplus (panel a) and fodder deficits (panel b) in case of no migration. Deficits are mainly found in the western areas of the Afar where high TLU densities are found, which makes migration for these areas prominent.

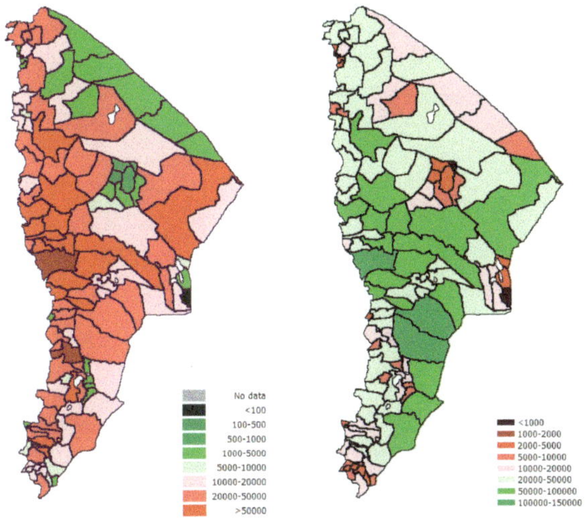

(a) (b)

Figure 6. Livestock distribution in TLU with sub-clan totals at own territory (panel **a**) and annual fodder supply expressed in TLU equivalents (panel **b**) (scale 1:6,500,000)

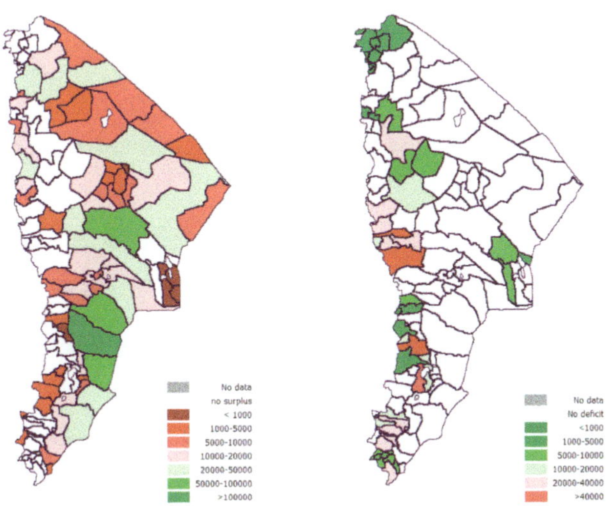

(a) (b)

Figure 7. Surplus (panel **a**) and deficit (panel **b**) of annual fodder supply in case of no migration expressed in TLU (scale 1:6,500,000).

3.2. Mapping the Migratory Routes by Sub-Clan and Season

For each of the 112 sub-clans, the routes followed by their herds have been modelled by season. The four seasons in the Afar are based on rainfall activity. During the Bega dry weather prevailed, Belig is the short rainy season, Keremt the main rainy season and Tsedaye the dry spell. There are important differences between sub-clans and seasons in the number of destinations visited and duration of stay within each sub-clan territory. Figure 8 depicts migratory routes for the Harkemella sub-clan in the four seasons, while Figure 9 concentrates on the Wadijma sub-clan. In the supplementary material, migratory routes for all 112 subclans are provided.

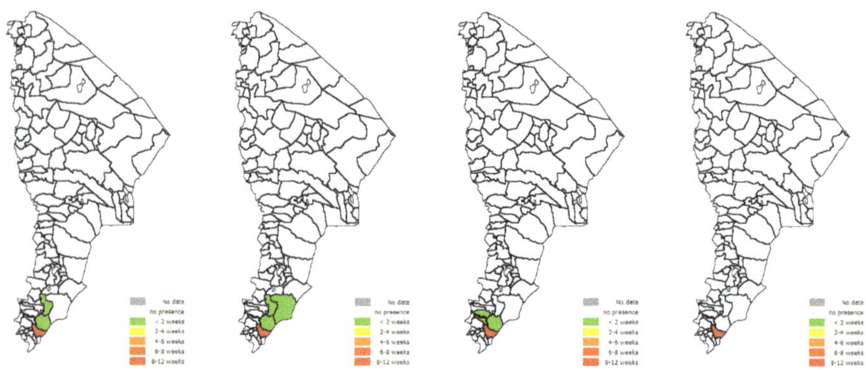

(**a**) (**b**) (**c**) (**d**)

Figure 8. Harkamella sub-clan, Belig (**a**), Keremet (**b**), Bega (**c**) and Tsedaye (**d**) seasons (scale 1:9,900,000).

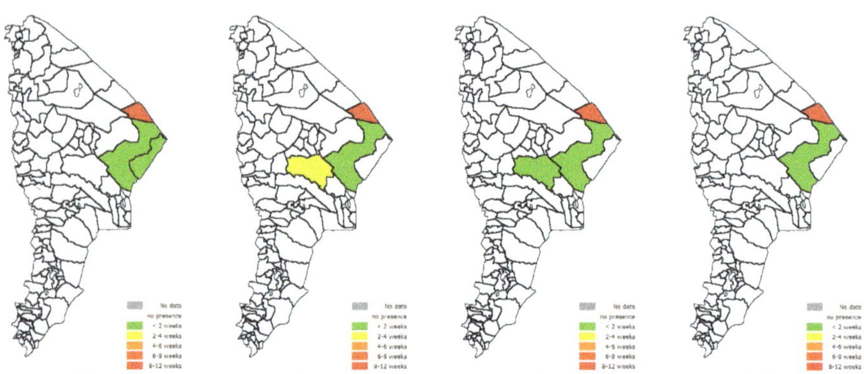

(**a**) (**b**) (**c**) (**d**)

Figure 9. Wadijma sub-clan, Belig (**a**), Keremet (**b**), Bega (**c**) and Tsedaye (**d**) seasons (scale 1:9,900,000).

3.3. The Full Migration Model

This section presents the results of the full migration model when all herds simultaneously migrate. Compared to the static situation (left side of Figure 6; panel a) we observe that at annual level the number of livestock after migration (left side of Figure 10; panel a) follows a more or less similar pattern. When zooming in on surpluses and deficits (left (panel a) and right (panel b) side of Figure 11), some special effects come to the fore. After migration, the deficits in the Northern Part are dissolved, the Central part, though, shows higher deficits. Going to the South we see a general decrease in deficits in areas and magnitude after migration.

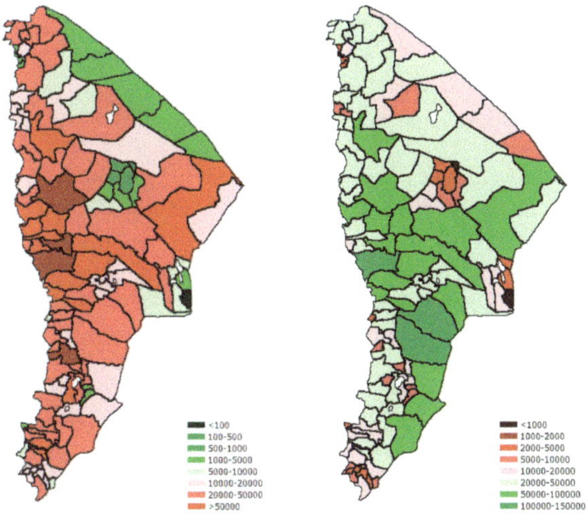

(**a**) (**b**)

Figure 10. Annual livestock presence in TLU after migration (panel **a**), annual fodder supply expressed in TLU equivalents (panel **b**) (scale 1:7,200,000).

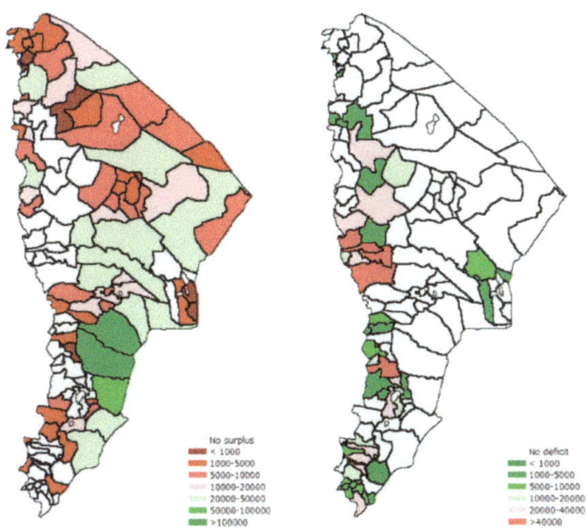

(**a**) (**b**)

Figure 11. Surplus (panel **a**) and deficit (panel **b**) of annual fodder supply expressed in TLU equivalents after migration (scale 1:7,200,000).

Figure 12 shows the ratio of livestock demand over fodder supply which confirms the earlier detected patterns of fodder deficits in the Western part. We define 'hot spots' areas where demand is two to three times larger than supply.

As our estimates on fodder supply are at the annual level we have to forgo a comparison of seasonal fodder demand-supply ratios. Yet, to obtain a seasonal assessment of hot spots it is interesting to compare the herd movements per season where it can be analysed which areas are visited most and which places are avoided. Figure 13 shows herd presence in case of no migration with seasonal movements during the Belig (a), Keremet (b), Bega (c) and Tsedaye (d) season. Remarkable movements are seen in the blue circle, where during the Belig and Keremet a concentration of herd presence is observed which drastically is reduced during the Bega and Tseday. The red circle also shows much activity, during the Belig high concentrations are found in the upper part while decreasing in the Keremet and Bega season, and then again increasing in the upper part during the Tsedaye. This is related to the substantial differences between the part of the herd that remains behind and the migrating share (Table 7).

Table 7 illustrates that there are also substantial seasonal differences between the part of the herd that remains behind and the migrating share. In the Keremet season, half of all herds are included in migratory movements, while in the Tsedaye season, over 90% of herds remain in their sub-clan territory. Anecdotal evidence suggests that the part of the herd that remains behind is made up of animals that are not yet able to join the migrating herd. The very low percentage of migrating herds in the Tsedaye (spring) season would be consistent with the assumption that new-born animals and their mothers stay behind.

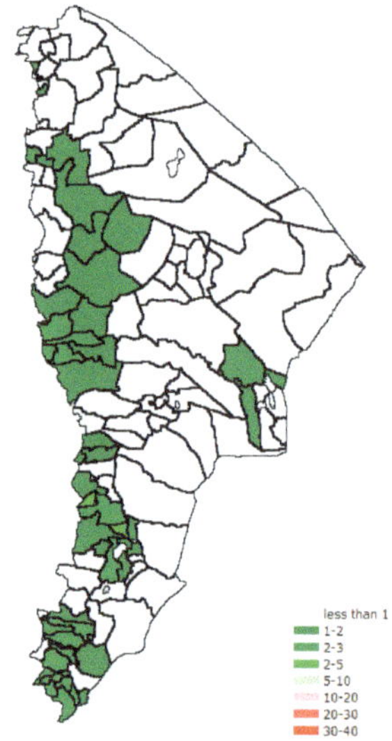

Figure 12. Ratio of livestock demand over fodder supply after migration (scale 1:7,200,000).

Figure 13. Comparing seasonal herd presence during the Belig (**a**), Keremet (**b**), Bega (**c**) and Tsedaye (**d**) seasons with the no migration option (**upper**) (scale 1:7,900,000).

Table 7. Percentage of herds migrating, by season.

	Belig	Keremet	Bega	Tsedaye
Average	30%	49%	20%	8%
Minimum	13%	30%	8%	3%
Maximum	36%	58%	34%	13%

4. Sensitivity Analysis

In this section we analyze the sensitivity of our assumed parameters on the identification accuracy of hot spots. We analyze the sensitivity for production assessment, livestock distribution, and forage consumption per TLU and relate these to the ratio feed demand over supply. Specifically, we use the highest and lowest estimates for total herd size (in TLU) from the different data sources, dry matter availability (in ton) using the highest and lowest production function estimates, and conversion rates from dry matter to TLU. Combined with the assumed values used in the baseline, this leads to 27 scenarios (one of which is the baseline itself), described in Table 8. Here, the 'low' estimate for total herd size is 2,491,037 TLU [67]; 'high' estimate is 3,662,078 TLU [64]. For total dry matter availability, the 'low' estimate is 3714 thousand ton [52] while 'high' equals 9652 thousand ton [55]. For TLU conversion the 'low' estimate equals 2.2 MT dry matter per year [71], while the 'high' equals 2.7 MT dry matter per year [52]. Base values are as described in Section 2 above.

Table 8. Scenario description.

	Herd Estimate	Dry Matter	TLU Conversion
Scenario 1	Low	Low	Low
Scenario 2	Low	Low	Base
Scenario 3	Low	Low	High
Scenario 4	Base	Low	Low
Scenario 5	Base	Low	Base
Scenario 6	Base	Low	High
Scenario 7	High	Low	Low
Scenario 8	High	Low	Base
Scenario 9	High	Low	High
Scenario 10	Low	Base	Low
Scenario 11	Low	Base	Base
Scenario 12	Low	Base	High
Scenario 13	Base	Base	Low
Baseline	Base	Base	Base
Scenario 14	Base	Base	High
Scenario 15	High	Base	Low
Scenario 16	High	Base	Base
Scenario 17	High	Base	High
Scenario 18	Low	High	Low
Scenario 19	Low	High	Base
Scenario 20	Low	High	High
Scenario 21	Base	High	Low
Scenario 22	Base	High	Base
Scenario 23	Base	High	High
Scenario 24	High	High	Low
Scenario 25	High	High	Base
Scenario 26	High	High	High

Results of all scenarios are available on request. Here we present the results on the 'Best' scenario where TLU assessments are low, fodder estimates high and fodder consumption by TLU low, the 'Baseline' scenario, referring to results reported in Section 3, while the 'Worst' Scenario has highest TLU, lowest fodder assessment and high fodder consumption by TLU.

Concerning the results of the corresponding feed demand over supply ration for the Afar at provincial level we depict the outcome in Figure 14. We observe that, especially, the 'Worst' scenario is very sensitive for negative outliers and results in serious deficits, state wide.

Outcomes of the scenarios by sub-clan territory are presented in Figure 15, where the baseline, the worst case outcome and the optimal outcome are depicted for each of the 112 sub-clan territories. We observe that the 'Best' and 'Baseline ' scenarios are more or less the same with absolute differences oscillating between 0.04 and 1.51 with average difference 0.42. Differences between 'Baseline' and 'Worst' scenarios are much larger, varying from 0.02 to 7.5 with an average of 2.09. Magnitude of feed surpluses and deficits for the three scenarios are illustrated in Figures 16 and 17, respectively. In the 'Best' scenario, surpluses are found in all sub-clan territories except for some areas in the Centre bordering the Highlands, where feed deficits all fall in the first class. The 'Worst' scenario shows some surpluses in the areas in the North East, bordering Eritrea, in the South-East near Somali Province and near the capital Semera. Yet, the amount of areas with deficits is alarming especially for the Central part of the Afar where deficits classes fall into the highest categories.

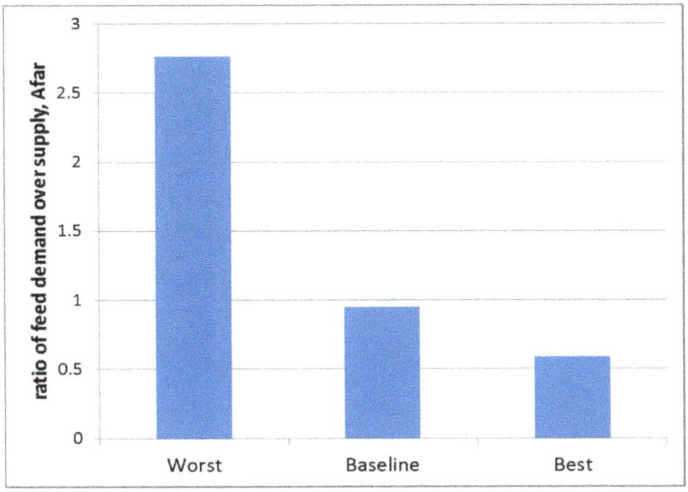

Figure 14. Ratio feed demand over supply for 'Worst, 'Baseline' and 'Best' scenario.

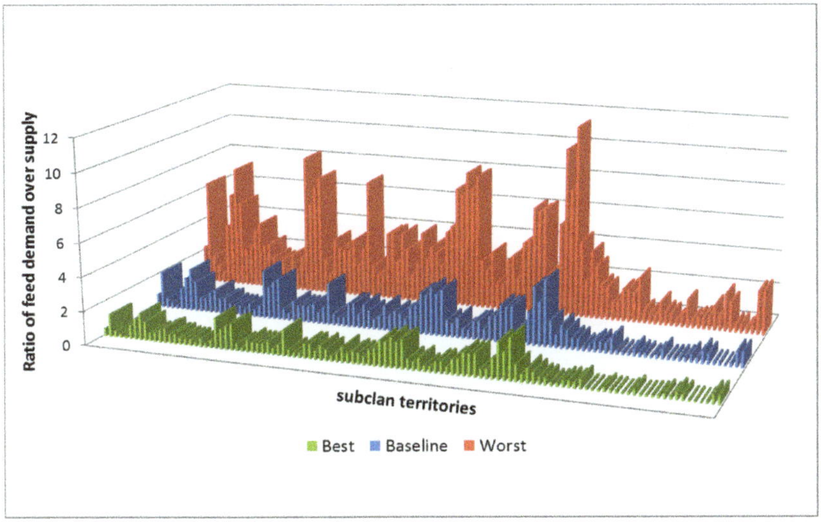

Figure 15. Results of sensitivity analysis. Ratio feed demand over supply for 'Best', 'Baseline' and 'Worst' scenario.

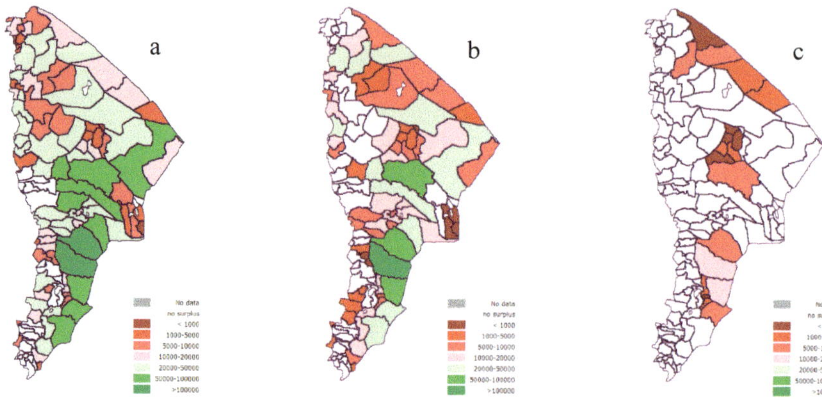

Figure 16. Surplus feed: 'Best' (panel **a**), 'Base' (panel **b**) and 'Worst' (panel **c**) scenario's (scale 1:7,200,000).

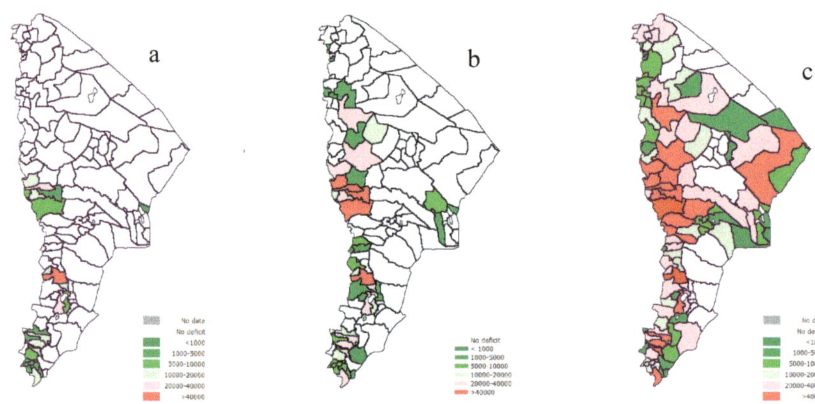

Figure 17. Feed deficits: 'Best' (panel **a**), 'Base' (panel **b**) and 'Worst' (panel **c**) scenarios (scale 1:7,200,000)

5. Conclusions

Using a migration model to represent herd movements over time and space, we found in most of Afar State a reasonable balance between annual fodder supply and demand. Most areas show a small surplus, except for the Western part of the Afar bordering the foot of the Highland slopes where in 'hot spots' demand exceeds supply by a factor two or three. The results may support the argument that by and large institutions that regulate the time and duration of visits of Afar pastoralists to other sub-clan territories favor the development of arid areas by maintaining a balance between demand and supply of fodder and water resources. All of the sub-clans practice migration mechanism though we found considerable differences in share of the herd that migrates and the distances that were covered by the various sub-clans. By itself the results of this study-dispatch an important message to local and regional authorities in that migration is needed for sustainable development of the Ethiopian dryland and that negative effects of land developments that interfere with traditional patterns should be minimized. For example authorities could guarantee safe corridors for pastoralists through planned biofuel plantations and the large sugar cane farm near the Tenaha dam.

Our research is a clear step forward in the impact analysis of migration patterns on the sustainable land development. The simultaneous evaluation of the presence of livestock in its spatial and temporal dimension gives an accurate representation of reality and allows policy makers to target their intervention geographically to support drought coping strategies of the pastoralists such as the design of the best spatial configuration of a system of groundwater pumps and forage storage points. When such "enclaves" are well regulated they could help pastoralists through dire periods and avoid overgrazing and land degradation of the few areas that are not yet affected by drought.

Yet, this study also showed that still large data gaps exist and much ground truthing need to be done to complete a sound empirical basis. A consolidated data base with detailed information on trekking routes, biophysical resources, land uses, market prices, conflict zones, household/pastoralist surveys, and narratives on coping strategies in appropriate spatial and temporal dimensions. The collection of these data at the appropriate level and its organization in a dynamic modelling environment is a big scientific challenge in the coming years.

Several improvements of the migration model are envisaged. First, using satellite information on start and end of season allows assessing seasonal variation of feed demand and representing fodder supply at a higher spatial and temporal resolution. A follow up study is planned to refine the Second, adding availability and access to water resources, a vital element will improve the explanation of migrating routes. Third, introducing prices and bringing the model under an optimization framework maximizes herders' income under various scenarios like alternative routings, climate change effects and improved rangeland management. Fourth, and finally, combining the model outcomes with results of a survey held among 180 pastoralists will deepen the understanding of regulations and agreements between sub-clans on sharing of common resources and indicate where institutional support can strengthen the resilience of Afar pastoralists to cope with the new challenges.

Acknowledgments: Authors gratefully acknowledge the financial support for this study from the OPEC Fund for Development under the grant R279 'Improving Pastoral System's Drought Resilience in Afar State, Ethiopia'.

Author Contributions: B.G.J.S. Sonneveld, project coordinator, involved in modelling and final editing, C.F.A. van Wesenbeeck involved in editing and modelling of migration scenarios. M.A. Keyzer, project leader, involved in editing and modelling of migration scenarios. K. Georgis, involved in editing, project design and supervision of field work. F. Beyene involved in project design, editing and supervised field work. F. Urbano, M. Meroni and O. Leo, were involved in biomass assessments and editing. M.K. Yimer prepared the sub-clan map, was involved in identifying migration routes and editing. M. Abdullatif was involved in preparation of sub-clan map, prepared information on migration routes and was involved in editing.

Conflicts of Interest: The authors declare no conflict of interest.

Appendix

Scatter plots of relative number of heads for camels (panels A1–A3) (Figure A1), cattle (panels B1–B3) (Figure A2), sheep and goats (panels C1–C3) (Figure A3), equine (panels D1–D3) (Figure A4) and TLU (panels E1–E3) (Figure A5) for CSA03 against REGA08 (panels A1, B1, C1, D1 and E1), CSA03 against REGA13 (panels A2, B2, C2, D2 and E2) and REGA08 against REGA13 (panels A3, B3, C3, D3 and E3).

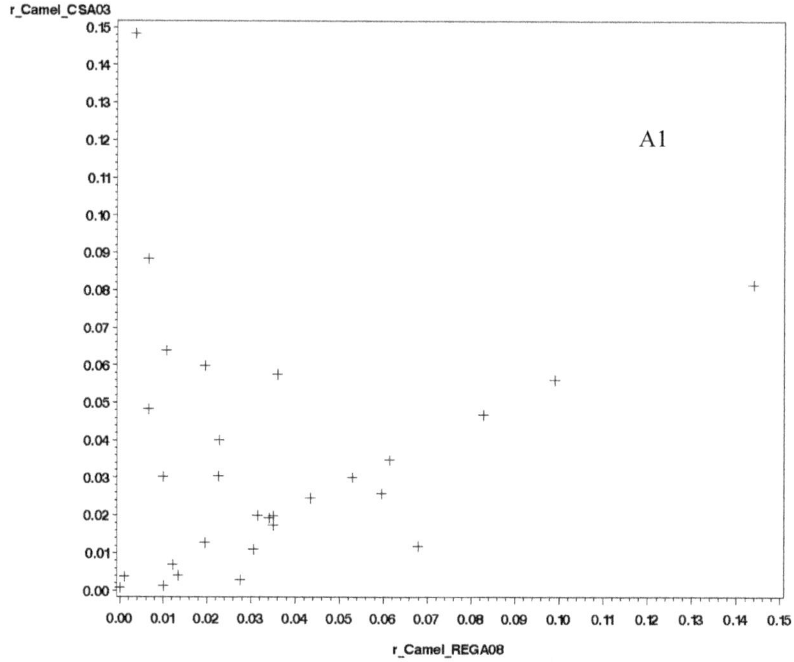

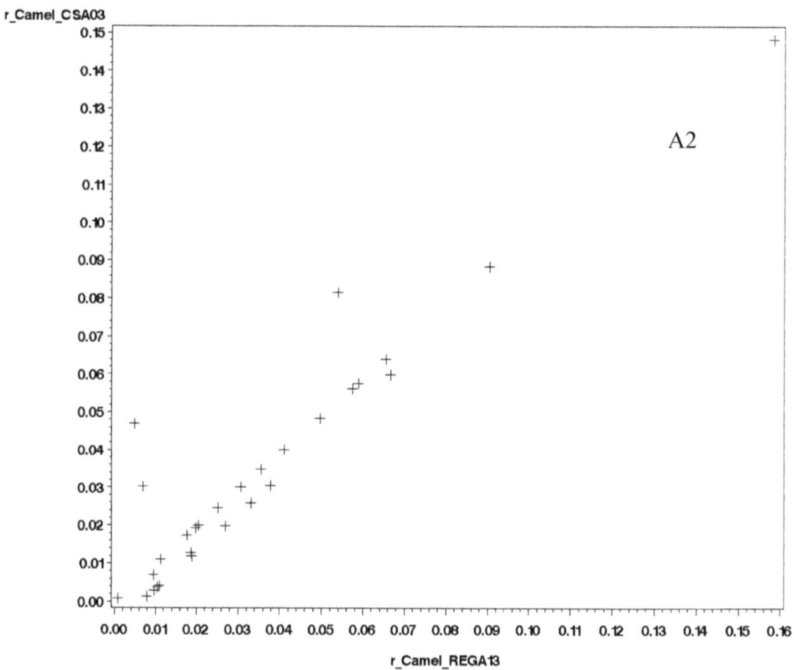

Figure A1. *Cont.*

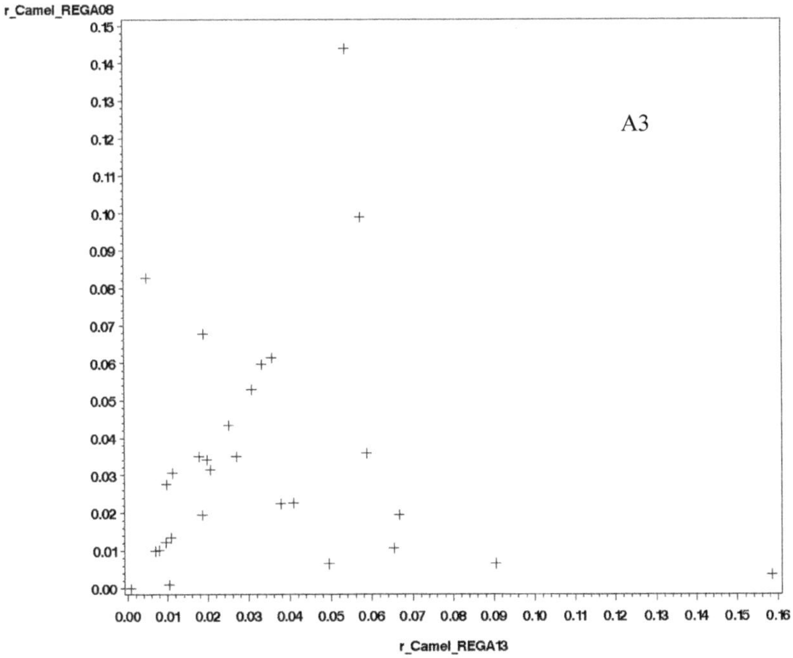

Figure A1. Scatter plots of relative number of heads for camels (A1–A3).

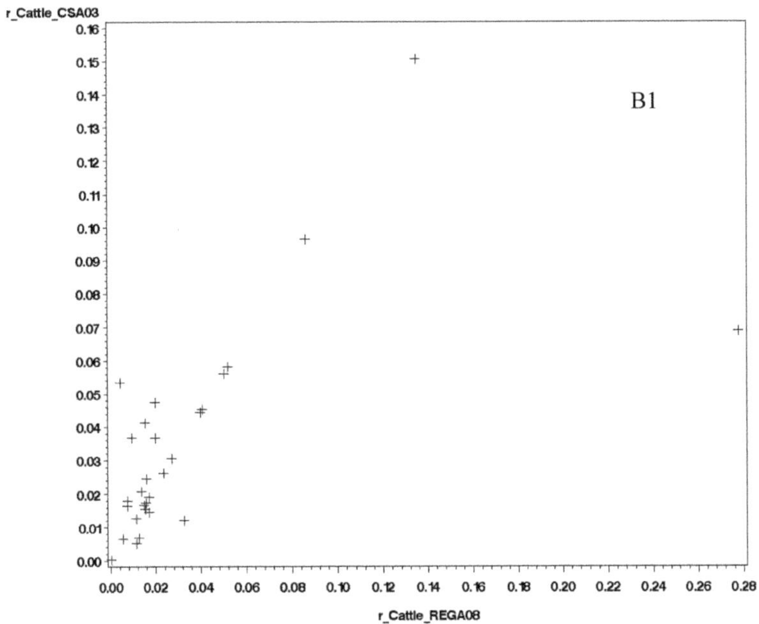

Figure A2. *Cont.*

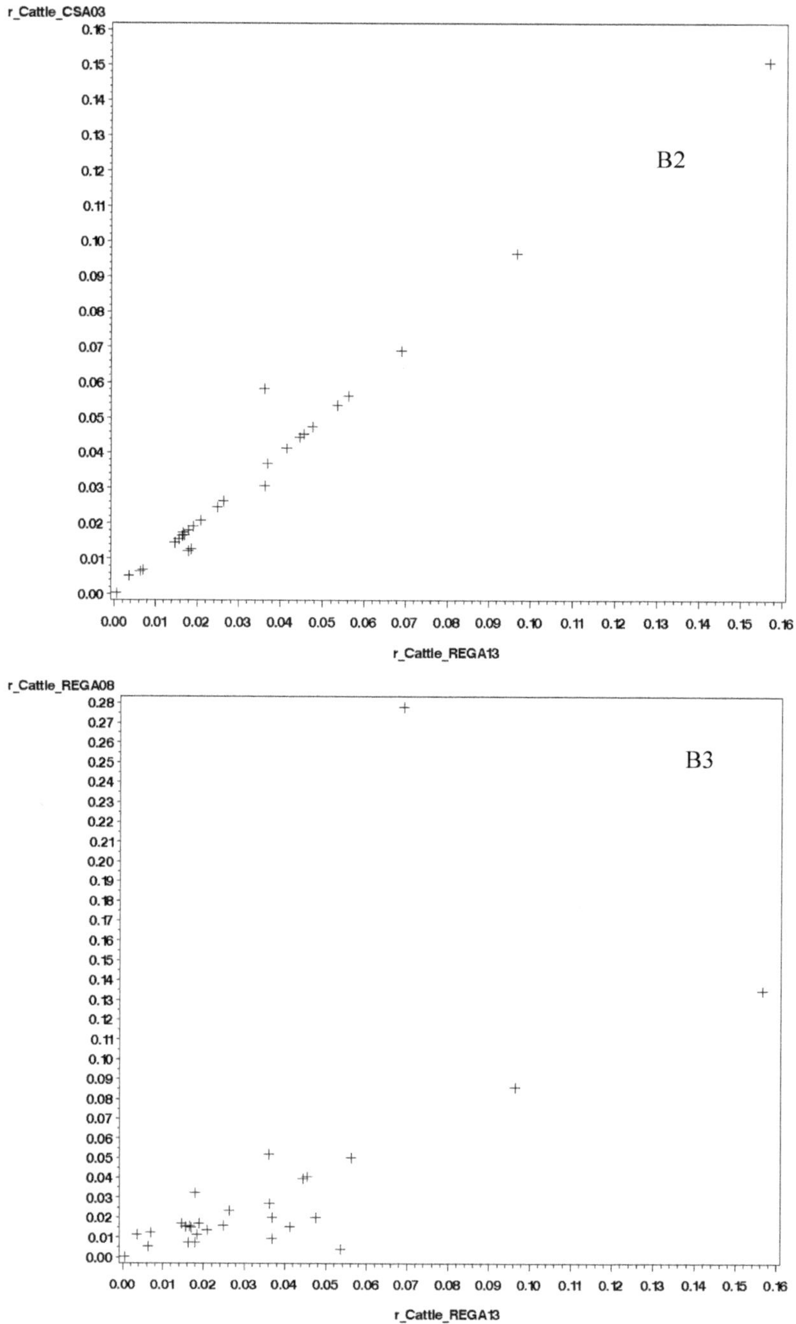

Figure A2. Scatter plots of relative number of heads for cattle (B1–B3).

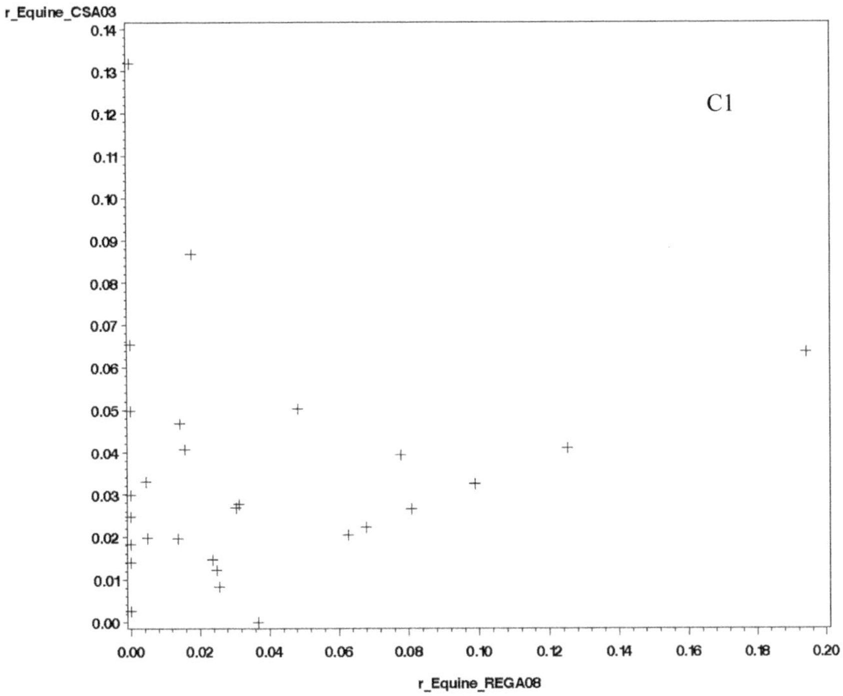

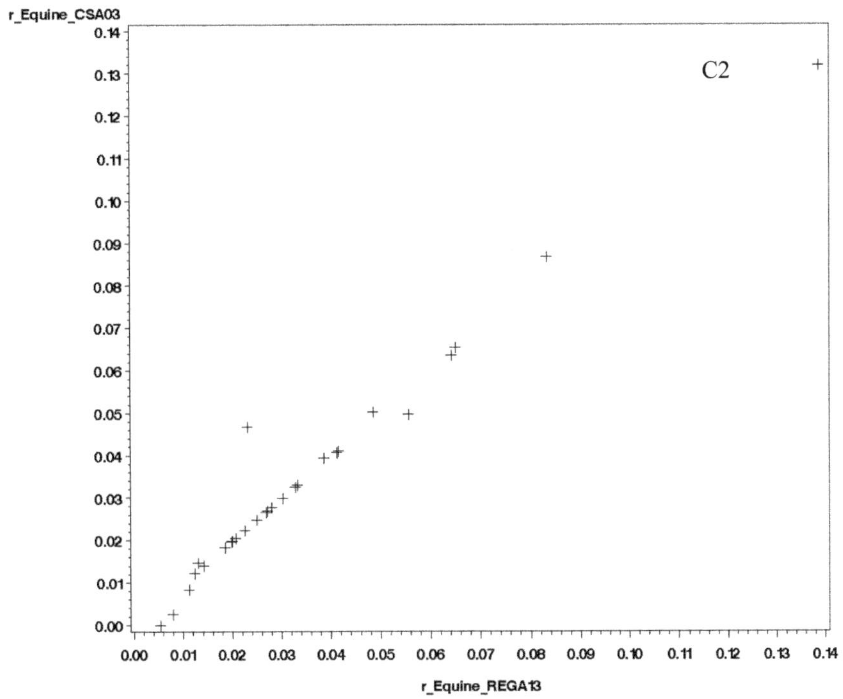

Figure A3. *Cont.*

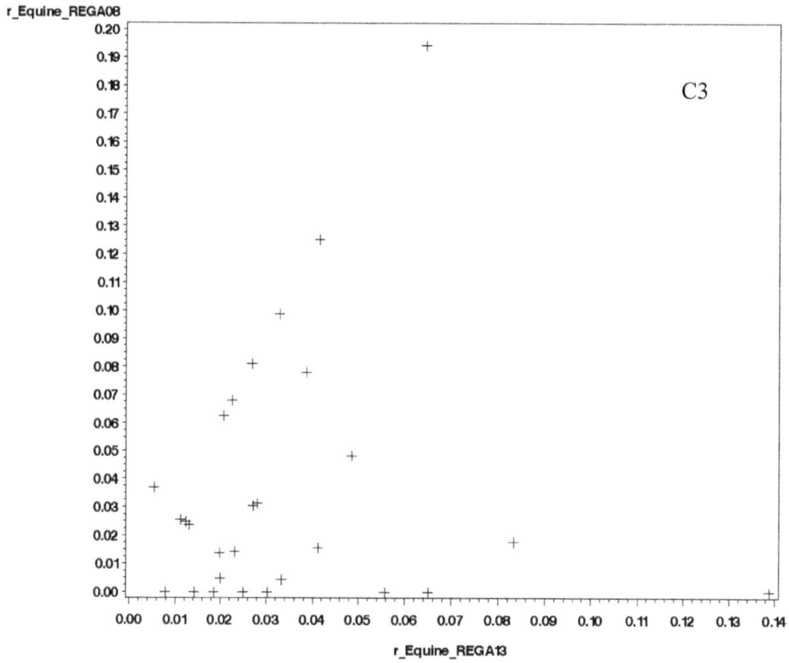

Figure A3. Scatter plots of relative number of heads for equine (C1–C3).

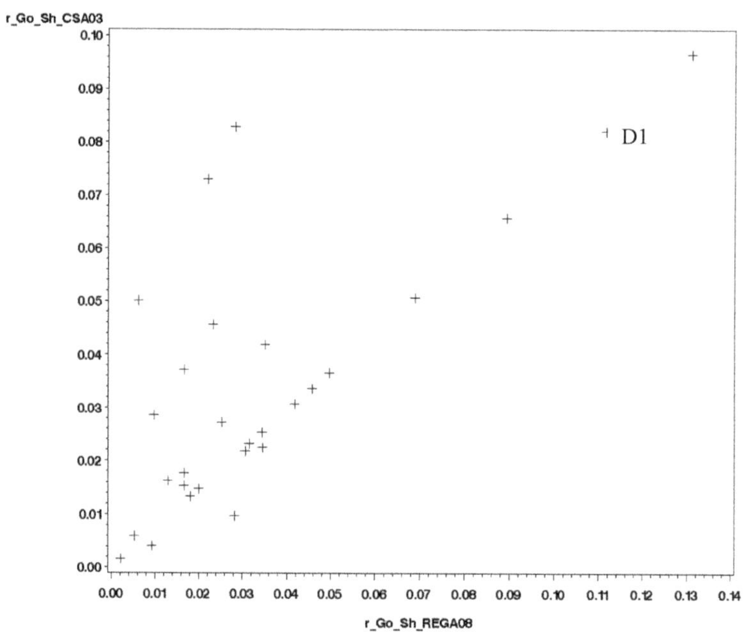

Figure A4. *Cont.*

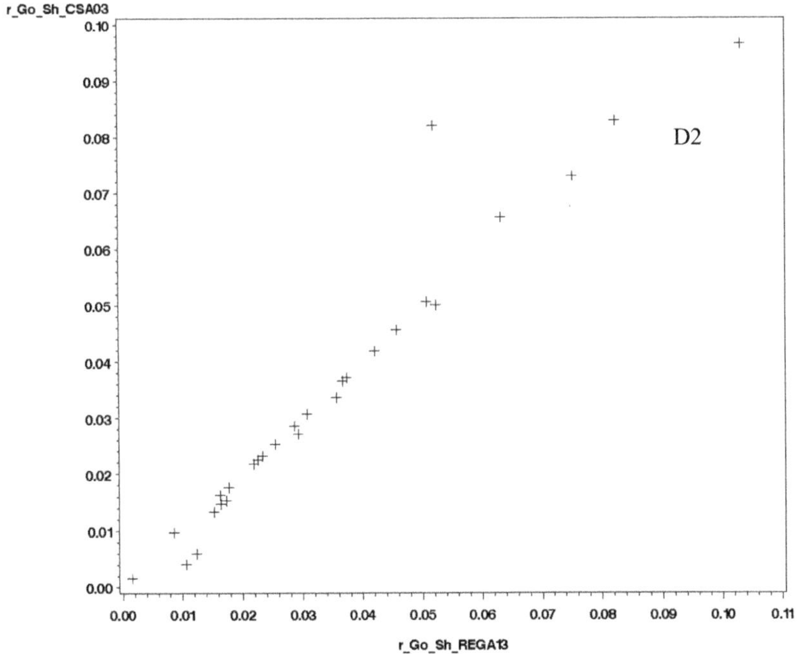

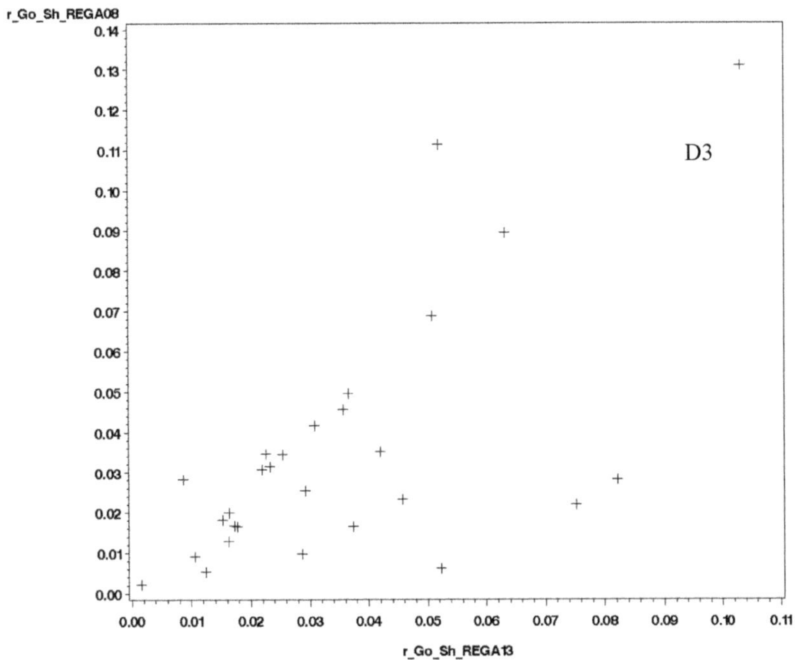

Figure A4. Scatter plots of relative number of heads for goats and sheep (D1–D3).

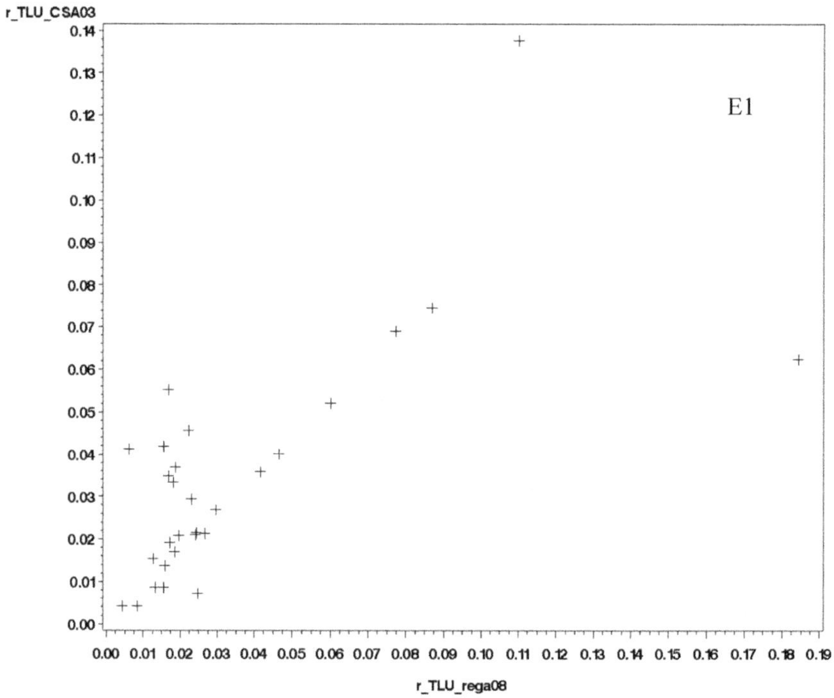

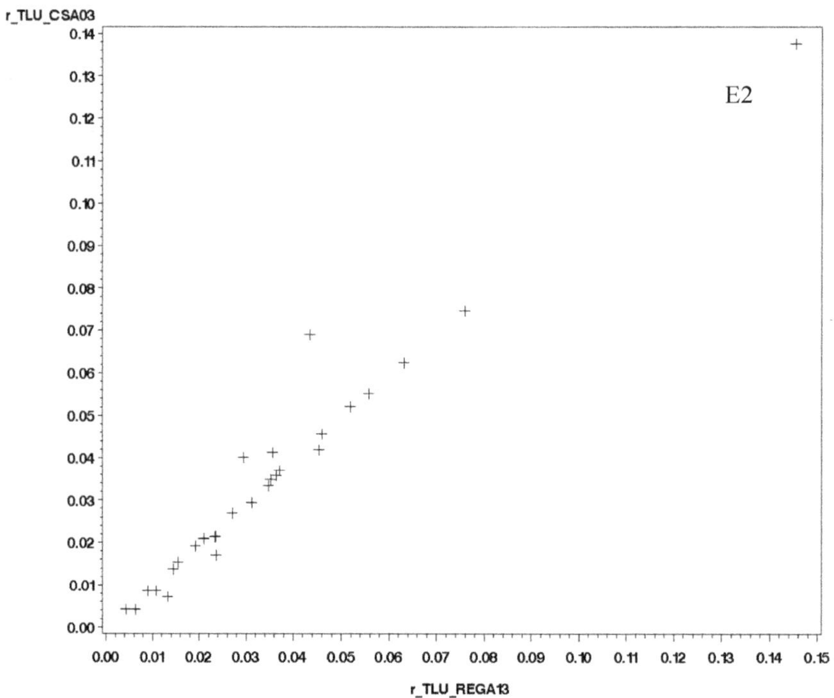

Figure A5. *Cont.*

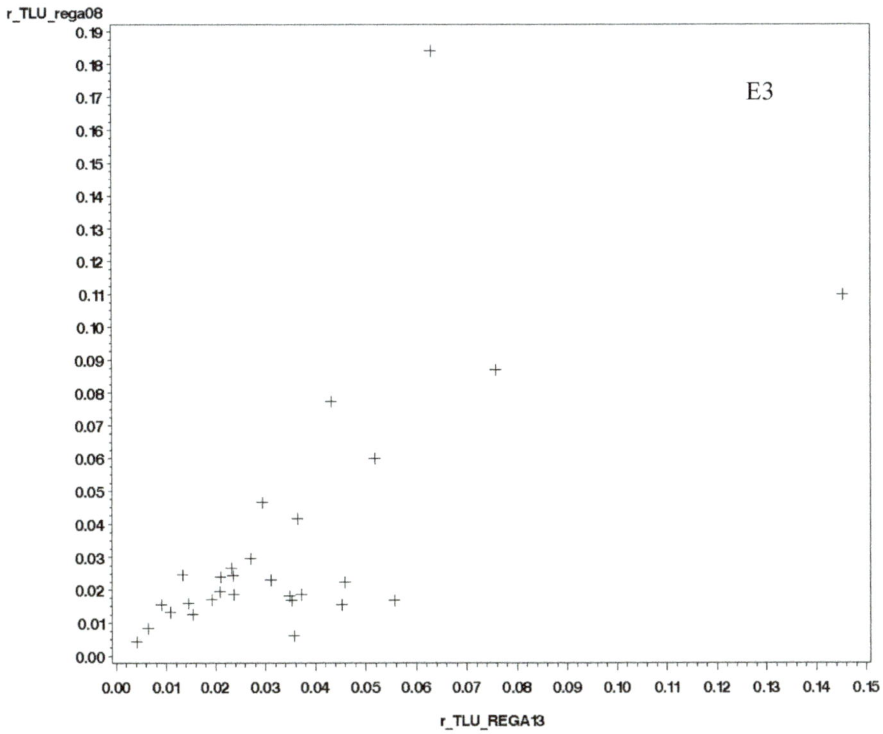

Figure A5. Scatter plots of relative number of heads for TLU (E1–E3).

References

1. UN. Global Drylands: A UN system-wide response Environment Management Group. 2011. Available online: http://www.unccd.int/Lists/SiteDocumentLibrary/Publications/Global_Drylands_Full_Report.pdf (accessed on 15 May 2017).
2. Dodd, J.L. Desertification and degradation in sub-Saharan Africa—The role of livestock. *BioScience* **1994**, *44*, 28–34. [CrossRef]
3. Homewood, K.; Rodgers, W.A. Pastoralism, conservation and the overgrazing controversy. In *Conservation in Africa: People, Policies and Practice*; Anderson, D., Grove, R., Eds.; Cambridge University Press: Cambridge, UK, 1987; 355p.
4. Mearns, R. *The Lie of the Land: Challenging Received Wisdom on the African Environment*; Leach, M., Ed.; James Currey Ltd.: Oxford, UK, 1996; 240p.
5. Lamprey, H.F. Pastoralism yesterday and today: The overgrazing controversy. In *Tropical Savannas. Ecosystems of the World*; Bourliere, F., Ed.; Elsevier: Amsterdam, The Netherlands, 1983; Volume 13, pp. 643–666.
6. Sinclair, A.R.E.; Fryxell, J.M. The Sahel of Africa: Ecology of a disaster. *Can. J. Zool.* **1985**, *63*, 987–994. [CrossRef]
7. Hardin, G. The tragedy of the commons. *Science* **1968**, *163*, 1243–1248. [CrossRef]
8. Behnke, R.H.; Abel, N.O.J. Revisited: The overstocking controversy in semiarid Africa. *World Anim. Rev.* **1996**, *87*, 3–27.
9. Ellis, J.E.; Swift, D.M. Stability of African pastoral ecosystems: Alternate paradigms and implications for development. *J. Range Manag.* **1988**, *41*, 450–459. [CrossRef]
10. Sandford, S. *Management of Pastoral Development in the Third World*; John Wiley and Sons: New York, NY, USA, 1983; 316p.

11. Sullivan, S. Towards a non-equilibrium ecology: Perspectives from an arid land. *J. Biogeogr.* **1996**, *23*, 1–5.
12. Sullivan, S.; Rohde, R. On non-equilibrium in arid and semi-arid grazing systems. *J. Biogeogr.* **2002**, *29*, 1595–1618. [CrossRef]
13. NIM. Nomads, the facts. New Internationalist Magazine. Available online: http://newint.org/features/1995/04/05/facts/ (accessed on 10 September 2016).
14. Fernandez-Gimenez, M.E. The role of mongolian nomadic pastoralists ecological knowledge in rangeland management. *Ecol. Appl.* **2000**, *10*, 1318–1326. [CrossRef]
15. Warioa, H.T.; Robab, H.G.; Kaufmann, B. Responding to mobility constraints: Recent shifts in resource use practices and herding strategies in the Borana pastoral system, southern Ethiopia. *J. Arid Environ.* **2016**, *127*, 222–234. [CrossRef]
16. Warren, A. Changing Understandings of African Pastoralism and the Nature of Environmental Paradigms. *Trans. Inst. Br. Geogr.* **1995**, *20*, 193–203. [CrossRef]
17. Desta, S.; Coppock, D.L. Pastoralism under Pressure: Tracking System Change in Southern Ethiopia. *Hum. Ecol.* **2004**, *32*, 465–486. [CrossRef]
18. Butt, B. Ecology, mobility and labour: Dynamic pastoral herd management in an uncertain world. *Rev. Sci. Tech.* **2016**, *35*, 461–472. [CrossRef] [PubMed]
19. Salpeteur, M.; Madella, M.; Patel, H.R.; Reyes-García, V. *Adaptation, Access to Resources and Mobility: From Contemporary Pastoral Systems to Ancient Societies. Nomadic Peoples*; The White Horse Press: Winwick, Cambridgeshire, UK, 2017; Volume 21, pp. 191–213.
20. Hundie, B. Property Rights among Afar Pastoralists of Northeastern Ethiopia: Forms, Changes and Conflicts. In Proceedings of the Survival of the Commons: Mounting Challenges and New Realities. Eleventh Conference of the International Association for the Study of Common Property, Bali, Indonesia, 19–23 June 2006.
21. Rass, N. *Policies and Strategies to Address the Vulnerability of Pastoralists in Sub-Saharan Africa*; PPLPI Working Paper No. 37; Pro-Poor Livestock Policy Initiative; Food and Agriculture Organization of the United Nations: Rome, Italy, 2006.
22. Biopact. ICRISAT launches pro-poor biofuels initiative in drylands. 2007. Available online: http://biopact.com/2007/03/icrisat-launches-pro-poor-biofuels.html (accessed on 15 May 2015).
23. Cotula, L.; Vermeulen, S.; Leonard, R.; Keeley, J. *Land Grab or Development Opportunity? Agricultural Investment and International Land Deals in Africa*; IIED/FAO/IFAD: London, UK/Rome, Italy, 2009; ISBN 978-1-84369-741-1.
24. Scoones, I. New Directions in Pastoral Development in Africa. In *Living with Uncertainty. New Directions in Pastoral Development in Africa*; Scoones, I., Ed.; Intermediate Technology Publications Ltd.: London, UK, 1994; pp. 1–36.
25. Deressa, T. *Measuring the Economic Impact of Climate Change on Ethiopian Agriculture: Ricardian Approach*; CEEPA Discussion Paper No. 25; Centre for Environmental Economics and Policy in Africa, University of Pretoria: Pretoria, South Africa, 2006.
26. Kohyani, P.T.; Bossuyt, B.; Bonte, D.; Hoffmann, M. Grazing as a management tool in dune grasslands: Evidence of soil and scale dependence of the effect of large herbivores on plant diversity. *Biol. Conserv.* **2008**, *141*, 1687–1694. [CrossRef]
27. Ebrahimi, A.; Milotic, T.; Hoffmann, M. A herbivore specific grazing capacity model accounting for spatio-temporal environmental variation: A tool for a more sustainable nature conservation and rangeland management. *Ecol. Model.* **2010**, *221*, 900–910. [CrossRef]
28. Keyzer, M.A.; Merbis, M.D.; Pavel, I.F.P.W.; Van Wesenbeeck, C.F.A. Diet shifts towards meat and the effects on cereal use: Can we feed the animals in 2030? *Ecol. Econ.* **2005**, *55*, 187–202. [CrossRef]
29. De Han, C. *Livestock Development: Implications for Rural Poverty, the Environment, and Global Food Security*; The World Bank: Washington, DC, USA, 2001.
30. CSA. Central Statistical Agency; Federal Democratic Republic of Ethiopia. 2016. Available online: http://www.csa.gov.et/images/documents/pdf_files/regional/Afar.pdf (accessed on 15 October 2016).
31. Rettberg, S. Der Umgang mit Risiken im Spannungsfeld zwischen Konflikten und Nahrungskrisen. Eine politisch-geographische Untersuchung in der Afar Region Ethiopien. Ph.D. Thesis, Bayreuth University, Bavaria, Germany, 2008.
32. Hundie, B. *Pastoralism, Institutions and Social Interaction: Explaining the Coexistence of Conflict and Cooperation in Pastoral Afar, Ethiopia*; Shaker Verlag GmbH: Aachen, Germany, 2008.

33. Hundie, B. Conflicts between Afar pastoralists and their neighbors: Triggers and motivations. *Int. J. Confl. Violence* **2010**, *4*, 135–148.
34. Ame, A. Cross-border livestock trade and small arms and conflict in pastoral areas of the horn of Africa: Case study from southern Ethiopia and northern Kenya. In Proceedings of the Survival of the Commons: Mounting Challenges and New Realities, the Eleventh Conference of the International Association for the Study of Common Property, Bali, Indonesia, 19–23 June 2006; Available online: http://dlc.dlib.indiana.edu/archive/00001823/00/Ame_Abdurahman.pdf (accessed on 1 April 2015).
35. Bekele, K. Ethiopia: Promoting Bio-Fuel. The Reporter (Addis Ababa). 2007. Available online: http://allafrica.com/stories/200711050688.html (accessed on 12 May 2015).
36. Eshetu, J. Overview of the potentials and constraints on export of meat and livestock. In Proceedings of the national workshop on managing animal health constraints to export marketing of meat and livestock, MoARD–FAO, Addis Ababa, Ethiopia, 27–28 April 2004.
37. LMA. *Meat Exports Market Study, MoARD*; Livestock Marketing Authority: Addis Ababa, Ethiopia, 2004.
38. Shapiro, B.I.; Gebru, G.; Desta, S.; Negassa, A.; Negussie, K.; Aboset, G.; Mechal, H. *Ethiopia Livestock Master Plan Roadmaps for Growth and Transformation A Contribution to the Growth and Transformation Plan II (2015–2020)*; Ministry of Agriculture Livestock Resources Development: Addis Ababa, Ethiopia, 2015.
39. The Ethiopian Herald. Ethiopia Eyeing 12 Million USD from Ramadan Meat Export: Institute. 2017. Available online: https://www.alleastafrica.com/2017/06/08/ethiopia-eyeing-12-million-usd-from-ramadan-meat-export-institute/ (accessed on 8 November 2017).
40. Thornton, P.K.; Jones, P.G.; Owiyo, T.; Kruska, R.L.; Herrero, M.; Kristjanson, P.; Notenbaert, A.; Bekele, N.; Omolo, A. *Mapping Climate Vulnerability and Poverty in Africa*; Report to the Department for International Development; ILRI: Nairobi, Kenya, 2006.
41. Hundie, B.; Padmanabhan, M. The Transformation of the Commons: Coercive and Non-Coercive Ways. In *Collective and Property Rights for Poverty Reduction: Lessonsfrom a Global Research Project*; IFPRI: Washington, DC, USA, 2012.
42. Beyene, F. Institutional arrangements in mutually beneficial grazing systems: An example from herding communities in Ethiopia. *J. Land Use Sci.* **2014**, *9*, 438–452. [CrossRef]
43. Beyene, F. Incentives and Challenges in Community-Based Rangeland Management: Evidence from Eastern Ethiopia. *Land Degrad. Dev.* **2015**, *26*, 502–509. [CrossRef]
44. Mulatu, A.; Bekure, S. The Need to Strengthen Land Laws in Ethiopia to Protect Pastoral Rights. In *Pastoralism and Development in Africa: Dynamic Change at the Margins*; Andy, C., Jeremy, L., Ian, S., Eds.; Routledge: Abingdon, UK; New York, NY, USA, 2013; pp. 186–194.
45. USAID. Ethiopia—Strengthening Land Administration Program (ELAP). 2012. Available online: http://www.moa.gov.et/.../Pastoral...pastoral+Land.../eba08eee6d8c-a8f (accessed on 12 December 2016).
46. Little, P.D.; Behnke, R.; McPeak, J.; Gebru, G. Policy Options for Pastoral Development in Ethiopia. 2010. Available online: http://www.researchgate.net/publication/257993033 (accessed on 2 April 2015).
47. Abbink, J. Land to the Foreigners: Economic, legal, and socio-cultural aspects of new land acquisition schemes in Ethiopia. *J. Contemp. Afr. Stud.* **2011**, *29*, 513–535. [CrossRef]
48. Sonneveld, B.G.J.S.; Pande, S.; Keyzer, M.A.; Georgis, K.; Seid Ali, A.; Takele, A. Land degradation and overgrazing in the Afar Region, Ethiopia: A spatial analysis using Rainfall Use Efficiency. In *Land Degradation and Desertification: Assessment, Mitigation and Remediation*; Zdruli, P., Pagliai, M., Cano, S.K.A.F., Eds.; Springer: Dordrecht, The Netherlands, 2010; pp. 97–110.
49. Abdulatife, M. *Training Report on Spatial Data Management*; Afar Pastoral and Agro-pastoral Research Institute (APARI): Semera, Ethiopia, 2016.
50. Sutcliffe, P. The Woody Biomass Project. *reCOMMEND* **2006**, *3*, 1–4.
51. GAEZ v3.0 Global Agro-Ecological Zones. Available online: http://www.gaez.iiasa.ac.at (accessed on 8 January 2015).
52. De Leeuw, P.N.; Tothill, J.C. The concept of carrying capacity in sub-Saharan Africa: Myth or reality? In *Range Ecology at Disequilibrium: New Models of Natural Variability and Pastoral Adaptation in African Savannas*; Behnke, R.H., Scoones, I., Kerven, C., Eds.; Overseas Development Institute: London, UK, 1993; pp. 77–88.
53. Western, D.; Grimsdell, J.J.R. *Measuring the Distribution of Animals in Relation to the Environment*; Handbook No. 2; African Wildlife Leadership Foundation: Nairobi, Kenya, 1979.

54. Too, D.K. Effects of Defoliation Frequency and Intensity on Production of Four Burned and Unburned Bushed Grassland Communities in South-Central Kenya. Master's Thesis, Texas A & M University, College Station, TX, USA, 1985.
55. Braun, H.M.H. Primary production in the Serengeti: Purpose, methods and some results of research. *Ann. Univ. d'Abidjan (E)* **1973**, *6*, 171–188.
56. Van Wijngaarden, W. *Elephants-Trees-Grass-Grazers: Relationships between Climate, Soil, Vegetation and Large Herbivores in a Semi-Arid Savanna Ecosystem*; No. 4; ITC Publ.: Enschede, The Netherlands, 1985.
57. Sinclair, A.R.E. The eruption of the ruminants. In *Serengeti, Dynamics of an Ecosystem*; Sinclair, A.R.E., Norton-Griffiths, M., Eds.; University of Chicago Press: Chicago, IL, USA, 1979; pp. 82–103.
58. Potter, H.L. Aspects of Climate, Herbage Growth and Animal Production in a Semi-Arid Area of Kenya. Ph.D. Thesis, University of Nairobi, Nairobi, Kenya, 1985.
59. Deshmuk, I.K. A common relationship between precipitation and grassland peak biomass for East and southern Africa. *Afr. J. Ecol.* **1984**, *22*, 181–186. [CrossRef]
60. Gebremeskel, K. *Rangeland Production and Carrying Capacity of Afar Regional State*; Consultancy Report; Werer Agricultural Research Center of the Ethiopian Institute for Agricultural Research: Werer, Ethiopia, 2012.
61. Le Houérou, H.N.; Hoste, C.H. Rangeland production and annual rainfall relations in the Mediterranean Basin and in the African Sahelo-Sudanian zone. *J. Rangel. Manag.* **1977**, *30*, 183–189. [CrossRef]
62. Dye, P.J.; Spear, P.T. The effects of bush clearing and rainfall variability on grass yield and composition in southwest Zimbabwe. *J. Agric. Res.* **1982**, *20*, 103–118.
63. De Leeuw, P.N.; Nyambaka, R. The prediction of rangeland primary 18 production from rainfall data in arid and semi-arid Eastern Africa. In *Proceedings of the Third PANESA Workshop: 260–268 ILCA*; Dzowela, B.H., Ed.; International Livestock Centre for Africa: Addis Ababa, Ethiopia, 1988.
64. CSA. *Agricultural Sample Survey 2010/11 [2003 e.c.] Volume ii. Report on Livestock and Livestock Characteristics (Private Peasant Holdings)*; Federal Democratic Republic of Ethiopia, Central Statistical Agency: Addis Ababa, Ethiopia, 2011.
65. CSA. *Pastoral Areas Livestock Enumeration, Results for Afar Region*; The Central Agricultural Census Commission; Central Statistical Authority: Addis Ababa, Ethiopia, 2003.
66. Behnke, R. *The Contribution of Livestock to the Economies of IGAD Member States. Study Findings, Application of the Methodology in Ethiopia and Recommendations for Further Work*; IGAD Livestock Policy Initiative Odessa Centre: Great Wolford, UK, 2010.
67. ESGPIP. Afar Region. The Ethiopia Sheep and Goat Productivity Improvement Program. Available online: http://www.esgpip.org/Afar.html (accessed on 15 June 2015).
68. LDMPS. *Livestock Development Master Plan Study for 2006 (Phase I, Volume B)*; Ministry of Agriculture, Government of Ethiopia: Addis Ababa, Ethiopia, 2007.
69. BoFED. *Regional Atlas 2: Afar National Regional State*; Finance and Economic Development Bureau: Semera, Ethiopia, 2009.
70. Abdullatif, M.; Semera, Ethiopia. Afar Livestock statistics from the Afar Pastoral & Agricultural Development Bureau. Personal communication, 30 January 2016.
71. Amsalu, T.; Addisu, S. Assessment of Grazing Land and Livestock Feed Balance in Gummara-Rib Watershed, Ethiopia. *Curr. Agric. Res. J.* **2014**. [CrossRef]

land

MDPI

Article

Berenty Reserve—A Gallery Forest in Decline in Dry Southern Madagascar—Towards Forest Restoration

Vanessa Winchester [1,*], Kate Hardwick [2], Hantanarina Rasamimanana [3], Sahoby M. Raharison [4], Anne Mertl-Millhollen [5], Holger Gärtner [6] and Janet McCrae [7]

1 School of Geography and the Environment, University of Oxford, Oxford OX1 3QY, UK
2 Royal Botanic Gardens, Kew, Wellcome Trust Millennium Building, Wakehurst, Ardingly, West Sussex RH17 6TN, UK; k.hardwick@kew.org
3 Ecole Normale Supérieur, University of Antananarivo, Antananarivo 101, Madagascar; hantani1@yahoo.fr
4 Lycée Ambohimalaza Miray, Cisco Antananarivo, Avaradrano, Antananarivo 101, Madagascar; raioby@gmail.com
5 Department of Anthropology, 128 University of Oregon, Eugene, OR 97403-1218, USA; hplam_1998@yahoo.com
6 Swiss Federal Research Institute WSL, Zürcherstrasse 111, 8903 Birmensdorf, Switzerland; holger.gaertner@wsl.ch
7 20 Aston Street, Oxford, OX4 1EP, UK; jemccrae12@btinternet.com
* Correspondence: vanessa.winchester@geog.ox.ac.uk; Tel.: +44-(0)-186-555-7600

Received: 22 October 2017; Accepted: 8 January 2018; Published: 15 January 2018

Abstract: Berenty Reserve, a fully protected gallery forest beside the Mandrare River is renowned for its lemurs, but the continuous canopy of the main forest is shrinking, fragmenting and degrading. The aim of this study, before any restoration can be considered, is to investigate why canopy-cover is declining and define the forest's vegetation status and composition. Our study includes analysis of tamarind age (the dominant species) and regeneration, forest extent, climate and soil. Measurement of trunk circumference and annual rings indicated a median age of 190 years, near the accepted maximum for tamarinds. There is no regeneration of tamarind seedlings under the canopy and an invasive vine, *Cissus quadrangularis* suffocates any regeneration on the forest margins. A vegetation survey, based on fifteen transects, broadly characterized three forest areas: continuous canopy near the river, transitional canopy with fewer tall trees, and degraded dryland; the survey also provided a list of the 18 most common tree species. Ring counts of flood-damaged roots combined with measurement to the riverbank show that erosion rates, up to 19.5 cm/year, are not an immediate threat to forest extent. The highly variable climate shows no trend and analysis of forest soil indicates compatibility with plant growth.

Keywords: degrading; tamarind age; regeneration; invasive vine; vegetation survey; erosion

---

1. Introduction

Madagascar is one of the richest countries on Earth in terms of biodiversity, endemism and range of habitats, but the combination of a rapidly growing human population coupled with extreme poverty constitutes a serious threat to its uniquely diverse flora and fauna. In the dry south, with the highest percentage of endemic plants on the island, forests are burnt for charcoal production, planted with maize or sisal or degraded by cattle pasturing [1]. Much has been written about the need for conservation [2–4], but there have been few surveys of the dryland vegetation [5] and little is known concerning the growth requirements of the local species. There is no information on successional patterns [6] or which species could be termed pioneer or climax and, excluding one reforestation project on the southwestern side of Madagascar focused on planting eucalyptus and tamarind trees for local use [7], regeneration of a gallery forest aiming at re-establishing its structure, productivity and species diversity has, to the authors' knowledge, never been attempted.

Gallery forests form as corridors along rivers or wetlands running through arid or semi-arid landscapes. The riparian zone offers fertile alluvial soil, a more reliable water supply at depth and better drainage than is available to the vegetation beyond the immediate river valley. As a result, the boundary between the gallery forest and the surrounding woodland or grassland is usually abrupt, with an ecotone only a few metres wide, as is the case at Berenty Private Reserve.

The reserve, embracing patches of forest on the banks of the Mandrare River and famous for its lemurs, is among the very few remaining securely protected gallery forests in southern Madagascar. However, since the 1980s researchers have noticed that the canopy of the dominant species, *Tamarindus indica* (tamarind), has been fragmenting and shrinking [8]: this is of particular concern since tamarinds are a keystone resource for lemurs. That the forest exists at all and has not been lost, as have most other gallery forests in Madagascar, is due to the foresight of the de Heaulme family, who set the forest aside as a conservation area 80 years ago and have protected it ever since. Although only 330 ha in area, the reserve has an importance far beyond its size: studies of its six species of lemur have been ongoing for the last 50 years and as a centre for eco-tourism and student education it has a preeminent position in spreading knowledge of this fragile, unique and potentially vanishing ecosystem [9].

Previous studies have shown that the main forest area, Malaza, includes at least three different vegetation zones [10], which can be further subdivided into six vegetation types [8]. Maps of the zones produced in 1973 and 1995 show that in 1973 the continuous canopy gallery forest formed a belt about 250 m at its deepest mainly along the western section of the Malaza riverbank, with patchy cover along the bank to the east. The south-east side is more open with occasional large trees surrounded by grass, scrub and thorny thickets. By 1995, the dense tamarinds near the river showed a 25% decrease in canopy cover, with the more open areas expanding. Fire is a minor problem; three small patches were set alight in Malaza, one east of the central north-south trail in 1998 and the other two on the western edge of the same trail in 2006, all were quickly extinguished by watchful guardians and reserve staff.

This present study sets out to provide a background for forest restoration. Several restoration methods have been developed and most approaches involve planting tree seedlings, but many tree planting projects fail, due to the planting of inappropriate species, inadequate planting techniques and post-planting maintenance regimes, particularly when using native tree species about which little is known. Although most of these applications have been in humid tropical zones, some of the lessons learned can be carried over to dryland reforestation: principally, investigation of the reasons for forest degradation, the importance of a preliminary survey of the target forest and the advisability of an initial planting program to trial different cultivation practices [11]. The first two of these reasons are the main objects of this study: an investigation of the reasons for forest decline and a vegetation survey to assess the existing balance of species in the gallery forest and identify what intervention may be needed to support or accelerate regeneration. The long-term aim is to arrest the decline of the continuous canopy forest and restore biodiversity levels to those typical of the remaining forest fragments, thus conserving key resources for the lemurs and other endangered fauna.

2. Methods and Materials

2.1. Study Area

Berenty lies 25 km from the southern coast (25°00′ 35.50°18′ 28.50 E) and west of the coastal range in a relative rain shadow with average annual precipitation of 500 mm (Figure 1). Cyclones affecting southern Madagascar occur intermittently [12], but the forest structure, with an approximately uniform canopy height and the deep taproots of the tamarinds, is well adapted to resist cyclone events.

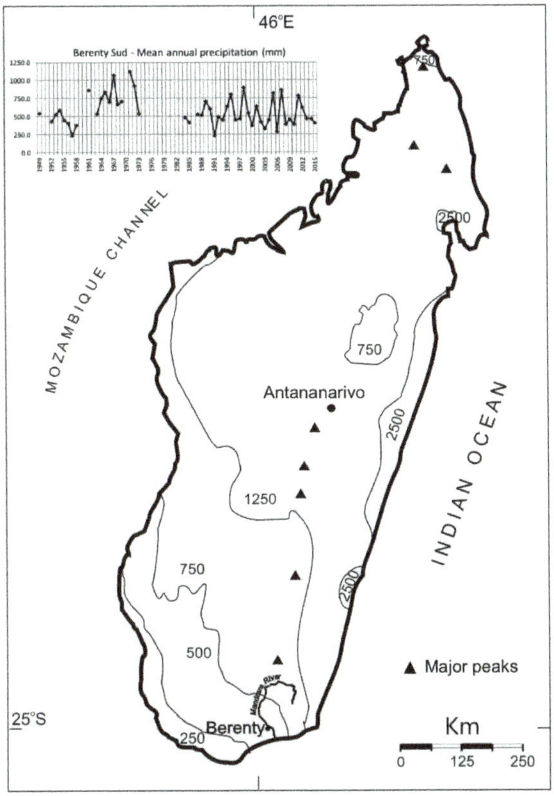

Figure 1. Precipitation map with isohytes (mm), with inset of Berenty precipitation data: note the exceptional flood in 1971. Data courtesy of Mr Rakotomalala, Berenty.

There are three main forest areas in the reserve, Malaza, Ankoba and Bealoka, and each have different management histories [13]: Malaza, covering approximately 100 ha is located on the floodplain on an outside bend of the Mandrare River where grazing has been excluded since 1936. A 10 m high ancient river terrace backed by spiny forest defines part of its southern boundary (Figure 2).

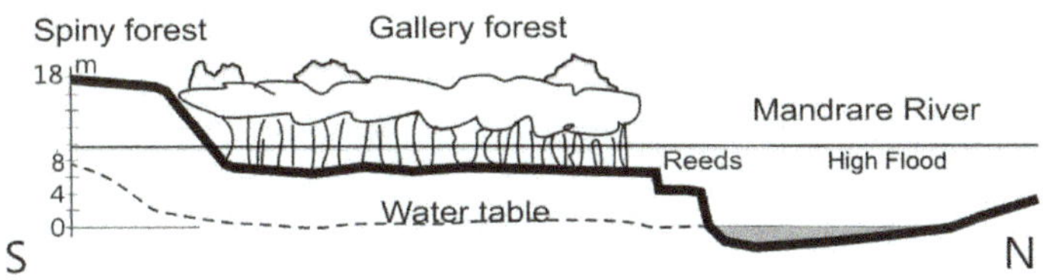

Figure 2. Generalised north-south cross section of Malaza forest showing the river at average annual low-flow, bank-side growth of reeds (*Phragmites mauritianus*) on a lower terrace frequently eroded by floods, height of intermittent high floods (in 1971 up to 1.5 m above forest floor) and an upper, ancient terrace rising up to 10 m above the gallery forest floodplain.

Tourist amenities and fields encompass the remaining perimeter. Dry ancient river channels and sandbars dissect the forest floor. Tree cover within the forest varies from dense tamarind forest, at its densest along the riverbank, to open scrub and grassland interspersed with impenetrable thickets of thorny scrub, with *Capparis sepiaria* and *Azima tetracantha* among the most common. Plant species beginning to invade the open areas include *Agave sisalana* and *Opuntia monacantha* [14], used as a defence to keep cattle out of the reserve, but the most prolific invasive species is the succulent African vine, *Cissus quadrangularis*, growing wherever there is sunlight, climbing in great suffocating waves to the tops of isolated trees.

Ankoba, joined by a forest corridor to Malaza, is a 15 ha patch of mature secondary-growth forest. It was originally farmed by the locals and then used as a sisal nursery. In the 1950s it was allowed to revert to forest, with the addition of exotic *Pithecellobium dulce* [15] (p. 34). Bealoka, where grazing has been excluded since 1985, lies 4.5 km to the northwest and covers around 110 hectares. It is separated from Ankoba by Berenty village and sisal fields. Not included in this study are Rapily/Analalava, a 60 ha disjointed strip of spiny forest south of Malaza and Anaramalangy, a 45 ha patch east of Malaza once abandoned to *Cissus*, but now being cleared.

2.2. Objectives

The field work has three main objectives: the first is to establish why the forest is shrinking; the second is to carry out a vegetation survey, including names of the most common trees for eventual restoration, and the third is to investigate riverbank erosion rates to determine if the forest riverside margin is threatened. A further objective is an analysis of the soil composition across the Malaza floodplain, since for any eventual forest restoration an understanding of the organic, mineralogical, pH and moisture content of the soils is essential.

2.3. Tree Age

A random sample of 48 trunk circumferences was measured in Malaza and Ankoba at 95 cm height; eleven stem cores were taken, at the same height, with a 10 mm increment-boring tool. Only three of the cores reached the central pith due to exceptional wood hardness (we broke two boring tools). The cores were dried, mounted and fine-polished until the annual ring widths were clearly visible. Ring widths were measured using the LINTAB system and TSAP-Win software [16,17]. To find tree age in the reserve, tamarind circumference measurements were converted to radii and each radius divided by the average ring-width value. This provides tree age above the coring position; to find true age, years of growth below the core must be added to the ring count: a mean growth rate was derived from nine seedlings and saplings from 3 to 337 cm tall, aged one to 18 years, taken from a range of environments across the reserve.

2.4. Vegetation Survey

The main vegetation areas were surveyed using the rapid site assessment approach for data collection recommended by Elliott et al. [18] (pp. 72–77; 124). The procedure, defining five stages outlining the health of a forest, is designed as a broad-brush approach for assessing levels of forest degradation to aid the determination of the most suitable restoration strategy. However, the five stages are designed for the analysis of forests that have suffered most from human intervention. At Berenty, Malaza the main forest, has been protected from outside disturbance since the 1930s and thus the decline in forest cover here may mainly be due to natural causes and only the first three of Elliott's stages were considered relevant. Saplings and small trees, under 30 cm girth but over 50 cm tall, are termed 'regenerants' by Elliott et al. [18] (p. 72):

- Stage 1: at least 25 regenerants per circle, few shrubs and grasses and a high number of tree seedlings.
- Stage 2: at least 25 regenerants per circle, a moderate to low cover of shrubs and tree cover that is frequently insufficient to shade out the grasses. Large-tree numbers are declining as are seedling numbers.

- Stage 3: fewer than 25 regenerants per circle, shrubs and grasses dominate, large trees are rare and seedling numbers are greatly reduced.

The vegetation survey was based on transects laid out on set compass bearings, with 10 m diameter circles established every 25 m. To avoid edge effects [19], wherever possible circles were initiated 20 m from any trail or path. The circles were defined using 5 m lengths of cord attached to a central pole. We aimed to include 10 circles per transect (i.e., 250 m), but a number of transects were shorter due to the thorny impenetrability of the undergrowth.

Within each circle, as well as regenerants, we recorded the number of trees with stems over 30 cm girth at breast height (large trees) and numbers of seedlings under 30 cm tall. Tamarind seedlings were counted separately as these have been of particular interest in previous studies [9,20]. Lianas and vines were classified as few, moderate or many (0–4; 5–9; 10<). The percent cover of shrubs and grasses was estimated as low, moderate or high, with their measured height range, and the percent cover of bare soil and numbers of dead trees. The short transect, T2, following a line of dead trees was included to see if this made a difference as regards forest growth.

2.5. Riverbank Erosion and Roots

The extent of canopy cover could also be threatened by riverbank erosion (Figure 3).

Figure 3. Bank-side tamarind roots exposed by floods at Ankoba. Flood dates and rates of erosion can be estimated from anatomical changes in roots damaged by catastrophic flooding.

Incipient bank erosion can be dated from the year root cells become more stem-like with smaller and more densely packed thicker cell walls, providing more rigid support (Figure 4) [21]. Sections cut from damaged roots similarly provide a date for exposure, with this defined by the ring dating the year of damage [22]. Average annual erosion rates were calculated by dividing the distance from the damaged root to the eroded bank by the number of years since damage occurred.

Fifteen damaged roots were sectioned and dried in the sun, fine-polished in the laboratory and examined under a microscope, with a camera attachment for details and a high-resolution scanner used to view whole sections. This is an important necessity since annual rings may only be seen in part of a section due to trauma or wound callusing and may be eccentric in relation to root centres. Micro-sections were cut from some of the roots using a GSL1 microtome [23].

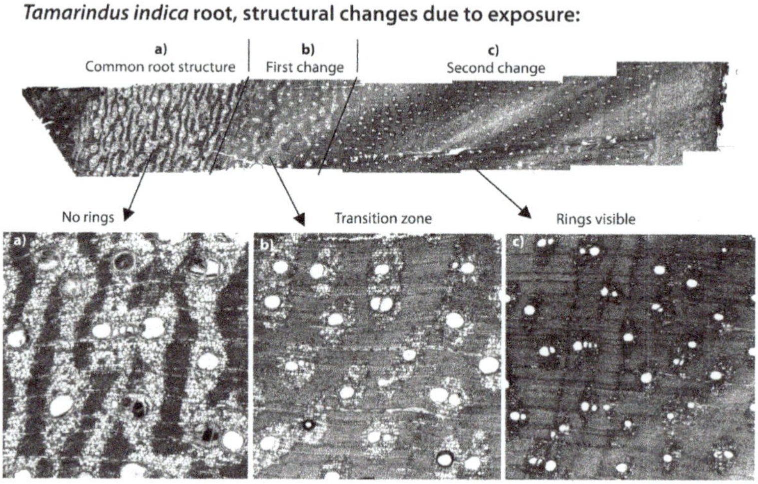

Figure 4. Root micro-section tamarind root sample from the eastern end of the Malaza riverbank. Structural changes due to gradual exposure: (**a**) Common root structure; (**b**) First structural change: root potentially close to the surface but not exposed; (**c**) Second structural change, root fully exposed.

2.6. Soil Analysis and Leaf Identification

A total of twenty-four soil samples were taken, at 25 cm below leaf litter, from the transects and also from the river bank, mid-forest and in the degraded area. These were analyzed in Oxford to provide information on the suitability of the forest soil for seedling transplantation. For chemical analysis, a Dionex IC DX500 Chromatograph was used for determination of organic and inorganic ions using 1 g soil in 100 mL water. Iron was analysed using atomic spectroscopy, with 2 g in 10 mL EDTA 0.05 m. The organic content was calculated using the loss on ignition (LOI) method, with a sample size of 2 g dry weight and exposure time of 5 h at 550 °C [24]. A Malvern Master-sizer 2000, with the programme Gradistat, was used for soil particle size analysis [25]. Three soil samples were taken, also at 25 cm depth, to provide moisture measurements.

Identification of tree species was a priority. Specimens of each species were collected and dried and given their local names on site by the chief forest guardian Mr Remanonja, while Latin names were provided by Professor Rasamimanana. The specimens were then compared with a photographic collection left by a previous researcher. Specimens were mounted in a book and given to botanists at the Kew Madagascar Conservation Centre in Antananarivo for future reference and a photographic record was sent to botanists at the Muséum National d'Histoire Naturelle (MNHN) in Paris.

3. Results

3.1. Tree Age and Regeneration

Tree ages was derived from a combination of two values: an average core ring-width measurement of 0.225 cm, with this value agreeing with that obtained from a nine-year direct growth study [26], and an average value of five years extrapolated from a mean growth rate of 18.7 cm/year, below coring height. A histogram of the age structure of the tamarinds shows that two of the trees from Ankoba, at the lowest end of the distribution, are 54 and 67 years old, while the central cohort sampled in Malaza is aged between 154 and 226 years, with a median age of 190 years (Figure 5); the four trees in the upper range are aged from 323–405 years. On the forest floor, there are large numbers of one-year-old tamarind seedlings, but no young saplings: this does not, however, extend to lower canopy species, which regenerate plentifully under tamarinds.

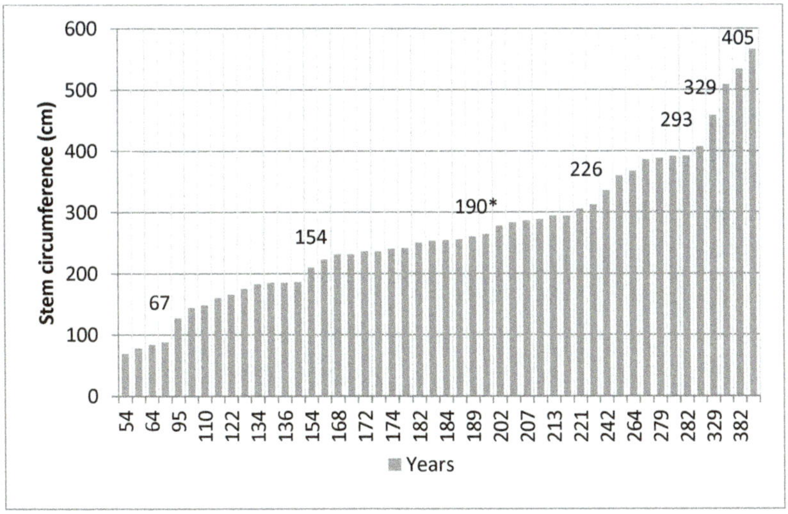

Figure 5. Tamarind tree age in Ankoba and Malaza. Note that exceptionally advantaged trees can attain ages between 300–400 years; whereas the usually accepted age is 200 years. 190* is the median age in Malaza of the central cohort (154–226 years).

3.2. *Vegetation Survey*

Figure 6 shows the positions of the transects carried out in 2014, together with the outlines of the 1973 vegetation zones, as drawn by Blumenfeld-Jones [8] (p. 73), superimposed over a Google Earth June 2013 view of the forest. Comparison of the 1973 zones with the 2013 vegetation suggests that, at the scale of the 2013 satellite view, zone 1 appears relatively unchanged, despite the 25% decrease in canopy cover described by Blumenfeld-Jones [8] (p. 81); by 2013 half of zone 2, and large areas of 3 and X have fragmented together with the southern end of zone 4. Zone 5 has expanded north into zone 4 and into the smaller zone 3 areas, with almost no big trees and an extensive cover of shrubs, vines and grasses.

The positions of the three areas burnt in 1998 and 2006 show that the vegetation has to some extent regenerated over the interval, with the scar left by the 1998 fire reduced in size (it formerly covered almost a hectare), the small 2006 fire nearest the river is recovering and only the 2006 fire in the centre of the forest is still open.

In Table 1, the transects, listed in descending order of the average number of large plus small (regenerant) trees, illustrate the three stages defining the health of the forest [18]. Regeneration is at its highest under the continuous canopy of stage 1, apart from an extraordinary number of small trees in T9, an old riverbed. Average small tree numbers are reduced under the stage 2 canopy, but the ratio of small to large trees is at its highest, with this signaling the transitional character of the vegetation, while shrub and grass cover is mostly in the low to moderate range. In stage 3 transects, the low ratio between small and large trees and their overall low numbers highlights the degraded character of the area, with shrub and grass cover moderate to high.

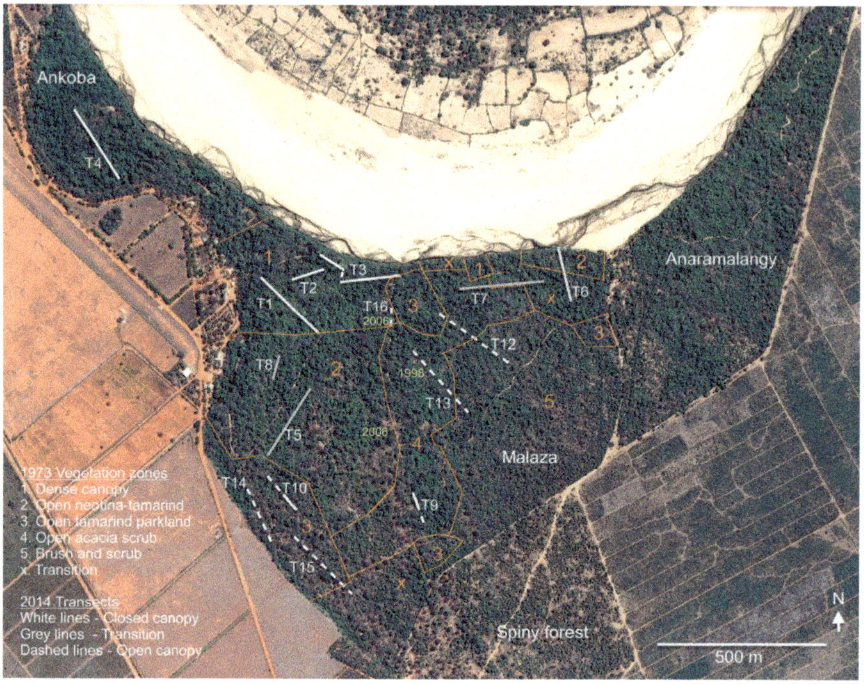

Figure 6. View of Ankoba and Malaza with the river at low water. The red lines show the positions of the vegetation zones as defined in 1973. The 2014 transects are numbered and dates mark the locations of the 1998 and 2006 burns. The 'dogleg' transect, T3, crosses the riverside trail where it touches the riverbank.

Table 1. Rapid site assessment data for transects, divided into 3 stages. The transects are arranged in descending order of average large plus small trees, defined as 'regenerants' by Elliott et al. [18]. Regenerants are saplings and small trees <30 cm girth. Short transect lengths are italicized and highlighted. Averages are based on counts per circle in each transect. Standard deviations (SD) indicate considerable variability between circles.

Malaza 1936				**Ankoba 1950**					**River bed**		**Bealoka 1985**		**Burnt 1998**		**River Bed**	**Burnt 2006**
Transect	1	3	3	4	5	**6**	7	**8**	**9**	**10**	11	**12**	**13**	14	15	**16**
Length (m)	250	250	250	250	250	***150***	250	***75***	***100***	***125***	250	250	250	***225***	250	***50***
	STAGE 1—continuous canopy				STAGE 2—transitional canopy							STAGE 3—open dryland				
Average large + regenerants − SD	80 21.0	61.7 12.8	58.9 20.6	55.7 11.5	46.8 21.9	46 12.2	40.9 12.6	37.6 1.9	37.2 27.1	35.2 14.9	33.0 13.9	13.1 9.3	14.7 12.4	14 9.3	13.8 9.2	9.5 3.5
Average large trees (>30 cm girth) − SD	7 5.4	9.7 4.0	7.3 2.9	6.9 3.0	4.0 2.4	4.5 1.6	3.1 2.0	2.6 0.9	1.5 0.5	3 2.6	4.4 2.0	2.9 5.4	1.8 2.7	1.9 1.8	2 1.7	2 2.0
Average regenerants (<30 cm girth) − SD	73.1 20.1	52 13.9	51.6 20.3	48.9 1.9	42.8 22.7	41.5 11.7	37.8 12.2	35 1.6	35.7 27.0	32.2 12.5	28.6 12.4	10.2 8.4	12.9 10.8	12.1 8.4	11.9 8.0	7.5 1.5
Average all seedlings (<50 cm ht) Average tamarind seedlings	22.9 Not counted	35.0 8.5	34.5 11.7	40 8.4	17.3 0.2	32.8 2.5	32 0.9	19.6 0.6	54.3 27	17.2 0.6	27.2 9.5	3.7 0.2	2.4 0.1	13 3.9	8.2 0.3	2.5 0
Shrub cover ht range (cm)	low 10–200	low 100–150	low 7–150	low 10–200	low/mod 15–200	low/high 130–800	low/mod 30–200	low/mod 200	low/mod 50–150	mod/high 200–250	low 20–200	mod 25–200	mod 30–200	mod/high 100–300	mod 100–200	low/high 70–150
Lianas/vines nos of circles. Few mod many	6 1 3	1 3	7 2 1	7 1 2	3 1 6	1 1 4	1 3 6	3	1 2 1	1 4	2 2 6	6 4	4 1 5	2 1 6	7 2 1	1 1
Grass cover % ht-range (cm)	low 5–10	low 0–10	low 5–10	low 5–20	low/mod 10–50	low/high 0–10	low/high 5–150	high 10	low/high 15–75	mod 15–100	low5 –15	high 100–150	high 5–500	low/mod 5–100	mod/high 20–280	high 30–40
Bare soil % cover	3.2	0	1	0	0.5	4	0	0.2	0.1	0.1	0.9	4.7	2	0.1	2.5	1
Dead trees	11	6	1	12	8	4	7	2	4	8	5	9	10	0	3	0

(i) On each transect, circles of 10 m diameter are spaced 25 m apart; (ii) Shrub and Grass cover %: low = 0–33%; moderate = 34–66%; high = 67–100%; (iii) Lianas and vines (principally *Combretaceae* and *Cissus*) in circles: few = 0–4; moderate = 5–9; many = 10+.

Comparison of the average number of seedlings in each of the three stages is also indicative of stage: highest in stage 1 areas, including T2, where a line of acacia trees was blown down in a wind storm in 1999 [27] and lowest in the degraded open dryland. In stage 2, the T9 riverbed is again an exception with an extraordinary number of seedlings of all species, but otherwise seedling numbers are variable. The average tamarind seedling numbers are generally moderate to low in the other stages. Of note is that despite the eight years difference between the two burns, covered by transects T13 and T16, they contain closely similar seedling counts, with only one tamarind seedling in T13 and none in T16. Grass and shrub cover is least in stage 1 but variable elsewhere, while percentages of bare soil follow no pattern.

Figure 7 presents the data as a graph showing averages of both small and large trees declining across the stages, as expected given the sampling criteria. Shrub and grass cover under the dense canopy of stage 1 is low as is shrub cover for half of the transitional stage 2, although the presence of grass here is highly variable. In stage 3, shrubs and grasses (mainly *Panicum maximum*) dominate. The incidence of dead trees, peaking in T2 and T10, is also highly variable showing no pattern. Vines and lianas could not be quantified due to sampling difficulties.

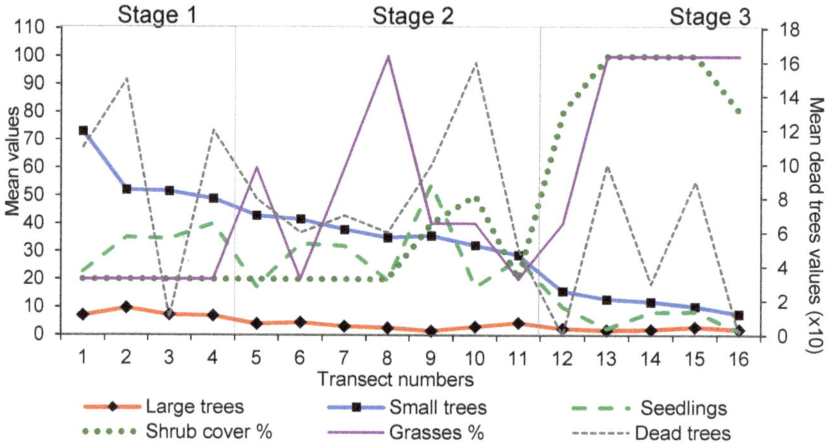

Figure 7. Graph of the data from Table 1 showing the inter-relationship of the six main factors across the three stages.

In Table 2, species are listed in descending order of percent occurrence in all transects in each vegetation type: continuous canopy, transitional and dryland. Other species are doubtless present, but are insufficiently common to occur in any of the transects (for example, there are a number of *Ficus* spp. found elsewhere in Malaza). The single record of *Cordia caffra* (a native from mainland Africa) suggests that the species is not yet a threat to the mainly endemic tree population in the forest and probably never will be since the presence of large specimens in the tourist precinct imply that the species has been in the area for some time. Table 2 also shows that *Euphorbia* does not grow under the continuous canopy and that *Salvadora angustifolia* is the only one of the listed species to grow solely in the dryland area. Not shown in the table, but also growing in the dryland, are several species from the spiny forest including sisal (*Agave sisalana*) from the nearby fields and a thick growth of tall grasses, thriving even beside the narrower trails wherever there are openings letting in light.

Table 2. Transect tree species classified according to habit and habitat. Short transects are highlighted.

Transect (T)	1	2	3	4	5	6	7	8	9	10	11	12	13	14	15	16	
DRYLAND		STAGE 1					STAGE 2							STAGE 3			%
*Enterospermum**	x	x		x	x	x	x	x	x	x		x	x	x	x	x	7
Strychnos madagascariensis	x	x	x	x		x	x	x	x	x	x	x	x	x			7
*Euphorbia**					x	x		x	x	x	x	x		x	x		5
Hazunta modesta	x	x		x	x		x	x			x	x			x		5
*Grewia**				x	x		x			x		x	x		x		4
Salvadora angustifolia												x	x	x	x		2
LOWER CANOPY																	
Celtis philippensis	x	x	x	x	x	x	x	x	x	x	x	x	x	x	x	x	8
Rinorea greveana	x	x	x	x	x	x	x	x	x	x	x	x	x	x	x	x	8
Crataeva excelsa	x	x	x	x	x	x	x	x	x	x	x	x	x	x	x	x	8
Quisivianthe papinae	x	x	x	x	x	x	x	x	x		x	x		x	x	x	7
Celtis bifida	x	x	x	x			x			x	x	x	x	x		x	6
Rubiaceae sp. **	x	x	x	x	x	x	x	x				x			x	x	6
Tricalysia sp. **	x	x	x	x								x		x		x	4
Celtis gomphophylla			x								x						1
Cordia caffra							x										-
UPPER CANOPY																	
Tamarindus indica **	x	x	x	x	x	x	x		x	x	x	x	x	x	x		7
Albizia polyphilla	x	x	x	x	x	x	x			x	x	x		x			6
Neotina isoneura	x	x	x	x	x			x		x			x	x	x		5
Acacia rovumae	x		x	x			x			x	x			x	x	x	5
TOTAL SPECIES																	
Dryland species	3	3	1	4	4	3	4	4	3	4	3	6	4	4	5	1	
Lower canopy	7	7	8	7	5	5	6	5	4	4	6	7	4	6	5	7	
Upper canopy	4	3	4	4	3	2	4	1	1	4	3	2	2	4	3	1	
AVERAGE spp./circle	1.4	3.2	1.3	1.5	1.2	1.6	1.4	3.3	2.0	2.4	1.2	1.5	1.1	1.4	1.3	4.5	

Average canopy height: dryland—10 m; lower canopy—12 m; upper canopy—20 m. * At least two species; ** Species that are native but probably not endemic; -*Cordia caffra* is non-native and should not be planted in forest; %Guide as to the relative percentage of species to plant to approximate the distribution of species in the natural; ecosystem

3.3. Riverbank Erosion

Root section ring counts show damage in three main flood years since 1998 (Figure 8). According to root damage, the bank towards the northern end of Ankoba is the most threatened, with maximum projected erosion of 19.5 cm/year, while on the Malaza bank erosion averages between 6 and 9 cm/year (Figure 9)

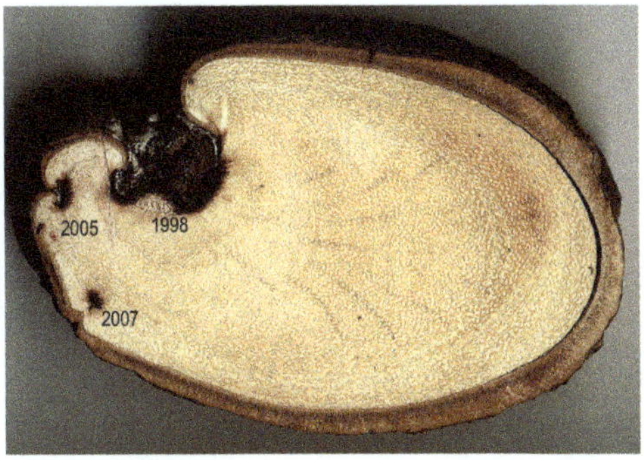

Figure 8. Tamarind root section from the Ankoba riverbank. The section shows changes in structure from root to stem-type cells after the damage that occurred in the 1998 flood (Figure 9) when the root was first exposed to the air and subsequently during the floods of 2005 and 2007.

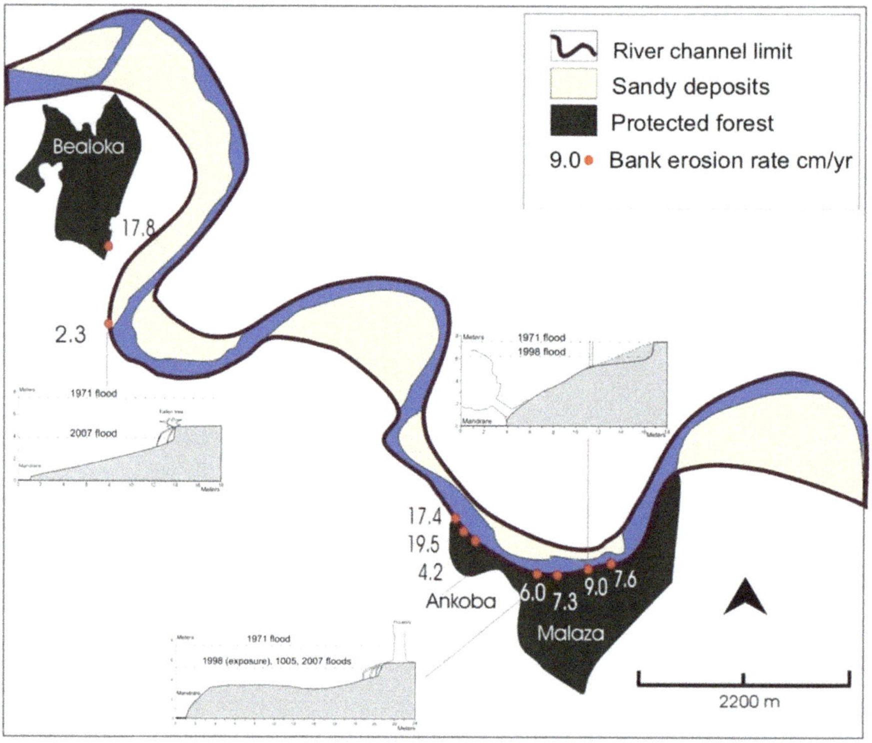

Figure 9. Berenty Reserve with the Mandrare river at low water and the range of different bank profiles with dated flood heights: 1971 was the highest flood on record. The local people practice transient cultivation on inside channels and bends.

3.4. Soil Analysis

The soils in Malaza can be divided into four domains (Figure 10): riverbank, south, centre-east and west. Near the river, the average sand percentage compared with silt is high and clay is low (73.1%, 25.5%, 1.4%). In two widely separated sites, one within 250 m of the river (T12) and the other near the southern margin in the open area east of the central trail (S3), the sand content is almost twice that of silt, and clay is moderately increased compared with the riverbank (62%, 36%, 2%). In the centre of Malaza, the sand to silt ratio is nearly even and the clay percentage is raised (49.5%, 47.5%, 3%), while on the west side of the forest the percentage of sand is reduced and silt predominates, with clay remaining the same as in the centre (40%, 57%, 3%).

The organic content of the soil averages 12%, with the notable exception of 38% at T9 in the old river channel running beside the central trail where leaf litter is particularly deep; here, as a consequence of the high organic content, cation values are high compared with the other transects influencing the soil's ability to hold onto essential nutrients [28]. Elsewhere anion values are variable from a low of 62 ppm in T11, a dry channel in Bealoka, to 1110 ppm at the start of T3 in Malaza. Other raised values are found in transects T5 and T16.

Levels of salinity are well below concentrations that could reduce growth. They are highest on the riverbank, T3, at 53.3 ppm and T7, at 45.4 ppm. Elsewhere they are within the range 23–43 ppm. Phosphorous at T16 is high (336 ppm compared with an average elsewhere of 55 ppm), with an anomalous pH of 4.4. At all other transects pH values are approximately neutral ranging from 6.5 to 7.6, with most nutrients optimally available and a range highly compatible with plant growth.

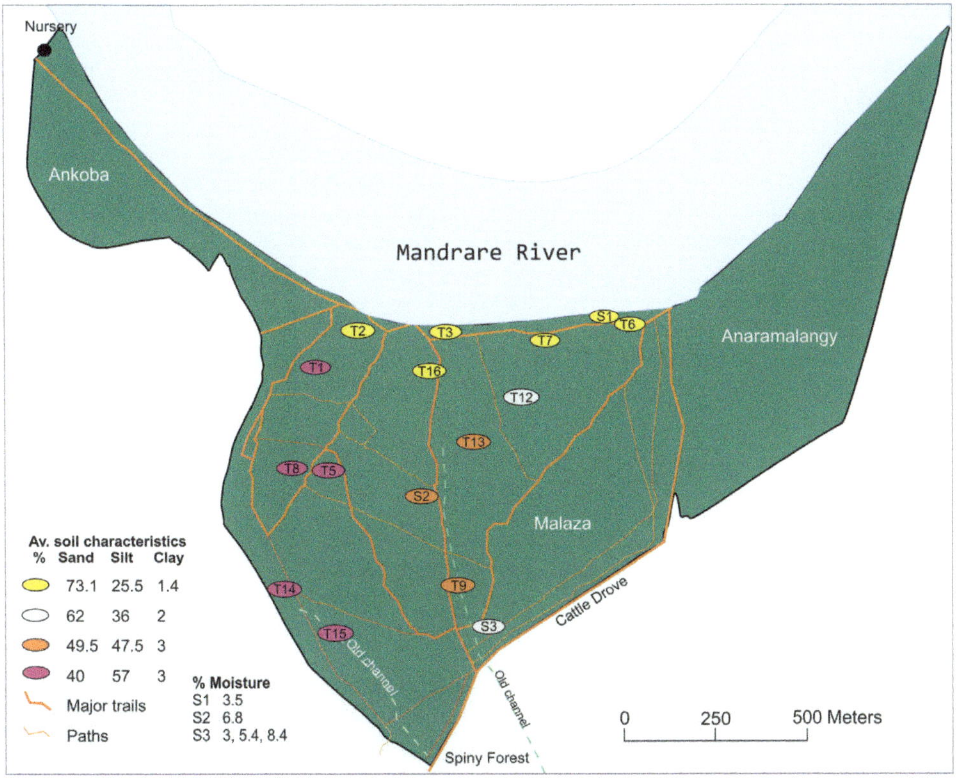

Figure 10. Soil sampling sites: The T samples are from the 2014 transects and the S1–S3 sites included soil moisture measurements.

Soil readings from T16, the small burnt patch, show an exceptionally low pH and a high phosphate and iron content, with this suggesting that the sample could contain urine; nevertheless, the high average number of tree species on the site, with as yet no upper canopy species, imply that this area is as suitable for seedlings as elsewhere.

Soil moisture measurements show that the driest site is near the river under the continuous canopy at S1, with high sand content. At S2, mid-forest where there was more silt than sand, moisture was over twice that of S1, while S3 showed highly variable results.

4. Discussion

4.1. Tamarind Age and Regeneration

The young ages of 54 and 67 years of the two trees from Ankoba are consistent with the known age of this secondary forest. However, the four trees between 329 and 405 years old (Figure 5) are a surprise, since the generally accepted age for tamarinds is 200 years. Evidently, these trees are growing in exceptionally advantageous locations. The oldest, growing on the edge of a large clearing in Ankoba, which measured 28 m tall, died in 2010 (Alison Jolly personal communication).

Blumenfeld-Jones [8], who detailed canopy changes over a 22-year period (1973–1995), suggests that the whole forest may be in transition to a dryer forest type, with the pattern of change in vegetation boundaries along the eastern section of the riverbank reflecting the overall drying of the forest due to changing dynamics of the river. However, our study of riverbank erosion rates reveals that there have been only small changes over the period, hence changes in river dynamics are unlikely to have affected the forest. Climate change might be implicated, but the precipitation data (Figure 1) show no trend, with droughts or floods occurring every few years, although warming temperatures may have contributed to drying.

Blumenfeld-Jones also carried out an extensive study of tamarind regeneration [8] (pp. 69–85), she found almost no tamarind seedlings (between 1–2 m tall) growing less than 15 m from large tamarinds and none in canopy gaps around dead tamarinds. She also found that tamarind abundance was uncorrelated with canopy cover or soil type and that young trees tended to grow in clusters in all forest types especially along the sides of the wider trails. Overall, she concludes that young tamarinds mostly survive in scattered patches not already overhung by mature tamarinds: this confirms our findings. The lack of tamarind saplings under mature trees suggests either a toxic effect [29–31] or some other negative plant–soil feedback [32].

Measurement of tamarind annual ring widths revealed widely differing patterns making comparison impossible; hence, neither age decline in ring-width nor climate-related trends could be determined due to ring eccentricity, a common growth feature of older tamarinds.

4.2. Cissus Quadrangularis, Pests and Pathogens

Cissus is a major problem in the reserve, stifling regeneration especially on the margins of the dense canopy [33,34]. There is no record of when *Cissus* originally arrived in Berenty; evidently it was well established by 1983 when a program of vine clearance was initiated, following which tree growth was found to improve [35]. Nevertheless by 2007, *Cissus* was widespread covering between 50% to 60% of Malaza [15], with this being particularly a problem for regeneration around dense canopy margins and wherever trees are dead or dying and there is full sunlight.

Pests and pathogens, particularly insects and fungi, form a vital part of the ecosystem in any forest [36], but at Berenty there are no obvious signs that they are causing forest decline; although the leaves of individual trees are frequently mined by insects, termites favour the smooth bark of *Neotina* and dying tamarinds are often associated with *Phellinus adamantinus*. This last is a species of wood-rot fungus identified for us by Leif Ryvarden (Botanical Institute University of Oslo) who affirmed that it is not usually pathogenic on living trees.

4.3. Climate

There are no local records of historical temperatures, but meteorological data from Taolagnaro (Fort Dauphin) indicate that since the 1950s annual temperatures have increased by 0.2 °C [37]. This small increase, together with highly variable local rainfall since the 1980s (Figure 1), suggests that if changes in forest cover are climate-related they are more likely linked to vegetation distribution and microclimatic variations than to any changes in regional climate.

River and water table levels vary seasonally and floods are likely to be important in the life cycle of the trees [38]. One of the growth variables for a riverine climate is distance from the river, with riverside vegetation providing shade, evapotranspiration and humid air that, cooling the microclimate, contrasts with the hot dry air over the open areas, the fields and the spiny forest fragments on Malaza's southern and eastern margins. In flood years the river can top its banks and flow over a meter deep across the forest floor (Figures 2 and 9) and although recurrent droughts can be disastrous for lemurs [13], droughts in terms of gallery forest survival are normal. In order to survive, many of the tree species have deep taproots, so that even when the river dries completely their roots can reach the saturated zone below the water table [39].

4.4. Growth Variables

Gallery forests survive as spatial and temporal mosaics at different successional stages, with complex inter-related patterns of cause and effect closely related to micro-environmental variations. The short transect of T2 was included to see if a line of dead trees made a difference to regeneration, but although *Capparis sepiaria* grows around the dead trunks, growth elsewhere is similar to that in the other stage 1 transects. Otherwise, growth variations are likely to be linked to minor topographic features, for although the forest floor is generally characterised as flat (Figure 2), former river-channels and sandbars have left numerous dips and gentle rises in the floodplain. These channels, containing more leaf-litter than elsewhere will, during high floods, retain moisture for longer than the flatter surroundings providing favourable seedbeds for species regeneration. This is illustrated by the high numbers of seedlings and saplings and few large trees in the T9 channel bed, with this strongly suggesting more frequent floods over recent decades than formerly (Figure 1). Growth in transect T15 along the other old channel further from the river does not appear to have been similarly advantaged.

An additional variable is provided on the western forest boundary by the planting of exotic species to provide cooling shade around the tourist facilities. The presence of this shelterbelt may be one reason why the neighbouring forest can be placed in stage 2 and has not declined to stage 3.

The 3–4 m wide forest trails may also be affecting the microclimate: Rambeloarivony and Jolly [13] remark that temperatures are noticeably higher by the wider trails than in the surrounding forest. In general, wherever the continuous canopy remains intact and 1 m wide trails through the forest are narrow and shaded, the penetration of hot air and light is reduced and vines, shrubs and grasses are largely excluded, but wide trails by providing channels for hot air may hasten forest fragmentation [40].

The generally high species presence in each transect, especially lower canopy species (Table 2), suggests that Malaza, left to regenerate on its own with the only intervention being *Cissus* removal, might eventually recover, but with a warming climate and under the pressure of drying as ageing trees die this could be a race against time.

4.5. Limitations

The choice of where to place transect lines and their lengths was partly controlled by the nature of the undergrowth; wide-scale clearance of prickly shrubs and strangling vines was impractical and thus the representation of the forest as a total ecosystem is incomplete.

Counts of the numbers of young trees in sampling circles were complicated by the growth of multiple stems in 'clumps' so that it was unclear if the stems were individual trees or shoots from a single initial stem. We counted all stems, but this could be a mistake since lemurs also eat seedling leaves possibly stimulating new shoots from the root-collar, with this subsequently buried by leaf litter and hidden by the expansion of the root system. The raised count of small trees in T1 may be due to this effect, but the similarities with the assessment data in the other stage 1 transects indicate that the findings are otherwise reasonably representative.

In Table 1, the addition of standard deviations shows the high degree of variability within transects. There was also high variability between circles in shrub and grass cover, but no standard deviations are given because it was impossible to provide realistic percentages due to patchy grass cover and the aerial cover of vines and lianas being often much greater than their ground cover.

A simple count of tamarind seedlings (Table 1) does not include seedling age; the limited size of seedlings suggests that they seldom thrive beyond a year (Figure 3), with this due to an unidentified negative effect. Whatever the cause, the death of tamarind seedlings under the dense canopy has implications for succession.

Cissus distribution is under-represented in the data. Although it was recorded wherever it occurred in the transects, we were unable to penetrate the heavily invaded areas and thus the data presented in Table 1, especially in stage 3 areas, do not give a true picture of its prevalence and for that reason *Cissus* is not included in Figure 7. Bare soil was also excluded as a factor of minor importance in Malaza.

Table 2 only shows the presence or absence of species, not their frequencies in each transect. Hence, to some extent this masks differences in forest character. For example, in Bealoka species diversity is similar to that in Malaza at T1 (excluding *Neotina* and *Celtis bifida*), but possibly due to grazing history, trees are more widely spaced resulting in this transect being placed at the lower limit of stage 2 (Table 1). The main difference is that in the Malaza bank-side forest mature tamarinds predominate compared with T4 in Ankoba and T11 in Bealoka where canopies are more open. However, the high number of tamarind seedlings in Bealoka suggests that, if this forest remains protected and is kept clear of *Cissus*, a continuous tamarind canopy might develop in time, with this signaled by the growth of tamarind saplings.

We classified tree growth in Table 2 as dryland, lower and upper canopy rather than ''pioneer'' and ''climax'' as recommended by Elliott et al. [18] (p. 46) since there is no information on the growth characteristics of the endemic species of the region. Hence, our definitions apply to habit and habitat rather than species position in succession.

4.6. Future Intervention and Work

In Malaza and Bealoka, the de Heaulme family have already established a programme of systematic clearance of *Cissus quadrangularis*, with work starting from *Cissus*-free areas and working towards the more infested areas. *Cissus* requires light and does not grow under a dense canopy. However, even in the 'clean' areas, these forests will need to be regularly searched to remove any new invasions since *Cissus* can re-sprout from the smallest fragments. In the long term, the battle can only be won by persistent clearance and the planting of trees to out-shade the vine.

Minimum intervention termed "accelerated natural regeneration" by Elliott et al. [18] (pp. 118–123) in stage 2 areas, could include weeding of grasses and clearance of competing shrubs. The list in Table 2 suggests areas where species are missing that could benefit from addition of seedlings. For example, tamarind (in T8), *Neotina* (T6, T7, T9, T10) and *Acacia* (T5, T6, T8, T9, T10). A further intervention that could moderate hot air invasion in Malaza could be to plant upper canopy species where these are lacking to shade the wider trails.

Other topics for future work could include a comparison of temperature differences beside wide and narrow trails in relation to vegetation and a controlled nursery study of the reasons for tamarind seedling inhibition under the tamarind canopy; measurement of seedling growth in the nurseries; and, if feasible, regular measurement of water table levels in the nursery water well for comparison with the precipitation data gathered at Berenty; and finally a more extensive survey of the forest's tree species and continuation of the specimen collection for future reference.

Large-scale reforestation in the degraded areas should be proceeded by an experimental planting trial lasting at least three years to investigate optimum planting distances and growing conditions, species inter-relationships and soil conditions for the native trees [18]. For the planting trial and eventual reforestation, two tree nurseries have been established with enough space to propagate tree seedlings: Table 2 provides guidelines as to the approximate percentages of seedlings to plant for reforestation.

5. Summarizing Remarks

Our 2014 survey enabled us to divide the forest into three vegetative stages: continuous canopy, transitional canopy and open dryland. The decline in forest canopy cover in Malaza, although largely protected from human interference, results from a combination of three major factors: senescence of tamarind, the dominant tree species; a lack of its regeneration under the canopy; and the invasion in open spaces of the exotic vine, *Cissus quadrangularis*.

- Firstly, almost 50% of the tamarinds in Malaza are nearing the accepted 200-year life span of the species, although this study finds that exceptionally advantaged specimens can live twice as long (Figure 5).
- Secondly, the loss of ageing tamarinds and lack of tamarind regeneration under the continuous canopy will lead to changes in vegetation character. Either lower-canopy species will predominate, with *Neotina*, *Acacia*, *Albitzia* and the occasional surviving tamarind providing a thin upper canopy (as in our Stage 2 zone); or

- thirdly, wherever there are large sunlit gaps *Cissus* will invade and overwhelm all seedlings and young trees, as is currently occurring in stage 3 areas and increasingly in stage 2 (Figure 6).

Table 1 illustrates the decline in canopy cover (average large plus small trees) across the three stages, with seedling survival particularly challenged in stage 3 areas. The high variability both within and between transects is likely related to microhabitat variations especially as regards dead tree locations, trails or former river channels, with T9 a particular example of an old channel with deep leaf litter that has recently become highly favourable for plant growth as witnessed by the exceptional number of saplings and seedlings, but few mature trees.

Table 2 supplies a list of the most common tree species for planting in a recommended three-year experimental planting trial. Such a trial is mandatory to avoid costly and time-consuming mistakes and to ensure that the integrity of the forest remains uncompromised before attempting any larger-scale reforestation. It is intended that the selection of common native trees will, in time, form a nucleus for the repopulation of the forest with ecological processes and functions matching those of the existing continuous canopy areas [18] (pp. 124–125).

Although riverbank erosion rates between 2.3 and 19.5 cm/year indicate that the forest is not immediately at risk from erosion, tamarind regeneration along the riverbank [41] provides a slim margin for re-growth at best, with seedlings and young trees caught between damaging flood events, embracing reeds and the negative effect of overhanging mature tamarinds (Figure 3). The problem is that the gallery forest at Berenty is limited to the existing area since riparian sites outside the reserve are used by the local people for transient cultivation and thus the forest cannot evolve along the riverbank over time. Hence, the urgent need for planting trials to provide a basis for forest restoration.

Soil characteristics, although variable, are supportive of plant growth. Sand predominates near the riverbank, where coarse particles are the first to be deposited by fast-flowing floodwaters, while the fine particles of silt and clay are held in suspension and deposited where flooding has lost its impetus over the western and southwestern reaches of Malaza. The open dryland areas to the east and around the central trail hold an intermediate soil position suggesting an intermediate position as regards flood deposition, with soil moisture levels reflecting soil composition rather than canopy cover.

According to Tadross et al. [37], climate warming in southern Madagascar is likely to increase, but modelling of temperature and precipitation trends is ill-constrained, being confounded by high inter-annual variability. It is foreseen that Indian Ocean warming is likely to result in fewer but more intense cyclones.

The data gathered in this study are intended to assist development of a restoration strategy in an ecoregion where there is very little information on species growth habits or requirements. If the dense canopy continues to shrink at the rate indicated by Blumenfeld-Jones et al. [8], in twenty years there may be very little forest left for the lemurs; ecotourism will collapse and the de Heaulme family, the local people and the wider world will lose a unique ecosystem.

Acknowledgments: Funding for Vanessa Winchester was provided by the Percy Sladen Memorial Fund and the Appleyard fund administered by the Linnean Society of London. The late Alison Jolly contributed to the expenses of Anne Mertl-Millhollen, Hantanarina Rasamimanana, Vanessa Winchester and the students Tolotra Randrianirina and Mirana Ratovolol from the École Normale Supérieure, Madagascar. James Tysell (USA), the students Camille Kaynor and Cassondra Gavazza (University of Oregon) and Dee Wallace (Portland State University), are all warmly thanked for their valiant help in the field as is Mona Edwards of the School of Geography and the Environment, Oxford, (UK) who carried out the soil analysis. Lastly, we thank the de Heaulme family for their support and encouragement during our work at Berenty. Two anonymous referees are thanked for their helpful comments.

Author Contributions: Vanessa Winchester conceived, researched and wrote the paper; Holger Gärtner sectioned and photographed the root samples, described the cell structures and assisted at the referee stage. Kate Hardwick and Anne Mertl-Millhollen, contributed ideas, context and wordsmithing. Hanta Rasamimanana, Sahoby Raharison and Janet McCrae all helped in the field work.

Conflicts of Interest: The authors declare no conflict of interest.

References

1. Waeber, P.O.; Wilmé, L.; Ramamonjisoa, B.; Garcia, C.; Rakotomalala, D.; Rabemananjara, Z.H.; Kull, C.A.; Ganzhorn, J.U.; Sorg, J.-P. Dry forests in Madagascar: Neglected and under pressure. *Int. For. Rev.* **2015**, *17*, 127–148. [CrossRef]
2. Vieira, D.L.M.; Scanot, A. Principles of natural regeneration of dry tropical forests for regeneration. *Restor. Ecol.* **2006**, *4*, 11–20. [CrossRef]
3. Aronson, J.; Vallauri, D.; Jaffré, T.; Lowry, I.I. PP 2005: Restoring dry tropical forests. In *Forest Restoration in Landscapes: Beyond Planting Trees*; Mansourian, S., Vallauri, D., Eds.; Springer Science + Business Media, Inc.: New York, NY, USA, 2005; pp. 285–290. [CrossRef]
4. Scales, I.R. The future of conservation and development in Madagascar: Time for a new paradigm? *Madag. Conserv. Dev.* **2014**, *9*, 5–12. [CrossRef]
5. Seddon, N.; Butchart, S.; Tobias, J.; Yount, J.W.; Rémi Ramanampamonjy, J.; Randrianizahana, H. Conservation issues and priorities in the Mikea Forest of south-west Madagascar. *Oryx* **2000**, *34*, 287–304. [CrossRef]
6. Sussman, R.W.; Rakotozafy, A. Plant Diversity and Structural Analysis of a Tropical Dry Forest in Southwestern Madagascar. *Biotropica* **1994**, *26*, 241–254. [CrossRef]
7. Kassi N'Dja, J.K.; Decocq, G. Successional patterns of plant species and community diversity in a semi-deciduous tropical forest under shifting cultivation. *J. Veg. Sci.* **2008**, *19*, 809–820. [CrossRef]
8. Blumenfeld-Jones, K.; Randriamboavonjy, T.M.; Williams, G.; Mertl-Millhollen, A.S.; Pinkus, A.; Rasamimanana, H. Tamarind recruitment and long-term stability in the gallery forest at Berenty, Madagascar. In *Ringtailed Lemur Biology*; Jolly, A., Sussman, R.W., Koyama, N., Rasamimanana, H., Eds.; Springer Science + Business Media, LLC: New York, NY, USA, 2006; pp. 69–85. ISBN 978-0-387-34126-2.
9. Jolly, A. Berenty Reserve, Madagascar: A long time in a small space. In *Long-Term Field Studies of Primates*; Kappeler, P.M., Watts, D.P., Eds.; Springer-Verlag: Berlin/Heidelberg, Germany, 2012; pp. 1–464. ISBN 978-3-642-22513-0.
10. Budnitz, N.; Dainis, K. Lemur catta: Ecology and Behaviour. In *Lemur Biology*; Tattersall, I., Sussman, R.W., Eds.; Plenum: New York, NY, USA, 1975; pp. 219–236.
11. Kuaraksa, C.; Elliott, S. The use of Asian *Ficus* species for restoring tropical forests ecosystems. *Restor. Ecol.* **2012**, *21*, 86–95. [CrossRef]
12. Fitchett, J.M.; Grab, S.W. A 66-year tropical cyclone record for south-east Africa: Temporal trends in a global context. *Int. J. Climatol.* **2014**, *34*, 3604–3615. [CrossRef]
13. Jolly, A.; Koyama, N.; Rasamimanana, H.; Crowley, H.; Williams, G. Berenty Reserve: A research site in southern Madagascar. In *Ringtailed Lemur Biology*; Jolly, A., Sussman, R.W., Koyama, N., Rasamimanana, H., Eds.; Springer Science + Business Media, LLC: New York, NY, USA, 2006; pp. 32–42. ISBN 978-0-387-34126-2.
14. Binggeli, P. Cactaceae, *Opuntia* spp., Prickly Pear. In *The Natural History of Madagascar*; Goodman, S.M., Benstead, J.P., Eds.; University of Chicago Press: Chicago, IL, USA, 2003; pp. 257–268.
15. Rambeloarivony, H.; Jolly, A. Berenty Reserve: Past, Present and Future. In *Leaping Ahead. Developments in Primatology: Progress and Prospects*; Masters, J., Gamba, M., Génin, F., Eds.; Springer Science + Business Media: New York, NY, USA, 2013; Volume 153, pp. 353–359.
16. Rinn, F. Eine neue Method zur Berechnung von Jahrringparametern. Auszug aus der Diplomarbeit. Ph.D. Thesis, Universität Heidelberg, Heidelberg, Germany, 1988; pp. 1–86.
17. Rinn, F. *TSAP-Win Time Series Analysis and Presentation for Dendrochronology and Related Applications*; Version 0.53 for Microsoft Windows; Rinn Tech.: Hedelberg, Germany, 2003; pp. 1–88.
18. Elliott, S.; Blakesley, D.; Hardwick, K. *Restoring Tropical Forests: A Practical Guide*; Kew Publishing: Royal Botanic Gardens, Kew, UK, 2013; pp. 72, 118–125. ISBN 978-1-84246-442-7. Available online: http://www.forru.org/en/content.php?mid=78 (accessed on 14 November 2017).
19. Harris, L.D. Edge effects and conservation of biotic diversity. *Conserv. Biol.* **1988**, *2*, 330–332. [CrossRef]
20. Mertl-Millhollen, A.S.; Rambeloarivony, H.; Miles, W.; Kaiser, V.A.; Gray, L.; Dorn, L.T.; Williams, G.; Rasamimanana, H. The influence of tamarind tree quality and quantity on *Lemur catta* behavior. In *Ringtailed Lemur Biology*; Jolly, A., Sussman, R.W., Koyama, N., Rasamimanana, H., Eds.; Springer Science + Business Media, LLC: New York, NY, USA, 2006; pp. 102–118. ISBN 978-0-387-34126-2.

21. Gärtner, H.; Schweingrube, F.H.; Dikau, R. Determination of erosion rates by analyzing structural changes in the growth pattern of exposed roots. *Dendrochronologia* **2001**, *19*, 81–91.
22. Gärtner, H. Tree roots. Methodological review and new development in dating and quantifying erosive processes. *Geomorphology* **2007**, *86*, 243–251. [CrossRef]
23. Gärtner, H.; Lucchinetti, S.; Schweingruber, F.H. New perspectives for wood anatomical analysis in Dendrosciences: The GSL1-microtome. *Dendrochronologia* **2014**, *32*, 47–51. [CrossRef]
24. Heiri, O.; Lotter, A.F.; Lemcke, G. Loss on ignition as a method for estimating organic and carbonate content in sediments: Reproducibility and comparability of results. *J. Paleolimnol.* **2001**, *25*, 101–110. [CrossRef]
25. Ryżak, M.; Bieganowski, A. Methodological aspects of determining soil particle-size distribution using the laser diffraction method. *J. Plant Nutr. Soil Sci.* **2011**, *174*, 624–633. [CrossRef]
26. Koyama, N.; Soma, T.; Ichino, S.; Takahata, Y. Home range of ringtail lemur troops. In *Ringtailed Lemur Biology*; Jolly, A., Sussman, R.W., Koyama, N., Rasamimanana, H., Eds.; Springer Science + Business Media, LLC: New York, NY, USA, 2006; pp. 86–101.
27. Rasamimanana, H.; Ratovonirina, J.A.; Jolly, A.; Pride, E. Storm Damage at Berenty Reserve. *Lemur News* **2000**, *5*, 7–8.
28. Hazelton, P.; Murphy, B. Interpreting soil test results: What do all the numbers mean? *Eur. J. Soil Sci.* **2007**, *58*, 1219–1220.
29. Parvez, S.S.; Parvez, M.M.; Fujii, Y.; Gemma, H. Allelopathic competence of *Tamarindus indica* L. root involved in plant growth regulation. *Plant Growth Regul.* **2003**, *41*, 139–148. [CrossRef]
30. Parvez, S.S.; Parvez, M.M.; Nishihara, E.; Gemma, H.; Fujii, Y. *Tamarindus indica* L. leaf is a source of allelopathic substance. *Plant Growth Regul.* **2003**, *40*, 107–115. [CrossRef]
31. Parvez, S.S.; Parvez, M.M.; Fujii, Y.; Gemma, H. Differential allelopathic expression of bark and seed of *Tamarindus indica* L. *Plant Growth Regul.* **2004**, *42*, 245–252. [CrossRef]
32. Mangan, S.A.; Schnitzer, S.A.; Herre, E.A.; Mack, K.M.; Valencia, M.C.; Sanchez, E.I.; Bever, J.D. Negative plant-soil feedback predicts tree-species relative abundance in a tropical forest. *Nature* **2010**, *446*, 752–755. [CrossRef] [PubMed]
33. Martínez-Izquierdo, L.; García, M.M.; Powers, J.S.; Schnitzer, S.A. Lianas suppress seedling growth and survival of 14 tree species in a Panamanian tropical forest. *Ecology* **2016**, *97*, 215–224. [CrossRef] [PubMed]
34. Addo-Fordjour, P.; Rahmad, Z.B.; Shahrul, A.M.S. Effects of human disturbance on liana community diversity and structure in a tropical rainforest, Malaysia: Implication for conservation. *J. Plant Ecol.* **2012**, *5*, 391–399. [CrossRef]
35. Crowley, H.M. *Berenty Reserve Management Plan*; Wildlife Preservation Trust International: Philadelphia, PA, USA, 1995.
36. Dietze, M.; Matthes, J.H. A general ecophysiological framework for modelling the impacts of pests and pathogens on forest ecosystems. *Ecol. Lett.* **2014**, *17*, 1418–1426. [CrossRef] [PubMed]
37. Tadross, M.; Randriamarolaza, L.; Rabefitia, Z.; Zheng, K.Y. *Climate Change in Madagascar; Recent Past and Future*; Climate Systems Analysis Group, University of Cape Town: South Africa; World Bank: Washington, DC, USA, 2008; pp. 1–17. Available online: http://www.csag.uct.ac.za/%7Emtadross/Madagascar%20Climate%20Report.pdf (accessed on 1 October 17).
38. Rasamimanana, N.; Ratsirarson, J.; Richard, A.F. Influence de la variabilité climatique sur la phénologie de la forêt de la Réserve Speciale de Beza Mahafaly. *Malagasy Nat.* **2012**, *6*, 67–82.
39. Duan, L.; Liu, T.; Wang, X.; Luo, Y.; Wang, W.; Liu, X. Water table fluctuation and its effects on vegetation in a semi-arid environment. *Hydrol. Earth Syst. Sci. Discuss.* **2011**, *8*, 3271–3304. [CrossRef]
40. Costello, J.F.; Leeder, J.; Strang, M. Drivers of the distribution of a dominant riparian tree species (*Eucalyptus coolabah*) on a dryland river system, Diamantina River, Lake Aire Basin. Paper 2667, Hydrology-ecology interactions, G9. In Proceedings of the 11th International Symposium on Ecohydraulics, Melborne, Australia, 7–12 February 2016; The University of Melbourne: Melborne, Australia, 2016.
41. Ranaivoson, T.; Brinkmann, K.; Rakouth, B.; Buerkert, A. Distribution, biomass and local importance of tamarind trees in south-western Madagascar. *Glob. Ecol. Conserv.* **2015**, *4*, 14–25. [CrossRef]

land

MDPI

Article

Land Cover Change in Northern Botswana: The Influence of Climate, Fire, and Elephants on Semi-Arid Savanna Woodlands

John Tyler Fox [1], Mark E. Vandewalle [2] and Kathleen A. Alexander [1,2,*]

[1] Department of Fish and Wildlife Conservation, Virginia Polytechnic Institute and State University, 1981 Kraft Drive (0321), Blacksburg, VA 24061, USA; jtylerfox@vt.edu

[2] CARACAL: Centre for Conservation of African Resources: Animals, Communities, and Land Use, Kasane, Botswana; caracal@btcmail.co.bw

* Correspondence: kathyalx@vt.edu; Tel.: +1-540-267-4947

Received: 28 July 2017; Accepted: 20 October 2017; Published: 25 October 2017

Abstract: Complex couplings and feedback among climate, fire, and herbivory drive short- and long-term patterns of land cover change (LCC) in savanna ecosystems. However, understanding of spatial and temporal LCC patterns in these environments is limited, particularly for semi-arid regions transitional between arid and more mesic climates. Here, we use post-classification analysis of Landsat TM (1990), ETM+ (2003), and OLI (2013) satellite imagery to classify and assess net and gross LCC for the Chobe District, a 21,000 km^2 area encompassing urban, peri-urban, rural, communally-managed (Chobe Enclave), and protected land (Chobe National Park, CNP, and six protected forest reserves). We then evaluate spatiotemporal patterns of LCC in relation to precipitation, fire detections (MCD14M, 2001–2013) from the Moderate Resolution Imaging Spectroradiometer (MODIS), and dry season elephant (*Loxodonta africana*) aerial survey data (2003, 2006, 2012, 2013). Woodland cover declined over the study period by 1514 km^2 (16.2% of initial class total), accompanied by expansion of shrubland (1305 km^2, 15.7%) and grassland (265 km^2, 20.3%). Net LCC differed importantly in protected areas, with higher woodland losses observed in forest reserves compared to the CNP. Loss of woodland was also higher in communally-managed land for the study period, despite gains from 2003–2013. Gross (class) changes were characterized by extensive exchange between woodland and shrubland during both time steps, and a large expansion of shrubland into grassland and bare ground from 2003–2013. MODIS active fire detections were highly variable from year to year and among the different protected areas, ranging from 1.8 fires*year^{-1}/km^2 in the Chobe Forest Reserve to 7.1 fires*year^{-1}/km^2 in the Kasane Forest Reserve Extension. Clustering and timing of dry season fires suggests that ignitions were predominately from anthropogenic sources. Annual fire count was significantly related to total annual rainfall ($p = 0.009$, adj. $R^2 = 0.50$), with a 41% increase in average fire occurrence in years when rainfall exceeded long-term mean annual precipitation (MAP). Loss of woodland was significantly associated with fire in locations experiencing 15 or more ignitions during the period 2001–2013 ($p = 0.024$). Although elephant-mediated damage is often cited as a major cause of woodland degradation in northern Botswana, we observed little evidence of unsustainable pressure on woodlands from growing elephant populations. Our data indicate broad-scale LCC processes in semi-arid savannas in Southern Africa are strongly coupled to environmental and anthropogenic forcings. Increased seasonal variability is likely to have important effects on the distribution of savanna plant communities due to climate-fire feedbacks. Long-term monitoring of LCC in these ecosystems is essential to improving land use planning and management strategies that protect biodiversity, as well as traditional cultures and livelihoods under future climate change scenarios for Southern Africa.

Keywords: Chobe; forest resources; ecosystem services; non-linear change; protected areas; disturbance; drought; sustainable livelihoods; ecotone; dryland; KAZA; Southern Africa

1. Introduction

Dryland ecosystems [1,2] occupy 41% of the global land surface and support nearly one third of the world's population, concentrated primarily in countries that share a significant portion of the global poverty burden [3–6]. Across Southern Africa, an estimated 150 million rural and urban residents support their livelihoods through extracting natural resources from semi-arid savannas [7], including firewood, food, fresh water, traditional medicine, thatching grass, reeds, poles, and other materials for building and crafts [8–10]. Scarcity of surface water and high rainfall variability make these systems particularly vulnerable to land degradation resulting in the potential loss of ecosystem services [11], with rural and impoverished communities being disproportionately impacted [12,13].

Savannas are defined by the coexistence of shade-intolerant and fire tolerant C4 grasses with tree cover maintained below climate-defined equilibrium values [14–16]. Proposed mechanisms for tree-grass coexistence in savannas fall broadly into two classes that emphasize either the role of niche partitioning and competition for resources [17–19] or demographic bottlenecks in tree recruitment mediated by disturbance [20–22]. Tree-grass ratios in savannas range from areas where grasses are dominant, to woodland savanna where tree cover may approach a closed-canopy state [17,23]. Water and nutrient limitations, together with individual and interactive effects of disturbance at different life-history stages appear central to tree persistence in savannas [15,22,24,25]. While rainfall and edaphic conditions tend to be the most important factors limiting maximum tree cover in arid savannas where mean annual precipitation (MAP) is less than 400 mm, in savannas where MAP exceeds 650 ± 134 mm, disturbance dynamics exert a greater influence on vegetation structure [21]. In these semi-arid and mesic savannas, the role of fire and herbivory becomes increasingly important for maintaining the coexistence of trees and grasses through positive feedback that prevents woodlands from attaining a closed canopy state [14,24].

The semi-arid savanna of northern Botswana provides an important example of a vegetation ecotone situated along a transitional zone between the arid (<400 mm MAP) Kalahari landscape to the south and the more mesic (>1000 mm MAP) Zambezian *Baikiaea* Woodland ecoregion to the north [26,27]. Unlike the Kalahari, northern Botswana possesses extensive woodlands composed primarily of Zambezi teak (*Baikiaea plurijuga*) and Mopane (*Colophospermum mopane*), commonly in association with *Pterocarpus angolensis*, and *Terminalia sericea*. These woodlands provide essential habitat for a wide range of taxa, along with important ecosystem services, including regulating decomposition processes and nutrient cycling, and improving water quality by limiting runoff of precipitation, soil, and contaminants [28–31].

Local land managers and communities frequently encounter varying and competing perceptions of land degradation, compounded by a lack of knowledge about the drivers and direction of LCC [32]. In Southern Africa, land degradation is frequently associated with changes in vegetation structure, particularly the loss of mature trees and encroachment of low-growing shrub species into savannas [33–37]. While expansion of cropland, livestock grazing, and urbanizing land use are frequently cited as driving LCC in sub-Saharan Africa [38–40], in northern Botswana wildlife and fire disturbance, together with human encroachment and climate change, are identified as the primary threats to woodland resources [41].

Remote sensing studies of LCC in northern Botswana have typically focused on relatively short time spans (<10 years), providing limited information about regional change dynamics and emergent management needs [42–50]. Global-scale LCC monitoring efforts such as the Global Land Cover Network's Africover project and the European Space Agency's Globcover mapping project are currently unavailable for much of Southern Africa or, where available, lack the necessary spatial resolution to detect important changes in land cover, information critical to resource management. Here, we use Landsat Thematic Mapper (TM), Enhanced Thematic Mapper Plus (ETM+), and Operational Land Imager (OLI) satellite images to classify multi-decadal (1990–2013) LCC and assess broad-scale changes in a semi-arid savanna across varying land uses and management. We evaluate LCC in relation to potential drivers including fire, climate, and growing elephant (Loxodonta africana) populations, and provide recommendations for future change monitoring and management needs in semi-arid savannas in Southern Africa.

2. Materials and Methods

2.1. Study Area

The Chobe District of northeastern Botswana (Figure 1) is a semi-arid dryland supporting a diverse assemblage of wildlife species and habitats of global conservation significance, including the largest elephant population in Africa [51,52]. The region has an aridity index between 0.20 and 0.50, calculated as the ratio between mean annual precipitation (MAP) and mean potential evapotranspiration [53], and is characterized by highly variable rainfall, with nearly the entire annual rainfall budget concentrated during the November—March wet season, followed by a general absence of precipitation from April-October. Wetter and drier periods frequently occur on the order of 5–10 years, with drought conditions associated with the strength of the prevailing El Niño-Southern Oscillation [54]. The Chobe River and its surrounding wetlands provide the only permanent source of surface water for the Chobe District. As a transboundary watershed, the Chobe River Basin is considered a core component of the Kavango-Zambezi (KAZA) Transfrontier Conservation Area, shared by Angola, Botswana, Namibia, Zambia, and Zimbabwe.

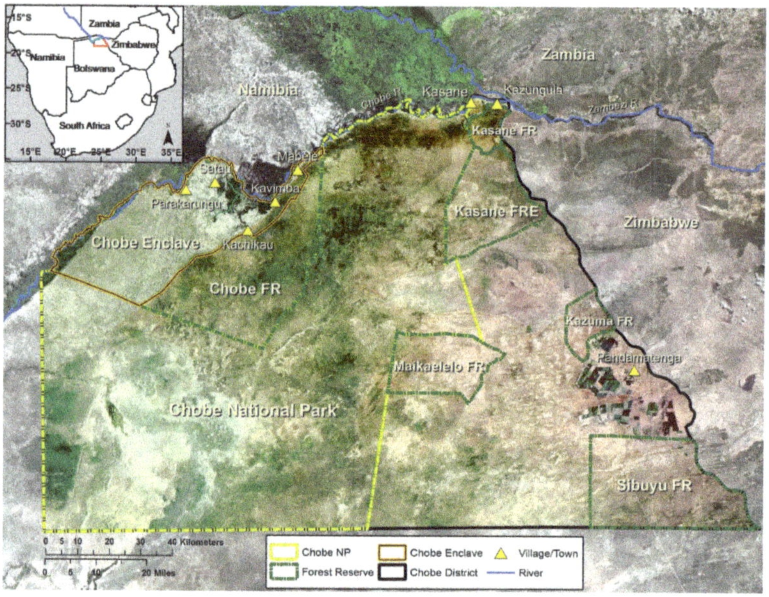

Figure 1. Landsat-based (2013 OLI) map of the Chobe District study area displayed using a natural color (4,3,2) band combination.

2.2. Land Use and Management

Land tenure in the Chobe District consists primarily of state land (national parks, forest reserves, and wildlife management areas), and tribal land (communal areas and commercial agriculture). Communal lands are typically utilized for subsistence livestock grazing and agriculture at a lower intensity than other areas in Botswana due to high wildlife conflict and barriers to cattle movement and production associated with endemic livestock diseases such as the virus that causes Foot-and-Mouth disease (*Aphthae epizooticae*). Protected areas account for approximately 58% of the District's total land area, of which, the Chobe National Park (CNP) occupies 11,700 km^2 (see Figure 1). Six protected forest reserves and extensions are located in the study site: Kazuma, Maikaelelo, Sibuyu, Chobe, Kasane Forest Reserve, and the Kasane Forest Reserve Extension, occupying a combined land area of approximately 4100 km^2. The town of Kasane is the regional government seat (est. 2011 population 9008) and, together with Kazungula (est. pop. 4133), are the largest urban settlements in the District [55]. The communally-managed Chobe Enclave (1690 km^2, Figure 1), located along the District's northern border with Namibia, is surrounded on all sides by protected areas. The Chobe River floodplain occupies a large portion of the Enclave, with forested areas confined to higher elevations. Human population density is generally low in the Enclave's five main villages, with a combined population of 7500 [55], approximately 30% of whom work outside of the Enclave [8]. Crop and livestock production and wage employment are the primary economic activities, with a majority of households relying on extraction of natural resources for subsistence purposes [10].

2.3. Data Sources

Landsat satellite images of the Chobe District were obtained from the U.S. Geological Survey with a spatial resolution of 30 m for visible, shortwave infrared (SWIR), and near infrared (NIR) bands, for the years 1990 (Landsat 5 Thematic Mapper; TM), 2003 (Landsat 7 Enhanced Thematic Mapper Plus; ETM+), and 2013 (Landsat 8 Operational Land Imager; OLI) (Figure 2). A decadal revisit period was chosen in order to focus on broad-scale LCC in the study area during a period of rapid growth of both elephant [52] and human [55] populations in the District. This time period also includes the implementation of Botswana's National Conservation Strategy (1990), which outlines fundamental ecosystem conservation goals for the country [56]. Timing of green-up and leaf-fall is highly variable in northern Botswana, so we limited our classifications to cloud-free images acquired from 8–28 April during each year of the analysis to minimize seasonal effects on land cover classifications (Table S1). Monthly active fire detections from the Moderate Resolution Imaging Spectroradiometer (MODIS; MCD14ML Collection 5.1) available for the period 1 January 2001–31 December 2013 were used to evaluate fire dynamics in the Chobe District. The MODIS sensor detects fires burning in 1 km pixels using a contextual algorithm to apply thresholds to observed mid-infrared and thermal infrared brightness temperature, and rejects false detections by comparing them to values from neighboring pixels [57]. Only fire pixels assigned high-confidence (>75%) were included in the analysis in order to exclude potential false detections [57–59].

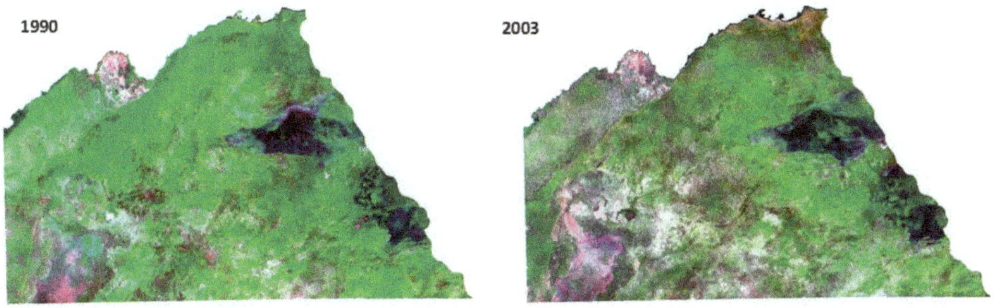

Figure 2. *Cont.*

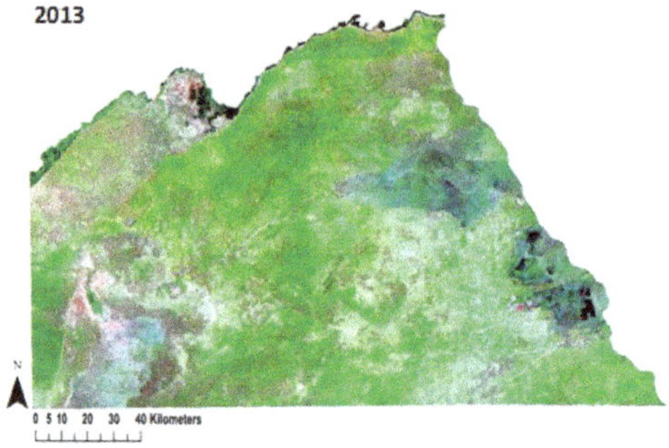

Figure 2. Landsat mosaic images of the study area taken 8–28 April in each year and displayed using a false color 7,4,2 band combination. Healthy vegetation appears green with darker hues indicating forested areas. Water is black to dark blue and pink and magenta indicate bare ground, while sand, soil, and sparsely-vegetated areas appear in lighter shades of pink and tan.

Climate and Wildlife Aerial Survey Data

Daily precipitation records measured at the Kasane Meteorological Station were provided by the Botswana Department of Meteorological Services for the years 1922–2013 and were used to evaluate rainfall patterns during the study period. Dry season aerial wildlife survey data provided by the Botswana Department of Wildlife and National Parks (DWNP) for 2003, 2006, 2012, and 2013 were used to examine spatial associations between elephants and land cover. Surveys were flown in late August following a stratified systematic transect sampling design and were analyzed using Jolly's method for sampling units of unequal size [60]. Elephant biomass is represented as Large Stock Units (LSU)/km^2, where one LSU is equal to the metabolic equivalent of a 454 kg cow [52,61]. Land cover values in the 2003 and 2013 classification maps were extracted within aerial survey units from the corresponding year to determine elephant presence in the different land cover classes.

2.4. Satellite Imagery Pre-Processing

Landsat images were standardized onto a common geographic and radiometric scale, and projected to Universal Transverse Mercator (UTM) WGS 84 Zone 34 South prior to classification and analysis. Landsat TM and ETM+ images were processed to surface reflectance using the Ecosystem Disturbance Adaptive Processing System (LEDAPS), while Landsat OLI surface reflectance was calculated using the USGS L8SR algorithm. Multi-spectral data often possess a high degree of correlation among adjacent spectral bands, particularly in semi-arid savanna systems where environmental and ecological variability create a mosaic landscape of diverse plant community assemblages [62]. In this study, we used principal components analysis (PCA) to transform the original multi-band Landsat images to reduce redundancy in spectral data in preparation for land cover classification and change detection. The first three principal components contained >98% of the original information in the multi-spectral datasets for 1990, 2003, and 2013, and were used for all classifications.

2.5. Land Cover Classification

We initially classified land cover using an unsupervised Interactive Self-Organizing Clustering (ISO Cluster) approach in ESRI ArcMap 10.2 (Environmental Systems Research Institute, Redlands, CA, USA). Image pixels were grouped into 45 spectral clusters using a sampling interval of 25 cells and a minimum class size of 150 cells, with a 3 × 3 majority filter applied to reduce noise and outliers. Spectral clusters from the unsupervised classification were then assigned to one of six general land cover classes (Table 1) using a supervised classification approach based on characteristics defined by the Food and

Agriculture Organization (FAO) of the United Nations 2010 Forest Resources Assessment (FRA) [63]. Final class assignments were based on examination of current and historical aerial orthophotography and expert knowledge of the study area and local conditions. Any obvious localized misclassifications were corrected prior to merging polygons of the same grid value in preparation for post-classification change analysis.

Table 1. Land-cover classifications based on 2010 FAO Forest Resources Assessment designations.

Class	Description
Water	Perennial and seasonally-flooded river channels, water-filled pans, reservoirs, and watering holes
Wet/irrigated vegetation	Permanent and seasonally-inundated wetland, floodplain, and riparian vegetation; irrigated agricultural land, and water-saturated vegetation surrounding ephemeral water pans.
Grassland	Land cover dominated by grasses with <10% tree and shrub cover.
Woodland	Land with tree cover ranging from open (>10%) tree savanna intermixed with a grass or shrub understory, to closed-canopy (>40% cover) dry deciduous forest.
Shrubland	Land cover dominated by mixed, short woody vegetation (usually <3 m tall) ranging from open shrub savanna (>10% cover) with a discontinuous understory cover of annual grasses, herbs, and occasional tall trees, to thickets (>50%) with little or no grass cover.
Bare/impervious	Natural sandy and bare soils; rock outcrops; and impervious surfaces including dirt and tar roads, parking lots, and rooftops.

2.6. Classification Accuracy Assessment

We assessed classification accuracy for 2013 land cover assignments using standard confusion matrix cross-tabulation methods [64]. A total of 928 reference sites were selected using a random point generator and existing cover was validated using a combination of field-based ground-truthing conducted in July 2015 and high-resolution orthophotography from 2013. As we were unable to assess classification accuracy for 1990 and 2003 directly due to a lack of suitable reference data, the 2013 training site data were used to calculate proportional error for each class by dividing the number of misclassified cells by land cover contingency table row totals for each time step. Proportional errors for each cover class were then multiplied by the nominal class areas, and summed along each column to derive an adjustment representing upper (omission) and lower (commission) classification uncertainty intervals. We calculated overall classification accuracy with binomial 95% confidence intervals for net LCC using the "caret" package in R by dividing the total number of correctly identified reference points by the total number of sample units in the matrix [64]. Individual class accuracies were computed in a similar manner and include measures of: sensitivity (producer accuracy), specificity (true negatives), detection rate and detection prevalence, and positive predictive (user accuracy) and negative predictive values [65]. Quantity disagreement and allocation disagreement between the 2013 land cover classification and reference data were calculated following methods described by Pontius Jr and Millones [66], along with omission and commission error for each class. Quantity disagreement derives from a proportional difference between the number of cells of a particular land class in two maps, while allocation is the additional disagreement resulting from a less than optimal match in the spatial allocation of the categories [67]. We also include an unweighted Cohen's Kappa statistic describing agreement of categorical data relative to what would be expected by chance, with values of 1 indicating perfect alignment.

2.7. Change Detection

Net and gross LCC were estimated for the Chobe District and communally-managed Chobe Enclave, and for seven protected areas: Chobe National Park, Sibuyu, Maikaelelo, Kazuma, Chobe, Kasane Forest Reserve (FR), and the Kasane Forest Reserve Extension (FRE) using change detection analysis in ENVI version 4.8 (Exelis Visual Information Solutions, Boulder, CO, USA) and in ArcMap v.10.2. Images were coregistered (UTM zone 34S, Datum WGS-84) prior to analysis and for each pixel in the initial state classification the land cover class in the final state image was recorded.

2.8. Statistical Analysis

All statistical analysis was conducted in the R open source integrated programming environment [68] unless otherwise noted. A two sample *t*-test of equal variances was used to test the hypothesis that mean fire occurrence was significantly different in alternating (even and odd) years, and ordinary least squares (OLS) regression was used to analyze the association between total annual rainfall and fire count. We evaluated the influence of elephants on woodland loss by quantifying the total number of pixels within aerial survey blocks from 2003, 2006, and 2012 that changed from woodland to grassland, shrubland, or bare/impervious from 2003–2013. Woodland loss values were normalized by survey block area (km^2) and elephant data for the three years were grouped into low (<10), medium (10.1–20), and high (>20) elephant biomass (LSU/km^2) for analysis. We assessed the relationship between woodland loss and fire in a similar manner by calculating the total number of woodland loss pixels within a 1 km radius buffer around MODIS active fire detections. Fire counts were grouped into low ($\leq$5), medium (6–14), and high ($\geq$15) classes for analysis.

We initially fit OLS regressions and tested for the presence of spatial autocorrelation in model residuals using a Global Moran's I statistic. Significant spatial autocorrelation was present in OLS model residuals comparing woodland loss and fire count (0.014, $p < 0.0001$) and woodland loss and elephant biomass (0.273, $p < 0.0001$). As result, we fit generalized least squares (GLS) regressions to these data in the R package "nmle" [69] using five autocorrelation structures: exponential, Gaussian, spherical, linear, and rational quadratic. The GLS methodology has been extensively applied to analyze and account for the influence of spatial correlation in a wide range of ecological data [70]. Akaike Information Criterion (AIC) estimation was used to select the best performing model based on lowest delta AIC. A square root transformation was applied to the woodland loss data to satisfy assumptions of normality and verified using normal qqplots.

In addition, we conducted an optimized hot spot analysis of fire clusters using Getis-Ord Gi* in ArcGIS (v. 10.3) following integration and event collection of fires occurring within a 10 km-radius neighborhood. Getis-Ord Gi* identifies statistically significant spatial clusters of high values (hot spots) or low values (cold spots) based on z-scores and *p*-values for each feature. Fires in the neighboring countries of Namibia, Zambia, and Zimbabwe were included in the hot spot analysis to assess potential fire spread into the Chobe District across international borders.

3. Results

3.1. Land Cover Classification Accuracy

Spectral clusters from the unsupervised classification were assigned to one of the six general land cover classes according to the definitions in Table 1. Overall disagreement between 2013 land cover classifications and the training data was 13%, the majority of which was due to differences in class allocation (Figure 3a). Omission error was higher than commission error for grassland and bare/impervious cover, indicating a tendency to underestimate these classes across the study area domain (Figure 3b). A full accounting of contingency table results for land cover classification accuracy compared to reference data and a summary of accuracy assessment statistics are included in Tables S2 and S3, respectively. Overall classification accuracy was 0.867 (95% CI 0.843, 0.888) with a Kappa coefficient of 0.832, indicating generally low misclassification error, with the highest confusion arising between woodland and shrubland.

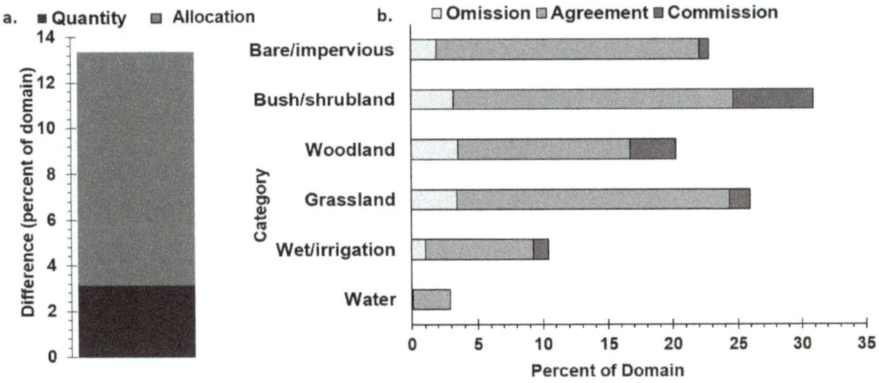

Figure 3. (**a**) Summary of overall disagreement between 2013 land cover classification and reference data, separated into the components of quantity and allocation. (**b**) Summary of agreement, omission disagreement, and commission disagreement between 2013 land cover classification and reference data by cover class as a percent of the study area domain.

3.2. *Summary of Net LCC*

3.2.1. Chobe District

The dynamic nature of LCC in northern Botswana is illustrated in thematic land cover maps for 1990, 2003, and 2013 (Figure 4). Net LCC during the study period is summarized in Figure 5 and was characterized by loss of woodland (1514 km^2, 16%) and expansion of shrubland (1304 km^2, 16%). Locations of woodland gains and losses from 1990–2013 are shown in Figure 6. Woodland losses during the first time step were followed by a small net gain from 2003–2013 (143 km^2, 2%), although overlap of uncertainty intervals for woodland indicate this gain may be attributable to errors in classification. Shrubland losses from 1990–2003 were followed by a large net gain of 2189 km^2 (29.4%) from 2003–2013, with shrubland becoming the dominant land cover in the Chobe District by 2013. Grassland gained 265 km^2 (20%) over the study period, despite a net loss of 1321 km^2 (45%) from 2003–2013, with the largest changes observed in the southwestern corner of the District around the Savuti Marsh, and in Mokororo Pan located to the south of Kasane FRE. Bare/impervious gains during the first time step were followed by a loss of 1023 km^2 (40%) from 2003–2013, with large reductions in bare ground around Savuti Marsh and in the Chobe Enclave. Water and wet/irrigated vegetation cover varied considerably, with the majority of change associated with seasonal flooding of the Chobe River and irrigation of commercial agricultural fields in southeastern portion of the District around the village of Pandamatenga.

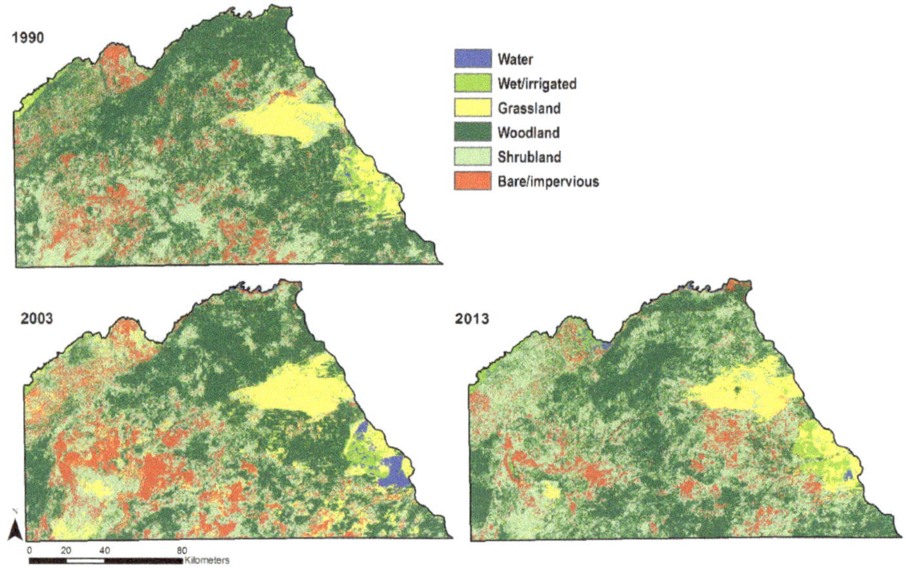

Figure 4. Land cover classification maps of Chobe District for 1990, 2003, and 2013.

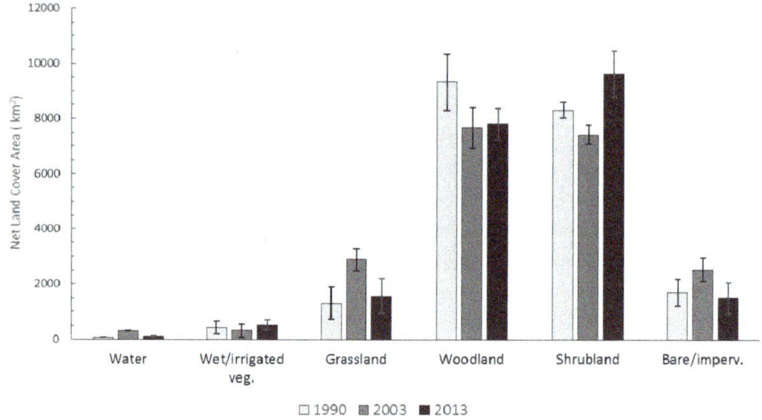

Figure 5. Estimated net area (km^2) for Chobe District land cover in 1990, 2003, and 2013. Bars show estimated error calculated using field and training site data. Estimated errors for consecutive time steps were less than net change for all classes except wet/irrigated in all three years, and woodland in 2003 and 2013.

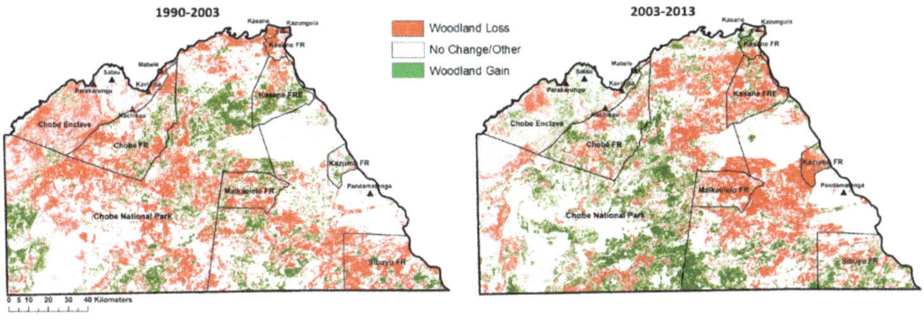

Figure 6. Maps of woodland change from 1990–2003 and 2003–2013. Woodland losses are shown in red and gains in green. White indicates areas of no change or land cover other than woodland.

3.2.2. Protected Areas and Communal Lands in the Chobe Enclave

Net LCC in protected areas and the Chobe Enclave are summarized in Figure 7. Woodland declined in five of the six protected forest reserves from 1990–2003, with net losses averaging 31%. While only three protected areas lost woodland from 2003–2013, average declines were 58%, driven by high losses in Maikaelelo (190 km^2, 70%) and Kazuma (91 km^2, 65%) FRs. In the CNP, woodland losses from 1990–2003 (618 km^2, 15%) were followed by a gain of 703 km^2 (20%) from 2003–2013. Shrubland expanded in all of the protected areas over the study period, with the smallest gains observed in the CNP and the largest gains in Kasane and Kazuma FR. Grassland also increased over the study period in a majority of the protected areas, with all but Kasane FR gaining grassland from 1990–2003 and losing grassland from 2003–2013. Kasane and Maikaelelo FR experienced the greatest increases in bare/impervious cover over the study period. Changes in water and wet/irrigated vegetation in the Sibuyu FR were related to seasonal flooding of nearby ephemeral pans along with flood irrigation in the agricultural fields bordering the reserve to the north.

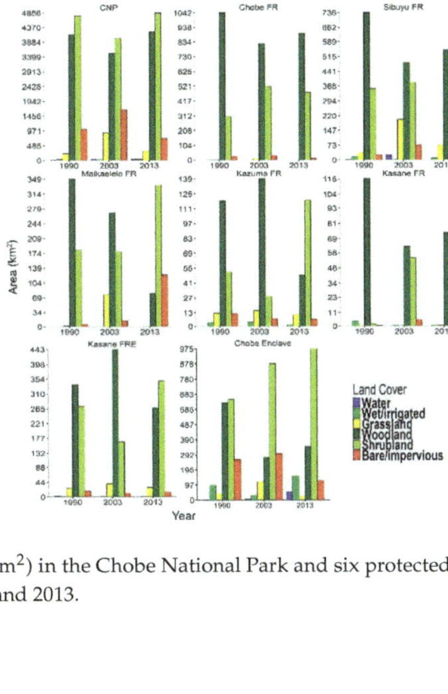

Figure 7. Land cover (km^2) in the Chobe National Park and six protected forest reserves, the Chobe Enclave for 1990, 2003, and 2013.

Land cover in the communally-managed Chobe Enclave changed from nearly equal representation of woodland and shrubland in 1990, to being shrubland-dominated in 2013. A large woodland loss from 1990–2003 (351 km^2, 56%) was followed by a gain of 71 km^2 (26%) from 2003–2013. In contrast, both grassland and bare/impervious cover expanded from 1990–2003, but subsequently declined from 2003–2013. Changes in water and wet/irrigated vegetation in the Enclave were primarily due to fluctuations in seasonal flooding of the Chobe River and associated wetlands.

3.3. Summary of Gross LCC

Estimates of gross change among the different cover classes highlight the complex and dynamic nature of LCC across the Chobe District (Table S4). The magnitude and direction of gross percent change among woodland, shrubland, grassland, and bare/impervious cover from 1990–2003 and 2003–2013 are summarized in Figure 8. Over the study period, woodland lost 4317 km^2 to shrubland, while gaining an estimated 2788 km^2 from shrubland in return (Table S4). Extensive woodland regeneration occurred in areas of bare ground, with 237 km^2 (15%) of bare/impervious changing woodland from 1990–2003 and 423 km^2 (17%) changing to woodland from 2003–2013. Grassland gains from 1990–2003 were driven predominately by the loss of shrubland (1088 km^2, 13%) and bare/impervious (229 km^2, 14%) cover, while shrubland encroachment drove grassland losses from 2003–2013. Grassland also lost 395 km^2 (14%) to woodland from 2003–2013. Shrubland gains intensively targeted bare ground, with around half of all bare/impervious losses attributable to expansion of shrubland during both time steps. Bare/impervious gains from 1990–2003, on the other hand, were predominately driven by loss of shrubland. Although not shown in Figure 8, the majority of change from water was to wet/irrigated vegetation, while wet/irrigated vegetation predominately changed to grassland Table S4).

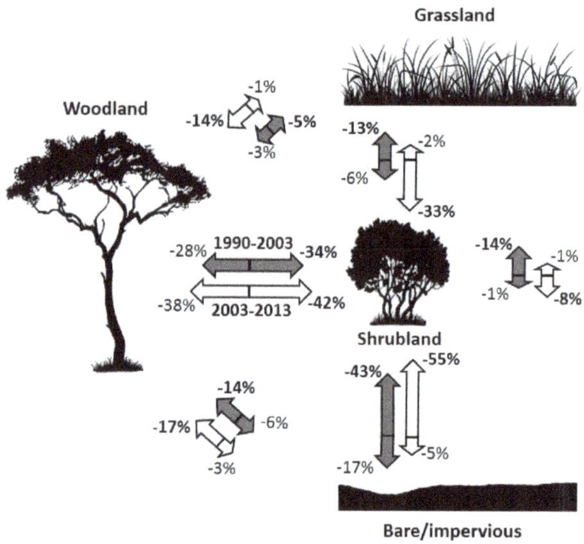

Figure 8. Dynamics of gross change among woodland, shrubland, grassland, and bare/impervious land cover in the Chobe District for 1990–2003 (gray arrows) and 2003–2013 (white arrows), represented as percent losses. Arrows indicate the direction of class losses, with arrow length scaled by change magnitude and largest transitions shown in bold. For example, from 1990–2003 34% of all woodland losses were to shrubland, while 28% of all shrubland losses were to woodland.

3.4. Rainfall

Analysis of precipitation records showed long-term (1922–2013) mean annual precipitation (MAP) in Kasane was 632 mm ± 181 mm with total annual rainfall ranging from 301–1205 mm. For the study period, MAP was 562 mm ± 177 mm. While not statistically significant, MAP differed importantly between the two time steps and boxplots of rainfall for the different time steps show the larger interquartile range from 2003–2013 (Figure 9). Between 1990 and 2003, total annual rainfall only exceeded the long-term MAP in 1991, compared to four years of above average rainfall from 2003–2013 (Figure 9).

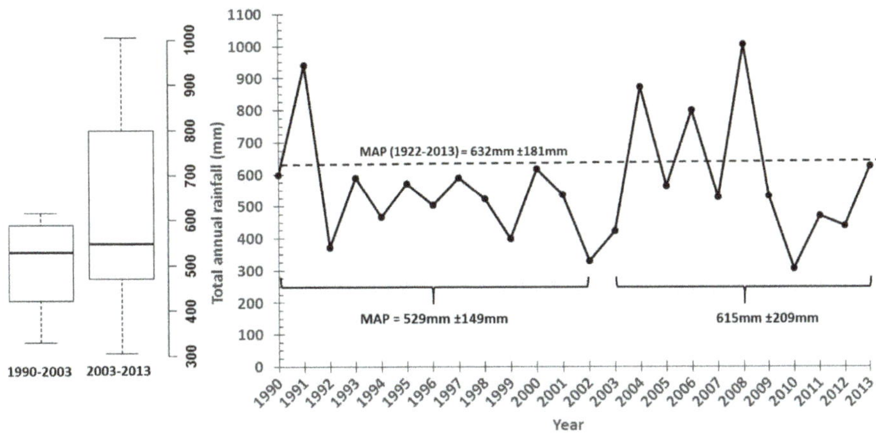

Figure 9. Total annual rainfall (mm) for the study period with long term (1922–2013) mean annual precipitation (MAP) and MAP for each time step with standard deviations shown, along with boxplots of total annual rainfall for the different time steps.

3.5. Spatiotemporal Analysis of Fire, Rainfall, and Woodland Loss

Fire was almost entirely a dry season phenomenon in the Chobe District with activity peaking during the period August-October (Figure 10a) when more than 85% of fires in the study area were detected. In contrast, only 11 fires were detected during the wet season from December-March. The timing and location of fires prior to the arrival of convective storms typically associated with lightning production, suggests the majority of fire ignitions were anthropogenic in origin. While the CNP had the most ignitions of any protected area, it had comparatively few fires per year by total land area (Table 2), while Kasane FRE and Kazuma FR had the highest fire incidence.

Fire counts were highly variable from year to year, with more active fires recorded in years with higher rainfall (Figure 10b). OLS regression comparing annual fire count and total annual rainfall over the period 2001–2013 was not significant at the alpha 0.05 level; however, regression residual versus leverage plots indicated 2010 and 2011 as potentially influential outliers, and the fitted regression model excluding these two years was significant (F(1,9) = 10.9, 95% CI (0.388, 2.08), $p = 0.009$) with an adjusted $R^2 = 0.50$, compared to an adjusted $R^2 = 0.04$ for the full model. High fire and low fire years tended to alternate biennially, with a significantly greater number of fires in even years (excluding 2010 and 2011) (df = 9, $t = 2.78$, $p = 0.02$). Higher total annual precipitation was associated with a 41% increase in mean fire occurrence in years when rainfall totals exceeded the long-term MAP of 632 mm.

Comparing the spatial distributions of woodland loss and fire data shows woodland loss increased from low to high fire groups (Figure 10c). However, gls regression results show woodland loss was significantly related to fire ($p = 0.024$) only in locations experiencing more than 15 fires during the observation period (Table 3).

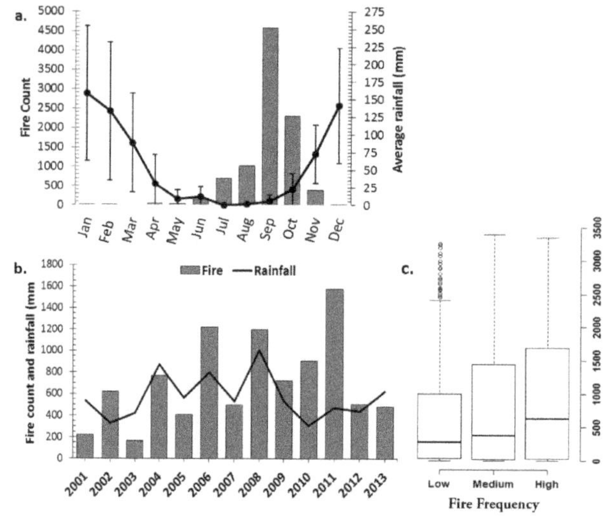

Figure 10. (**a**) MODIS fire detections for the Chobe District from 2001–2013 by month (n = 9288, s = 1376) with long-term (1922–2013) average total monthly rainfall (mm) with standard deviations (**b**) total annual rainfall (mm) and active fires in the Chobe District from 2001–2013. (**c**) boxplots of woodland loss and fire frequency grouped by low (≤5), medium (6–14), and high (≥15) classes.

Table 2. Summary of fire frequency for the Chobe District, Enclave, and protected areas. Mean detection confidence levels for each location are included, along with an area-weighted annual fire index (AFI), calculated by dividing average annual fire frequency by total land area (fires*year^{-1}/km^2), representing a straightforward way of comparing fire incidence in analysis units of varying size.

Location	Area (km^2)	Mean Detection Confidence	Total Fires	Average Annual Fires	AFI
Chobe District	21,153	91	9288	714	3.4
Chobe Enclave	1677	89	1128	87	5.2
Protected Area					
Chobe NP	10,168	92	2835	218	2.1
Chobe FR	1382	89	326	25	1.8
Sibuyu FR	1169	92	726	56	4.8
Kasane FRE	660	90	613	47	7.1
Maikaelelo FR	538	91	310	24	4.4
Kazuma FR	194	91	151	12	6.0
Kasane FR	123	92	33	3	2.1

Table 3. Generalized least squares regression results.

Variable	Woodland Loss	
Intercept	10.77 **	19.94 **
df	846	1257
Medium elephant biomass	0.002	-
High elephant biomass	0.008	-
Medium fire	-	1.58
High fire	-	**5.31**
Variogram Form	Rational quadratic	Exponential
Range	0.002	0.03
Nugget	<0.01	0.66
Residual standard error	6.49	17.07

Significance of variable coefficients: $p \leq 0.001$ **, $\boldsymbol{p \leq 0.05}$.

3.6. Geospatial Analysis of Fire Hot Spots

Fire density was high in unprotected land between the Kazuma and Maikaelelo FRs with significant fire hot spots clustered temporally and spatially in areas where harvest of thatch materials is permitted from September to October (Figure 11). Another large cluster of fire hot spots was centered near the agricultural fields of Pandamatenga and were associated with clearing of agricultural land. Fire density was also high in several protected areas (Kazuma, Maikaelelo, and Kasane FRE), but was generally low in the central and southern CNP and Chobe FR, with numerous fire cold spots indicating significantly fewer fires than predicted compared to neighboring land. Fire hot spots along the Botswana side of the Chobe River were predominately confined to within ~5 km of the river course and decreased in number towards the interior of the Enclave. In contrast, significant fire hot spots on the Namibian side of the river extended into the Bamunu and other communal land conservancies surrounding Mudumu National Park in Namibia, but declined abruptly along the western border of Salambala Conservancy, as well as in neighboring portions of the Enclave.

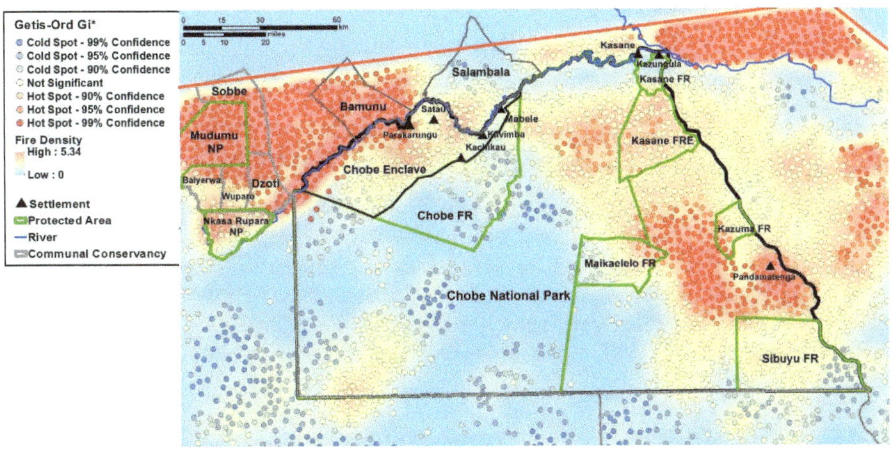

Figure 11. MODIS active fires (2001–2013, n = 33,631) for Chobe District and neighboring countries, showing kernel density (fires/10 km^2) overlaid with Getis-Ord Gi* hot and cold spot confidence level bins. The red line marks the spatial extent of MODIS fire data used in the analysis.

3.7. Spatial Analysis of Woodland Loss in Relation to Elephant Biomass in Aerial Survey Units

Associations between elephants in dry season aerial survey units and 2003 and 2013 land cover are shown in Figure 12a. Survey units with elephants totaled 5077 km^2 in 2003 and 5452 km^2 in 2013. Elephants were predominantly associated with woodland and shrubland in both 2003 and 2013 (Figure 12a), however, we found no significant difference in woodland loss among survey blocks with low, medium, or high elephant biomass (Table 3), despite evidence of elevated woodland loss in survey blocks with high elephant biomass (Figure 12b).

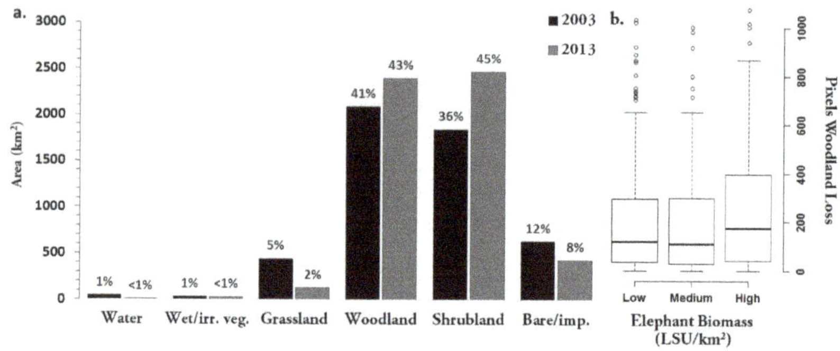

Figure 12. (**a**) Area (km^2) and percent land cover in 2003 and 2013 aerial survey units with elephants (**b**) boxplots of elephant biomass (LSU/km^2) in aerial survey blocks from 2003, 2006, and 2012 grouped into low (<10), medium (10.1–20), and high (>20) classes vs. woodland loss from 2003–2013.

4. Discussion

4.1. Chobe District Net and Gross LCC

Spatial distributions of tree, shrub, and grass communities varied considerably in the climatically unpredictable, semi-arid savanna of northern Botswana, influenced by interdependent environmental and anthropogenic forcings over decadal time periods. Rapid land cover transitions within savanna biomes can be driven by complex nonlinear feedbacks between biophysical and human systems at local, regional, and even global scales [14,71]. For example, recent evidence suggests increasing levels of atmospheric CO_2 may play a significant role in the expansion of woody plants in savannas by favoring regrowth following injury from fire or browsing [72,73] and inducing plant water-saving [74]. Changes in fire frequency due to increasing rainfall variability related to global climate warming, as well as human activity, may greatly impact the distribution of trees and grasses in Southern African savannas [54,75].

Our analysis found net LCC in the Chobe District from 1990–2013 was characterized by a decrease in woodland cover accompanied by expansion of shrubland. The observed woodland decline of 16% (1514 km^2) was in line with United Nation's FAO estimates that Botswana lost approximately 20% of its forested area from 1990–2015 [76]. Woodland declined at a faster rate across the District from 1990–2003 compared to 2003–2013, driven primarily by losses to shrubland (Figure 8). Shrubland underwent extensive expansion during the second time step, with the majority of gains occurring in bare/impervious and grassland (Figure 8). In contrast, grassland gained across the District during the first time step when conditions were drier, but subsequently lost 1321 km^2 (942 km^2, 33% to shrubland) from 2003–2013 (Figure 8, Table S4). Bare/impervious cover also varied considerably over the different years of the analysis. Large bare/impervious gains from 1990–2003 in the southwest of the study area around the Savuti marsh were followed by a large decrease from 2003–2013, likely associated with changes in water flow into the Savuti channel and surrounding wetlands, which only began receiving water again in 2010 after an extended dry period [77]. Bare/impervious gains around the towns of Kasane and Kazungula in the northeast of the District, on the other hand, were associated with urban expansion and infrastructure development driven by a 32% increase in population between 2001 and 2011 [55]. Net changes in open water and wet/irrigated vegetation were relatively small compared to total District land area, but were highly variable and primarily related to seasonal flooding of the Chobe River and associated wetlands.

Gross (class) changes at the landscape level were dominated by exchange between woodland and shrubland (Figure 8, Table S4), further illustrating the nonlinear nature of LCC in the study area. Exchange occurs when a pair of pixels is classified as category "A" in the first map and as category

"B" in the second map, and vice-versa [67]. Woodland losses to shrubland were higher during both time steps, although the rate of exchange among the two classes increased by approximately 10% from 2003–2013 when conditions were wetter. Such high rates of exchange between shrubland and woodland over decadal time periods may be a common feature of transitional savannas situated between arid and more mesic climate zones. In semi-arid savanna, woody vegetation growing on nutrient-poor soils maintains a large fraction of biomass below ground [78,79]. *C. mopane* in particular, which is widespread in the south of the Chobe District, may coppice or persist in low shrub form under unfavorable drought conditions or following disturbance, quickly returning to a taller, more closed-canopy state once conditions improve [79]. Observations of woody vegetation growing on Kalahari sand in the study area have noted large areas of shrub savanna have developed into woodland since the 1990s [42,48]. Despite a net loss from 1990–2003, our results indicate woodlands in the Chobe District may undergo expansion even during prevailing drought conditions in locations where fire disturbance remains low to moderate. Woodland gains in bare/impervious cover provide additional evidence of the rate of successful establishment of tree seedlings following fire or other significant disturbance, with 15% of bare ground becoming woodland on average during each 10-year time step (Figure 8).

4.2. LCC in Protected Areas

Woodland losses were generally higher in the protected forest reserves (FRs) compared to the Chobe District and Chobe National Park (CNP), where woodland grew by 20% (702 km^2) from 1990–2013, possibly reflecting differences in natural resource management strategies. For example, while the CNP is considered a "no access" area in regards to natural resource extraction, limited subsistence and commercial harvesting of forest resources is allowed in FRs through a government permit system. Similar results were observed for protected areas in East Africa, where only national parks increased their forest area between 2001 and 2009, while forest reserves, nature reserves, and game parks were all more likely to lose forest cover [80]. Wilcock et al. [81], in contrast, found that deforestation in Tanzania shifted to net increases in forest cover once legal protection had been established. However, all the protected areas examined in their study were established more than 50 years prior to the final time step of the analysis (2000), allowing more time for forests to recover.

With the exception of Kasane FR and the CNP, which were created in the late 1960s, the protected areas in the Chobe District were all established after 1980. De-gazetting of protected land during the late 1990s for residential and commercial expansion likely contributed to woodland losses in Kasane and Kazuma FRs. High woodland losses observed in several FRs from 1990–2003 may also be related to commercial logging, which began in the 1930s and greatly reduced the areal coverage of mature trees, particularly Bloodwood (*Pterocarpus angolensis*) and Zambezi teak (*Baikiaea plurijuga*) [42,82]. Growing concerns about economic sustainability, along with a lack of concessionaire adherence to contractual agreements, led the government to suspend commercial timber harvesting operations in the forest reserves in 1994 [63,83]. Woodland area also declined in the communally-managed Chobe Enclave over the period 1990–2003, but grew from 2003–2013. A Community-Based Natural Resources Management (CBNRM) framework was implemented in the Enclave beginning in the mid-90s, and has increasingly focused on preserving forests due to their significance to rural livelihoods [84].

4.3. Drivers of Land Cover Change

4.3.1. Rainfall and Fire

Short-term climate variability at inter-annual time-scales, together with strong climate-plant couplings and non-linearities inherent in the system can greatly influence savanna plant community structure and composition, and its response to external forcings [24,85–88]. Total annual rainfall had a significant positive influence on fire occurrence in the Chobe District from 2001–2013. High fire and low fire years tended to alternate biennially in the Chobe District, a pattern found to contribute to significantly higher mean fuel loads in long-term burn experiments conducted in well-wooded savanna in South Africa [89]. Greater plant biomass in high rainfall years can facilitate dry season fire spread when moisture content of vegetation is lowest [89] and we observed a 41% increase in fires during years when total rainfall exceeded the long-term MAP of 632 mm (Figure 10b). Loss of woodland was significantly higher in locations experiencing 15 or more fires from 2001–2013, but did not differ significantly among areas that burned less frequently, suggesting that a threshold exists below which woodlands may be able recover from fire disturbance in the region. Unprotected land between the Maikaelelo and Kazuma FRs burned at a significantly greater frequency than surrounding land, with many fires coinciding temporally with dry season harvesting of thatch materials permitted only from September to October. Increased plant biomass during high rainfall years improves the condition of thatch, but may also contribute to fire spread due to accidental ignitions from escaped campfires used by grass cutters. Similarly, fire hot spots in communal land along the Chobe River in the communally-managed Enclave appeared to be influenced by the large concentration of fires across the border in Namibia's Caprivi Strip where people frequently burn reeds and floodplain vegetation to clear land for livestock. Moderate climate change scenarios for the Chobe District and the greater KAZA region predict annual rainfall to decrease by more than 100 mm by the year 2080, while interannual rainfall variability is predicted to significantly increase [90,91]. Given the dynamic feedback between rainfall and fire, climate change impacts on woodlands and the ecosystem services they provide to local communities are likely to be complex.

4.3.2. Woodland Change and Elephants

Over-browsing by elephants is frequently cited as one of the most important drivers of woodland degradation in the Chobe District [92], particularly within the riparian corridor of the Chobe River where elephant and other water-dependent wildlife congregate at high densities during the dry season [45–47,49]. Ben-Shahar [48] found that while the proportion of woody plants utilized by elephants increased with proximity to the Chobe River, browsing damage was extremely patchy with some sites little used by elephants and patches varying in size by plant species diversity. Our analysis showed that elephants were predominately associated with woodland and shrubland in dry season aerial survey units (Figure 12a). However, we did not detect any significant differences in woodland loss in aerial survey units with higher versus lower elephant biomass for the years 2003, 2006, and 2012. These results suggest that elephant populations are not contributing significantly to the loss of interior woodlands, despite average annual increases in the regional elephant population of between 5.5% and 7% since the 1960s [93,94].

4.4. Fire Management Implications

The vast majority of fires in the Chobe District occurred late in the winter dry season (Figure 10a) prior to the arrival of convective storms typically associated with lightning production, which suggests the majority of fire ignitions were from anthropogenic sources. Experimental studies have shown that frequent fires in the late dry season can lead to large reductions in the amount of woody vegetation cover compared to those occurring in wet season [95]. Frequent dry season fires are also associated with an increase in multi-stemmed coppicing contributing to the dominance of trees of lower stature [89]. High grass yield in high rainfall years can lead to intense fires as grasses become desiccated during the dry season, which may in turn impose a strong demographic bottleneck on tree establishment and recruitment by increasing mortality of seedlings and saplings [96]. In addition to identifying the primary anthropogenic causes of fire, careful implementation of controlled fuel load reduction fires or thinning in areas identified as significant fire hot spots may present a useful tool in the management of the Chobe District's forest resources. In high rainfall years, shifting a proportion of fires to the late summer wet season or earlier in the dry season could be another means to reduce the frequency and intensity of dry season fires and limit fire spread in managed forest reserves [89,97]. Natural resource managers will need to carefully consider the impacts of fire season, frequency, and intensity, in their fire policy development, as these factors can significantly influence vegetation cover at a range of growth stages, in addition to affecting other aspects of biodiversity [89,95]. Understanding of the complex spatial interactions between fire, vegetation dynamics, human and wildlife activities, and climate drivers in dryland savanna will be essential to developing greater predictive capacity and improving adaptive fire management strategies.

4.5. Limitations

In this study, we used a hybrid unsupervised ISO Cluster supervised maximum likelihood approach to classify land cover based on contextual knowledge [98]. Our use of PCA helped reduce spectral overlap and redundancy of image data, which has been shown to improve model discrimination between vegetation and bare ground in savannas [99]. While the overall accuracy of our classification was 86.7% (95% CI 0.843, 0.888) with a Kappa coefficient of 0.832, it is important to note that any land cover classification will contain error, and this is particularly true for semi-arid savanna systems where environmental variability creates a mosaic landscape of diverse plant community assemblages [62]. Our classification targeted broad scale LCC patterns at decadal time scales; hence, we may have failed to detect patterns occurring at a smaller scales, such as shifts in woodland community structure or functional diversity. To overcome these limitations, very high resolution satellite data may be used to extend and improve the existing analysis [100], particularly within the narrow riparian zone of the Chobe River. Use of spectral unmixing methods in combination with hyperspectral data has the potential to improve classification accuracy in structurally complex landscapes [101]. We did not examine potential interactions between elephants and fire, variations in fire intensity, or specific anthropogenic forcings on woodland loss, and future research may focus on integrating multi-disciplinary approaches, such as analysis of stakeholder interviews and additional field observations [102] to provide further information on potential drivers and impacts of LCC in northern Botswana.

5. Conclusions

We classified multi-decadal changes in land cover in the semi-arid Chobe District in northeastern Botswana using post-classification analysis of Landsat datasets from 1990, 2003, and 2013. We observed a long-term trend of decreasing woodland cover and increasing shrubland. However, LCC was not strictly linear in nature, with substantial spatiotemporal variation in land cover trajectories across the study area. Grassland and bare/impervious cover, in particular, increased from 1990–2003 but declined substantially during the subsequent time step. MODIS active fire detections varied greatly year to year and among the different protected areas, with the forest reserves and communally-managed lands tending to burn more frequently than the Chobe National Park. Fire frequency increased in years of high rainfall, particularly when annual rainfall exceeded the long-term MAP of 632 mm. However, loss of woodland was significantly associated with fire only in locations experiencing 15 or more ignitions during the period 2001–2013. Although elephants are often cited as a major cause of woodland degradation in northern Botswana, we did not detect significantly higher woodland losses in areas of high elephant biomass estimated from aerial surveys.

Changes in fire frequency as a result of increased climate variability across the Southern African region may have profound effects on the distribution of woodlands in savanna systems. The Chobe District's transitional location between more arid and mesic climate zones makes it an important model system for the study of the role of disturbance in mediating tree-grass dynamics. Our analysis contributes to improving understanding of long-term land cover changes and savanna ecosystem dynamics, in particular, fire effects on woodland cover stability. The wealth of ecosystem services provided by semi-arid savannas in Southern Africa, coupled with their vast land area, further increases the importance of satellite remote-sensing for monitoring disturbance and change in both naturally-functioning and human-impacted ecosystems. Given the dynamic nature of LCC in the region, land managers should take a long-term approach to understanding change dynamics, in order to correctly identify and isolate emergent threats to vital natural resources and rural livelihoods.

Supplementary Materials: The following are available online at www.mdpi.com/2073-445X/6/4/73/s1, Tables S1–S4.

Acknowledgments: We would like to thank the Government of Botswana's Department of Wildlife and National Parks and Department of Meteorological Services for providing wildlife aerial survey and rainfall data used in this research. We further acknowledge Stephanie Joos-Vandewalle for her assistance in the field. Funding for this work was provided by the Forest Conservation Botswana (grant agreement number FCB00028) and the National Science Foundation Dynamics of Coupled Natural and Human Systems (Award #1518486).

Author Contributions: J.T.F., K.A.A., and M.E.V. conceived and designed the study. J.T.F. analyzed the data. J.T.F., K.A.A., and M.E.V. wrote the paper.

Conflicts of Interest: The authors declare no conflict of interest.

References

1. Smith, R. *Global Forest Resources Assessment 2000 Main Report*; Food and Agriculture Organization: Rome, Italy, 2001.
2. Unccd, Z.N.L.D. United Nations Convention to Combat Desertification. Available online: http://www2.unccd.int/ (accessed on 4 April 2014).
3. Safriel, U.; Adeel, Z. *Dryland Systems.|| in Ecosystems and Human Well-Being: Current State and Trends*; Hassan, R., Scholes, R., Ash, N., Eds.; Island Press: Washington, DC, USA, 2005; Volume 1, pp. 623–662.
4. Safriel, U.; Adeel, Z. Development paths of drylands: Thresholds and sustainability. *Sustain. Sci.* **2008**, *3*, 117–123. [CrossRef]
5. Huber-Sannwald, E.; Palacios, M.R.; Moreno, J.T.A.; Braasch, M.; Peña, R.M.M.; de Alba Verduzco, J.G.; Santos, K.M. Navigating challenges and opportunities of land degradation and sustainable livelihood development in dryland social-ecological systems: A case study from Mexico. *Philos. Trans. R. Soc. B Biol. Sci.* **2012**, *367*, 3158–3177. [CrossRef] [PubMed]
6. Scholes, R.; Walker, B. *Nylsvley: The Study of an African Savanna*; Cambridge University Press: Cambridge, UK, 1993.
7. Ryan, C.M.; Pritchard, R.; McNicol, I.; Owen, M.; Fisher, J.A.; Lehmann, C. Ecosystem services from Southern African woodlands and their future under global change. *Philos. Trans. R. Soc. B* **2016**, *371*, 20150312. [CrossRef] [PubMed]
8. Joos-Vandewalle, S. *The Effects of Urbanisation on Non-Timber Forest Product Dependencies: A Case Study of Three Settlements in the Chobe District of Northern Botswana*; University of Cape Town: Cape Town, South Africa, 2015.
9. Schuyt, K.D. Economic consequences of wetland degradation for local populations in Africa. *Ecol. Econ.* **2005**, *53*, 177–190. [CrossRef]
10. Jones, B.T. Chobe enclave. In *IUCN/SNV CBNRM Support Programme*; Printing and Publishing Company: Botswana, Gabarone, 2002.
11. Scholes, R. Syndromes of dryland degradation in Southern Africa. *Afr. J. Range Forage Sci.* **2009**, *26*, 113–125. [CrossRef]
12. Dregne, H.; Kassas, M.; Rozanov, B. A new assessment of the world status of desertification. *Desertification Control Bull.* **1991**, *20*, 6–18.
13. Wily, L.; Mbaya, S. *Land, People, and Forests in Eastern and Southern Africa at the Beginning of the 21st Century: The Impact of Land Relations on the Role of Communities in Forest Future*; No. 7; IUCN: Gland, Switzerland, 2001.
14. Staver, A.C.; Archibald, S.; Levin, S. Tree cover in sub-saharan Africa: Rainfall and fire constrain forest and savanna as alternative stable states. *Ecology* **2011**, *92*, 1063–1072. [CrossRef] [PubMed]
15. Sankaran, M.; Ratnam, J.; Hanan, N.P. Tree-grass coexistence in savannas revisited-insights from an examination of assumptions and mechanisms invoked in existing models. *Ecol. Lett.* **2004**, *7*, 480–490. [CrossRef]
16. Ratnam, J.; Bond, W.J.; Fensham, R.J.; Hoffmann, W.A.; Archibald, S.; Lehmann, C.E.; Anderson, M.T.; Higgins, S.I.; Sankaran, M. When is a 'forest' a savanna, and why does it matter? *Glob. Ecol. Biogeogr.* **2011**, *20*, 653–660. [CrossRef]

17. Walker, B.H.; Ludwig, D.; Holling, C.S.; Peterman, R.M. Stability of semi-arid savanna grazing systems. *J. Ecol.* **1981**, *69*, 473–498. [CrossRef]
18. Walter, W.G.; Bottman, R.P. Microbiological and chemical studies of an open and closed watershed. *J. Environ. Health* **1967**, *30*, 157–163.
19. Casper, B.B.; Jackson, R.B. Plant competition underground. *Annu. Rev. Ecol. Syst.* **1997**, *28*, 545–570. [CrossRef]
20. Beckage, B.; Platt, W.J.; Gross, L.J. Vegetation, fire, and feedbacks: A disturbance-mediated model of savannas. *Am. Nat.* **2009**, *174*, 805–818. [PubMed]
21. Sankaran, M.; Hanan, N.P.; Scholes, R.J.; Ratnam, J.; Augustine, D.J.; Cade, B.S.; Gignoux, J.; Higgins, S.I.; Le Roux, X.; Ludwig, F. Determinants of woody cover in African savannas. *Nature* **2005**, *438*, 846–849. [CrossRef] [PubMed]
22. Higgins, S.I.; Bond, W.J.; Trollope, W.S. Fire, resprouting and variability: A recipe for grass-tree coexistence in savanna. *J. Ecol.* **2000**, *88*, 213–229. [CrossRef]
23. Scholes, R.; Archer, S. Tree-grass interactions in savannas. *Annu. Rev. Ecol. Syst.* **1997**, *28*, 517–544. [CrossRef]
24. Bond, W.; Midgley, G.; Woodward, F.; Hoffman, M.; Cowling, R. What controls South African vegetation—Climate or fire? *S. Afr. J. Bot.* **2003**, *69*, 79–91. [CrossRef]
25. Baudena, M.; Dekker, S.C.; van Bodegom, P.M.; Cuesta, B.; Higgins, S.I.; Lehsten, V.; Reick, C.H.; Rietkerk, M.; Scheiter, S.; Yin, Z. Forests, savannas, and grasslands: Bridging the knowledge gap between ecology and dynamic global vegetation models. *Biogeosciences* **2015**, *12*, 1833–1848. [CrossRef]
26. Scholes, R.; Dowty, P.; Caylor, K.; Parsons, D.A.B.; Frost, P.G.H.; Shugart, H. Trends in savanna structure and composition along an aridity gradient in the Kalahari. *J. Veg. Sci.* **2002**, *13*, 419–428. [CrossRef]
27. Vetter, S. Zambezian Baikiaea Woodlands (at0726). Terrestrial Ecoregions Profiles. World Wildlife Fund. Available online: http://www.worldwildlife.org/wildworld/profiles/terrestrial/at/at0726_full.html (accessed on 4 April 2014).
28. Lowrance, R.; Altier, L.S.; Newbold, J.D.; Schnabel, R.R.; Groffman, P.M.; Denver, J.M.; Correll, D.L.; Gilliam, J.W.; Robinson, J.L.; Brinsfield, R.B. Water quality functions of riparian forest buffers in chesapeake bay watersheds. *Environ. Manag.* **1997**, *21*, 687–712. [CrossRef]
29. Anderson, H.W.; Hoover, M.D.; Reinhart, K.G. *Forests and Water: Effects of Forest Management on Floods, Sedimentation, and Water Supply*; USDA Forest Service: Berkely, CA, USA, 1976.
30. Swift, M.J.; Heal, O.W.; Anderson, J.M. *Decomposition in Terrestrial Ecosystems*; University of California Press: Oakland, CA, USA, 1979; Volume 5.
31. McCulley, R.L.; Archer, S.; Boutton, T.; Hons, F.; Zuberer, D. Soil respiration and nutrient cycling in wooded communities developing in grassland. *Ecology* **2004**, *85*, 2804–2817. [CrossRef]
32. Bunting, E.; Steele, J.; Keys, E.; Muyengwa, S.; Child, B.; Southworth, J. Local perception of risk to livelihoods in the semi-arid landscape of Southern Africa. *Land* **2013**, *2*, 225–251. [CrossRef]
33. Cui, X.; Gibbes, C.; Southworth, J.; Waylen, P. Using remote sensing to quantify vegetation change and ecological resilience in a semi-arid system. *Land* **2013**, *2*, 108–130. [CrossRef]
34. Southworth, J.; Rigg, L.; Gibbes, C.; Waylen, P.; Zhu, L.; McCarragher, S.; Cassidy, L. Integrating dendrochronology, climate and satellite remote sensing to better understand savanna landscape dynamics in the Okavango delta, Botswana. *Land* **2013**, *2*, 637–655. [CrossRef]
35. Herrero, H.V.; Southworth, J.; Bunting, E. Utilizing multiple lines of evidence to determine landscape degradation within protected area landscapes: A case study of Chobe national park, Botswana from 1982 to 2011. *Remote Sens.* **2016**, *8*, 623. [CrossRef]
36. Vogel, M.; Strohbach, M. Monitoring of savanna degradation in Namibia using Landsat TM/ETM+ data. In Proceedings of the 2009 IEEE International Geoscience and Remote Sensing Symposium, Cape Town, South Africa, 12–17 July 2009; pp. III-931–III-934.
37. Lehmann, C.E.; Archibald, S.A.; Hoffmann, W.A.; Bond, W.J. Deciphering the distribution of the savanna biome. *New Phytol.* **2011**, *191*, 197–209. [CrossRef] [PubMed]
38. Wessels, K.; Prince, S.; Frost, P.; Van Zyl, D. Assessing the effects of human-induced land degradation in the former homelands of northern South Africa with a 1 km AVHRR NDVI time-series. *Remote Sens. Environ.* **2004**, *91*, 47–67. [CrossRef]
39. Brink, A.B.; Eva, H.D. Monitoring 25 years of land cover change dynamics in Africa: A sample based remote sensing approach. *Appl. Geogr.* **2009**, *29*, 501–512. [CrossRef]

40. Aleman, J.C.; Blarquez, O.; Staver, C.A. Land-use change outweighs projected effects of changing rainfall on tree cover in sub-saharan Africa. *Glob. Chang. Biol.* **2016**, *22*, 3013–3025. [CrossRef] [PubMed]
41. Ministry of Environment Wildlife & Tourism. *Forest Policy, Ministry of Environment, Wildlife, and Tourism*; Botswana Government Printer: Gabarone, Botswana, 2011.
42. Skarpe, C.; du Toit, J.; Moe, S.R. *Elephants and Savanna Woodland Ecosystems: A Study from Chobe National Park, Botswana*; John Wiley & Sons: Hoboken, NJ, USA, 2014.
43. Simpson, C.D. A detailed vegetation study on the Chobe River in north-east Botswana. *Kirkia* **1975**, *10*, 185–227.
44. Child, G. *An Ecological Survey of Northeastern Botswana*; Food and Agricultural Organization of the United Nations: Rome, Italy, 1968.
45. Moe, S.R.; Rutina, L.P.; Hytteborn, H.; du Toit, J.T. What controls woodland regeneration after elephants have killed the big trees? *J. Appl. Ecol.* **2009**, *46*, 223–230. [CrossRef]
46. Rutina, L.P.; Moe, S.R. Elephant (loxodonta africana) disturbance to riparian woodland: Effects on tree-species richness, diversity and functional redundancy. *Ecosystems* **2014**, *17*, 1384–1396. [CrossRef]
47. Teren, G.; Owen-Smith, N. Elephants and riparian woodland changes in the Linyanti Region, Northern Botswana. *Pachyderm* **2010**, *47*, 18–25.
48. Ben-Shahar, R. Patterns of elephant damage to vegetation in Northern Botswana. *Biol. Conserv.* **1993**, *65*, 249–256. [CrossRef]
49. Mosugelo, D.K.; Moe, S.R.; Ringrose, S.; Nellemann, C. Vegetation changes during a 36-year period in northern Chobe National Park, Botswana. *Afr. J. Ecol.* **2002**, *40*, 232–240. [CrossRef]
50. Fullman, T.J.; Bunting, E.L. Analyzing vegetation change in an elephant-impacted landscape using the moving standard deviation index. *Land* **2014**, *3*, 74–104. [CrossRef]
51. Blanc, J.J. *African Elephant Status Report 2007: An Update from the African Elephant Database*; No. 33; IUCN: Gland, Switzerland, 2007.
52. DWNP. *Aerial Census of Wildlife and Some Domestic Animals in Botswana*; Department of Wildlife and National Parks. Monitoring Unit, Research Division: Gaborone, Botswana, 2014.
53. Secretariat of the convention on biological diversity. The Convention on Biological Diversity. In *Global Biodiversity Outlook*, 1st ed.; Secretariat of the Convention on Biological Diversity: Montreal, QC, Canada, 2001; pp. 119–136.
54. Pricope, N.G.; Binford, M.W. A spatio-temporal analysis of fire recurrence and extent for semi-arid savanna ecosystems in Southern Africa using moderate-resolution satellite imagery. *J. Environ. Manag.* **2012**, *100*, 72–85. [CrossRef] [PubMed]
55. Republic of Botswana. *Population and Housing Census, 2011*; Statistics Botswana: Gaborone, Botswana, 2011.
56. Republic of Botswana. *Botswana, National Conservation Strategy: National Policy on Natural Resources Conservation and Development*; Gaborone Government Printer: Gaborone, Botswana, 1990.
57. Giglio, L.; Kendall, J.; Mack, R. A multi-year active fire dataset for the tropics derived from the TRMM VIRS. *Int. J. Remote Sens.* **2003**, *24*, 4505–4525. [CrossRef]
58. Justice, C.; Giglio, L.; Korontzi, S.; Owens, J.; Morisette, J.; Roy, D.; Descloitres, J.; Alleaume, S.; Petitcolin, F.; Kaufman, Y. The MODIS fire products. *Remote Sens. Environ.* **2002**, *83*, 244–262. [CrossRef]
59. Giglio, L.; Descloitres, J.; Justice, C.O.; Kaufman, Y.J. An enhanced contextual fire detection algorithm for MODIS. *Remote Sens. Environ.* **2003**, *87*, 273–282. [CrossRef]
60. Jolly, G. Sampling methods for aerial censuses of wildlife populations. *East Afr. Agric. For. J.* **1969**, *34*, 46–49.
61. Boshoff, A.; Kerley, G.; Cowling, R. Estimated spatial requirements of the medium-to large-sized mammals, according to broad habitat units, in the Cape Floristic Region, South Africa. *Afr. J. Range Forage Sci.* **2002**, *19*, 29–44. [CrossRef]
62. Kiage, L.; Liu, K.B.; Walker, N.; Lam, N.; Huh, O. Recent land-cover/use change associated with land degradation in the Lake Baringo catchment, Kenya, East Africa: Evidence from Landsat TM and ETM+. *Int. J. Remote Sens.* **2007**, *28*, 4285–4309. [CrossRef]
63. Food and Agriculture Organization (FAO). *Global Forest Resources Assessment 2010: Main Report*; Food and Agriculture Organization of the United Nations: Rome, Italy, 2010.
64. Congalton, R.G.; Green, K. *Assessing the Accuracy of Remotely Sensed Data: Principles and Practices*; CRC Press: Boca Raton, FL, USA, 2008.

65. Kuhn, M. The Caret Package. R Foundation for Statistical Computing: Vienna, Austria. Available online: https://cran.r-project.org/package=caret (accessed on 21 September 2012).
66. Pontius, R.G., Jr.; Millones, M. Death to kappa: Birth of quantity disagreement and allocation disagreement for accuracy assessment. *Int. J. Remote Sens.* **2011**, *32*, 4407–4429. [CrossRef]
67. Pontius, R.G., Jr.; Santacruz, A. Quantity, exchange, and shift components of difference in a square contingency table. *Int. J. Remote Sens.* **2014**, *35*, 7543–7554. [CrossRef]
68. Team, R.C. *R: A Language and Environment for Statistical Computing [Internet]*; R Foundation for Statistical Computing: Vienna, Austria, 2014.
69. Pinheiro, J.; Bates, D.; DebRoy, S.; Sarkar, D.; R Core Team. *NLME: Linear and Nonlinear Mixed Effects Models*, R Package Version 3.1-117; 2014. Available online: http://CRAN.R-project.org/package=nlme (accessed on 20 September 2014).
70. Griffis, V.; Stedinger, J. The use of GLS regression in regional hydrologic analyses. *J. Hydrol.* **2007**, *344*, 82–95. [CrossRef]
71. Mayer, A.L.; Khalyani, A.H. Grass trumps trees with fire. *Science* **2011**, *334*, 188–189. [CrossRef] [PubMed]
72. Bond, W.J.; Midgley, G.F. A proposed CO_2-controlled mechanism of woody plant invasion in grasslands and savannas. *Glob. Chang. Biol.* **2000**, *6*, 865–869. [CrossRef]
73. Bond, W.J.; Midgley, G.F. Carbon dioxide and the uneasy interactions of trees and savannah grasses. *Philos. Trans. R. Soc. Lond. B Biol. Sci.* **2012**, *367*, 601–612. [CrossRef] [PubMed]
74. Lu, X.; Wang, L.; McCabe, M.F. Elevated co2 as a driver of global dryland greening. *Sci. Rep.* **2016**, *6*. [CrossRef] [PubMed]
75. Milly, P.C.; Dunne, K.A.; Vecchia, A.V. Global pattern of trends in streamflow and water availability in a changing climate. *Nature* **2005**, *438*, 347–350. [CrossRef] [PubMed]
76. Food and Agricultural Organization (FAO). *Global Forest Resources Assessments 2015*; FAO Forestry Paper No. 1; The Food and Agricultural Organization of the United Nations (FAO): Rome, Italy, 2015.
77. Darkoh, M.B.; Mbaiwa, J.E. Okavango Delta-A Kalahari Oasis under environmental threats. *J. Biodivers. Endanger. Species* **2014**, *2*. [CrossRef]
78. Guy, P. Changes in the biomass and productivity of woodlands in the Sengwa Wildlife Research Area, Zimbabwe. *J. Appl. Ecol.* **1981**, *18*, 507–519. [CrossRef]
79. Cowling, R.M.; Richardson, D.M.; Pierce, S.M. *Vegetation of Southern Africa*; Cambridge University Press: Cambridge, UK, 2004.
80. Pfeifer, M.; Burgess, N.D.; Swetnam, R.D.; Platts, P.J.; Willcock, S.; Marchant, R. Protected areas: Mixed success in conserving East Africa's evergreen forests. *PLoS ONE* **2012**, *7*, e39337. [CrossRef] [PubMed]
81. Willcock, S.; Phillips, O.L.; Platts, P.J.; Swetnam, R.D.; Balmford, A.; Burgess, N.D.; Ahrends, A.; Bayliss, J.; Doggart, N.; Doody, K. Land cover change and carbon emissions over 100 years in an African biodiversity hotspot. *Glob. Chang. Biol.* **2016**, *22*, 2787–2800. [CrossRef] [PubMed]
82. Lepetu, J.; Alavalapati, J.; Nair, P. Forest dependency and its implication for protected areas management: A case study from kasane forest reserve, Botswana. *Int. J. Environ. Res.* **2010**, *3*, 525–536.
83. Kgathi, D.; Sekhwela, M. Sustainability of the commercial exploitation and management of the chore forest reserves in Botswana. *S. Afr. Geogr. J.* **2003**, *85*, 26–34. [CrossRef]
84. Thakadu, O. Success factors in community based natural resources management in Northern Botswana: Lessons from practice. *Nat. Resour. Forum* **2005**, *29*, 199–212. [CrossRef]
85. Archibald, S.; Bond, W.; Stock, W.; Fairbanks, D. Shaping the landscape: Fire-grazer interactions in an African savanna. *Ecol. Appl.* **2005**, *15*, 96–109. [CrossRef]
86. Staver, A.C.; Levin, S.A. Integrating theoretical climate and fire effects on savanna and forest systems. *Am. Nat.* **2012**, *180*, 211–224. [CrossRef] [PubMed]
87. Vanacker, V.; Linderman, M.; Lupo, F.; Flasse, S.; Lambin, E. Impact of short-term rainfall fluctuation on interannual land cover change in sub-saharan Africa. *Glob. Ecol. Biogeogr.* **2005**, *14*, 123–135. [CrossRef]
88. Zeng, N.; Neelin, J.D. The role of vegetation-climate interaction and interannual variability in shaping the African savanna. *J. Clim.* **2000**, *13*, 2665–2670. [CrossRef]
89. Govender, N.; Trollope, W.S.; Van Wilgen, B.W. The effect of fire season, fire frequency, rainfall and management on fire intensity in savanna vegetation in South Africa. *J. Appl. Ecol.* **2006**, *43*, 748–758. [CrossRef]

90. Rogelj, J.; Meinshausen, M.; Knutti, R. Global warming under old and new scenarios using IPCC climate sensitivity range estimates. *Nat. Clim. Chang.* **2012**, *2*, 248. [CrossRef]
91. Intergovernmental Panel on Climate Change. *Climate Change 2014—Impacts, Adaptation and Vulnerability: Regional Aspects*; Cambridge University Press: Cambridge, UK, 2014.
92. Nichols, C.; Vandewalle, M.; Alexander, K. Emerging threats to dryland forest resources: Elephants and fire are only part of the story. *For. Int. J. For. Res.* **2017**, *90*, 473–484. [CrossRef]
93. Calef, G.W. Maximum rate of increase in the African elephant. *Afr. J. Ecol.* **1988**, *26*, 323–327. [CrossRef]
94. Vandewalle, M. Historic and Recent Trends in the Size and Distribution of Northern Botswana's Elephant Population. In *Effects of Fire, Elephants and Other Herbivores on the Chobe Riverfront Ecosystem Proceedings of a Conference Organised by the Botswana-Norway Institutional Cooperation and Capacity Building Project (BONIC)*; Government Printer: Gaborone, Botswana, 2003.
95. Smit, I.P.; Asner, G.P.; Govender, N.; Kennedy-Bowdoin, T.; Knapp, D.E.; Jacobson, J. Effects of fire on woody vegetation structure in African savanna. *Ecol. Appl.* **2010**, *20*, 1865–1875. [CrossRef] [PubMed]
96. Favier, C.; Aleman, J.; Bremond, L.; Dubois, M.A.; Freycon, V.; Yangakola, J.M. Abrupt shifts in african savanna tree cover along a climatic gradient. *Glob. Ecol. Biogeogr.* **2012**, *21*, 787–797. [CrossRef]
97. Booysen, P.D.V.; Tainton, N.M. *Ecological Effects of Fire in South African Ecosystems*; Springer Science & Business Media: Berlin, Germany, 2012; Volume 48.
98. Phiri, D.; Morgenroth, J. Developments in Landsat land cover classification methods: A review. *Remote Sens.* **2017**, *9*, 967. [CrossRef]
99. Almeida, T.; Filho, D.S. Principal Component Analysis applied to feature-oriented band ratios of hyperspectral data: A tool for vegetation studies. *Int. J. Remote Sens.* **2004**, *25*, 5005–5023. [CrossRef]
100. Dons, K.; Smith-Hall, C.; Meilby, H.; Fensholt, R. Operationalizing measurement of forest degradation: Identification and quantification of charcoal production in tropical dry forests using very high resolution satellite imagery. *Int. J. Appl. Earth Obs. Geoinf.* **2015**, *39*, 18–27. [CrossRef]
101. Meyer, T.; Okin, G. Evaluation of spectral unmixing techniques using modis in a structurally complex savanna environment for retrieval of green vegetation, nonphotosynthetic vegetation, and soil fractional cover. *Remote Sens. Environ.* **2015**, *161*, 122–130. [CrossRef]
102. Kiruki, H.M.; Zanden, E.H.; Malek, Ž.; Verburg, P.H. Land cover change and woodland degradation in a charcoal producing semi-arid area in kenya. *Land Degrad. Dev.* **2017**, *28*, 472–481. [CrossRef]

land

MDPI

Article

Making Sense of Past, Present and Future. Images of Modern and Past Pastoralism among Nyangatom Herders in South Omo, Ethiopia

Jill Philine Blau

School of Social Sciences (Development Studies), Friedensau Adventist University, An der Ihle 19, 39291 Möckern, Germany; jill.blau@thh-friedensau.de; Tel.: + 49-(0)-3921-9160

Received: 10 March 2018; Accepted: 13 April 2018; Published: 19 April 2018

Abstract: This article asks how Nyangatom pastoralists currently make sense of the past, present and future of their pastoralist livelihood. Nyangatom pastoralists, like all agro-pastoralist groups in southern Ethiopia, are faced with enormous structural changes in their immediate surroundings, primarily due to large-scale industrial agriculture and a government policy encouraging them to be sedentary. While the impacts have been discussed elsewhere, thus far little focus has been placed on what images of the past, present and future these changes create among the Nyangatom. This article pays attention to these changes by highlighting the results of a larger qualitative study. It becomes evident that discourses of modernity and culture are translated into the everyday lives of Nyangatom herders. While the past is constructed as a cultural/traditional time by the older generation, an image of modernity shapes the present life of younger generations. The administration plays a contradictory role in transmitting modernity ideals. The future of the Nyangatom is envisioned as a modern pastoralism, yet there is general pessimism with regards to pastoralism persisting.

Keywords: pastoralism; Ethiopia; South Omo; Nyangatom

1. Introduction

Pastoralism in Ethiopia is practiced on more than 62% of the national land categorized as arid or semi-arid rangeland [1]. There are about 10 million pastoralists and agro-pastoralists in Ethiopia, which is about 15% of the Ethiopian population, with their numbers rising and their rangelands decreasing [2]. Most of them live in the lowlands of Ethiopia.

Globalization and its inherent commercialization and commoditization processes are affecting pastoralist livelihoods in Ethiopia in numerous ways. Even though pastoralists have always been involved in trading, exchanging and bartering, the pressure to diversify their livelihood has never been as strong as now [3]. This is likely to increase in the coming years in Ethiopia due to the government strategy to encourage pastoralists to "pursue sedentary life with diversified and sustainable income" [4] as well as pressures related to large-scale agricultural development. Yet there are no numbers on how many pastoralists receive some form of income from their livelihood and how many are still self-sufficient.

Scholars have argued that increasing commoditization and privatization processes are changing concepts and systems of production, consumption and distribution in pastoralist societies. Anthropologists highlight how the symbolic meanings of objects are also changing due to the changes in the pastoralist livelihood [5].

The Nyangatom woreda, home to the Nyangatom pastoralists in Ethiopia, is located in the Lower Omo Valley which is situated in south-western Ethiopia in the Southern Nations and Nationalities Peoples Region (SNNPR). The SNNPR borders South Sudan in the west, Kenya in the south, the Ilemi triangle (a region claimed by Ethiopia, South Sudan and Kenya) in the south-west and the Ethiopian regions Gambela and Oromia in the north-west and north and east.

As the name proclaims, the SNNPR is the most ethnically and culturally diverse region of Ethiopia, and 56 ethnicities live within it. Of these, 17 groups are pastoralists and agro-pastoralists. Originating in the central Ethiopian plateau, the Omo River runs across the SNNPR before spilling into Kenya's Lake Turkana [6]. The Omo River is very important for the pastoralists and agro-pastoralists in the region, structuring wandering paths and flood retreat cultivation in accordance with its annual flooding. The pastoralists and agro-pastoralists living in the SNNPR include Hamer, Kara, Mursi, Nyangatom, Bodi, Tsemai and Dassanech, among others.

The Nyangatom live mainly in South Sudan and Ethiopia along the northern part of the Ilemi triangle. In Ethiopia, they inhabit the eastern side of the Omo River and the northern part of the seasonal Kibish River. Their population is estimated to be 30,000 in Ethiopia and a similarly high number in South Sudan [7]. The administrative center for the Nyangatom woreda is Kangaten, which is located on the eastern bank of the Omo River and has just been connected to the other side by a bridge and is urbanizing rapidly. Their mother tongue is the Nyangatom language.

Ethiopia wants to stop importing sugar. It also wants to increase its production seven-fold in order to eventually export half of the total amount of sugar. All of the "additional land" for sugar plantations is situated in South Omo and alongside the Gibbe III dam by which it shall also be irrigated [8]. The Ethiopian Omo–Kuraz Project [9] states: "The project will have a total area of 175,000 hectares of land for its sugarcane cultivation. This wide area will get its water supply from the Omo River through a diversion weir which is under construction with 381 m width and 22.4 m height." Sugar cane is a crop that requires a lot of water. The sugar plantations will, therefore, further exacerbate water scarcity issues. The currently constructed and future plantations are all located on sites that are crucial for local pastoralists, including the Nyangatom pastoralists. The Gibbe III dam is a highly disputed and politicized megaproject. It is the third in a series of hydro-dams constructed in recent years. Upon completion, it will be the tallest dam in Africa, with the biggest hydroelectric power plant attached to it. The tentative completion date of the dam was 2015, but heavy delays have characterized the project thus far. Gibe III was designed to produce electricity and export half of the power to neighboring countries. Construction began in 2006, and as the NGO International River writes, included "flagrant violations of Ethiopia's own laws on environmental protection and procurement practices, and the national constitution" [6]. The Gibe III dam is also a cross-border issue, as Kenya's World Heritage Site Lake Turkana is affected. According to estimates, the water level of the world's largest desert lake will drop by two meters in the first year already, its salinity levels will increase and negatively impact the fish stocks of the lake [6]. Therefore, a political dispute between Ethiopia and Kenya has arisen, with Kenya asking for compensation for the impact of Gibe III. This dispute also affects Ethiopia's pastoralists such as the Nyangatom. As Ethiopia prefers not to challenge Kenya on the cross-border pastoralist conflicts between the Nyangatom and the Turkana, the Nyangatom are left without governmental recognition of cases of cattle raiding and killings which happen on a regular basis. Faced with the very rapid transformations of their livelihood highlighted above, Nyangatom pastoralism in Ethiopia is under real threat.

What has not been looked at thus far is how—confronted with these facts—the Nyangatom pastoralists create different images that characterize the way they make sense of their past and their future. After presenting the methods that lay the foundation of this article, I discuss how the past is constructed as the "cultural times". This will be followed by ideas about the future, "modernity" and how that links to Nyangatom culture. Afterwards, I discuss the Nyangatom's hope for a compromise between modern and cultural identities. I also discuss the role played by the employees of the local administration, which still consider themselves pastoralist. This includes the analysis of symbolic meanings of objects and practices that are changing due to the changes in the pastoralist livelihood.

2. Methods

This article is based on empirical qualitative research done throughout my time as a PhD fellow. Qualitative data analysis is an iterative and self-reflexive process which was supported by atlas.ti to analyze interviews and develop codes and categories. In my empirical research I combined open-ended interview techniques with participatory observation techniques. I did formal and informal interviews, used exploratory research techniques including informal focus-group discussions, and analysed research diaries made in the course of my field work. I kept adjusting my ideas in the course of the process, re-interpreting my work as I went along. The basis for this article are eight formal interviews with elderly male and female Nyangatom and four interviews with members of the Nyangatom administration, all male, as well as numerous background and informal participatory observations and informal interviews. I concentrate on the inclusion of field material (interview segments) that was collected and analysed between 2011 and 2015 in the Nyangatom woreda.[1] All names are anonymized in order to protect the privacy of my interviewees.

In the analysis, I focused on the categories that evolved naturally from my empirical material. Here, I used classical data analysis approaches such as that of Mayring [10] and Quinn Patton [11], where categories are built through the inductive analysis of the data. Mayring's data analysis classically comprises 8 steps which, however, I did not follow strictly but instead combined with grounded theory analysis techniques [12] where I gradually viewed my data and related the content to these categories, which often led to a re-adaptation of the categories or the creation of new categories altogether [13,14].

With regards to language, I was not able to translate the interviews myself. Even though I did learn some local vocabulary in order to engage in polite small talk, no interview was possible without a translator. My translator was a local educated Nyangatom working for an international organization who had already accompanied some researchers in their field work and was thus experienced with different research techniques as well as translation-related issues. We reflected on each interview afterwards, which served to clarify the content of the interviews as well as ensuring a cooperative working atmosphere. Even though, or maybe because, his English was not the strongest, I always felt like he did his best to translate as close to Nyangatom as possible as well as explaining or paraphrasing symbols that I would not understand after the interview. In the transcription, I left the English the way it was, as that seemed most authentic to me and, upon editing request, adapted small parts for general understanding. I then used footnotes for clarification.

With regards to his influence on the work, I had the impression that he was respected by the pastoralists and also did not have any relevant political associations, as he was working for an international organization at that time. Yet with his modern clothing and education he represented a pastoralist who has "left the culture". Needless to say, matters might have been said differently to me if I had conducted the interviews by myself.

Underlying the idea that power and knowledge are inseparable, science has, on many occasions, not been neutral but has served to legitimize power structures within society [15,16]. Gender studies has spent a significant amount of time developing methodologies that place the analysis of power structures, e.g., between "the researched" and "researcher" within research as opposed to rendering it invisible [16]. Accordingly, the way the researcher moves within the research field is also political. Even the manner in which research is conducted influences and often legitimizes already existing dimensions of power.

As Do Mar Castro Varela [17] argues, the researcher tends to gain prestige by conducting research, while the "objects of study" generally suffer a loss in terms of energy and time invested. Being shaped by these strands of research, I was acutely aware of aspects of power and prestige within my research procedures due to European and white privilege. Needless to say, I sometimes felt uncomfortable in

[1] Some of the interview excerpts can also be found in my dissertation Blau, J. (2018: forthcoming).

the interview situation with pastoralists confronted with my own positionality and the issues with it. Amina Mama writes about intensified objectification here:

> "The danger of such work is that of our intensified objectification, to use Fanon's term. By this he meant the process by which our reality is constructed by others (oppressors in the colonial and neocolonial contexts), to serve their psychological and political needs, and then projected on to us and internalized. In a context where our own interpretations and accounts are as yet largely unpublished, others become the experts on us, and this monopoly on knowledge about us must be seen as imperialistic. While some of this knowledge may be well researched, much of it is partial and particular, and does not serve our interests as a group, or our psychological and intellectual development. All that imperial feminism has meant here, is that it is European (and North American) women's preoccupations rather than those of men which have come into vogue [18].

Below I outline a few dealings I found with these issues. I was also influenced by the research of Nadig [19], who writes about how transferences by the community of research can always be informative about certain conflicts within the community of research and vice versa. In that sense, I attempted to use all the projections that were made towards me for my research and vice versa to assess my own projections.

In line with Tuana's [20] concept of "location for the text", which argues that one always needs to critically assess whom a text addresses and whom it speaks about, I emphasize the incorporation of many segments of my interviews in the hope of transmitting how the larger sense of pastoralists losing their livelihood to modernity translates into the everyday. While there remains a strong and fundamental asymmetry between me as a white researcher using my findings for an international academic audience, I intentionally include as many 'voices' as possible as well as making my discursive interactions transparent. Keeping in line with Schutte's cautioning that perfect translations are not possible between culturally differentiated subjects, I have refrained from structural language corrections within the translations of interviews that were made by the translator that I worked with and insist that element of incommensurability always remains [21]. This includes sticking to terms such as "cultural" and "modern" not because of academic discourses but because they emerged out of the data. It implies making texts as understandable yet keeping them as authentic as possible.

3. The Past Was the Cultural Time

For the Nyangatom pastoralists, modern life is seen as something coming "from the outside".[2] Outside here refers to a blurry mix of Ethiopian highland urban culture and "Western" practices. By contrast, the Nyangatom speak about their customary habits as cultural habits. The word for cultural in Nyangatom language is "Etal", which my translator insisted to be the same as culture even though in some instances traditional seems fitting as well. Upon asking him what the difference is between cultural and traditional, he said:

> **Translator**: "Traditional doesn't fit to the meaning of the Nyangatom. And I don't really know the real definition of traditional, I only know it when it goes with the sentences, for example: 'do it in a traditional way' so is this the same as 'do it in a cultural way?' Or they are different? That is why I don't see the real meaning and the difference."

In trying to stay as close as possible to the local language and its translation, I am therefore using the word cultural. Just like the idea of modernity, my emphasis is on the way the Nyangatom use and interpret this word. While members of the older age sets[3] describe themselves and their practices

---

2 "Brackets" refer to codes that were developed during data analysis, many of them in vivo codes.

3 The Nyangatom organize themselves according to age sets which structures roles, responsibilities and rights within the community.

as cultural, they portray the younger generations as tempted by modernity. This coincides with any description of life in the past being highlighted as the "cultural time" and the future foreseen as a "modern life". Customary habits remain, in that sense, with the elderly generations. The past is also constructed as the time of the "real Nyangatom life" and the future as a distortion of that through modernity. This is often illustrated through a depiction of the everyday.

> **Lokamusio** (male, elder, pastoralist): In the past we did things differently. Normally the Nyangatom, they have a special way of sleeping. You sleep and then the stick[4] will make you sleep a deep sleep but also wake up in the night and see what is going on. But nowadays the children do not want to take this one anymore. They are sleeping deep sleep like the women do (…).

The past is seen as a time in which even sleep was "better, more pastoralist". Waking up in the night or sleeping lightly is a responsibility of the male herders in order to watch out for theft and threats by wild animals. Modernity is described as changing this everyday practice with men not taking this responsibility as seriously anymore. Older generations become the practitioners of pastoral obligations and therefore need to make a point of preferring it and thereby safeguarding it. This can be seen in how Nakaale finished her interview with me.

> **Jill:** *Thank you so much. I am finished with my questions. Is there anything else that you would like to add?*
>
> **Nakaale:** (female, elder, pastoralist): No ... there is not … (silence) Hmmm … . (silence)
>
> You know there is something that I would like to add. Most of the time what I like is the cultural things. I don't want to cook my porridge with the modern pot, I want with the cultural one. Even if you don't put salt or something, it tastes good with the cultural pot. And there is the transportation, if we want to go far, far away we want to put our luggage on the donkey. And then these things are good. You know instead of sleeping on the ground, I saw that yesterday he (Note of author: referring to a young male herder) was sleeping on a mat. That is not good for me, it is burning me. It is good to sleep with the skin. And then you look good, like a pastoralist, and then you put your leather here and butter there and it is a good look, the cultural look.

Nakaale also stresses how sleeping in her way is still good for her. She highlights how one looks beautiful as a real pastoralist and how being connected to the ground actually makes one look like a proper Nyangatom. This is a very literal depiction of something coming "in between" the former and the current livelihood of pastoralism. Toro also speaks of another everyday practice, eating. Milk, the essence of cattle herding, features heavily in the diet and is sometimes referred to as the "pastoralist gold".

> **Toro** (female, elder, pastoralist): They (author's note: pastoralists in the past) had a good-looking style, a good posture because of the milk. A real pastoralist has to eat porridge with milk, has to eat porridge with butter, has to eat several types of meat, like there is the fresh meat, there is the dried meat, there is the meat mixed with butter, in the pastoral way of life. But now if you could travel somewhere … then there is porridge, but there is no salt, there is no butter, just simply … they will eat simply like that. This is why I generally say the past pastoralist was better. Now it is mixed up, with the changes. Now the things are changed with Nyangatom.

---

[4] Some Nyangatom sleep with a stick in their hands and/ or with their head on a wooden stool.

The disorder within livelihood and cultural practices is perceived and described through the changing of recipes. Older pastoralists speak negatively about modern foods and also consider the weight gain of the younger generation as a consequence of new eating styles. Just like the sleeping habits, transformations in eating as an intrinsic everyday practice characterizes the depth of the structural changes and the way they are being evaluated. Some interviewees like Toro also make a particular point of highlighting that they "used to look different" and in fact were more beautiful.

Jill: *How would you describe the situation of women specifically?*

Toro: You know, I compare again with the past. When we were married, luckily I was married in the good times, when a married woman had four leather skirts reserved, she had decorations to put over her head, and she had feathers, rings everywhere, and butter here, good and shiny beads.[5] Then she was healthy, she could jump. Whenever she moved, you could say: "There is a real Nyangatom woman". You know there was something, if you make a hole here (author's note: by the teeth) and then you put decoration on. And then she would have the small chains on the ears (author's note: earrings).

Then even their husbands would have metals up to here (author's note: bracelet), they would be very nice, because this showed that he was a rich man. And then whenever he would come home, he would do this (shows a movement of bracelets rattling), so that he comes home. The woman, they could work with a lot of things, milking the cow, a lot of joy with the women. And while she would do that, the husband would make a sound.

Everyone, even the children, were happy.

But right now, the men are moving without having this one (author's note: bracelet), the women stopped wearing skirts—look at me now, I am not wearing one and nothing could make me decorate myself with the skirt, with the cattle. Now everyone is moving about without having these cultural attributions.

And now even the husbands of the Nyangatom, they don't wear feathers, they don't wear skirts, nothing.

You know, this is because they were right, they had cattle, their cattle were productive, they could produce butter and milk, and so they were sitting as a rich person, the past ones, but now ... Maybe you will interview some of them, they are working hard just to survive. They don't have any decorations, they don't even look like pastoralists. So these clothing styles are mixed. I can say that the past and now is not uniform.

This quote of Toro's illustrates many in-depth matters which can be examined. The "cultural look" is what made the Nyangatom "real and beautiful", and this realness is in turn related to gender. As also stressed in the first interview segment, "real" men and women are those older ones, those that still know how to be pastoralist. This is in line with a societal organization in which the oldest have the most authority and responsibility. But it also means that young people's gender roles are weakening or degenerating in the eyes of the older generations. The richness of wearing cultural attire is, furthermore, connected to an understanding of wealth: whereas the older generations see being a real pastoralist—to look, sleep and eat like a pastoralist—as a wealthy man/woman, the younger generations seek wealth elsewhere and start to view pastoralists as poor and deprived in a way that was not evident before. In that sense, it is a confrontation of value systems in which the modern choices of the younger people are not seen as relevant for their parent generation. The mix of clothing, like the mix of recipes, symbolizes a mix-up of livelihood strategies. Yet it is clear for everyone that the "modern life" will be the dominating paradigm for the future.

5 Butter is also put on the bodies for skincare and beauty.

4. Modernity is 'Coming Inside' Pastoralism

Jill: *How do you see the future of your children?*

Lokamusio: Nowadays the children have their own thoughts. Sometimes I compare the life and say: Maybe this life is pushing the people towards modern life. People like Etuk (note of author: points to my translator) go to the village and then everybody wants to be like him. So in this case, I do not know. (silence)

So I can say, like my wife told you, some of my children are in school. Sometimes in the break they come back home, and then the ones that are left at home will start mumbling: "I should go to school". So I must say: "If they quit pastoralism, they quit". Then me as an old person I have to take care of the cattle. And then maybe the cattle will be disappearing. Because I will be dead and they will be in school and there will be no one to take care of the cattle.

As Lokamusio describes, modernity is perceived to be like a magnet, symbolized by certain people who encourage others to go to school and quit their previous lifestyle. It is understood through items such as mobile phones and clothing like jeans and shirts. It attracts the younger generation away from the cattle. Having no one to take care of the cattle any longer means the end of pastoralism.

Nakawo (male, elder, pastoralist): But here in Nyangatom we are backward. Now everyone wants to be in the new things, fast fast. Nobody gave us a background in these things. Now the fashion came and everyone wants to be modern.

The Nyangatom, older and younger, who are still practising pastoralism, tend to label themselves as backward. By doing so and constructing the past as the time of the real pastoralist, they are situating themselves in the past; thereby foreseeing the end of their livelihood and devaluing it. "Fast, fast" here indicates the quick changes of a livelihood strategy from one generation to the next. Fashion is hereby very symbolic, as it is a literal depiction of the physical attributions of Nyangatom becoming invisible. With Western clothing, it is harder for locals to understand which pastoralist group/ethnicity people in South Omo identify with.

The key driver of modernity, however, is not perceived in fashion or food but in school.

Ngaye (male, middle-age, pastoralist): But I fear that our children will be educated, some of them will be even missionaries, some of them will be this or that. So I do not know about our cattle. I am afraid that this pastoralist life, it will not continue. But maybe we will see with time. If their time is good, let it be. But what I see myself is that school is put into the mind of the people.

Education plays a special role in the immersion of the Nyangatom pastoralists into sedentary livelihoods. Training for older age sets is provided by the government as well as elementary and high school for the youth where more than half the Nyangatom population now attends. Once a child is sent to school, they "will become modern", meaning that they will adopt the habits discussed here. For a while it was predominately boys that were sent to school, but female attendance is increasing.

Jill: *So how do you see the future of the Nyangatom?*

Toro: You know, the school, let me start with the school ... the school is built beside the village ... in every village you can find a school. And then the government officials, they would come and say: "Take your child to school". Most of them are in school now. There is a lot of training being given to the pastoralists. Like for example, us now, we are going to have a "harmful traditional practices training". And you know, it's not only me, it is all of the pastoralists. The training will continue like this. And the school is there. And our trained officials, they are dressed well and they want to educate themselves. So in the future, I do not think pastoralism will exist.

The government is rapidly building schools and enforcing training programs in the Nyangatom woreda with lasting impacts. The children's school education prevents them from acquiring cattle-herding skills in as much depth as their parents' generation did. As Toro highlights above, it is also about introducing new value sets and practices among people, such as eradicating some of their habits labelled as harmful/traditional. Linking the words harmful and traditional also seems to coincide with the idea that traditional practices are backward, as mentioned earlier. They create a devaluation in which the traditional gets associated with harmfulness and thus is in need of transformation

> **Lokeyn** (male, middle-age, pastoralist): During our fathers, our fathers are very great in looking after the cattle. During the time, they transfer this to us. They have been keeping things like that. But nowadays the school thing came and everyone is going to the school. Even my children. They refuse suddenly . . . and go to school. Whenever you want to take him back, the government says: "No, let him stay". So I think either the animals, sheep and goats will be finished by selling or finish with this environment, because no one will take care of them. So this is Etuk's (author's note: my translator) generation. If they are happy to look after the cattle, they are happy. If they are happy to go to the school, this is their problem. But in our case, we have been taking care of our cattle since our fathers. Etuk's generation, they want to go to school, they want to be educated. So I do not think the cattle will be surviving. They will not survive. Even the animals will not survive.

As one can see in this quote, the relationship between schooling and the end of pastoralism is clearly stressed. Yet the Nyangatom also have visions for compromises in which their livelihood can continue somewhat, despite the challenges to it.

5. Hoping for a Compromise, a Mix between Modern and Pastoral

Many of the Nyangatom I interviewed nevertheless dream of a future in which they can take the best of both past and future livelihoods. This is described as a life in which modernity and the "cultural" combine to benefit the pastoralists. A lot of parents have compromised by sending half their children to school and keeping the other half "in the village".

> **Nakawo:** In fact I was in Yabello [6]. There were a lot of pastoralists from the world that shared a lot of information. So I heard in the world there is someone who builds this kind of house (note of author: makes a drawing in the air that is meant to illustrate a more urban house), but he has cattle, sheep and goats and camel. And then the person who was like this was a herder, and educated. He goes to school and then looks after the animals. So they are sharing the culture, the traditional and the modern one.

> **Lokeyn:** The reason I put one in school and one in herding is because I thought that the school-goer will help me with two things, with the modern things. For example, if my parents are sick, then they will help taking them to the hospital. Whenever my cattle are sick, he will buy me the drugs for the animals. So I will use him as a help in town. If we are hungry, he can buy food, he can assist me with the money. Then the pastoralist son can help me with the cattle and with the cultural things. So both of them are useful for me. This is why I split them up, my sons.

> **Lkirito** (male, middle-age, pastoralist): I have two options. One: putting them in school; Two: sending them far away to mind the animals. I will be in the middle. For me, no matter for me, I will give them a role, benefitting from this one and then from this one.

6 Here, Nakawo is referring to an international pastoralist meeting that took place in Yabello, Ethiopia in 2006.

What becomes evident is that the idea of sending one child to school is nevertheless often evaluated from the standpoint of the pastoralist: An educated child could have money for medicines for the cattle. One could try to benefit from both, in order to maintain pastoralism. Pointing to this imagined future is not to undermine the hardship that the effort of even creating a vision creates for the Nyangatom pastoralists in the face of their current challenges. The interventions of modernity, be it through changes in food, sleep patterns, fashion, formal education or government policies in general, are experienced as an intrusion—as something that destroys the Nyangatom livelihood from "inside".

> **Lolamai** (male, pastoralist, herder, middle-age): "You know, it is hard for me. Because cattle is everything for me. And if I lose cattle, I do not know. Maybe I will advise my children not to lose the cattle. To have more than one life. To have modern life and pastoralist life together. Not to lose all the pastoralist life. Just to have it together. This is what I advise my children [. . .] You see, you are interviewing me. I am a herder. It is good. If you come again and interview my child, you will not get the information I am giving you. Never. Because the government is coming inside the pastoralists. Inside. This fashion, this modernization thing, is coming inside the pastoralists. Everyone wants to have mobile, wants to eat *injera* and foods like this, wants to wear clothes, everything is coming inside. So I am afraid that your child will not get enough information about herding from my child. I am afraid that my child will be, you know ... he will be in school, he will be somewhere. And then pastoralism is over, I can say like that. Maybe there will be someone. Maybe one of my children will listen to me and they remain as a pastoralist in a far place, far place. But most of them, I think, they will not continue this life.

In that sense, the hope of finding a way to adopt the modern life while at the same time sustaining what is important to the Nyangatom pastoralists must be understood as a compromise; as a hope in the face of challenges that are not perceived as resistible—and as a response to: "I am afraid that pastoralist life, it will not continue". Some Nyangatom have also moved to South Sudan because they believe it is easier to continue a pastoralist life there than in Ethiopia.

6. Even the Administrators are Pastoralists

With the increasing number of educated Nyangatom men who work in the administration and are thereby seen as representing modernity, it was interesting to see whether they still considered themselves pastoralists.

> **Jill:** *Do you have cattle also?*
>
> **Administrator 1** (male, middle-age, formally-educated): Yes, I have. In my father's place. I have cattle, sheep, donkey.
>
> **Jill:** *Who is taking care of them?*
>
> **Administrator 1:** My father. You know, because my father has five women, he married five women, my brothers are there. They take care for that. They come to me if they want medicine, anything for the car, I buy for them.
>
> **Jill:** *Now I understand a little bit better. One thing that would be interesting for me, would you call yourself a pastoralist?*
>
> **Administrator 2** (male, formally-educated, middle-age): Really. Definitely I am a pastoralist.
>
> **Jill:** *So what makes you a pastoralist?*
>
> **Administrator 2:** You know, I know even, not only here. From here to another side, we are known by pastoralists. I have animals like cattle, goats, sheep. So, in this case I am really a pastoralist.

Even formally educated Nyangatom possess cattle and small ruminants which some family members in rural areas are taking care of. They also consider themselves pastoralist. Yet, at the same time, those working in the administration (must) replicate the idea of "changing" the pastoralists.

> **Jill:** *And how do you see the future of Nyangatom land use?*
>
> **Administrator 4** (male, Nyangatom, formally educated)**:** According to my personal aim, I wish, if I wish that all the Nyangatom area would turn to farming. Because as I told you in the first question, our aim is to change our community to live in a good situation. So, farming can change their lives. It can decide if they eat once, twice or three times a day. This is my personal belief. According to the government strategy plan, this is also their one point, their one aim for change. Also to use the Nyangatom land for investment, not by casting them from their area, not by ignoring their cattle: side by side. The government can use half of the land for investment and for factories and half of the land for pastoralists' animals.

The idea of change is justified and explained by an improvement of the situation for the Nyangatom. However, from what has been described earlier, many of those still practicing herding see the change as defined by the governments' strategy as a sign of more and not less poverty. The villagization strategy of the Ethiopian government is directly aimed at transforming the livelihood of pastoralists towards becoming settled, by "wanting to change them". The strategy was still in its early stages during the time of my research. It was beginning to take form in the Nyangatom woreda through four established "villagisation" villages, with the relevant infrastructure officially consisting of traditional houses, access to water, schools and medical facilities. With these villages, each Nyangatom household is allocated their own piece of land of up to five hectares and additional land for cattle. The Nyangatom sometimes stressed how they were still waiting to see any changes themselves, as Toro and Lobuti illustrate:

> **Toro:** So, we didn't experience the sugar and some of the infrastructure. But we are waiting to see how they change. But the government officials, they say that it will change you. So we are waiting for it, whether it will change or not.
>
> **Lobuti** (female, middle-age, pastoralist): You know, for me they have been telling me that whenever there is the sugar factory here, there will be given grazing land and irrigation. So if it is like this, we accept. Because if the sugar factory is not touching grazing land, and if there is additional land given to us for cultivation and irrigation, I accept that idea.

Whereas both the official policy of the Ethiopian government and the interviewed public servants reason that these changes will bring an overall improvement for the local population, the "Lands of the future network"—a network involved in critical pastoralist land-use research especially in the Horn of Africa—evaluates the Ethiopian government's strategy as a voluntary, but actually forceful, policy and relates it directly to the sugar plantations as follows: "A good example of how not to go about the displacement and resettlement of pastoralists is provided by the river-basin development in Ethiopia's Omo Valley. The Omo has long been seen as a river with an excellent hydropower and irrigation potential. It has been estimated, for example, that, once the river's highly seasonal flow has been regulated by hydropower dams in its upper and middle basins, over 50,000 ha will become available in its lower basin for reliable large-scale irrigation development [...]. In fact, the Ethiopian Sugar Corporation has now begun developing an area more than three times this size, in anticipation of the completion by 2015 of the Gibe III hydropower dam. Known as the "Kuraz Sugar Development Project", this will eventually equal the entire area currently irrigated in Kenya. Thousands of agro-pastoralists are already being evicted, by government fiat, from their most valuable agricultural land along the banks of the Omo [22]. As stressed in this depiction, the participation level in any of the changes involving the lives of the Nyangatom is minimal and coerced. The administration employees would rather insist on all the advantages being brought about. For example, they emphasized that the infrastructure was one of the main advantages of the pastoralists living in the villages established through villagization.

> **Administrator 1:** Actually in our woreda, we have the strategy to have the Nyangatom people come to one place. This is what we call villagization or re-establishment. Because of that, there is a kind of movement in our people. So they move from place to place. In this case, we want to have these people in one permanent place. But for the cattle and the goats, they can move like this, no problem. Because they have the young boys like this. So for the old women, they must stay in one place. Because when they stay in one place, the government will build the school there, a health centre there. Then they will give them the land for farming, yeah. The government will build all the infrastructure, even water, there. So this is our fast plan or strategy for them.

While my emphasis here is not on the assessment of the various "change agendas" taking place, I wish to highlight how the administration often spoke in contradictions. They would stress how Nyangatom cannot continue to live as they do, and yet make sure to reclaim their own pastoralist identity. They would speak about "these people" and yet insist that they were one of them. They would speak about themselves as part of the government and then insist on a distinction between their personal and the government's strategy—as the job comes with a lot of conflicting demands on the individuals who, themselves, grew up herding. The administrators would promise prosperity through sedentarization, yet the pastoralists hold a different understanding of prosperity.

The underlying proposition of the public servants would often be that, whatever the government wanted, the local Nyangatom would eventually understand because it was "reasonable". "Eventually they would like" whatever changes would be brought about, one just needed to "discuss".

> **Jill:** *How did they react when you spoke about the village?*

> **Administrator 3** (male, Nyangatom, formally educated)**:** Yes, they accepted very much, because, as I told you, we have a lot of infrastructure. You know, they cannot walk 100 km, 200 km, so, for them, we want them stay in one place. We let the cattle outside, we want the young people to move, no problem for them because of grass and water.

What seems important to highlight in this context is that the villagization villages were seen by the local population as something provisional. This becomes evident through phrasing which depicted other places as their home while officially living in a villagization village. Home was specifically the place where their cattle was located, as illustrated by this interview:

> **Amareng** (young, female, pastoralist): You know, from X I moved first to Kangaten[7] first because I needed food. So I and my husband we settled here in some place. And then from here we shifted to Y (Note of author: villagization village).

> **Jill:** *Why did you move there?*

> **Amareng:** You know the people say that if you build a house, you will get food. And remember I came from X looking for food. So I say okay, there is something here. So let me move to these people and then make my house so I can get food.

> **Jill:** *So if you compared your life in X and in Y, what is good about life in X, and what is good about the life in Y?*

---

[7] Capital of Nyangatom woreda.

Amareng: It's different here in Y. Back in X, we watered the animals, and some of the neighbors are different. Also the activities that I used to do there in X in the home, it is not normal to do those here.

Jill: *So maybe you can tell me about a normal day in Y from morning until evening?*

Amareng: You know, in the morning I wake up early and grind. Look at my hands, the hard skin. (laughter). You know you grind early in the morning, you fetch water, you make food for your husband. Then you collect some firewood. You make the house clean, you go from here to there, this is our normal work.

Jill: *But this is also what you did in X, right?*

Amareng: Yes, it is the same.

Jill: *So what is different now?*

Amareng: The one thing missing is the watering of animals, making fences for the animals, milking them. The activities with the animals are the ones missing.

Jill: *So do you like living in the new village?*

Amareng: I am not happy about living here. I will go back to my village.

Jill: *Why?*

Amareng: Or I will make two houses. One with my animals, so I can see how they are doing, and one here to this village.

Jill: *So what makes you not happy about the village?*

Amareng: Just animals. It is still living. No problem with that.

What Amareng describes is a situation in which a home is seemingly created in one place in order to receive food aid. But the place is not really accepted as a home or seen as a long-term residence, especially because it is not the place where animals can do well. This inner image can also be interpreted as an act of resistance, as it shows that government agendas are complied with for their own purposes/needs but not internalized or interpreted as the government would like it.

7. Discussion

As I have shown, the Nyangatom are in a process of deep structural change with regards to their life. This impacts the way they present and define themselves just as much as these identifications are fluid and contradictory. Lange, when looking at the philosopher Dussel, writes that: "modernity justifies an irrational praxis of violence, despite its ideal of a discursive community in which coercion is unacceptable". First and foremost, Europe understands itself as more developed, its civilization superior to others [23]. While the idea of Europe as the "modernity haven" is not applicable in a context in which the sedentary Ethiopian society and a more abstract idea of "Western" represents modern, what is applicable is the colonially charged discourse between modernity and how it is used discursively as a justification by the state to implement new ways of living as more "reasonable". What has also been shown is how this co-exists with notions of superiority of the modern, sedentary ideal. This notion in itself is contradictory, as coercion to the more "reasonable" will lead to subordination and, eventually, a sedentary poor, as is the case for many urbanized pastoralists [24]. Settling comes along with an arguably increased state of dependency on the state and the cash economy.

Barker speaks about the enlightenment ideal as an ideal of innocent knowledge: "an ideal that masks the instrumental role that development has played in maintaining global structures of neocolonialism and dependency. Instead of progress and prosperity, much of the world has experienced profound poverty, growing income inequality, high debt burdens, and environmental degradation" [25]. In the Ethiopian case it can, however, also be seen that the extent to which the ideals about a somewhat enlightening modernity are internalized vary greatly and next to individual opinions, age plays a key role. Whereas some of the younger generation has internalized modern notions of prosperity and wealth, elder pastoralists reject them and actually view current and forthcoming lifestyles as poor, less beautiful, less cultural and even as reducing the possibility to be a "real man or real woman". Furthermore, pastoralists sometimes go along with administrative policies because they have no choice, yet reject their meaning—such as the meaning of home.

While one should not fall into the hasty dualism of the "good, premodern, traditional world (...) and the bad, modern, industrialized world and the thoughtless" [26], this article has shown how both notions get internalized in contradictory ways within the everyday of a livelihood in transition. The everyday experience here is the site in which larger ideals intersect and compete.

8. Conclusions

The Nyangatom pastoralists of South Omo are faced with drastic structural changes in their lives and livelihoods. In this article, I have focused not on the transformational process of large-scale-agriculture or the government's sedentary strategy itself, but rather on the images that float especially among elder herders about the past, present and future of the Nyangatom. By showing these through segments of my interviews, I show how radical structural changes are translated into the everyday of eating, fashion and sleeping habits. The Nyangatom pastoralists already now position themselves in the past by highlighting that "real pastoralism" does not exist anymore. Modern life, especially through education, has come "inside pastoralism". The governmental strategy works discursively and actively by introducing discourses that devalue pastoralism, which are internalized, especially by the younger generation, but also show up in contradictory ways. Yet hopes and ideas of compromises between modern and pastoral persist and the Nyangatom already employ strategies to benefit best from both livelihoods.

Acknowledgments: For the conduct of this research, I have been funded by the Heinrich Böll Foundation as well as the Freie Universität Belin. Further acknowledgement goes to the Adventist University Friedensau for supporting my current research activities and Denise Erdmann for sensible editing.

Conflicts of Interest: The founding sponsors had no role in the design of the study; in the collection, analyses, or interpretation of data; in the writing of the manuscript, and in the decision to publish the results.

References

1. Kassahun, A.; Snyman, H.; Smit, G. Impact of rangeland degradation on the pastoral production systems, livelihoods and perceptions of the Somali pastoralists in Eastern Ethiopia. *J. Arid Environ.* **2008**, *72*, 1265–1281. [CrossRef]
2. IRINNEWS Pastoralism against the Odds. Available online: http://www.irinnews.org/report/90373/ethiopia-pastoralism-against-odds (accessed on 1 February 2016).
3. Ridgewell, A.; Flintan, F. *Gender and Pastoralism Vol. II: Livelihoods and Income Development in Ethiopia*; SOS: Sahel, Ethiopia, 2007.
4. Federal Democratic Republic of Ethiopia. *Ministry of Finance and Development (Growth and Transformation Plan 2010–2014)*; Main Text; Federal Democratic Republic of Ethiopia: Addis Ababa, Ethiopia, 2010; p. 59.
5. Homewood, K. *Ecology of African Pastoralist Societies*; James Currey: Oxford, UK, 2008.
6. International Rivers Ethiopia's Gibe III Dam: Sowing Hunger and Conflict. 2011. Available online: https://www.internationalrivers.org/sites/default/files/attached-files/gibe3fs_051509final.pdf (accessed on 1 February 2015).
7. Glowacki, L.; Wrangham, R. Warfare and reproductive success in a tribal population. *Proc. Natl. Acad. Sci. USA* **2015**, *112*, 348–353. [CrossRef] [PubMed]

8. Oakland Institute. Understanding land investment deals in Africa: Half a million lives threatened by land development for sugar plantations in Ethiopia's Lower Omo Valley. *Land Deal Brief* **2011**.
9. Ethiopian Omo-Kuraz Project (2015). Available online: http://ethiopiansugar.com/index.php/en/projects/kuraz-sugar-development-project (accessed on 1 May 2015).
10. Mayring, P. *Qualitative Inhaltsanalyse, Grundlagen und Techniken*; Auflage: Beltz Verlag, Basel, 2008.
11. Patton, Q. *Qualitative Research and Evaluation Methods*; Sage Publications: London, UK, 2002.
12. Pfitzner, J. Unser Kalender funktioniert nicht mehr. In *Ernährungssicherung und Anpassung im Zeichen des Klimawandels: Ein Fallbeispiel von Agropastoralisten in Südäthiopien. Diplomarbeit*; Geographisches Institut der Rheinischen Friedrich Wilhems Universität: Bonn, Germany, 2010; unpublished work.
13. Glaser, B.; Strauss, A. *The Discovery of Grounded Theory: Strategies for Qualitative Research*, 1st ed.; Aldine: Chicago, IL, USA, 1967.
14. Foucalt, M. *Discipline and Punish*; Pantheon: New York, NY, USA, 1977.
15. Said, E. *Orientalism*; Vintage Books: New York, NY, USA, 1979.
16. Naples, N. Feminist methodology and its discontents. In *The SAGE Handbook of Social Science Methodology*; Outhwite, W., Ed.; SAGE Publications: London, UK, 2007.
17. Do Mar Castro Varela, M. *Unzeitgemäße Hoffnungen: Migrantinnen Zwischen Selbsterfindung Und Gelehrter Hoffnung*; Transcript Verlag: Bielefeld, Germany, 2007.
18. Mama, A. In: Taiwo, O. Feminism and Africa: Reflections on the poverty of theory. In *African Women and Feminism: Reflecting on the Politics of Sisterhood*; Oyewumi, O., Ed.; Africa World Press: Trenton, NJ, USA, 2003.
19. Nadig, M. *Die Verborgene Kultur der Frau*; Psyche: Fischer, Germany, 1992.
20. Tuana, N. *Women and the History of Philosophy*; Paragon Issues in Philosophy; Centre for Digital Philosophy: London, ON, Canada, 1992.
21. Schutte, O. Cultural alterity: Cross-cultural communication and feminist theory in North-South Contexts. *Hypatia* **1998**, *13*, 53–72. [CrossRef]
22. Abbink, J.; Askew, K.; Dori, D.F.; Fratkin, E.; Gabbert, E.C.; Galaty, J.; La Tosky, S.; Lydall, J.; Mahmoud, H.A.; Markakis, J.; et al. Transforming pastoral lands and livelihoods in eastern Africa. In *Lands of the Future*; Working Paper 154; Max Planck Institute for Social Anthropolog: Halle, Germany, 2014.
23. Lange, L. Burnt offerings to rationality: A feminist reading of the construction of indigenous peoples in Enrique Dussel's theory of modernity. *Hypatia* **1998**, *13*, 132–145. [CrossRef]
24. Gertel, J.; Rottenburg, R.; Calkins, S. (Eds.) *Disrupting Territories: Land, Commodification and Conflict in Sudan*; Boydell & Brewer Ltd.: Suffolk, UK, 2014.
25. Barker, D. Dualisms, Discourse and Development. *Hypatia* **1998**, *13 Pt 2*, 83–94. [CrossRef]
26. Harcourt, W. *Feminist Perspectives on Sustainable Development*; Zed Books: London, UK, 1994.

land

MDPI

Review

Environmental Conservation in the Tibetan Plateau Region: Lessons for China's Belt and Road Initiative in the Mountains of Central Asia

J. Marc Foggin [1,2]

1 Mountain Societies Research Institute, University of Central Asia, Bishkek 720001, Kyrgyzstan; marc.foggin@ucentralasia.org; Tel.: +996-(0)770-822456

2 Institute of Asian Research, University of British Columbia, Vancouver, BC V6T 1Z2, Canada

Received: 15 March 2018; Accepted: 10 April 2018; Published: 17 April 2018

Abstract: Environmental conservation has developed significantly in China over the past 20 years, including more collaborative approaches and recent advances in establishing a national parks system. This study reviews the development of protected areas in the headwaters of the Yangtze River, drawing lessons from experiences of community development and co-management approaches. Community engagement and participation in developing localized plans for natural resource utilization and conservation have been critical features of successful ventures. Government programs and policies, the emergence of grassroots civil society, and the development of herders' cooperatives and protected areas, are all tracked, each pointing towards the significant value of *inclusive biodiversity conservation* approaches for meeting broadly agreed development agendas, such as achieving the Sustainable Development Goals by 2030. Observations from the vast, high, arid, and semi-arid lands of the Tibetan Plateau are then considered in light of China's Belt and Road Initiative, which is bringing vast financial and technical resources to the world. Special attention is given to applying the lessons that have been learned in China to the mountains of Central Asia, globally recognized as a biodiversity hotspot and a water tower for large downstream populations. Keeping local people *at the heart of conservation* is deemed fundamentally important.

Keywords: China; Tibetan Plateau; Sanjiangyuan region; social–ecological systems; pastoralism; partnerships; co-management; national parks; Belt and Road Initiative; mountains of Central Asia

---

1. Introduction

With a global population now exceeding 7.6 billion people and estimated to reach over 11 billion by 2100 [1], humanity consumes significantly more resources than our planet can provide in the long term, and our patterns of consumption and the waste we produce cannot be absorbed by the world's natural systems. In short, our use of natural resources is stretching beyond regenerative capacities [2–5], or even more succinctly, the world has reached planetary limits [6].

The need for sustainability is now agreed by nearly all countries in the world and on this basis a suite of Global Goals (or Sustainable Development Goals) have been adopted to help end poverty, protect the planet, and ensure prosperity for all (see UN Resolution 70/1, *Transforming our world: the 2030 Agenda for Sustainable Development* [7]). The 17 Global Goals are further developed as 169 targets that we corporately aim to achieve over the next 12 years, and for which everyone will need to play their part: all governments, the private sector, civil society, and citizens alike. Now, with global sustainability perspectives agreed in principle, and many national commitments in place, one final question remains: *scale*. Of course, it should be at all scales; but therein lies the great challenge. Who will be the winners, and who the losers? In particular, no party should advance "local sustainability" purely through an externalization of costs. Conversely, "regional sustainability" should not be achieved through an inequitable transfer of environmental or development burdens between geographic regions, especially if shifting burdens toward those less powerful, remote, or unseen populations. For this reason, comprehensive assessments with multi-stakeholder dialogues and inclusive decision-making processes and governance arrangements are fundamentally important.

With increasing regional and international connectivity (cf. globalization), the approaches adopted and decisions made in one place now almost always impact communities and situations elsewhere, often well beyond the original intended areas or sectors of focus. Additionally, sustainability-oriented goals alone rarely drive all decisions. Political, economic, and environmental factors bring trade-offs such that, at best, we can hope for *science-informed*, if not actually evidence-based, decisions. Nonetheless, learning from experience and then subsequently feeding the lessons learned into development processes are of paramount value.

With more than 1.4 billion people, China is coping with some of the world's greatest development challenges. It also is demonstrating great technological and scientific innovations and, over the last decade or two, it is become increasingly recognized as a global leader in research and development in numerous fields and disciplines [8–10]. Its growing economic and military strength and increasing global engagement, e.g., through the unprecedented *Belt and Road Initiative* (in this review paper referred to as BRI, but also known as *The Silk Road Economic Belt and the 21st Century Maritime Silk Road*, or for short the *One Belt, One Road* (OBOR) Initiative), also mark China as emerging leading agency in international affairs and global development trends, bringing its own geopolitical dimension to aid [11,12]. In light of its growing influence and geographic reach, how China has been dealing with the governance and management of its natural resources and the mechanisms it has adopted in recent years for achieving conservation and sustainability are very important aspects to consider, as these may be leveraged more widely to increase the overall likelihood of long-term success wherever China invests its time and resources, at home and abroad.

Within China, the Tibetan Plateau encompasses one-quarter of the national territory, and provides many benefits nationally and internationally, especially as ecosystems goods and services from the headwaters of Asia's major rivers. This vast, high altitude mountain landscape is recognized for such values, and much attention and financial resources have been given for environmental protection. However, conserving the natural environment is not contingent only on increased knowledge about nature or ecology or earth systems. It is equally dependent on appropriate understanding of the human or social dimensions of people's livelihoods and the utilization of natural resources, including who decides how resources are used (governance), as well as how they are used (management). The past two decades have seen particularly significant developments in how natural resources are utilized and conserved in the Tibetan Plateau region, including the emergence of community conservation initiatives and the development of a network of protected areas. Building on these experiences, China is now in process of establishing its first national parks, to be launched formally in 2020.

This review paper seeks to follow the development of conservation initiatives, along with local community development within emerging protected areas, over the past couple decades in western China. From this, it aims to draw out some key lessons that could help to inform future programming. How China engages with local communities and how it seeks to balance the overlapping needs and interests of nature and people when undertaking environmental action, will affect not only short-term returns on conservation investment, but also more generally the viability and long-term sustainability of projects and initiatives. Lessons gained from recent experiences in the Sanjiangyuan region of China (referring to the "three rivers sources", i.e., the headwaters of the Yellow, Yangtze, and Mekong rivers) may be applied well beyond protected areas—national parks or otherwise—and should include landscape level conservation and development strategies across western China, as well as neighboring mountainous regions in Central Asia encompassed in the scope of China's BRI.

With its rapidly growing influence around the world, strategic decisions and approaches adopted by China clearly will affect global sustainability. This review article aims to make some of the recent experiences in environmental management and conservation more widely known outside China, and to render their links with regional and global sustainability more explicit.

2. Methods

Many research methods have been employed over the course of the two decades that it has taken to learn and compile the information, insights, and recommendations presented herein.[1] Methods have included both qualitative and quantitative approaches: longer time-series, as well as shorter-duration or "snapshot" perspectives; larger sample size household surveys, as well as more in-depth interviews with selected informants and small focus groups; desk reviews of published literature; participatory action research; direct participant-observation; horizontal policy analysis; and other approaches spanning the natural and social sciences. These have been further complemented by a process of continuous reflection on past successes and failures, and renewed consideration of the foundations of sustainable development (e.g., the role of culture as the "fourth pillar" of sustainability [17,18]. More details about the interrelated social and ecological contexts encompassed in this review and about the methods adopted in this long-term endeavor are noted elsewhere (e.g., [13,19–24]).

As outlined in a recent documentary about the project work and affiliated research conducted over the past 20 years, upon which the majority of the reflections and recommendations for conservation and sustainable community development presented in this article are based,[2] the key roles played by the author and his team have included (i) *bridging* between a range of stakeholders (especially local versus external parties, each bringing different sociocultural, ideological, or economic ideals), which in turn has required (ii) *translation* between widely differing perspectives and/or worldviews (including scientific versus traditional languages, i.e., ways of seeing the world) and (iii) *trialing* new ideas or approaches to help address specific problem areas or to more effectively attain common goals in sustainable development, including the conservation of biodiversity.[3]

The analytic frameworks employed in this review paper are multiple, given the breadth of academic disciplines and perspectives, academic and otherwise, that have a bearing on environmental conservation outcomes. Most fundamentally, though, a socioecological systems (SES) framework is adopted, not least because of the wide chasm noted in China, now reducing but still significant, between the natural and social sciences and the differing policy recommendations that ensue. Greater mutual understanding across disciplines through SES-based analyses of the major challenges and benefits of emerging conservation models may lead to more robust syntheses.

The experiences and lessons learned are then further compared and embedded within the wider academic and development literature about community-based conservation and the governance of common property resources, including "co-management" of natural resources and protected areas [26–29]; about indigenous and community conserved areas (ICCAs) [30,31]; and about the need for mainstreaming biodiversity and protected areas within regional development planning [32,33]. This is further complemented by consideration of the wider experiences in conservation and sustainable development in western China [34–38] and in mountainous areas of Central Asia [22,39,40].

[1] The author is the founding director of Canadian non-government organization (NGO) Plateau Perspectives, established in 1998, through which the majority of learning opportunities described in this article have taken place. The grassroots organization Upper Yangtze Organization (UYO) was established the same year by Tashi Dorje Hashil. Many of the author's and Tashi Dorje's conservation and development activities were undertaken together or in coordinated fashion, with many learnings taking place in unison. The founding directors' parallel histories and their mutual support and influence on each other are noted elsewhere [13–16].

[2] See http://plateauperspectives.org/documentary/.

[3] These three roles (i.e., bridging divides, translating between the different "languages" of culture and science, and providing support through trialing of new ideas or approaches) hold significant overlap with key functions identified for the Chinese environmental NGO Shanshui Conservation Center [25], which has been engaged in conservation programs in the Sanjiangyuan since 2007—namely mediation, translation, and coordination. Both NGOs have sought to enhance local participation in ecosystem management and collaboration amongst actors.

Insights about China's BRI stem from published literature and media, as well as personal exchanges and interactions with Chinese and non-Chinese colleagues, and participation in recent BRI-related workshops [41–45].

3. Background

3.1. Major Geographies

This review of the recent history of conservation efforts in China is focused on the Tibetan plateau region, specifically, the headwaters of the Yangtze River in the renowned vast Sanjiangyuan region. Administratively, the Sanjiangyuan is situated in southern Qinghai Province, one of China's largest provincial level units. As with most of western China, climatic conditions are generally arid, human population density is very low (on average < 10 people/km^2) (Figure 1), and most strategic decisions are made in light of a perceived need to strengthen social stability, and security perspectives related to proximity to borderlands. A major transition in development priorities took place in China around 2000, with the country shifting its focus from developing its eastern coastal cities to developing the expansive western hinterlands under the auspices of the Great Western Development Program. Since then, urbanization has continued on massive scale. Over half of the world's population now resides in urban centers, with similar national figures and even faster rates of urbanization occurring in China [46]. The peoples and communities who manage the mountains, grasslands and desert resources in western China, however, remain primarily rural, and often are mainly ethnic minorities within China.

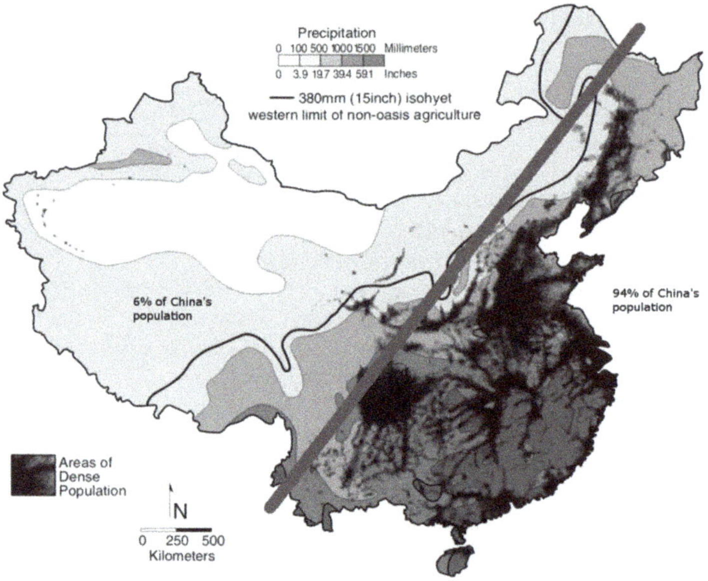

Figure 1. Precipitation and human population in China; showing precipitation isopleths (shading) and highly concentrated areas of human population. Areas north and west of the solid line cannot support agriculture in the absence of irrigation, and hence qualify as "western China." The majority of China's people resides in the eastern half of the country. Adapted from [34], used with permission.

Encompassing around one-quarter of China's land area—or 2.5 million km^2—and with an average elevation over 4000 m above sea level (asl), the Tibetan Plateau is the largest and highest mountain area in the world. While China itself is recognized as one of the most mega-diverse countries in the world [47], the diversity of biogeographic realms and ecosystems that meet or are found in the Tibetan Plateau region make it particularly noteworthy for conservation of global biodiversity. Mountain ranges both bind and transect the Tibetan Plateau, including the Himalayas, Hengduan, Tanggula, Qilian, Kunlun, and Karakoram ranges, which also connect, in turn, with the Pamirs, Hindu Kush, and Tian Shan ranges (Figure 2). Considered together, these contiguous mountainous areas—characterized for millennia by nomadic pastoralism, most prominently yak husbandry [22,48,49]—are more broadly known as High Asia or Greater Central Asia [50–53].

For its part, the Tibetan Plateau *per se* is often referred to as the Third Pole of the world, because of its vast extent and high altitude [54], also as the Water Tower of Asia, because of the six major Asian rivers that originate on the Plateau [55]. Around 3 billion people live on or downstream of the Plateau [56], and its physical presence and annual snowpack conditions influence the global climate system and Asian monsoon characteristics, respectively. Four major ecosystem types are identified: alpine meadow, alpine steppe, alpine desert, and alpine shrub [57]. Taken together, these vegetative communities form one of the most extensive grazing social–ecological systems in the world [58], home to a unique, diverse assemblage of wildlife [59] and Tibetan livestock that, until recently, provided at least basic sustenance for the majority of herders in this vast highland region [60]. Altogether, around 1.3 million km^2 of the Tibetan plateau is classified as steppe (the entire plateau covers around 2.5 million km^2, or approximately one-quarter of China's land area), including alpine meadow (45%), alpine steppe (28%), alpine desert (5%), and other types [61].

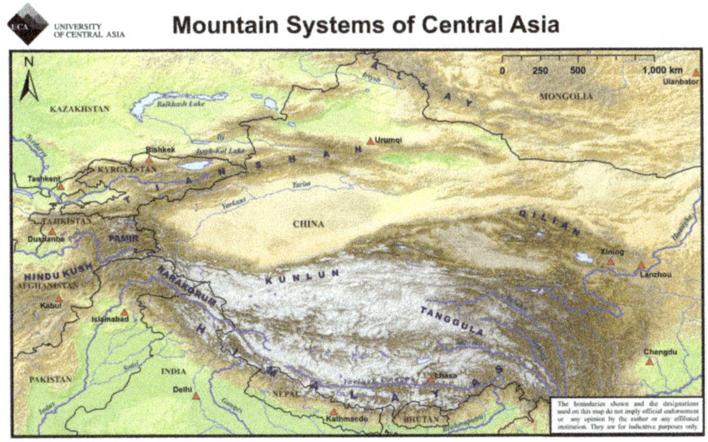

Figure 2. The mountains of western China and Central Asia (Mountain Societies Research Institute, University of Central Asia).

Returning again to the focal area of this review, the Sanjiangyuan region is situated in the heart of the Tibetan Plateau, encompassing the headwaters of the Yangtze, Yellow, and Mekong Rivers (see [13,14,55,62,63]). It comprises 363,000 km^2 or around half the land area of Qinghai Province [64] (Figure 3). As an area of great ecological significance to China as a whole, through its biodiversity and its critical water regulatory functions, a large portion of the region was designated as a National Nature Reserve in 2000, encompassing 153,200 km^2, or the area of England and Wales combined. Most recently, the decision was made in December 2015 to convert a large area of the nature reserve into China's first national park, to be administered by the central government—the Sanjiangyuan National Park—covering an area of 120,000 km^2 [65]. With initial trial operations underway since April 2016, the park management aims to hire at least 10,000 wardens from local herding families to oversee the vast region and to serve as community liaisons, bringing together the dual goals of conservation and socioeconomic development.

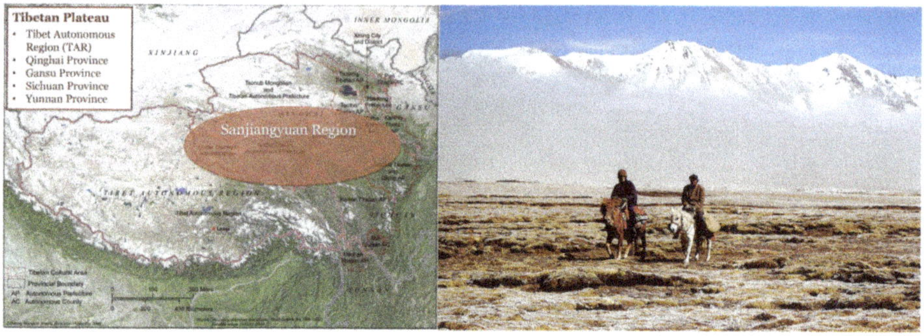

Figure 3. Map of the Tibetan Plateau and Sanjiangyuan (adapted from [66]), and Tibetan pastoralists riding their horses in the high altitude rangeland *circa* 4600 m asl near the source of the Mekong River.

However, as elsewhere across the Tibetan Plateau, and in the high mountains of Central Asia, the Sanjiangyuan area and its resident pastoral communities remain vulnerable to a range of factors originating outside its own geography, including the influence of climate change, such as glaciers shrinking and changing seasonality and intensity of precipitation patterns [63,67–70].

3.2. Protected Areas in China

China is a mega-diverse country with nearly 35,000 vascular plants (ranking third in the world) and 6445 vertebrate species (7 percent of the world's total) [71,72]. To protect its unique wildlife—half of China's terrestrial vertebrates are endemic, occurring only in China—the country has established a vast network of nature reserves since 1956, constituting the backbone of China's protected area system. Currently, a total of 2729 nature reserves are spread across the country, encompassing around 15 percent of the territory, with nationally administered reserves covering two-thirds of this area [73]. Altogether, more than 10,000 protected areas (including nature reserves) cover about 18 percent of the country, including forests, geological parks, wetland parks, world natural and cultural heritage sites, and scenic zones [74].

Since 1997, at least seven major protection plans have been developed at national level in China. Developing related laws and regulations, identifying priority areas for biodiversity conservation and key ecological function zones across the country [72], and participation in major conventions, are all constituent parts of these plans. Many positive steps have thus been made in the past few decades for the conservation of China's biodiversity, even while there are some limitations and challenges that clearly remain [34,72,75].

One of the challenges with nature reserves has been to find ways of balancing conservation needs with local communities' socioeconomic aspirations. The development of tourism in protected areas has been trialed in many instances; especially over the past couple decades. However, this approach has brought a range of inherent challenges revolving mainly around the sharing of tourism benefits and commercialization of reserves [76–78], with most instances of tourism development in protected areas being to finance the nature reserves, rather than local development. The overarching goal of nature reserves has remained as "conservation", with most reserve management authorities' interactions with and valuing of resident communities (e.g., for implementing conservation action, such as environmental monitoring or reducing human–wildlife conflict) being mainly pragmatic.

Most recently, China has proposed a new approach to conservation, in line with growing recognition—nationally and globally—of promoting the value of protected areas to not only protect biodiversity, but also to secure life-supporting ecosystem services *for human benefit*.[4] Aiming to protect special habitats and regulating services while integrating human activities, a new type of protected area is now being introduced in China: the "national park" system (see [71,79]).

Official plans for the national park system clearly indicate that "national parks will be set up... to achieve a combination of ecological protection and sustainable development" [80] and "the biggest difference between a nature reserve and a national park is that the latter offers more social and economic benefits... There should be a balance between protection and utilization in any successful national park, so that it can offer both development opportunities and effective protection" [81].[5] The national park system also aims to be a showcase for "ecological civilization", the government's vision for a sustainable relationship with the environment [82,83].

In the first national park—the Sanjiangyuan National Park—"herders and farmers [are expected to] be major forces behind environmental protection [and new] work is expected to provide jobs, boost farmers' incomes and give them an incentive to protect the environment" [80]. According to the *Overall Programme for Construction of a National Park System* published last year by China's highest level of decision-making bodies, altogether 10 new national parks are being trialed, and will officially be launched in 2020 [82]. As stressed by senior management in China's first national park within this select group (i.e., Sanjiangyuan National Park [65,84,85]), national parks shall have two distinct, parallel, reinforcing goals: environmental conservation, along with socioeconomic development of resident communities (personal communication, August 2016). The first 10 pilot national parks currently being developed in China [79,86,87] are introduced in Table 1.

Table 1. National Parks (NP) in trial development phase in China's new national parks system.

Name	Province	Description	Total Area
Sanjiangyuan NP	Qinghai	The first and largest of the 10 pilot national parks, encompassing the headwaters of 3 major rivers: the Yellow, Yangtze, and Mekong rivers [88]	123,100 km^2
Great Wall NP	Hebei	The only NP in China that is being built around cultural heritage, [1] including the Ming Dynasty Tombs and Badaling Great Wall [89]	60 km^2
Pudacuo NP	Yunnan	Renowned for its primitive landscape including lakes, wetlands, forests, and meadows, as well as nearly 100 endangered wildlife species. Very rich in biodiversity, it contains over 20 percent of China's plant species and around 1/3 of its mammals and birds. Home to black-necked cranes, orchids, and Himalayan Yew. Tibetan culture and customs also are component parts of the NP.	1313 km^2
Qianjiangyuan NP	Zhejiang	Includes habitat of threatened endemic species such as Elliot's pheasant and tufted deer. Known for seed plants.	252 km^2
Mount Wuyi NP	Fujian	Known for deep canyons, dense subtropical forests, waterfalls, and wildlife. One of China's most famous, beautiful landscapes. A World Heritage Site. [2]	983 km^2

[4] China has been protecting wildlife for many decades through a system of nature reserves and other protected areas. However, the value (or benefit) of such protected areas *for human wellbeing* has not always been considered explicitly. Until China's leadership recognized the integral relationship between people and nature even within protected areas, the "national park" system could not have developed. (Note: The term *protected area* is a high level generic category for all forms of *de jure* and *de facto* protection of the natural environment. *Nature reserves* constitute one form of protected areas, which prioritize the protection of wildlife and habitats, but without direct consideration of human wellbeing. On the other hand, *national parks* have a dual mandate to protect nature as well as to benefit people.) In China, a deeper understanding of "the environment" as coupled *socio*-ecological systems (SES) has emerged only in recent years. In the case of Sanjiangyuan region, an SES perspective has emerged as partners have jointly explored more collaborative models of conservation. Ultimately, it was decided that a national park model of conservation—with its dual mandate—was the most appropriate form to adopt for addressing both the ecological significance (and need for protection) and human development interests in the region.

[5] Both national parks and natures reserves are legal entities in China. Development of the national parks system aims to centralize and "rationalize" the planning and management of nationally prioritized areas. On the other hand, the legal home (government agency) and modes of operation of nature reserves is much more diverse. As for "community conserved areas", this term is more descriptive than legal in China; but is now well recognized by the International Union for the Conservation of Nature (IUCN) and other agencies internationally.

Table 1. *Cont.*

Name	Province	Description	Total Area
Shennongjia NP	Hubei	Well known for vast primeval forests, large numbers of endangered species and varieties of medicinal herbs, and legends such as the mysterious "Wild Man". Situated near China's Three Gorges. Largely karstic, with many caves. A World Heritage Site. [3]	1170 km^2
Mount Nanshan NP	Hunan	Stopping point for tens of thousands of migratory birds.	619 km^2
Siberian Tiger and Siberian Leopard NP	Jilin and Heilongjiang	The park covers one-quarter of the area of the Siberian tiger and Siberian leopard (*hubao*) habitats in China and over 75% of the animals' wild population [82,90,91].	14,600 km^2
Giant Panda NP	Sichuan, Gansu, and Shaanxi	More than 80 protected areas will be incorporated into this park, home to most of the world's giant pandas (*daxiongmao*), along with much other wildlife.	27,000 km^2
Qilian Mountains NP	Gansu and Qinghai	High altitude mountain range (at 4000 to 6000 m asl) with meadows, forests, glaciers, and rugged landscape. Home to snow leopard, white-lipped deer, and other rare wildlife species.	50,000 km^2

[1] http://whc.unesco.org/en/list/438. [2] http://whc.unesco.org/en/list/911. [3] http://whc.unesco.org/en/list/1509.

The emerging perspective in China about the multiplicity of roles and the major objectives of protected areas (i.e., beyond strictly protectionist approaches to conservation, as exhibited in the development of the new national park system) is consistent with the country's long-term commitment to the UN Convention on Biological Diversity (CBD) and the more recent UN Sustainable Development Goals (SDGs). A strategic mainstreaming of ecological protection across all development sectors was first apparent in China's "ecological red line" approach [74,82,92,93], reinforced by programs such as the Global Environment Facility-supported (GEF) suite of biodiversity- and protected areas-strengthening projects at provincial and national levels. About half of the total land area in western China is encompassed within the 25 national key ecological function zones (Figure 4), each prioritizing conservation functions for water source areas, soil and water, windbreaks and sand fixation, or biodiversity [94].

The value of protected areas for achieving the SDGs is now well established [95–98], as is China's commitment to the SDGs [99,100] and the tightly connected Paris Accord [101].

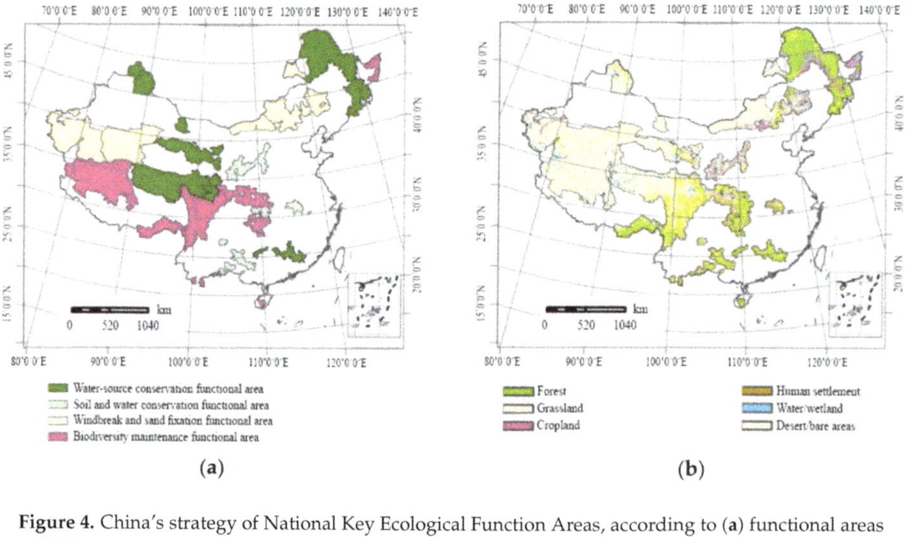

Figure 4. China's strategy of National Key Ecological Function Areas, according to (**a**) functional areas and (**b**) major ecosystems [91]. Used with permission.

4. Results

4.1. Recent Developments in Conservation and Sustainable Use of Tibetan Rangelands (Case Study)

Across the Sanjiangyuan region in southern Qinghai Province, traditional nomadic pastoralism, and to a lesser extent, subsistence hunting, have been practiced for at least 5000 years [19,55]. Moving from tribal feudalism (until *circa* 1958) to collectivism (*circa* 1958–1978) to quasi-privatization of land and privatization of livestock (from 1978 to present), local forms of pastoralism have seen major transformations in less than a lifetime. Nomadic pastoral practices have, by and large, given way to more sedentary lifestyles, and in many instances to people's permanent relocation and urbanization (Figure 5). Many elements of the former seasonal mobility of livestock grazing and flexibility in the resource management system are now largely gone, with reduced social resilience to shocks and stressors, including climate change. That being said, local sociocultural, political, and economic opportunities and local adaptations remain dynamic in the region; as the author has witnessed on multiple occasions in the Yangtze River headwaters, from 1998 to present [13,19,55,102].

(a) (b) (c) (d) (e) (f)

Figure 5. The last two decades have seen major changes in pastoral practices and people's livelihoods in the Sanjiangyuan region, shifting from traditional nomadic pastoralism on the vast grasslands up until the 1990s and early 2000s; to sedentarization, with construction of winter homes and introduction of agricultural practices (mostly for winter fodder) in the mid-2000s; to full relocation and settlement of a large portion of the population in new villages built under the eco-migration policy. Photo descriptions: (**a**) Tibetan family and yak hair tent on the grasslands; (**b**) pastoralist family moving to summer pastures; (**c**) winter dwelling of a Tibetan pastoralist family, including winter home, livestock shelter, and agricultural field; (**d**) family living in converted livestock shelter; (**e**) "nomads without pastures" in a recently built relocation village; (**f**) Tibetan "nomad" family, formerly livestock herders, now living in a relocation village.

Locally and around the world, traditional practices and perspectives are now increasingly recognized as converging with so-called more modern or scientific approaches [103–105]. In many places, an appreciation for complementary ways of seeing and knowing the world around us is becoming increasingly mainstream, and dialogue is starting to occur across former sociocultural divides. In the Tibetan plateau region, climate change is now being noted, and sometimes even documented, by herding communities. Much potential also exists for further development of "citizen science" to strengthen dialogue and collaborative studies, with partners coming from diverse educational and sociocultural backgrounds. Both recognizing and maximizing benefits of different "ways of knowing" have long been at the heart of this case study, whereby the author and colleagues have proactively sought to bridge different worlds [106] and to champion for effective restoration of "sense of ownership" to local communities [107].

Since 1998, the following developments have been particularly noteworthy for the management and conservation of the grassland social–ecological systems highlighted here, including their biodiversity and the social and livelihood practices of local communities. Transformations also have been taking place at the science–society interface. This is especially evident with the meeting of "scientific development" from Chinese national planners and government agencies, on one hand, and more traditional views and practices of herding communities, on the other hand, through a collaborative approach generally known as "community co-management" of natural resources and protected areas.

The following major developments from the past two decades together constitute a high-level review of recent history of conservation and environmental management in western China, as expressed in the particulars of the Sanjiangyuan region of the Tibetan Plateau.

4.1.1. Community Conserved Areas and the Development of Local Civil Society

Five local protected areas were created in 1998, established by community members in the Suojia district (former commune) in western Zhiduo County to protect endangered wildlife [13,19] (Figure 6). Although later incorporated into the Sanjiangyuan National Nature Reserve (and ultimately, the Sanjiangyuan National Park), their origins and ongoing operations have largely remained with the local herding communities themselves, who began to monitor local wildlife before any official protected area was established. The species focus built on a local cultural appreciation of wildlife as *jewels of the land* [6] including the snow leopard (*Panthera uncia*), Tibetan antelope (*Chiru hodgsoni*), wild ass or kiang (*Equus kiang*), wild yak (*Bos mutus*), and black-necked crane (*Grus nigricollis*).

Development of these community conserved areas occurred simultaneously with the establishment of the grassroots non-government organization (NGO), Upper Yangtze Environment and Development Organization, or Upper Yangtze Organization (UYO) for short, established in 1998. This NGO brought together more than 100 local community members for common purpose, with special emphasis on wildlife conservation and so-called "green development" (i.e., environmentally sound development) [13,19,103]. Embedded within a sociocultural ethic of conservation and a longer local history of activism—from the work of Sonam Dorje and the Wild Yak Brigade [108–110], to community-wide mobilization led by the UYO [13,21,33] and the establishment and operation of village-level tent schools and health centers (see [111,112] and subsequent annual reports,[7] also [23,24]; Figure 7)—the various activities and approaches initially adopted in Muqu community contributed to setting the scene for community based conservation in the Yangtze River headwaters. Under the leadership of Tashi Dorje Hashil, work undertaken by Snowland Great Rivers Environmental Protection Association (SGREPA [15]) has been instrumental in taking these and other community approaches even further, scaling them up and extending across the province.

---

6 For participatory video created by Muqu community, see http://www.marcfoggin.com/participatory-video/.
7 To download Plateau Perspectives' Annual Reports 2005–17, see http://plateauperspectives.org/downloads/.

Following establishment of the UYO, several other communities followed suit and established their own organizations or associations, often encouraged by the charismatic leadership of a few individuals through trusted relationships built over time. Most recently, a network of more than 30 local innovative thinkers and conservation leaders across the province have been recognized as conservation heroes for their sustained efforts [16,113].

Figure 6. Endangered wildlife protected by local herding communities includes (**a**) snow leopard, (**b**) wild yak, (**c**) black-necked crane, (**d**) wild ass, and (**e**) wetland birds, including (**f**) bar-headed geese.

Figure 7. The Muqu village tent school was established *circa* 2000, shown here (**a**) with first intake of students and (**b**) in its final location, including permanent buildings, with prime snow leopard habitat in the background.

4.1.2. Sanjiangyuan National Nature Reserve and Community Co-Management

Shortly after the above community-based protected areas were established, the Sanjiangyuan National Nature Reserve was announced, formally established in 2000. However, local Tibetan communities now found themselves living within a nature reserve, with a new suite of official regulations affecting their lives, yet having had very little, if any, involvement in its planning and establishment. With initially no provision for involvement in its management, the herders residing in the Yangtze River headwaters became deeply concerned about their future. It is only with an intentional process of multi-stakeholder dialogue in the first few years of the reserve's existence that the management authorities came to recognize both the potential roles and multiple benefits that could be derived from developing partnerships with herding communities. Thus was initiated what has become the main *modus operandi* of the reserve, "community co-management" (Figures 8 and 9).

Figure 8. Jiongqu village leader at a co-management planning meeting on the grassland, *circa* 2007.

Figure 9. Co-management meeting with nature reserve and village representatives, October 2007.

The decision to adopt a co-management approach was made for four main reasons: (i) to compensate for a shortage of field staff, (ii) to capitalize on local herders' knowledge and abilities, (iii) to increase environmental awareness, and (iv) to reduce illegal poaching. In addition to these four reasons, the author and his team also sought to advance this approach because (v) such dialogue and partnership would help empower and build the capacities of the local community, and (vi) actual demonstration of partnership-in-action could be used to highlight to policy makers how local communities may be partners in conservation, i.e., seen as part of the solution, rather than assumed (wrongly) to be primarily detrimental in areas of high biodiversity value.

As the national nature reserve's core zones were geographically overlapping with the community-based conservation areas, there was high risk that the national plan for conservation would in fact negate the community's plans and goodwill for contributing to environmental monitoring, raising environmental awareness, conducting anti-poaching patrols, etc.—all heretofore undertaken on voluntary basis by the community members.

At a planning workshop in October 2005 that included nature reserve management authorities, community representatives, local government, and selected international experts familiar with protected areas and their management, grassland ecology, and wildlife research [14], an arrangement was agreed between the nature reserve, the international NGO Plateau Perspectives and Muqu community, whereby co-management would be trialed through a joint monitoring and conservation project revolving around community based snow leopard monitoring, research, and conservation [14,114–117]. Training workshops and data collection by local community members subsequently took place over several years, leading to a clear demonstration of the value of sustained partnerships for conservation purposes [21,118].[8] Earlier direct observations of snow leopard and their sign by community members was subsequently extended with use of camera traps as collaborative projects intensified [66,116,119,120].

In another herding community near the source of the Yangtze River, village level regulations were agreed as early as 1998 to limit illegal poaching by local and external hunters—several years before creation of the Sanjiangyuan National Nature Reserve. Over time, this led to the establishment of a local organization, Friends of Wild Yak Association. This community venture was subsequently incorporated into the "contract conservation" model that was trialed by the nature reserve in its broader community co-management framework [25,118,121]. In this approach, a contract is agreed between the local community and reserve authorities, whereby local community rangers monitor wildlife populations (in this instance, wild yak) in exchange for agreed payments into a local community-managed development fund. Both this and the aforementioned snow leopard monitoring and conservation work were explicitly referenced in the early preparatory phase of the GEF-supported UNDP Qinghai Biodiversity Conservation Project, implemented by the Qinghai Forestry Department (2013–2017) (see [122]) as the positive experiences of co-management that should be scaled up by the project and integrated into future operations.

Through these collaborative approaches, threats such as poaching have greatly diminished, wildlife observations are reported to government conservation authorities, local awareness and pride in wildlife have increased through school-based and other outreach activities, and local communities' sense of empowerment has extended beyond environmental conservation action, *per se*. Fortunately, beyond these social outcomes (which remain critically important and the primary focus of this review paper), biodiversity benefits also have accrued over time, with increasing wildlife populations being reported (personal communications, repeated from several community members, as well as several field staff and leaders of the Sanjiangyuan nature reserve and national park).

---

[8] See *Snow Leopard Community Conservation Project*, http://plateauperspectives.org/en/project/snowleopard/.

4.1.3. Community Associations and Cooperatives

Just beyond the boundary of the nature reserve, another herding community also was mobilized in 2010. Encouraged by two young charismatic community members, the Kegawa Herders Cooperative (named after a local sacred mountain) was established in 2010 (see [102,117,123–126]), in addition to a local community association that successfully trialed the operation of a revolving development trust fund [123,124,127,128]. Both the county and prefecture governments provided the enabling policy environments, but actual development of these efforts has been largely endogenous.

The launch of the Kegawa Herders Cooperative marked the beginning of what has been one of the most empowering development activities seen in the region since the author first visited in 1998. Starting with around 40 families, now over 90 households regularly participate in the projects and activities organized by the cooperative. These activities include regular monitoring of grassland conditions, development of handicrafts, operation of a small shop, and ongoing development of a community-based tourism venture. Long-term partnership also has developed with an international clothing company, kora, which produces high quality thermal clothing from fine yak wool. Through this partnership, all members of the cooperative have the option to sell their yak wool at guaranteed upper fair market value. The cooperative makes major decisions at annual general meetings and operates activities through working groups. Membership is purely voluntary, but requires some financial contribution, and annual payment of dividends (from profits generated over the year) is proportional to each member's contribution, whether financial or in-kind, e.g., livestock. General meetings are held annually, and ensure that every member has a voice, based on a principle of one person, one vote. Working groups have been established to oversee regular operations and facilitate ongoing communication amongst members, including in ecological husbandry, grassland monitoring, environmental awareness, garbage collection, tourism development, and other topics. Some profit generated from activities is reinvested into the core fund, and some is earmarked for social needs. Most recently, in late 2015, the cooperative has decided to develop its own yak herd, together with designated pastureland. Sale of milk, butter and cheese is planned, targeting mainly the expanding urban population in the county capital.

A separate community association with a revolving fund mechanism to support small businesses was also established in a nearby village, providing another way by which to strengthen local communities in natural resource governance, business development, and community-wide negotiations. Through provision of micro loans, three rounds of ventures were successfully trialed. Beyond the individual activities funded, the broader success of this community revolving fund project was the mobilization of community members to jointly discuss, agree, and implement development action. Specifically, external capital was provided to create a trust fund, and members of the association were trained in project planning, monitoring, multi-stakeholder meetings, financial management, and other skills necessary to operate this venture. Formation of an association was necessary to provide the legal basis of operations and to enable user groups to convene and establish clear and transparent operational procedures (Figure 10). After running community management trainings and project operations for several years, the 100% payback of the short-duration loans instilled confidence to launch follow-up revolving fund projects in additional pastoralist communities in the Yangtze River headwaters (personal communication, Du Fachun, August 2016).

For the government, both approaches outlined above demonstrate the viability of community-based management of resources, including financial resources. This could become even more important in the future as new options for transfer payments are considered (i.e., eco-compensation) for local pastoralists, in exchange for protection and maintenance of critical ecosystems services in China's rangelands. The county government already has assimilated many of the learnings from these experiences and has integrated them into its broader poverty alleviation work and economic development planning for rural areas (personal communication, Gongbo Tashi, August 2016). In a world that is increasingly "flattened" by globalization, which tends to encourage homogeneity over sociocultural diversity, empowerment and sense of identity are themselves inherently important ingredients for people's wellbeing.

(a) (b)

Figure 10. Two community ventures: (**a**) Lari community development revolving fund management group (photo: Du Fachun) and (**b**) tourism development and glacier monitoring by Kegawa Herders Cooperative (photo: Jigme Rabden).

4.1.4. Development of China's New National Park System

The dual overarching purposes of national parks in China have already been highlighted—aiming to conserve the environment while simultaneously promoting community development and wellbeing. As such, developing national parks (as compared to nature reserves, with their stricter conservation focus) has come as a natural progression, incorporating local interests and concerns of both parties, government and communities. The Sanjiangyuan National Park will open formally in the year 2020, with other national parks to follow soon after [85,129]. While guided and overseen centrally, the daily operations of national parks will be delegated to local government, along with other partners if co-management approaches are maintained.

Over the past couple decades, the major contributions made to the development of the national parks system in China has been the trialing of community co-management, thereby integrating a focus on community-level development needs and interests into hitherto primarily conservation-only deliberations for protected area management. This has included, inter alia, community involvement in wildlife monitoring, awareness and outreach including community wildlife festivals, support for the provision of rural health and education services, mitigation of human–wildlife conflict (especially by brown bear), support for local development mechanisms such as community level associations and cooperatives and the brokering of international partnerships with them (e.g., with kora [9]), and the organization of strategic study tours to Mongolia, Canada, and Nepal (each with strong emphasis on development of national parks, including co-management approaches, the need for monitoring and research, and opportunities and challenges related to tourism in the framework of protected area) (see [111] and subsequent annual reports; also [23,24,130,131]). Through these many areas of collaboration between the nature reserve, local community members and international partners, and the valuable deliberations they have enabled from 1998 to present, a new vision began to emerge about the possibilities offered by a national park model, understood to be more inclusive of community needs and interests.

Moving forward, it will be especially important to find ways to strengthen the greatest asset in the Sanjiangyuan National Park—its 10,000 strong "work force" of community-based field staff.

[9] See the high performance technical clothing company *kora*, http://www.kora.net/row/our-story/.

Notably, these staff will serve not only as environmental monitors, but more broadly as park wardens and as liaisons to the wider communities of which they are a part. As such, they should be empowered and strengthened to assist their own communities in matters of socioeconomic development, as well as reaching the conservation goals of the national park. Their capacities should be developed through trainings, and appropriate tools should be provided to enable them to accomplish all the functions assigned or adopted by them. In regard to the management objectives of the park, particularly the management of its wildlife and habitat resources, the collection of data will necessarily remain of paramount importance, as resources cannot be managed if their status and trends are unknown. Given the vastness of the Sanjiangyuan National Park and the large but dispersed population, the management authorities are presently considering how best to meet the need for large repeated-measure datasets, and how the new community wardens could assist in this. Particular attention is now being given especially to the potential use of the ubiquitous smartphone, and increasing local cell phone coverage, along with emerging information and communication technologies (ICT) such as "citizen science"-enabled digital field guides (e.g., Lapis Guides as a data collection tool [132]).[10] The development of long-term partnerships between local community members and national park management authorities for joint action could, with suitable tools and "translation" mechanisms (between community-based observations and traditional knowledge, on the one hand, and scientists and science-informed managers, on the other hand), provide substantial returns on investment for formal (i.e., government) conservation authorities, through creation of "big data" (personal communication, Gongbo Tashi, February 2018) as well as the empowerment of local communities as partners in conservation [22,133,134].

4.2. Lessons Learned for Environmental Conservation, 1998–2018

In the previous section, this article has tracked themes of biodiversity conservation and sustainability on the Tibetan Plateau, from 1998 to present, through the lens of community conservation ventures, the development of a unique network of protected areas, and affiliated community mobilization and socioeconomic development mechanisms. We can learn much from comprehensive, retrospective analysis of such unique development experiences, even beyond the intentional "participatory action research" orientation set out at the beginning. Moreover, these valuable lessons, summarized in Table 2, can be applied to strategic planning and decisions about the design and management of protected areas and natural resources in the future.

Furthermore, as China expands the scope of its influence, as is now occurring through the high stakes Belt and Road Initiative, "putting its best foot forward" will benefit the neighboring mountain regions and societies in Central Asia, as well as mark China more favorably on the global scene.

What has been learned from the above experiences? What key lessons can be captured and shared?—both for the strengthening of China's new national park system, now in development, and also for transferring into China's wider spheres of engagement, such as with neighboring countries through its multi-national Belt and Road Initiative (BRI).

Below are a series of key lessons that have been learned since 1998 through practical, hands-on experience working in partnership with Tibetan herding communities in the Sanjiangyuan region in western China. Each lesson represents an important element to be considered and encouraged as a core principle of community development and sustainable livelihoods (see e.g., [135–139]).

---

[10] For smartphone application to develop citizen science field guides, see *Lapis Guides*, http://lapisguides.org/.

Table 2. Lessons learned from recent conservation history in the Sanjiangyuan region, 1998–2018.

Major Lessons	for National Parks and Other Protected Areas	for BRI
Comprehensive perspectives are essential	√	√
Conservation champions are major players	√	
Community ownership increases commitment	√	√
Community participation increases effectiveness	√	√
Community co-management helps to build trust	√	√
Cooperatives and associations are effective tools	√	√
Citizen science can serve as a connecting bridge	√	

4.2.1. Comprehensive, Integrated Perspectives Must Be Adopted

An integrated perspective of environmental conservation and development goals across multiple user groups and other stakeholders, with simultaneously conflicting and concurring outlooks and purposes, will allow better assessment of the major opportunities and trade-offs for community development. Overall, a more nuanced understanding and appreciation of all different stakeholder groups' perspectives are necessary precursors for developing genuine, viable and lasting partnerships. An appreciation that people, communities, and cultures are to some extent "tied to the land" and work as part of broader socioecological systems, rather than as external agencies, also will aid in development of viable models for more sustainable, community-oriented development [140].

4.2.2. Conservation Champions Are Major Players in Bringing Change

Change is most effective when someone within the community, or with prior or current connection to the community, is convinced that some change is needed and is committed to mobilizing his or her fellow community members to take action. Such persons are generally known as "champions" for a cause, and can play an invaluable role. In many instances, engaging with and building the capacity of a charismatic conservation champion who is accepted by the community may be the most effective strategy that can be adopted to promote change. This person will play both bridging and translation roles, bringing different parties together and helping them to understand each other's perspectives, needs, limitations, and desires. A successful community conservation champion serves as a catalyst, often introducing innovative new ideas and bringing people together in constructive dialogue.

4.2.3. Community Level Sense of Ownership and Commitment are Essential

The value of personal and community-level commitment to environmental conservation cannot be overestimated. In some instances, an appreciation and commitment to sustainable use of resources is already present. Where it is not present—or has waned over time—early efforts should be placed on communications and outreach activities, as the fundamental purposes of any project intervention will not last much beyond a project's lifespan if such purposes are not suitably internalized. It is only with a sense of local ownership of decisions and actions that the durability of community members' involvement will be ensured. The mobilization of people for common purpose is often necessary, and this is usually most effective when calling on long-standing sociocultural norms and/or personal or community gain.

4.2.4. Community Participation Increases Sense of Responsibility and Effectiveness

Beyond inherent commitment to an idea or ideal, people also need to be involved in a project or activity for them to take on a fuller ownership of specific project outcomes, and conversely, to feel responsible when project outcomes are not being achieved. The level of community participation in a project also can feed back into people's commitment, which often can grow as dialogue amongst stakeholders improves and as communication and understanding increase. Participation generally is found in association with empowerment, improved management, and conservation success.

4.2.5. Community Co-Management Approaches Strengthen Actions and Build Trust

In the context of natural resource management, including management of protected areas, "community co-management" has a proven track record of bringing together a wide range of stakeholders to the figurative table and strengthening partnerships for common purposes. More voices and perspectives can be heard, with wisdom often emerging from diversity. Although negotiation is always necessary, collective decisions and subsequent actions taken are generally more resilient and lasting. With rich evidence in the literature, Canada has made co-management with indigenous communities one of the mainstays of the federal national park system, and in Norway, co-management extends beyond natural resources and conservation to the governance of virtually all aspects of socioeconomic development. With sufficient time allowed for trust and understanding to develop, as witnessed in the Tibetan plateau region, more participatory models of development (collaborative management), including provision of social services (especially health and education; see [23,24]) have developed as local examples of wider application of co-management principles.

4.2.6. Cooperatives and Associations Are Effective Tools for Common Action

One of the richest forms of "participation" is when communities themselves decide and drive their own development and conservation programs. They may then choose to organize through the development of associations (mostly related to common purposes, interest groups, or professions) or by establishing cooperative enterprises that help members raise capital, coordinate their efforts, and maximize opportunities through collective action. Community cooperatives and associations have been shown in many instances also to serve as effective tools, inter alia, for the mobilization of both people and resources, strengthening of organizational capacities, and accessing of financial resources. In the Tibetan plateau region, recent prime examples include the strengthening of value chains (direct sale of yak wool to international markets) and alternative uses of natural aesthetic and cultural resources (development of community-beneficial tourism, such as ecotourism and adventure tourism operated by local community members).

4.2.7. Citizen Science Can Serve as a Bridge between Communities and Government

As government and public alike increasingly wish to adopt more evidence-based decision-making, the need for more information and data is evident everywhere. For managing natural resources—including wildlife and habitats in national parks—regular, repeated collection of data is necessary. However, limited financial and human resources preclude this possibility in nearly all instances; but, with the advent of new information and communications technologies (ICT), such as smartphones and dedicated applications, and also building on the recognized value of community participation in conservation, larger scale data collection than ever before is now possible. Collecting "big data" will help move decision-making from guesswork and five-year plans toward more real-time responsive assessments and strategic interventions. Involving grassroots stakeholders in deciding what should be monitored and researched, as well as in subsequent analysis and decision-making, will further increase people's sense of ownership, empowerment, and commitment to common conservation goals. A new era of science with society may be brought to fruition with use of ICTs, where political will is present.

From a conservation perspective, and expanding on models of "inclusive protected area management" developed elsewhere [33], the above principles and approaches may be summed up as being the key components of *inclusive biodiversity conservation*.

5. Discussion

5.1. China's Belt and Road Initiative

Building on the legacy of the ancient Silk Road through Central Asia that for generations allowed the movement and trade of goods (along with ideas, philosophies, religion) between China and

Europe—and extending these to encompass many other routes and destinations, now including well over 60 countries with 70% of the world's population and nearly 30% of the global GDP [141,142] (Figure 11)—China has set the stage for a reshaping of international relations and globalization that is unprecedented in recent times. And, these new developments are here to stay. Not only has China recently enshrined *Xi Jinping Thought on Socialism with Chinese Characteristics for a New Era* into its constitution, but a lesser known amendment also was introduced with the statement that China will now be "following the principle of achieving shared growth through discussion and collaboration, and pursuing the Belt and Road Initiative" [143,144]. A new era has arrived and is now here to stay.

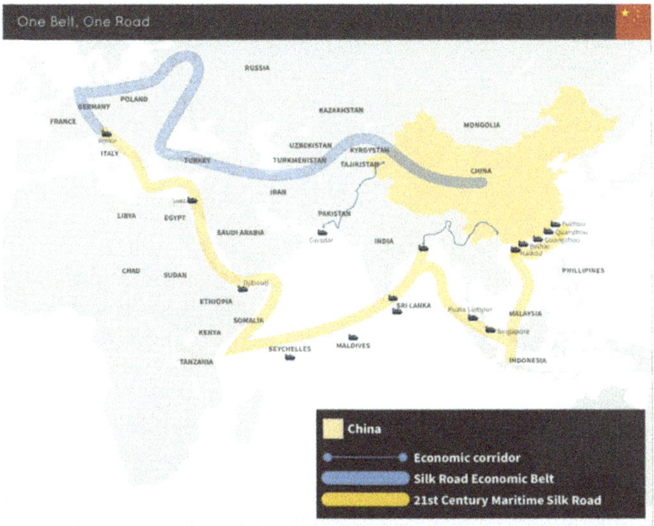

Figure 11. Geographic scope China's Belt and Road Initiative. Map by Peter Cai [141], reproduced with permission of the Lowy Institute, <https://www.lowyinstitute.org/publications/understanding-belt-and-road-initiative>.

The rationale for BRI is manifold, appearing either as "entirely a mercantile endeavor, designed to fortify China's economic interests around the world and open business opportunities for Chinese companies enduring a slowdown at home" [145]; or as a means for China "to address ... deepening regional disparity as [its] economy modernizes [with the hope that] its transnational infrastructure building program will spur growth in [its] underdeveloped hinterland and rustbelt" and as "platform to address the country's chronic excess capacity" [141]; or for China "to consolidate its position at the centre of global supply and manufacturing networks" [146]. At home, BRI is presented altruistically as "grounded in a philosophy of nations cooperating for mutual benefit, [with] no better strategy for worldwide growth, peace and stability" [147].

Whether used to "assert its leadership through a vast program of economic integration" [141] or seeking to achieve mutual benefit in ways that accommodate "the interests and concerns of all parties involved" through market operations [148] and with "promise of prosperity" [149], one way or another, China is bringing fresh geopolitical dimensions to international aid and development [11]. A range of BRI projects are already underway in Xinjiang Uyghur Autonomous Region, and many Chinese provinces are competing for BRI resources, and thus, have developed BRI-related plans [141]. Internationally, however, the potential opportunities and benefits along with challenges and costs of BRI for recipient countries beyond China's borders are much less obvious, having received significantly less attention to date (but see [41,45,150]).

The Central Asian countries of Kazakhstan, Kyrgyzstan, and Tajikistan already are highly dependent on China for trade and the movement of goods, along with very high proportions of foreign direct investment coming from China (i.e., 72% and 37% for Tajikistan and Kyrgyzstan, respectively, in 2016 [151]). Afghanistan is a country of substantial interest to all others in the region, due to security concerns, as well as substantial mineral and other natural resources. For its part, Pakistan already has a longer history of formal cooperation, through the development of the China–Pakistan Economic Corridor (CPEC). Further study and exploration of the local and regional opportunities offered by BRI, as well as potential impacts—including sociocultural, economic, and environmental—are just now starting to take place in greater depth (e.g., as introduced by [152], and further advanced at the *Silk Roads crossing the mountains of Central Asia: Ancient routes and new challenges in times of global change* workshop held in Dushanbe, Tajikistan on 3–4 October 2017, with recommendations presented in the *Dushanbe Declaration* [41]).[11]

Recognized as one of the world's biodiversity hotspots, the Mountains of Central Asia are now set to receive substantial investment from the Critical Ecosystem Partnership Fund (CEPF) supported by the European Union, the World Bank, MacArthur Foundation, and others [153] (Figure 12). As a meeting point of several vast biogeographic realms and with complex mountain topographies, this ecoregion has a unique, rich biodiversity. It also has a great diversity of peoples and cultures with well adapted rural livelihoods, a characteristic that is especially important in the context of climate change. What remains little known, though, is how China's BRI will intersect with local economies, ecological conditions, and cultures. Environmentally, there is some sense of hope [44,154,155], but some skepticism remains. Ultimately, "whether the new Silk Road brings environmental devastation or a new era of Chinese global resource stewardship and sustainable development will depend heavily on how China approaches the Belt and Road Initiative" [43].

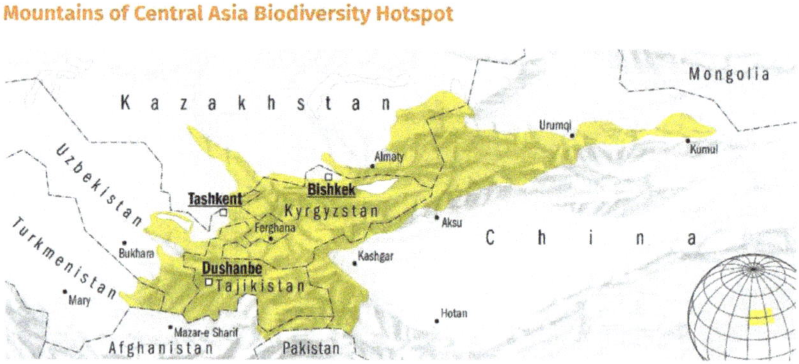

Figure 12. Mountains of Central Asia biodiversity hotspot [153]. Used with permission.

Ancient routes are being revived and recreated, and new approaches to international development and global relations have recently been launched in China—introducing a "new era" for a large part of the world and its population. For this reason, now is the time to revisit, consider and assess, and as appropriate to introduce lessons about environmental conservation and about sustainable community development—especially those learned in China, as presented herein—and begin to apply these to the international contexts beyond China's western borders, where it now is serving as one of the world's largest development partners.

[11] See the *Silk Roads in the Mountains of Central Asia* workshop proceedings and the resulting *Dushanbe Declaration* at http://ucentralasia.org/Research/Item/1625 and http://ucentralasia.org/Research/Item/1626, respectively.

5.2. The Way Forward

In China, the government has recently launched a new policy paradigm, ecological civilization, that calls on everyone at all levels to stop ignoring the environment in favor of the economy, and in fact, to pay much more attention to comprehensive environmental protection (see [156,157]). "The [Communist] party already accepts that [China's] environmental conditions need urgent improvements... The concept envisions better planning and carrying out future development within China's ecological capacity and rectifying degradation ... [prioritizing] pollution reduction, efficient use of natural resources, food security [and] climate change mitigation and adaptation" [158].

Clarifying and highlighting some of the key lessons learned over the past couple decades in China, especially in matters of use and protection of natural resources, both within and beyond protected areas, have been amongst the fundamental objectives of this article. These lessons have a potential application not only for China's emerging national parks system, but also for the ecosystems and landscapes in which protected areas are embedded. As China's global influence increases, particularly with its BRI programs in neighboring mountainous areas of Central Asia, how it views the environment and how such perspectives are integrated (or not) into development interventions is critically important for conservation. As rightfully queried, "Does China's intention to go through a 'green shift' domestically [also] resonate with these new transborder infrastructure development mega-projects?" [159]. How China chooses to operate, at home and abroad, will serve the interests or be detrimental to the interests of very large regions and enormous populations around the world.

More inclusive forms of biodiversity conservation in China—as outlined above—are encouraging, even if not perfectly developed, and consistent with the country's commitments as a signatory of the major global conventions (e.g., the Convention on Biological Diversity), and most recently, the UN Sustainable Development Goals (SDGs). The application of SDG principles and targets to its vast western mountainous regions has provided China with significant opportunity to consider various models and approaches—the learnings of which are now being incorporated, inter alia, in the trial development of its new national park system. Greater participation of communities with government authorities in the operations and to some extent the oversight of protected areas, in the monitoring and management of natural resources, and local socioeconomic development, are enabling new steps to be taken toward achieving the SDG goals—all 17 of the goals—but especially for *Goal 15, Life on Land*. These advances in more inclusive conservation should now be advocated also for wider application, both in and beyond protected areas, and wherever China has opportunity to bring its support [29,33,44,98,160–162].

Specifically, the lessons learned "at home" in China should be leveraged to inform and guide future developments not only in the Sanjiangyuan region, but also more widely across the entire Tibetan Plateau environment [54], the vast grasslands and arid lands in western China [34,35,104,133], and even extending into the mountains of Central Asia, where China, through its BRI, is now playing a decisive role in shaping the future across multiple sectors and in many countries [41,45,162].

Although mountains often may be viewed through the lens of geographic or politico-administrative borders and boundaries, they also can be special regions with not only ecological but also substantial sociocultural similarities [163,164]. For example, in the case of the Kirghiz and Wakhi people residing in four neighboring countries in the Pamir region, the similarities clearly outnumber their socioeconomic differences [53]. Likewise, there are many notable commonalities amongst yak herding communities across High Asia based on their mountain environments and livelihoods that exhibit a unifying effect (cf. sociocultural identity) extending far beyond formal language, nationality, or religion [49,165,166].

Through BRI, China is now expanding its geographic scope and socioeconomic and political influence in unprecedented ways. Neighboring mountain areas in Central Asia are amongst the first locations to be considered and receive major BRI inputs, yet little is known about BRI impacts on the receiving side; most studies to date have focused mainly on global considerations or on motivations from China's perspective [45]. The introduction and application of *lessons learned* in China that are relevant to these receiving geographies and the potential impacts of BRI on the environment and local people and communities' socioeconomic development and wellbeing are, thus, highly pertinent. The recommendation is therefore hereby made that the lessons learned—as outlined above—should be considered carefully and incorporated into BRI planning and operations, both in western China and beyond its borders, with special reference to greater Central Asia with its cultural and linguistic diversity as well as its renowned *Mountains of Central Asia* biodiversity hotspot [153]. How China chooses to address environmental, sociocultural, and development concerns is critically important.

Three broad development themes finally emerge from the case study materials presented above, including the key lessons that have been extracted from experiences in the Sanjiangyuan region:

- Mountain livelihoods: Pastoralism as sustainable food system
- Protected area networks and regional/landscape conservation
- "Access and Benefit Sharing" through community co-management

These important thematic foci should in future be developed in even more detail, with the aim of informing development planning, including BRI, in more inclusive and environmentally friendly ways.

5.2.1. Pastoralism Is a Sustainable Food System

Pastoralism is not just an economic occupation; it is food production system, a livelihood, and often, a cultural identity [133]. Traditionally, pastoralism has been the primary means by which communities have transformed rangeland resources into basic sustenance. Pastoralism is, de facto, one of the most sustainable food systems on the planet, as it has proven by its very existence over many centuries, despite some long-standing entrenched prejudices and pressures from agricultural and urban populations [167]. The adaptive capacities of pastoralists have led them, through trial and error, to choose flexible and responsive decision-making processes and seasonally mobile land use patterns, allowing them to respond pragmatically to variable, often unpredictable, climatic conditions. This does not mean that misuse (including overuse) of natural resources and ensuing environmental degradation does not ever occur in rangelands, mediated, at least in part, by pastoral communities. However, where such land degradation occurs, it tends to have been influenced significantly by changing patterns of natural resource use, often due to an imposition (whether active or passive) of new sociopolitical structures or a diminishing of some features (or "design principles") that are commonly associated with a sustainable management of common pool resources [168–170]. Adopting more integrated perspectives on the interconnected social, economic, and ecological realities of rangelands and mountain ecosystems is increasingly recognized as of paramount importance to achieve sustainability [171,172]. Under management that properly links these complementary dimensions, pastoral food production systems can yet continue to serve the interests of both humanity and global biodiversity.

The United Nations Environment Programme (UNEP) and the International Union for the Conservation of Nature (IUCN) also highlight the role of pastoralists as stewards of more than a quarter of the world's total land area. Pastoralism is practiced by between 200 and 500 million people globally, including nomadic communities, transhumant herders, and agropastoralists. Yet, despite the large numbers, four important facts about pastoralism are widely overlooked [173]:

- Mobility of livestock (and wildlife) is essential to maintain the health of dryland ecosystems—for carbon sequestration, watershed protection, and biodiversity conservation.
- Intensive livestock systems can degrade the environment by producing high levels of carbon dioxide and methane, polluting watercourses, and causing land degradation.
- Sustainable pastoral systems are more efficient, productive, and resilient than more sedentary agricultural systems in rangelands, when all factors and environmental benefits are counted.
- Pastoralism is a universal issue, as pastoralists in both the developing and developed countries share many environmental and economic challenges and opportunities.

Additionally, the benefits derived from the sustainable use and preservation of rangelands can have regional and sometimes global facets, in particular, through upstream–downstream water linkages [22]. Pastoralism may also be recognized not only as a means to harvest resources, but in itself a cultural asset of mountain and rangeland social–ecological systems, as well as being provider of ecosystem services [172]. As outlined in the Convention on Biological Diversity's *Good Practice Guide: Pastoralism, Nature Conservation and Development* [174], "vegetation maintained through grazing activities...captures carbon, reduces erosion, maintains soils, facilitates water holding capacity and provides [wildlife] habitat. Most pastoral systems [also] are steeped in cultural practices and indigenous knowledge, 'cultural services' which are highly valued and often irreplaceable". Thus, even beyond the provision of direct goods and services, such as meat, milk, and fibers, pastoralism in the high altitude rangelands of the world may yet be a viable livelihood option, and "rather than abandoning pastoralism, the revitalisation of traditional practices and indigenous knowledge is vital to secure sustainable livelihoods for millions of pastoralists [worldwide] and to maintain rangeland biodiversity and ecosystem services" [175].

5.2.2. Protected Area Networks and Regional/Landscape Level Conservation

Not all conservation occurs within protected areas. Rather, long-term regional conservation can only be achieved when landscape level approaches are adopted, recognizing that a matrix of land use types is always present (including but not limited to protected areas), with many different stakeholder groups holding different perspectives and priorities. That said, protected areas do still have an important role to play, especially with a broadening of our understanding of the governance and management options available, including with indigenous and community conserved areas [28,176–183].

Strengthening the roles of protected areas in regional development planning and, more generally, mainstreaming the value of biodiversity across sectors have been advanced over the past decade through the work of organizations such as the United Nations Development Programme (UNDP), with support of the Global Environment Facility (GEF) and other financial instruments, including in Qinghai Province and Xinjiang Uyghur Autonomous Region of China, as well as Kyrgyzstan, Tajikistan, and elsewhere in Central Asia [184]. High level government support has been leveraged successfully in recent years, with a focus placed on one charismatic species, the snow leopard, as an icon of high mountain ecosystems. In this way, the Global Snow Leopard and Environmental Protection (GSLEP) program has clearly articulated many of the priority needs and opportunities for conservation of snow leopard landscapes, inclusive of local herding and nearby farming communities' interests and aspirations [184,185]. Key transboundary aspects of wildlife management in mountain regions, including illegal trade, also are highlighted in international political and academic discourses [186], and reflected in ongoing projects, such as the Hindu Kush Pamir Karakorum Transboundary Landscape Conservation project undertaken by the International Centre for Integrated Mountain Development (ICIMOD), the University of Central Asia (UCA), local protected areas, and a range of other partners in four countries [187].

In China, official plans state that national parks will be set up to achieve a combination of ecological protection and sustainable development goals, with a fine balance to be found between protection and utilization. Development opportunities should be available to local people within national parks, which will be possible, in part, because herders and farmers are now beginning to be recognized as major forces behind environmental protection, with "conservation" expected to provide jobs as well as increase local farmers' and herders' incomes [80]. Adopting such parallel and reinforcing perspectives will enable conservation to take place more widely, last longer (beyond single project cycles), and be more effective (through greater level of community engagement and "ownership" of plans and activities).

Broad landscape-level conservation perspectives, incorporating but not limited to national parks and other protected areas, ought to be applied more widely in western China (see [34,104,133]), and should equally be introduced as a fundamental pillar in China's BRI, in which biodiversity is becoming recognized more explicitly as core value [44].

5.2.3. Access and Benefit Sharing (ABS) through Partnership and Co-Management

Finally, access to natural resources and the ability to benefit from biodiversity—including land, water, and other natural resources—must not be compromised for local communities and governments with the arrival of new external stakeholders in the region. The development of genuine partnerships and community co-management approaches represent an ideal way forward, yet their application in reality is not always straightforward. Negotiations and compromises are always needed, and not every group has equal power or voice. In many places and ways in China, local voices are increasingly heard—but not uniformly or entirely, and not in every aspect of the new national parks systems. In the core areas of national parks, for example, local farmers and herders will need to relocate; for obvious reasons, reactions are mixed.[12] When faced with relocations, herders in the Sanjiangyuan region have, themselves, struggled with reactions ranging from hopes for a better future for the next generation aided by new educational opportunities, to deep concern about how they should respond to such government policies (e.g., whether to sell their livestock and transition into a new way of life, or to "hold out" and lend their animals for a time to other family members or neighbors who plan to remain on the grassland), to sadness at anticipated loss of culture and traditional pastoral practices, to a general apprehension about all the change and challenges yet unknown [13,23,24,55,60,64]. At the same time, relative to the more strictly "protectionist" nature reserves in China (with their primary focus on nature conservation), the recent shift toward more collaborative approaches in management of protected areas and particularly the development of the national parks system in China (with its dual mandate, i.e., including both conservation and socioeconomic development) is promising in the trend that it represents. The positive elements identified in this review of environmental conservation in China over the past couple decades do not imply that all is perfect and improvements are not needed, but they do provide a focus for our corporate attention about some of the important foundations available, upon which to build future development programs and strategies, including especially China's BRI.

In short, conscientious attention must be given to ensure that infrastructure developments and business opportunities, as well as the management of protected areas, and in fact, any use of natural resources in the wider landscape, always be advanced in ways that bring greatest advantage to local communities and regions, and not be selected or promoted primarily on the basis of external drivers. Niche mountain products, tourism development, and value chains can also all be built to local and regional advantage. Special attention should equally be given to ensure that traditional knowledge and agrobiodiversity are preserved, both as protective adaptive mechanisms in the context of climate change, and to help anchor local communities and maintain cultural continuity in an era of rapid globalization [188–192]. Through such community-friendly approaches, greater equity in access to and sharing of the benefits derived from natural resources and other local assets can be maintained.

---

[12] Although such mixed reactions are extremely important *and must not be ignored*, they are not the focus of this paper—which instead has purposefully given most attention to the positive examples emerging from recent experiences in environmental conservation, such as the adoption of more integrated approaches and inclusive perspectives in natural resources management. The challenges associated with relocations can be enormous and deserve dedicated attention. For the present purposes, it shall suffice for now that this critical issue is noted as requiring further study, which should draw on local and global experiences surrounding "local communities" and their involvement in governance and management of natural resources and protected areas. In the present study, initial observations about perceptions and responses to relocations are based on numerous discussions with a wide range of actors over a period extending more than a decade, including local herders, community organizations, and national park field staff and leaders, as well as discussions with both natural science and social science specialists in China and internationally. Not surprisingly, perspectives and responses vary with people's positions and areas of specializations, with no single consensus emerging across all stakeholders.

6. Conclusions

In all three broad thematic areas just outlined—i.e., pastoralism in a globalizing world, protected areas and landscape-level conservation, and sharing of the benefits of biodiversity and ecosystems—many valuable lessons have been learned in the past two decades in the Sanjiangyuan region, which could beneficially be applied more widely in western China, both within and outside protected areas, as well as internationally through the country's ambitious Belt and Road Initiative. How China now refines its broad plans and long-range ambitions across continents, especially in regard to their social and environmental impacts—whether as intended development benefits, or unforeseen side-effects of large-scale projects—these early months and years of the BRI will likely set the stage for years to come, and will both define China's position and, to some extent, its degree of acceptance in the global development community, and determine the long-term viability of its investments in infrastructure, transport and trade, agriculture, and other business opportunities.

The complexity of all the interrelated conservation and development issues and the multiple scales that are considered, from community to continental levels, clearly render any conclusions or recommendations tentative at best. However, broad outlines have emerged from China's own recent experiences with collaborative approaches in environmental conservation. These emerging views should now be considered in greater detail and integrated into the country's broader development programming at home and abroad. Communities, economies, and the environment all stand to gain.

Acknowledgments: No funding was received to conduct this review paper. Project grants for the conservation and development work undertaken by Plateau Perspectives in the Sanjiangyuan region, as outlined above, were received from multiple sources (more details are available on Plateau Perspectives' website, e.g., annual reports). No funds were received from any grant, personal or institutional, to cover the cost to publish in open access. For professional collaboration and other forms of support, special thanks are due to Marion Foggin and Alistair Foggin; to Tashi Dorje Hashil; to Gongbo Tashi, Jigme Rabden, and the Plateau Perspectives field team; and to many colleagues and advisors, donors, and government partners, along with a host of community members—too many to name individually, but all enormously appreciated. Additionally, for dialogues and support in better understanding the plans and possible implications of China's Belt & Road Initiative, thanks are given to Troy Sternberg, Long Ruijun, Wu Ning, Yuri Badenkov, Hermann Kreutzmann, and others. In Central Asia, I also wish to thank my colleagues at University of Central Asia and especially at the Mountain Societies Research Institute, who introduced me over the past four years to Central Asia—its opportunities and challenges, as well as its people, politics, and environment. Finally, I express my genuine gratitude to Matthew Emslie-Smith, Chris Hergarten, and Stefanos Xenarios, who each have in different ways also have supported this writing project.

Conflicts of Interest: The authors declare no conflict of interest.

References

1. United Nations Department of Economic and Social Affairs. *World Population Prospects: The 2017 Revision*; United Nations: New York, NY, USA, 2017.
2. Ding, Y. Impacts of Affuence and Overexploitation of Natural Resources. In *Environment and Development*; Encyclopedia of Life Support Systems (EOLSS); Eolss Publishers Co. Ltd.; UNESCO: Oxford, UK, 2009; Volume 1.
3. Harris, P.G. *Environmental Policy and Sustainable Development in China: Hong Kong in Global Context*; University of Bristol Policy Press: Bristol, UK, 2012.
4. Mancini, M.S.; Galli, A.; Coscieme, L.; Niccolucci, V.; Lin, D.; Pulselli, F.M.; Bastianoni, S.; Marchettini, N. Exploring Ecosystem Services Assessment through Ecological Footprint Accounting. *Ecosyst. Serv.* **2018**. [CrossRef]
5. Magalhães, P.; Steffen, W.; Bosselmann, K. The Need for an Integrated Assessment Framework to Account for Humanity's Pressure on the Earth System. In *The Safe Operating Space Treaty: A New Approach to Managing Our Use of the Earth System*; Cambridge Scholars Publishing: Newcastle, UK, 2016; pp. 213–245.
6. Steffen, W.; Richardson, K.; Rockström, J.; Cornell, S.E.; Fetzer, I.; Bennett, E.M.; Biggs, R.; Carpenter, S.R.; Vries, W.; Wit, C.A.; et al. Planetary Boundaries: Guiding Human Development on a Changing Planet. *Science* **2015**, *347*, 1259855. [CrossRef] [PubMed]

7. United Nations General Assembly. Resolution 70/1: Transforming Our World: The 2030 Agenda for Sustainable Development. 2015. Available online: http://undocs.org/A/RES/70/1 (accessed on 27 October 2016).
8. Baker, S. China's Universities: Significant Progress, but More to Do. *Times Higher Education (THE)*. Available online: https://www.timeshighereducation.com/news/chinas-universities-significant-progress-more-do (accessed on 12 April 2018).
9. China Power Team. How Does Education in China Compare with Other Countries? Available online: https://chinapower.csis.org/education-in-china/ (accessed on 15 March 2018).
10. High Quality Research Booming in China. Available online: https://www.natureindex.com/news-blog/chart-chinas-top-ten-research-universities (accessed on 15 March 2018).
11. Kharas, H.; Rogerson, A. *Global Development Trends and Challenges: Horizon 2025 Revisited*; Overseas Development Institute: London, UK, 2017; p. 52.
12. Lin, Y.; Wang, Y. China's Growing Role in Global Development. *China Daily*. 12 May 2017. Available online: http://www.chinadaily.com.cn/opinion/2017-05/12/content_29312577.htm (accessed on 15 March 2018).
13. Foggin, J.M. Highland Encounters: Building New Partnerships for Conservation and Sustainable Development in the Yangtze River Headwaters, Heart of the Tibetan Plateau. In *Innovative Communities: People-Centred Approaches to Environmental Management in the Asia-Pacific Region*; United Nations University (UNU) Press: Tokyo, Japan, 2005; p. 341.
14. Foggin, J.M. *Promoting Biodiversity Conservation and Community Development in the Sanjiangyuan Region: Proceedings of Conservation Planning Meeting*; Plateau Perspectives: Xining, China, 2005.
15. Tashi Dorje's Redemption. *China Tibet Online*. 9 November 2011. Available online: http://chinatibet.people.com.cn/96069/7639089.html (accessed on 15 March 2018).
16. Liu, J. *Tibetan Environmentalists in China: The King of Dzi*; Rowen, I.; Hui, C.K.; Yeh, E.T., Translators; Lexington Books: Lanham, MA, USA, 2015.
17. Foggin, J.M.; Phillips, J. Horizontal Policy Analysis: A Tool to Promote Sustainable Livelihoods Development; with Implications for Ecological Resettlement and Other Major Development Programs in the Tibetan Plateau Region. In *Pastoralism in Contemporary China: Policy and Practice*; Social Science Academic Press: Beijing, China, 2013; pp. 3–30.
18. Soini, K.; Birkeland, I. Exploring the Scientific Discourse on Cultural Sustainability. *Geoforum* **2014**, *51*, 213–223. [CrossRef]
19. Foggin, J.M. *Biodiversity Protection and the Search for Sustainability in Tibetan Plateau Grasslands*; Arizona State University: Tempe, AZ, USA, 2000.
20. Foggin, J.M. *Building New Partnerships for Conservation & Sustainable Development in the Tibetan Plateau Region: Recent Experiences in the Yangtze River Headwaters*; Plateau Perspectives: Xining, China, 2004; p. 14.
21. Foggin, M. Pastoralists and Wildlife Conservation in Western China: Collaborative Management within Protected Areas on the Tibetan Plateau. *Pastor. Res. Policy Pract.* **2012**, *2*, 17. [CrossRef]
22. Foggin, J.M. Conservation Issues: Mountain Ecosystems. In *Reference Module in Earth Systems and Environmental Sciences*; Elsevier: Toronto, ON, Canada, 2016; ISBN 978-0-12-409548-9. Available online: https://doi.org/10.1016/B978-0-12-409548-9.09199-5 (accessed on 15 March 2018).
23. Foggin, M.; Torrance-Foggin, M. How Can Social and Environmental Services Be Provided for Mobile Tibetan Herders? Collaborative Examples from Qinghai Province, China. *Pastor. Res. Policy Pract.* **2011**, *1*, 21. [CrossRef]
24. Foggin, J.M.; Torrance-Foggin, M. Pastoralism, Development, and the Future of Tibetan Rangelands: Experiences in the Development and Provision of Social Services and Environmental Management. In *Tibetan Pastoralists and Development: Negotiating the Future of Grassland Livelihoods*; Reichert: Wiesbaden, Germany, 2017; pp. 175–186.
25. Shen, X.; Tan, J. Ecological Conservation, Cultural Preservation, and a Bridge between: The Journey of Shanshui Conservation Center in the Sanjiangyuan Region, Qinghai-Tibetan Plateau, China. *Ecol. Soc.* **2012**, *17*, 38. [CrossRef]
26. Agrawal, A. Common Property Institutions and Sustainable Governance of Resources. *World Dev.* **2001**, *29*, 1649–1672. [CrossRef]
27. Ballet, J.; Koffi, K.J.-M.; Komena, K.B. Co-Management of Natural Resources in Developing Countries: The Importance of Context, Abstract. *Econ. Int.* **2009**, *120*, 53–76.

28. Borrini-Feyerabend, G.; Dudley, N.; Jaeger, T.; Lassen, B.; Pathak, B.N.; Phillips, A. *Governance of Protected Areas: From Understanding to Action*; Best Practice Protected Area Guidelines Series No. 20; IUCN: Gland, Switzerland, 2013; p. 124.
29. Du, W.; Penabaz-Wiley, S.M.; Njeru, A.M.; Kinoshita, I. Models and Approaches for Integrating Protected Areas with Their Surroundings: A Review of the Literature. *Sustainability* **2015**, *7*, 8151–8177. [CrossRef]
30. Kothari, A.; Corrigan, C.; Jonas, H.; Neumann, A.; Shrumm, H. *Recognizing and Supporting Territories and Areas Conserved by Indigenous Peoples and Local Communities: Global Overview and National Case Studies*; Technical Series no. 64; Secretariat of the Convention on Biological Diversity, ICCA Consortium, Kalpavriksh, and Natural Justice: Montreal, QC, Canada, 2012; p. 160.
31. Borrini-Feyerabend, G.; Hill, R. Governance for the Conservation of Nature. In *Protected Area Governance and Management*; ANU Press: Canberra, Australia, 2015; pp. 169–206.
32. Ervin, J.; Sekhran, N.; Dinu, A.; Gidda, S.; Vergeichik, M.; Mee, J. *Protected Areas for the 21st Century: Lessons from UNDP/GEF's Portfolio*; UNDP; CBD: Montreal, QC, Canada, 2010; p. 132.
33. Foggin, J.M. Managing Shared Natural Heritages: Towards More Participatory Models of Protected Area Management in Western China. *J. Int. Wildl. Law Policy* **2014**, *17*, 130–151. [CrossRef]
34. Harris, R.B. *Wildlife Conservation in China: Preserving the Habitat of China's Wild West*; M.E. Sharpe: London, UK, 2008.
35. Squires, V.; Hua, L.; Zhang, D.; Li, G. (Eds.) *Towards Sustainable Use of Rangelands in North-West China*; Springer: Dordrecht, The Netherlands, 2010; p. 353.
36. Liu, J.; Ouyang, Z.; Pimm, S.L.; Raven, P.H.; Wang, X.; Miao, H.; Han, N. Protecting China's Biodiversity. *Science* **2003**, *300*, 1240. [CrossRef] [PubMed]
37. McBeath, J.; McBeath, J.H. Biodiversity Conservation in China: Policies and Practice. *J. Int. Wildl. Law Policy* **2006**, *9*, 293–317. [CrossRef]
38. Liu, J.; Diamond, J. China's Environment in a Globalizing World. *Nature* **2005**, *435*, 1179. [CrossRef] [PubMed]
39. Wehrli, A. Why Mountains Matter for Sustainable Development. *Mt. Res. Dev.* **2014**, *34*, 405–409. [CrossRef]
40. Schmidt-Vogt, D.; Foggin, M.; Hergarten, C. Strengthening Mountain Societies in Central Asia in a Context of Multidimensional Change. *Mt. Res. Dev.* **2016**, *36*, 380–383. [CrossRef]
41. Foggin, J.M. *Proceedings of International Workshop. Silk Roads in the Mountains of Central Asia: Ancient Routes & Modern Challenges in Times of Global Change*; Mountain Societies Research Institute, University of Central Asia: Bishkek, Kyrgyzstan, 2018; Available online: http://ucentralasia.org/Research/Item/1625 (accessed on 10 April 2018).
42. International Centre for Integrated Mountain Development (ICIMOD). *Building Partnerships for Conservation and Development in HKH: Exploring Complementary Opportunities for Improving HKPL Transboundary Cooperation from One Belt, One Road Initiative, 1–12 December 2016, Lanzhou, China*; ICIMOD: Kathmandu, Nepal, 2017; p. 39.
43. Pike, L. Will China's New Silk Road Be Green? Available online: https://www.chinadialogue.net/blog/9775-Explainer-Will-China-s-new-Silk-Road-be-green-/en (accessed on 15 March 2018).
44. Lechner, A.M.; Chan, F.K.S.; Campos-Arceiz, A. Biodiversity Conservation Should be a Core Value of China's Belt and Road Initiative. *Nat. Ecol. Evol.* **2018**, *2*, 408–409. [CrossRef] [PubMed]
45. Sternberg, T.; Ahearn, A.; McConnell, F. Central Asian 'Characteristics' on China's New Silk Road: The Role of Landscape and the Politics of Infrastructure. *Land* **2017**, *6*, 55. [CrossRef]
46. Maimaitiming, A.; Xiaolei, Z.; Huhua, C. Urbanization in Western China. *Chin. J. Popul. Resour. Environ.* **2013**, *11*, 79–86. [CrossRef]
47. Butler, R.A. The Top 10 Most Biodiverse Countries. Mongabay. Available online: https://news.mongabay.com/2016/05/top-10-biodiverse-countries/ (accessed on 12 April 2018).
48. Wiener, G.; Jianlin, H.; Ruijun, L. *The Yak*, 2nd ed.; RAP. Publication; Regional Off. for Asia and the Pacific, Food and Agriculture Organization of the United Nations: Bangkok, Thailand, 2003.
49. *Yak on the Move: Transboundary Challenges and Opportunities for Yak Raising in a Changing Hindu Kush Himalayan Region*; Wu, N.; Yi, S.; Srijana, B.; Bisht, N. (Eds.) Special Publication; International Centre for Integrated Mountain Development (ICIMOD): Kathmandu, Nepal, 2016.
50. Rowntree, L.; Lewis, M.; Price, M.; Wycoff, W. *Globalization and Diversity*; Pearson: London, UK, 2016.

51. Swanström, N.; Institute for Security and Development Policy; Silk Road Studies Program; Paul, H.; Nitze School of Advanced International Studies; Central Asia-Caucasus Institute. *China and Greater Central Asia: New Frontiers*; Silk Road Studies Program; Institute for Security and Development Policy: Stockholm, Sweden, 2011.
52. Squires, V.; Lu, Q. *Sustainable Land Management in Greater Central Asia | An Integrated and Regional Perspective | Taylor & Francis Group*, 1st ed.; Routledge: London, UK, 2017; p. 310.
53. Kreutzmann, H. *Pamirian Crossroads: Kirghiz and Wakhi of High Asia*; Harrassowitz Verlag: Wiesbaden, Germany, 2015.
54. Yao, T.; Thompson, L.G.; Mosbrugger, V.; Zhang, F.; Ma, Y.; Luo, T.; Xu, B.; Yang, X.; Joswiak, D.R.; Wang, W.; et al. Third Pole Environment (TPE). *Environ. Dev.* **2012**, *3*, 52–64. [CrossRef]
55. Foggin, J.M. Depopulating the Tibetan Grasslands. *Mt. Res. Dev.* **2008**, *28*, 26–31. [CrossRef]
56. Klein, J.A.; Yeh, E.; Bump, J.; Nyima, Y.; Hopping, K. Coordinating Environmental Protection and Climate Change Adaptation Policy in Resource-Dependent Communities: A Case Study from the Tibetan Plateau. In *Climate Change Adaptation in Developed Nations*; Advances in Global Change Research; Springer: Dordrecht, The Netherlands, 2011; pp. 423–438.
57. Long, R.; Ding, L.M.; Shang, Z.H.; Guo, X.H. The Yak Grazing System on the Qinghai-Tibetan Plateau and Its Status. *Rangel. J.* **2008**, *30*, 241–246. [CrossRef]
58. Miller, D. Grasslands of the Tibetan Plateau. *Rangelands* **1990**, *12*, 159–163.
59. Schaller, G.B. *Wildlife of the Tibetan Steppe*; University of Chicago Press: Chicago, IL, USA, 1998; p. 392.
60. Gruschke, A. Nomads without Pastures? Globalization, Regionalization, and Livelihood Security of Nomads and Former Nomads in Northern Khams. *J. Int. Assoc. Tibet. Stud.* **2008**, *4*, 1–40.
61. Sheehy, D.P.; Miller, D.; Johnson, D.A. Transformation of Traditional Pastoral Livestock Systems on the Tibetan Steppe. *Secheresse* **2006**, *17*, 142–151.
62. Shao, Q.; Jiyuan, L.I.U.; Lin, H.; Jiangwen, F.A.N.; Xinliang, X.U.; Junbang, W.; Quanqin, S.; Jiyuan, L.I.U.; Lin, H.; Jiangwen, F.A.N.; et al. Integrated Assessment on the Effectiveness of Ecological Conservation in Sanjiangyuan National Nature Reserve, Integrated Assessment on the Effectiveness of Ecological Conservation in Sanjiangyuan National Nature Reserve. *Geogr. Res.* **2013**, *32*, 1645–1656.
63. Yan, J.; Li, H.; Hua, X.; Peng, K.; Zhang, Y. Determinants of Engagement in Off-Farm Employment in the Sanjiangyuan Region of the Tibetan Plateau. *Mt. Res. Dev.* **2017**, *37*, 464–473. [CrossRef]
64. Du, F. Ecological Resettlement of Tibetan Herders in the Sanjiangyuan: A Case Study in Madoi County of Qinghai. *Nomadic Peoples* **2012**, *16*, 116–133. [CrossRef]
65. Xinhua. China to Invest 1 b Yuan in Sanjiangyuan National Park Construction. Available online: http://www.chinadaily.com.cn/china/2017-02/06/content_28117036.htm (accessed on 15 March 2018).
66. Foggin, J.M. Snow Leopard Landscape Conservation: Partnering with Local Communities in Western China and Central Asia for Conservation and Development. Available online: https://bit.ly/2Hvx2Gg (accessed on 12 April 2018).
67. Xu, J.; Grumbine, R.E.; Shrestha, A.; Eriksson, M.; Yang, X.; Wang, Y.; Wilkes, A. The Melting Himalayas: Cascading Effects of Climate Change on Water, Biodiversity, and Livelihoods. *Conserv. Biol.* **2009**, *23*, 520–530. [CrossRef] [PubMed]
68. Kang, S.; You, Q.; Flugel, W.; Pepin, N.; Yan, Y.; Xu, Y.; Huang, J. Review of Climate and Cryospheric Change in the Tibetan Plateau. *Environ. Res. Lett.* **2010**, *5*, 15101. [CrossRef]
69. Li, X.-L.; Gao, J.; Brierley, G.; Qiao, Y.-M.; Zhang, J.; Yang, Y.-W. Rangeland Degradation on the Qinghai-Tibet Plateau: Implications for Rehabilitation. *Land Degrad. Dev.* **2013**, *24*, 72–80. [CrossRef]
70. Manandhar, S.; Xenarios, S.; Schmidt-Vogt, D.; Hergarten, C.; Foggin, J.M. *Climate Vulnerability and Adaptive Capacity of Mountain Societies in Central Asia*; MSRI Research Report Series; Research Report 1; Mountain Societies Research Institute, University of Central Asia: Bishkek, Kyrgyzstan, 2018.
71. Li, J.; Wang, W.; Axmacher, J.C.; Zhang, Y.; Zhu, Y. Streamlining China's Protected Areas. *Science* **2016**, *351*, 1160. [CrossRef] [PubMed]
72. Cao, M.; Peng, L.; Liu, S. Analysis of the Network of Protected Areas in China Based on a Geographic Perspective: Current Status, Issues and Integration. *Sustainability* **2015**, *7*, 15617–15631. [CrossRef]
73. Guo, Z.; Cui, G. Establishment of Nature Reserves in Administrative Regions of Mainland China. *PLoS ONE* **2015**, *10*, e0119650. [CrossRef] [PubMed]

74. China Global Television Network (CGTN). How China's Ecological 'Red Line' Fits into a Much Larger Plan. *CGTN America*. Available online: https://america.cgtn.com/2017/03/10/how-chinas-ecological-red-line-fits-into-a-much-larger-plan (accessed on 12 April 2018).
75. Yeh, E. Do China's Nature Reserves Only Exist on Paper? Available online: https://www.chinadialogue.net/article/show/single/en/6696-Do-China-s-nature-reserves-only-exist-on-paper- (accessed on 15 March 2018).
76. Wang, G.; Innes, J.L.; Wu, S.W.; Krzyzanowski, J.; Yin, Y.; Dai, S.; Zhang, X.; Liu, S. National Park Development in China: Conservation or Commercialization? *Ambio* **2012**, *41*, 247–261. [CrossRef] [PubMed]
77. Li, W.; Han, N. Ecotourism Management in China's Nature Reserves. *AMBIO J. Hum. Environ.* **2001**, *30*, 62–63. [CrossRef]
78. Eagles, P.; McCool, S.; Haynes, C. *Sustainable Tourism in Protected Areas: Guidelines for Planning and Management*; Best Practice Protected Area Guidelines Series; Best Practice Guidelines No 8; IUCN: Gland, Switzerland, 2002.
79. Xu, W.; Xiao, Y.; Zhang, J.; Yang, W.; Zhang, L.; Hull, V.; Wang, Z.; Zheng, H.; Liu, J.; Polasky, S.; et al. Strengthening Protected Areas for Biodiversity and Ecosystem Services in China. *Proc. Natl. Acad. Sci. USA* **2017**, *114*, 1601–1606. [CrossRef] [PubMed]
80. Zheng, J. First National Park Receiving Strong Legal and Financial Support. Available online: http://english.gov.cn/news/top_news/2017/03/09/content_281475589549379.htm (accessed on 15 March 2018).
81. The Case for a National Park System in China. *China Daily*. Available online: http://www.chinadaily.com.cn/china/2017-02/10/content_28165369.htm (accessed on 12 April 2018).
82. Liu, Q. China Overhauls Its National Parks. *China Dialogue*. Available online: https://www.chinadialogue.net/article/show/single/en/10152-China-overhauls-its-national-parks- (accessed on 12 April 2018).
83. China Council for International Cooperation on Environment and Development (CCICED). *Ecological Civilization Shaping China's New Era*; CCICED 2017 Issues Paper; CCICED: Beijing, China, 2017; p. 31.
84. Xinhua. China to Set Up National Park System. Available online: http://english.gov.cn/policies/policy_watch/2017/09/27/content_281475888517430.htm (accessed on 15 March 2018).
85. Xinhua. China's First National Park to Open in 2020. Available online: http://www.xinhuanet.com/english/2018-01/18/c_136905975.htm (accessed on 15 March 2018).
86. China's National Parks: Natural, Scenic and Wild. *Chinadaily Forum*. 2017. Available online: http://bbs.chinadaily.com.cn/forum.php?mod=viewthread&tid=1888641 (accessed on 15 March 2018).
87. Xiao, X. Discover China's 10 Pilot National Park. Available online: http://discover.china.org.cn/trip-ideas/discover-chinas-10-pilot-national-park/ (accessed on 15 March 2018).
88. Yan, W. China's First National Park, an Experiment in Living with Snow Leopards. Available online: https://news.mongabay.com/2017/05/chinas-first-national-park-an-experiment-in-living-with-snow-leopards/ (accessed on 15 March 2018).
89. People's Daily Online. Protected Areas Identified for Construction of Great Wall National Park-People's Daily Online. Available online: http://en.people.cn/n3/2017/1220/c90000-9306790.html (accessed on 15 March 2018).
90. Xinhua. Administration for Tiger and Leopard National Park Inaugurated. *China Daily*. Available online: http://www.chinadaily.com.cn/china/2017-08/20/content_30851114.htm (accessed on 12 April 2018).
91. Ying. China Focus: China Plans National Park for Siberian Tiger, Amur Leopard. Available online: http://www.xinhuanet.com/english/2017-03/02/c_136097346.htm (accessed on 12 April 2018).
92. China Council for International Cooperation on Environment and Development (CCICED). *Progress in Environment and Development Policies in China and Impact of CCICED's Policy Recommendations (2013–2017)*; Progress Report; CCICED: Beijing, China, 2017; Available online: http://www.cciced.net/cciceden/Events/AGM/2017/file/201712/P020171204411638362626.pdf (accessed on 15 March 2018).
93. Xinhua. China Approves 15 Regional Plans for Ecological "Red Lines". Available online: http://www.globaltimes.cn/content/1089447.shtml (accessed on 15 March 2018).
94. Li, Y.; Lu, C.; Lu, U.; Deng, O. Ecological Characteristics of China's Key Ecological Function Areas. *J. Res. Ecol.* **2015**, *6*, 427–432.
95. International Union for the Conservation of Nature (IUCN). *Sustaining Development: IUCN and the Sustainable Development Goals*; IUCN: Gland, Switzerland, 2017.
96. International Union for the Conservation of Nature (IUCN). *Building the Sustainable Development Goals on the Aichi Biodiversity Targets*; Sustainable Development Goals Policy Briefs Series; IUCN: Gland, Switzerland, 2017.

97. Secretariat of the Convention on Biological Diversity (SCBD). *Biodiversity and the 2030 Agenda for Sustainable Development Technical Note*; Technical Note; Secretariat of the Convention on Biological Diversity: Montreal, QC, Canada, 2017.
98. Dudley, N.; Stolton, S.; Belokurov, A.; Krueger, L.; Lopoukhine, N.; MacKinnon, K.; Sandwith, T.; Sekhran, N. *Natural Solutions-Protected Areas Helping People Cope with Climate Change*; Report Funded and Commissioned by IUCN-WCPA, TNC, UNDP, WCS; World Bank; WWF: Gland, Switzerland; Washington, DC, USA; New York, NY, USA, 2009; p. 130.
99. Ministry of Foreign Affairs (MFA), People's Republic of China. *China's Position Paper on the Implementation of the 2030 Agenda for Sustainable Development*; Ministry of Foreign Affairs, PRC: Beijing, China, 2016. Available online: http://www.fmprc.gov.cn/mfa_eng/wjdt_665385/2649_665393/t1357701.shtml (accessed on 15 March 2018).
100. China Releases National Plan on Implementation of the 2030 Agenda for Sustainable Development. Available online: http://www.fmprc.gov.cn/mfa_eng/zxxx_662805/t1405596.shtml (accessed on 15 March 2018).
101. Leong, A. How the Paris Agreement and the SDGs Work Together. Available online: http://17goals.org/paris-agreement-sdgs/ (accessed on 15 March 2018).
102. Foggin, J.M. Kegawa Herders' Cooperative: Enhancing Community Resilience and Environmental Conservation in the Yangtze River Headwaters. 2016. Available online: https://www.slideshare.net/marcfoggin/kegawa-herders-cooperative (accessed on 15 March 2018).
103. Breivik, I. The Political Ecology of Grassland Conservation in Qinghai Province. Master's Thesis, Norwegian University of Life Science, Ås, Norway, 2007. Available online: https://tibetanplateau.wikischolars.columbia.edu/The+Political+Ecology+of+Grassland+Conservation+in+Qinghai+Province (accessed on 16 April 2018).
104. Shen, X.; Li, S.; Lu, Z. A New Approach to Conservation in Western China. In *Green China: Chinese Insights on Environment and Development*; Keeley, J., Yisheng, Z., Eds.; International Institute for Environment and Development: London, UK, 2011; pp. 118–135.
105. Shen, X.; Lu, Z.; Li, S.; Chen, N. Tibetan Sacred Sites: Understanding the Traditional Management System and Its Role in Modern Conservation. *Ecol. Soc.* **2012**, *17*, 13. [CrossRef]
106. Kassam, K.-A. *Biocultural Diversity and Indigenous Ways of Knowing: Human Ecology in the Arctic*; University of Calgary Press: Calgary, AL, Canada, 2009.
107. Salsberg, J.; Macridis, S.; Garcia Bengoechea, E.; Macaulay, A.C.; Moore, S. Engagement Strategies That Foster Community Self-Determination in Participatory Research: Insider Ownership. through Outsider Championship. *Fam. Pract.* **2017**, *34*, 336–340. [CrossRef] [PubMed]
108. Larson, C. A Model for Conservation in China? The Tibetan Antelope. *Christian Science Monitor*, 28 August 2012.
109. Xiao, H. Hoh Xil Heroes. *China Pictorial*. 8 August 2017. Available online: https://www.pressreader.com/china/china-pictorial-english/20170808/282458529077467 (accessed on 15 March 2018).
110. NewsChina. The Antelope and the Interloper. Available online: http://www.newschia/newschina/articleDetail.do?article_id=2585&section_id=&magazine_id= (accessed on 15 March 2018).
111. Plateau Perspectives. *Plateau Perspectives Annual Report 2005: Supporting Grassroots Initiatives in Conservation & Sustainable Development on the Tibetan Plateau*; Annual Report; Plateau Perspectives: Saint-Lambert, QC, Canada, 2005; p. 24.
112. Plateau Perspectives. *Plateau Perspectives Annual Report 2006: Supporting Community Development & Conservation on the Tibetan Plateau*; Plateau Perspectives: Saint-Lambert, QC, Canada, 2006; p. 16.
113. Wei, Z. Tibetans Fight Back Against Declining "Reverence for Nature". Available online: https://www.chinadialogue.net/article/show/single/en/5638-Tibetans-fight-back-against-declining-reverence-for-nature- (accessed on 15 March 2018).
114. Plateau Perspectives. *Plateau Perspectives Annual Report 2007: Strategic Conservation, Building a Better Future*; Plateau Perspectives: Saint-Lambert, QC, Canada, 2007; p. 24.
115. Plateau Perspectives. *Plateau Perspectives Annual Report 2008: People & Nature: Working Together for a More Sustainable Future on the Tibetan Plateau*; Plateau Perspectives: Saint-Lambert, QC, Canada, 2008; p. 16.
116. Plateau Perspectives. *Plateau Perspectives Annual Report 2009: Hope for the Future*; Annual Report; Plateau Perspectives: Saint-Lambert, QC, Canada, 2009.

117. Plateau Perspectives. *Plateau Perspectives Annual Report 2010: Rebuilding Yushu after the Earthquake;* Annual Report; Plateau Perspectives: Saint-Lambert, QC, Canada, 2010.
118. Foggin, J.M. ICCAs as 'Contract Conservation' in the Chinese Tibetan Plateau Region. In *Bio-Cultural Diversity Conserved by Indigenous Peoples & Local Communities—Examples & Analysis;* Companion Document to IUCN/CEESP. Briefing Note 10: Strengthening What Works; ICCA Consortium, CENESTA, GEF, SGP, GTZ, IIED, IUCN/CEESP: Gland, Switzerland, 2010.
119. Plateau Perspectives. *Snow Leopard Research and Conservation: Trialing New Forms of Collaborative Management in the Sanjiangyuan National Nature Reserve;* Plateau Perspectives: Xining, China, 2011.
120. Plateau Perspectives. *The Muqu Snow Leopard Conservation Project: Successful Trial of Collaborative Management in the SNNR;* Plateau Perspectives: Xining, China, 2011.
121. Foggin, J.M. Local Communities and Conservation on the Tibetan Plateau: Two Case Studies of Collaborative Management in the Sanjiangyuan Region. In Proceedings of the International Symposium on the Human Dimensions of Ecological Conservation in the Tibetan Plateau Region, Xining, China, 21–26 August 2011; pp. 50–60.
122. United Nations Development Programme (UNDP), China. Strengthening the Effectiveness of the Protected Area System in Qinghai Province, China to Conserve Globally Important Biodiversity. Available online: http://www.cn.undp.org/content/china/en/home/operations/projects/environment_and_energy/strengthening-the-effectiveness-of-the-protected-area-system-in-.html (accessed on 15 March 2018).
123. Plateau Perspectives. *Plateau Perspectives Annual Report 2012: Building Community Resilience through Development of New Opportunities in the Tibetan Plateau Region;* Annual Report; Plateau Perspectives: South Surrey, BC, Canada, 2012.
124. Plateau Perspectives. *Plateau Perspectives Annual Report 2013;* Annual Report; Plateau Perspectives: South Surrey, BC, Canada, 2013.
125. Plateau Perspectives. *Plateau Perspectives Annual Report 2016;* Annual Report; Plateau Perspectives: South Surrey, BC, Canada, 2016.
126. Foggin, J.M. Tibetan Herders Diversify Their Options. *Farming Matters*. 19 December 2016. Available online: https://www.ileia.org/2016/12/19/locally-rooted-ideas-initiatives-field-17/ (accessed on 15 March 2018).
127. Plateau Perspectives. *Plateau Perspectives Annual Report 2014;* Annual Report; Plateau Perspectives: South Surrey, BC, Canada, 2014.
128. Plateau Perspectives. *Plateau Perspectives Annual Report 2015;* Annual Report; Plateau Perspectives: South Surrey, BC, Canada, 2015.
129. Holtz, M. Following Damage Caused by Economic Rise, China Tackles Ambitious Conservation Experiment. *Christian Science Monitor*, 20 August 2017.
130. Worthy, F.R.; Foggin, J.M. Conflicts between Local Villagers and Tibetan Brown Bears Threaten Conservation of Bears in a Remote Region of the Tibetan Plateau. *Hum. Wildl. Confl.* **2008**, *2*, 200–205.
131. Foggin, J.M.; Rabden, J. Trial Use of Electric Fencing to Prevent Intrusions by Tibetan Brown Bear. *Int. Bear News* **2010**, *19*, 15–18.
132. Plateau Perspectives. *Plateau Perspectives Annual Report 2017;* Annual Report; Plateau Perspectives: South Surrey, BC, Canada, 2017.
133. Hodges, J.; Foggin, M.; Long, R.; Zhaxi, G. Globalisation and the Sustainability of Farmers, Livestock-Keepers, Pastoralists and Fragile Habitats. *Biodiversity* **2014**, *15*, 109–118. [CrossRef]
134. Foggin, J.M.; Hergarten, C. Empowering Communities through Citizen Science: Giving Voice to under-Represented Societies through the Practice of Participatory Conservation and Environmental Monitoring. In Proceedings of the Citizen Science 2015, San Jose, CA, USA, 12–16 February 2015.
135. United Nations, Social Commission of the Economic and Social Council (Eleventh Session). Concepts and Principles of Community Development and Recommendations on Further Practical Measures to Be Taken by International Organizations. *Ekistics* **1957**, *4*, 92–96.
136. Sanders, I. Theories of Community Development. *Rural Sociol.* **1958**, *23*, 1–12.
137. Scoones, I. *Sustainable Rural Livelihoods: A Framework for Analysis;* IDS Working Paper Series; Institute of Development Studies: Brighton, UK, 1998; p. 22.
138. Brocklesby, M.A.; Fisher, E. Community Development in Sustainable Livelihoods Approaches—An Introduction. *Commun. Dev. J.* **2003**, *38*, 185–198. [CrossRef]

139. Naku, D.W.C.; Afrane, S. Local Community Development and the Participatory Planning Approach: A Review of Theory and Practice. *Curr. Res. J. Soc. Sci.* **2013**, *5*, 185–191.
140. Stafford-Smith, M.; Griggs, D.; Gaffney, O.; Ullah, F.; Reyers, B.; Kanie, N.; Stigson, B.; Shrivastava, P.; Leach, M.; O'Connell, D. Integration: The Key to Implementing the Sustainable Development Goals. *Sustain. Sci.* **2017**, *12*, 911–919. [CrossRef]
141. Cai, P. *Understanding China's Belt and Road Initiative*; Lowy Institute for International Policy: Sydney, Australia, 2017.
142. Campbell, C. Ports, Pipelines, and Geopolitics: China's New Silk Road Is a Challenge for Washington. *Time.* 23 October 2017. Available online: http://time.com/4992103/china-silk-road-belt-xi-jinping-khorgos-kazakhstan-infrastructure/ (accessed on 15 March 2018).
143. Shepherd, W. Why China Just Added the Belt and Road Initiative to Its Constitution. Available online: https://www.forbes.com/sites/wadeshepard/2017/10/25/why-china-just-added-the-belt-and-road-initiative-to-its-constitution/#973bece42abc (accessed on 15 March 2018).
144. Xinhua. "Belt and Road" Incorporated into CPC Constitution. Available online: http://www.xinhuanet.com/english/2017-10/24/c_136702025.htm (accessed on 15 March 2018).
145. Djankov, S.; Miner, S. *China's Belt and Road Initiative: Motives, Scope & Challenges*; PIIE Briefing 16-2; Peterson Institute for International Economics: Washington, DC, USA, 2016; Available online: https://piie.com/system/files/documents/piieb16-2_1.pdf (accessed on 15 March 2018).
146. Sajjanhar, A. *China's Belt and Road Initiative: Prospects and Pitfalls*; Institute for Defence Studies and Analyses. Available online: https://idsa.in/idsacomments/china-belt-and-road-initiative_asajjanhar_281117 (accessed on 15 March 2018).
147. South China Morning Post. Beijing's 'Belt and Road Initiative' Will Benefit the World. Available online: http://www.scmp.com/comment/insight-opinion/article/2132943/beijings-belt-and-road-initiative-will-benefit-world (accessed on 15 March 2018).
148. State Council of the People's Republic of China (SCPRC). *Full Text: Action Plan on the Belt and Road Initiative*; Action Plan; State Council of the People's Republic of China: Beijing, China, 2015. Available online: http://english.gov.cn/archive/publications/2015/03/30/content_281475080249035.htm (accessed on 15 March 2018).
149. Ying, F. China's New Silk Road Promises Prosperity across Eurasia. *Huffington Post.* 31 July 2015. Available online: https://www.huffingtonpost.com/fu-ying/china-silk-road-eurasia_b_7899236.html (accessed on 15 March 2018).
150. Dörre, A.; Kreutzmann, H.; Schütte, S. *Pamirs at the Crossoads: Changing Challenges and Perspectives*; Berlin Geographical Papers; Center for Development Studies: Berlin/Heidelberg, Germany, 2016.
151. RICA Network. Available online: http://www.rica.network/publications/monthly-reports (accessed on 15 March 2018).
152. Badenkov, Y. Closing Statement and Remarks on Future Perspectives. In *Pamirs at the Crossroads: Changing Challenges and Perspectives*; Springer: Berlin/Heidelberg, Germany, 2016; pp. 89–100.
153. Critical Ecosystems Partnership Fund (CEPF). *Mountains of Central Asia Biodiversity Hotspot*; Ecosystem Profile; CEPF: Arlington, VA, USA, 2017; p. 184.
154. Nichols, S. China's New Silk Road Offers Environmental Promise. Available online: http://research.msu.edu/yesterdays-silk-road-tomorrows-environmental-superhighway/ (accessed on 15 March 2018).
155. Nichols, S. Yesterday's Silk Road could be Tomorrow's Environmental Superhighway. Available online: http://msutoday.msu.edu/news/2016/yesterdays-silk-road-could-be-tomorrows-environmental-superhighway/ (accessed on 15 March 2018).
156. Zhang, C. China's New Blueprint for an "Ecological Civilization". *The Diplomat*, 30 September 2015.
157. Xiao, L.; Zhao, R. China's New Era of Ecological Civilization. *Science* **2017**, *358*, 1008–1009. [PubMed]
158. Loh, C. Green Policies in Focus as China's Rise to an Ecological Civilisation Continues Apace. *South China Morning Post.* 11 October 2017. Available online: http://www.scmp.com/comment/insight-opinion/article/2114748/green-policies-focus-chinas-rise-ecological-civilisation (accessed on 15 March 2018).
159. Tracy, E.F.; Shvarts, E.; Simonov, E.; Babenko, M. China's New Eurasian Ambitions: The Environmental Risks of the Silk Road Economic Belt. *Eurasian Geogr. Econ.* **2017**, *58*, 56–88. [CrossRef]

160. Neupane, S. Community Engagement in the SDG Era and the Role of Development Donors. Available online: http://www.thinktankinitiative.org/blog/community-engagement-sdg-era-and-role-development-donors (accessed on 15 March 2018).
161. Howard, J.; Wheeler, J. What Community Development and Citizen Participation Should Contribute to the New Global Framework for Sustainable Development. *Commun. Dev. J.* **2015**, *50*, 552–570. [CrossRef]
162. Laruelle, M. (Ed.) *China's Belt and Road Initiative and Its Impact in Central Asia*; The George Washington University, Central Asia Program: Washington, DC, USA, 2018.
163. Debarbieux, B.; Rudaz, G. Linking Mountain Identities throughout the World: The Experience of Swiss Communities. *Cult. Geogr.* **2008**, *15*, 497–517. [CrossRef]
164. Aga Khan, H.H. Speech by His Highness the Aga Khan. Available online: http://ismaili.net/timeline/2000/000605/pakspeech06001.html (accessed on 15 March 2018).
165. Carralero Benitez, S. *World Yak Herders Association: Community Dialogues in High Asia. Report for FAO Pastoralism Knowledge Hub*; YURTA Association: Malaga, Spain, 2017.
166. Tsering, S. Whither the Tsampa Eater. *Himāl* **1993**, *6*, 8–11.
167. Krätli, S.; Schareika, N. Living off Uncertainty: The Intelligent Animal Production of Dryland Pastoralists. *Eur. J. Dev. Res.* **2010**, *22*, 605–622. [CrossRef]
168. Anderies, J.; Janssen, M.; Ostrom, E. A Framework to Analyze the Robustness of Social-Ecological Systems from an Institutional Perspective. *Ecol. Soc.* **2004**, *9*, 18. [CrossRef]
169. Ostrom, E. A General Framework for Analyzing Sustainability of Social-Ecological Systems | Science. *Science* **2009**, *325*, 419–422. [CrossRef] [PubMed]
170. Ostrom, E. Background on the Institutional Analysis and Development Framework: Ostrom: Institutional Analysis and Development Framework. *Policy Stud. J.* **2011**, *39*, 7–27. [CrossRef]
171. Sukhdev, P.; Bishop, J.; Ten Brink, P.; Gundimeda, H.; Karousakis, K.; Kumar, P.; Neßhöver, C.; Neuville, A.; Skinner, D.; Vakrou, A.; et al. *The Economics of Ecosystems & Biodiversity: An Interim Report*; European Communities: Wesseling, Germany, 2008.
172. McGahey, D.; Davies, J.; Hagelberg, N.; Ouedraogo, R. *Pastoralism and the Green Economy—A Natural Nexus?* International Union for the Conservation of Nature (IUCN); United Nations Environment Programme (UNEP): Nairobi, Kenya, 2014; p. 58.
173. Davies, J.; Hagelberg, N. *Sustainable Pastoralism and the Post 2015 Agenda: Opportunities and Barriers to Pastoralism for Global Food Production and Environmental Stewardship*; United Nations Environment Programme (UNEP); International Union for the Conservation of Nature (IUCN): Gland, Switzerland, 2015; p. 2.
174. Secretariat of the Convention on Biological Diversity (SCBD). *Pastoralism, Nature Conservation and Development: A Good Practice Guide*; Secretariat of the Convention on Biological Diversity: Montreal, QC, Canada, 2010.
175. Seid, M.A.; Kuhn, N.J.; Fikre, T.Z. The Role of Pastoralism in Regulating Ecosystem Services: -EN-FR-Le Rôle Du Pastoralisme Dans La Réglementation Des Services Écosystémiques -ES- La Función Del Pastoreo En La Regulación de Los Servicios Ecosistémicos. *Revue Scientifique et Technique de l'OIE* **2016**, *35*, 435–444. [CrossRef] [PubMed]
176. Borrini-Feyerabend, G.; Pimbert, M.; Farvar, T.; Kothari, A.; Renard, Y. *Sharing Power: Learning by Doing in Co-Management of Natural Resources throughout the World*; IIED and IUCN/CEESP: Gland, Switzerland, 2004.
177. Borrini-Feyerabend, G.; Kothari, A. *Recognising and Supporting Indigenous & Community Conservation-Ideas & Experiences from the Grassroots*; CEESP Briefing Note 9; IUCN: Gland, Switzerland; p. 28.
178. Corrigan, C.; Granziera, A. *A Handbook for the Indigenous and Community Conserved Areas Registry*; UNEP-WCMC: Cambridge, UK, 2010.
179. Lockwood, M.; Worboys, G.; Kothari, A. (Eds.) *Managing Protected Areas: A Global Guide*; Routledge: London, UK, 2006.
180. Worboys, G.L.; Kothari, A.; Pulsford, I.; Lockwood, M.; Feary, S. *Protected Area Governance and Management*; ANU Press: Canberra, Australia, 2015.
181. Dinerstein, E.; Olson, D.; Joshi, A.; Vynne, C.; Burgess, N.D.; Wikramanayake, E.; Hahn, N.; Palminteri, S.; Hedao, P.; Noss, R.; et al. An Ecoregion-Based Approach to Protecting Half the Terrestrial Realm. *BioScience* **2017**, *67*, 534–545. [CrossRef] [PubMed]
182. Baldwin, R.F.; Trombulak, S.C.; Leonard, P.B.; Noss, R.F.; Hilty, J.A.; Possingham, H.P.; Scarlett, L.; Anderson, M.G. The Future of Landscape Conservation. *BioScience* **2018**, *68*, 60–63. [CrossRef] [PubMed]

183. Infield, M.; Entwistle, A.; Anthem, H.; Mugisha, A.; Phillips, K. Reflections on Cultural Values Approaches to Conservation: Lessons from 20 Years of Implementation. *Oryx* **2017**, *52*, 1–11. [CrossRef]
184. Paxton, M.; Scott, T.; Watanabe, Y. *Silent Roar: UNDP. & GEF in the Snow Leopard Landscape*; United Nations Development Program: Washington, DC, USA, 2016; p. 48.
185. Jackson, R.M. HWC Ten Years Later: Successes and Shortcomings of Approaches to Global Snow Leopard Conservation. *Hum. Dimens. Wildl.* **2015**, *20*, 310–316. [CrossRef]
186. Vasilijević, M.; Zunckel, K.; McKinney, M.; Erg, B.; Schoon, M.; Rosen Michel, T. *Transboundary Conservation: A Systematic and Integrated Approach*; Groves, C., Phillips, A., Eds.; International Union for Conservation of Nature (IUCN): Gland, Switzerland, 2015.
187. International Centre for Integrated Mountain Development (ICIMOD). *Hindu Kush Karakoram Pamir Landscape Conservation and Development Initiative*; ICIMOD: Kathmandu, Nepal, 2016.
188. Johns, T.; Powell, B.; Maundu, P.; Eyzaguirre, P.B. Agricultural Biodiversity as a Link between Traditional Food Systems and Contemporary Development, Social Integrity and Ecological Health. *J. Sci. Food Agric.* **2013**, *93*, 3433–3442. [CrossRef] [PubMed]
189. Mijatovic, D.; Van Oudenhouven, F.; Eyzaguirre, P.; Hodgkin, T. The Role of Agricultural Biodiversity in Strengthening Resilience to Climate Change: Towards an Analytical Framework. *Int. J. Agric. Sust.* **2012**, *11*, 95–107. [CrossRef]
190. Van Oudenhouven, F.; Mijatovic, D.; Eyzaguirre, P. *Bridging Managed and Natural Landscapes*; Bioversity International: Rome, Italy, 2009; p. 1.
191. Zimmerer, K.S. Agriculture, Livelihoods, and Globalization: The Analysis of New Trajectories (and Avoidance of Just-so Stories) of Human-Environment Change and Conservation. *Agric. Hum. Values* **2007**, *24*, 9–16. [CrossRef]
192. Amend, T.; Brown, J.; Kothari, A.; Phillips, A.; Stolton, S. (Eds.) Protected Landscapes and Agrobiodiversity Values. Values of Protected Landscapes and Seascapes. Available online: https://www.cbd.int/doc/pa/tools/Values%20of%20Protected%20Landscapes%20and%20Seascapes.pdf (accessed on 15 March 2018).

land

MDPI

Article

Mapping Land Cover and Estimating the Grassland Structure in a Priority Area of the Chihuahuan Desert

Alberto Rodríguez-Maturino [1], José Hugo Martínez-Guerrero [2], Isaías Chairez-Hernández [3], Martín Emilio Pereda-Solis [2], Federico Villarreal-Guerrero [4], Marusia Renteria-Villalobos [4] and Alfredo Pinedo-Alvarez [4,*]

1 Programa Institucional de Doctorado en Ciencias Agropecuarias y Forestales, Universidad Juárez del Estado de Durango, Durango 34307, Mexico; maturino_al@yahoo.com
2 Facultad de Medicina Veterinaria y Zootecnia, Universidad Juárez del Estado de Durango, Durango 34305, Mexico; che_hugo1@hotmail.com (J.H.M.-G.); conplandg@hotmail.com (M.E.P.-S.)
3 Instituto Politécnico Nacional, CIIDIR, Durango 34220, Mexico; ichairez@hotmail.com
4 Facultad de Zootecnia y Ecología, Universidad Autónoma de Chihuahua, Chihuahua 31453, Mexico; fvillarreal@uach.mx (F.V.-G.); mrenteria@uach.mx (M.R.-V.)
* Correspondence: apinedo@uach.mx; Tel.: +52-614-434-0363 (ext. 15); Fax: +52-614-434-0345

Received: 5 September 2017; Accepted: 13 October 2017; Published: 20 October 2017

Abstract: A field characterization of the grassland vegetation structure, represented by the coverage of grass canopy (CGC) and the grass height, was carried out during three years (2009–2011) in a priority area for the conservation of grasslands of North America. Landsat Thematic Mapper (TM5) images were selected and the information of reflectance was obtained based on the geographical location of each field-sampling site. Linear models, constructed with field and satellite data, with high coefficients of determination for CGC (R^2 = 0.81, R^2 = 0.81 and R^2 = 0.72) and grass height (R^2 = 0.82, R^2 = 0.79 and R^2 = 0.73) were obtained. The maps showed a good level of CGC (>25%) and grass height (>25 cm), except for the year 2009, which presented the lowest values of grass height in the area. According to the Kappa Index, a moderate concordance among the three CGC maps was presented (0.49–0.59). Conversely, weak and moderate concordances were found among the grass height maps (0.36–0.59). It was observed that areas with a high CGC do not necessarily correspond to areas with greater grass height values. Based on the data analyzed in this study, the grassland areas are highly dynamic, structurally heterogeneous and the spatial distribution of the variables does not show a definite pattern. From the information generated, it is possible to determine those areas that are the most important for monitoring to then establish effective strategies for the conservation of these grasslands and the protection of threatened migratory bird species.

Keywords: remote sensing; modelling; coverage; grass height; Cuchillas de la Zarca

---

1. Introduction

Given their great biodiversity and the environmental services they provide; grasslands are ecosystems of great importance. According to Adams et al. [1] grasslands account for 40.5% of the Earth's surface and therefore play an important role in the global carbon cycle [2]. Currently, these ecosystems are undergoing significant deterioration due to overgrazing, urbanization, land use change, the presence of invasive species and habitat fragmentation [3–5]. Indeed, Hoesktra et al. [6] considered that grasslands are the most threatened ecosystems on Earth. Consequently, many species of migratory birds that use grasslands as habitat in the North American deserts have declined their populations continuously and steadily [7,8]. Previous studies on grasslands have focused on herbaceous productivity trends and on the factors that may affect the herbaceous coverage [9,10]. For instance, from studies on grasslands of the Chihuahuan Desert, which is distributed in northern

Mexico and the southern United States of America, some studies have focused on listing the plant species, on studies of its fauna and on its ecosystem fragmentation [11,12]. However, more research is needed on the spatio-temporal variation of the coverage of grass canopy (CGC) and grass height. This topic is particularly relevant because these variables are strongly related to the habitat quality for migratory birds and because the grasslands immersed in the Chihuahuan Desert do not escape from the issues and pressures that characterize this ecosystem [13,14].

One of the tools widely employed for monitoring habitat quality is remote sensing. This tool offers the possibility of studying extensive areas with a wide temporal margin at a low cost [15,16]. Data from the thematic mapper (TM) sensor (Landsat 5 satellite) provides the capability to perform synoptic monitoring. The Landsat archive provides detailed and consistent data about the change dynamics experienced in the terrestrial ecosystems [17–20]. The spectral data provided by the sensor have allowed the determination of the structural characteristics and composition of plants in various ecosystems [21–23]. Some of the parameters that have been correlated to habitat quality are CGC and grass height, parameters that have been employed to estimate the net annual biomass productivity of grasslands in the field. However, these kinds of methodologies require a significant investment of time and economic resources [24]. Thus, the use of remote sensing tools represents an alternative for the determination of grassland variables in a more economical and faster manner [25].

Several studies have shown the condition of winter habitat as a determining factor in the decline of some grassland avian species. In fact, the abundance and distribution of the native species of the Chihuahuan Desert, such as *Ammodramus bairdii*, *A. savannarum*, *Anthus spragueii*, *Calcarius ornatus* and *Pooecetes gramineus*, are strongly related to the structure of the grasslands, as well as to variables such as CGC and grass height [14,26]. Therefore, it is very important to obtain information about the dynamics of the grassland where the structural variables are highly susceptible for drastic changes in relatively short periods [27]. The objective of this study was to determine the spatial and temporal variations of CGC and grass height during a period of three years in an area under conservation called Cuchillas de la Zarca, located within the Chihuahuan Desert in Mexico.

2. Materials and Methods

The area of Cuchillas de la Zarca (CUZA) is a priority region for the conservation of grasslands and is located within the Chihuahuan Desert. This desert grassland is the most extensive in North America and is recognized for having high species richness despite being a dry area. In CUZA we can find both transition areas dominated by grasslands and extensive plains covered by them [28]. The area of CUZA is located between the coordinates 105°04′30.86″ W and 24°41′41.28″ N, as well as 104°57′07.48″ W and 25°27′04.24″ N. It comprises the northern part of the state of Durango and the southern part of the state of Chihuahua, Mexico. It has an area of 11,600 km^2 (Figure 1). The vegetation consists mainly of induced pasture, natural grassland, microphyll desert scrub, chaparral areas and pine-oak forest. In each of these vegetation types, grass species of the genera *Bouteloa*, *Aristida*, *Buchloe*, *Andropogon*, *Melinis*, *Muhlenbergia*, *Sporobolus*, *Heteropogon* and *Pleuraphis* may be present [29,30].

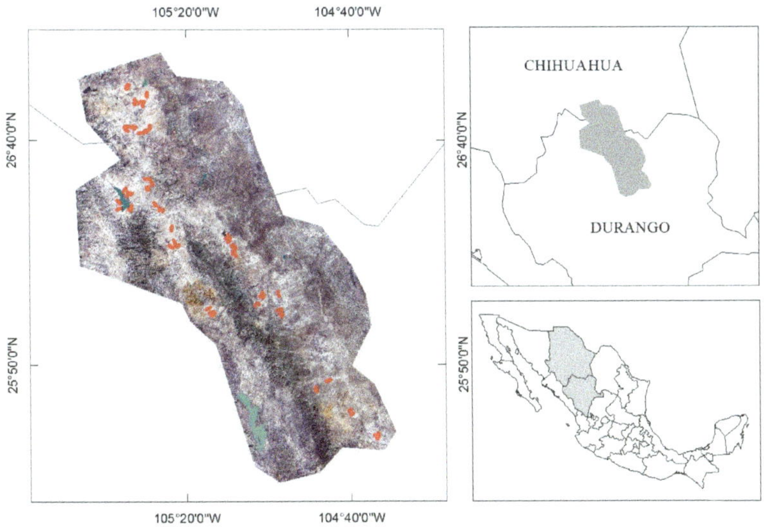

Figure 1. Study area. Sampling sites are marked in red.

2.1. Samples and Processing Date

The grassland structure, represented by CGC and grass height, was characterized during January and February by 96 transects in 2009, as well as 102 transects in 2010 and 2011. The transects of one kilometer of length were located according to a grid of squares with side lengths of 18 km covering the study area (Figure 2). The grid design was based on information from vector files of land use and roads network [31]. From all the 17 quadrants, only those with roads of acceptable condition were selected for easy access to the sampling points (red marks, Figure 1), which were located at every 500 m in each quadrant, as was established in a previous study [14]. For vegetation sampling, only the first three points of each quadrant were selected. In them, two sampling transects, perpendicular to the path, were established. Along the transects, circles of 5 m of radius were located each 100 m. In these circles, the area covered by grass canopy was visually assessed and the heights of three plants were measured at the ground level. For the height measurements, a small, a medium and a tall plant were chosen. From the three values, an average was estimated. All the measurements were made by only one person. For a summary of field data, the average CGC and grass height were obtained.

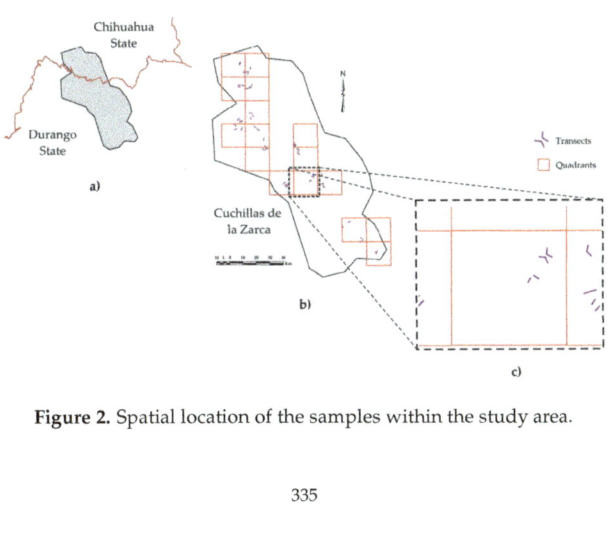

Figure 2. Spatial location of the samples within the study area.

Two scenes (path/row 31/41 and 31/42) of Landsat TM5 were acquired for each of the studied years from the United States Geological Survey (USGS) Global Visualization Viewer (GloVis, https://www.usgs.gov). The Landsat images correspond to the date of ground measurements, 5th February in 2009, 8th February in 2010, and 11th February 2011. For each year, the scenes were merged into a single mosaic.

To establish the relationship between the variables of vegetation structure and spectral values, the spectral information of the digital number (DN), extracted from the bands 1–5 and 7 of the sensor, the radiometrically-corrected values of reflectance and three synthetic bands, generated for the tasseled cap (TC), were correlated with CGC and grass height. The obtained bands of TC corresponded to the components of brightness (B), greenness (G) and wetness (W) [32]. The radiometric calibration was made by converting the gross value of each pixel (DN) into values of absolute spectral radiance. To reduce the variability among scenes, the spectral radiance was converted into exoatmospheric top of the atmosphere (TOA) reflectance. Formulae and coefficients provided by Chander et al. [33], represented by Equations (1) and (2), were used to perform the radiometric corrections applied to the TM sensor.

$$L_\lambda = \left(\frac{Lmax_\lambda - Lmin_\lambda}{QCALmax - QCALmin}\right) \times (QCAL - QCALmin) + Lmin_\lambda \quad (1)$$

$$\rho_\lambda = \frac{\pi \times L_\lambda \times d^2}{ESUN_\lambda \times \cos\,\theta_s} \quad (2)$$

where QCAL is DN; QCALmin and QCALmax are the minimum and maximum quantized calibrated pixel values, respectively; $Lmin_\lambda$ is the spectral radiance scales to QCALmin; $Lmax_\lambda$ is the spectral radiance scales to QCALmax; d is the distance from the earth to the sun; $ESUN_\lambda$ is the mean solar exoatmospheric irradiance; and θ_s is the solar zenith angle.

The atmospheric correction seeks to reduce or eliminate the image distortions resulting from the interaction of the atmosphere components with the sensor [34]. Such correction allows the standardization of image data and thus permits a comparison of images from different dates. From each of the circles of vegetation sampling located in the transects, the value of reflectance and TC components for each band were obtained from the corresponding pixel. Although some bushes and forests exist in the area, when sampling of vegetation in areas dominated by pasture, the sensor data were mainly associated to the spectral signatures of pastures. The bands that best explained the variables of CGC and grass height were selected to feed the models.

2.2. Statistical Analysis

The normality of the variables was proven with the Shapiro–Wilk test. After checking it, an analysis of variance was applied to determine if the differences among the quadrants were significant. Subsequently, multiple regression models [35] were run to explore the relationships of CGC and grass height with the spectral data. The CGC and grass height were established as response variables while the spectral data were used as independent variables. The spectral data included the DN, reflectance values and the values from the tasseled cap. The models were produced separately by dependent variable and by year. Once the models were obtained, the Gauss–Markov assumptions, which include normality, independence and homogeneity of variance, were verified on the data. To verify normality, the test was applied to the residuals of each model; to check independence, it was determined whether the correlation between the residual and predicted values were zero; and, to verify the homogeneity of variance, it was verified whether the correlation between the absolute value of the residuals and predicted values were close to zero ($p > 0.05$).

2.3. Maps of Cover of Grass Canopy and Grass Height

The estimators of the selected full models were used to construct maps of the variables CGC and grass height. For the case of CGC (%), their continuous values were reclassified in the following five classes: water bodies, 0–25 (low), 25–50 (medium), 50–75 (high) and 75–100 (very high). For the case of grass height (cm), the classes were water bodies, 0–25 (low), 25–50 (medium), 50–75 (high) and >75 (very high). To make a comparison between the resulting maps, a cross-tabulation of images was made. Such a cross-tabulation determines the frequency of pixels that stayed constant during the evaluation period and serves to make inferences about the dynamics of CGC and grass height through time in the study area. Through this operation, the categories of a raster image were compared with those of a second image; both images had to have the same number of categories and the same data type. The operation casts the Kappa Index (KI) as a measure of agreement. The KI ranges from zero, indicating that there is no agreement, and up to one, indicating perfect agreement [36,37].

2.4. Meteorological Data

Given that information about precipitation in the study area was scarce, vector files from the National Meteorological Service, providing information on drought events occurring in the country, were used. These data are based on obtaining and interpreting indicators such as the standardized precipitation index (SPI), precipitation percent anomaly, satellite vegetation health index (VHI), NOAA/CPC leaky bucket soil moisture model, the normalized difference vegetation index (NDVI), the average temperature anomaly, the water availability percentage in the country's dams and the contribution of local experts. These indicators are displayed in the form of layers in a geographic information system and, by consensus, the drought-affected regions are determined according to a scale of intensities ranging from abnormally dry, moderate drought, severe drought, drought extreme and exceptional drought. As a result of the consensus, polygons for the intensity of drought were delineated, generating shape files. When the polygons correspond to the mid-month analysis (issued on the 15th day of each month) they are used to quantify the drought on the domestic territory. Likewise, when the polygons correspond to the final evaluation of the month, they complement the regional or continental map of the North American Drought Monitor [38]. Weather information linked to drought events was used to explain the different scenarios shown by the maps, considering that these events are related to precipitation and, in turn, rains occurring in a given year can largely explain the grassland conditions [39].

3. Results

3.1. Field Data

The average CGC ranged between 50%–75% and there were no differences between the three years ($p > 0.05$). Conversely, significant differences in grass height (cm) were found among the years studied ($p < 0.05$). The lowest values were recorded in 2009 (Table 1).

Table 1. Average (or mean) of field data for Coverage of Grass Canopy (CGC) and grass height in the quadrants measured in Cuchillas de la Zarca.

QUADRANT	CGC (%)			Grass Height (cm)		
	2009	2010	2011	2009	2010	2011
1	65.5	57.8	54.3	9.0	13.5	12.2
2	63.0	58.8	59.9	15.8	25.9	30.8
3	50.7	49.8	44.1	13.4	8.2	7.3
4	68.1	61.2	66.0	16.9	28.9	27.6
5	57.7	45.4	55.6	19.8	15.8	12.5
6	49.8	52.7	46.1	8.7	10.6	9.4
7	67.1	69.5	67.8	19.6	30.2	32.7
8	60.6	51.2	55.2	8.3	15.4	11.3
9	60.6	58.3	61.9	9.2	25.5	20.0
10	65.1	65.3	60.2	13.2	20.1	17.4
11	70.9	62.9	65.5	11.9	21.3	23.5
12	60.5	53.8	57.5	8.7	15.9	20.0

Table 1. *Cont.*

QUADRANT	CGC (%)			Grass Height (cm)		
	2009	2010	2011	2009	2010	2011
13	66.5	41.8	50.8	11.2	22.6	15.9
14	49.5	51.2	53.7	12.4	20.6	21.7
15	41.3	26.6	27.5	7.0	9.7	9.8
16	36.0	16.8	22.3	10.4	15.3	19.8
17	-	64.0	64.0	-	27.3	24.4

3.2. Spectral Data

The digital numbers and reflectance values followed the same pattern over the three years evaluated; bands 5, 7 (middle-infrared) and 4 (near-infrared) registered the highest values. A decline of reflectance was observed for the quadrants located in the central and southeast parts of the study area, particularly from the bands 1–3 (blue, green and red, respectively). In 2011, the values from the bands 2–4 and 7 showed a slight difference compared to the other two years (Figure 3).

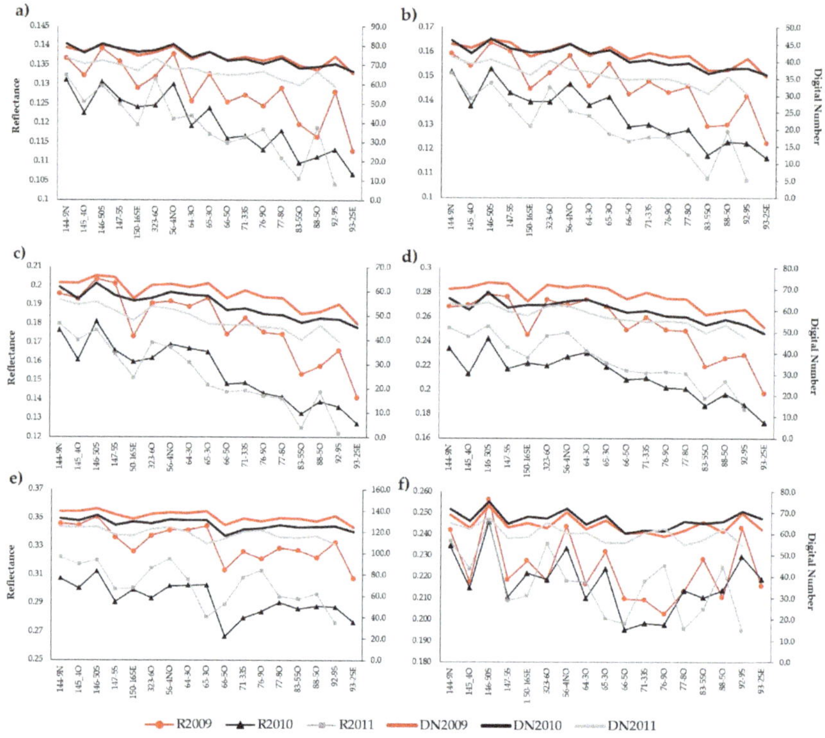

Figure 3. Spectral signature average expressed as reflectance and the digital number for each quadrant ((**a**) band1, (**b**) band2, (**c**) band3; (**d**) band4, (**e**) band5, (**f**) band7).

Forty-two models for CGC and grass height were obtained. They corresponded to three years, two dependent variables, three elements of TC with the original bands, radiometrically-corrected bands and three transformations of dependent with independent variables. Six of the models were selected, one for each variable per year. The selected models met the Gauss–Markov assumptions, and had the lowest statistical deviance information criterion, Akaike information criterion, Bayesian information criterion, (DIC, AIC, BIC) and the Root Mean Square Error (RMSE). In all the cases, the selected models only used images from bands into original digital values or in combination with an element of the TC. The coefficients of determination for the six models were high and with a significant probability ($p < 0.01$). In addition, the validation tests allowed us to verify the usefulness of each model, since the values of the mean square error of validation were slightly lower than those of calibration (Table 2).

Table 2. Coefficients of determination of linear regression models ($p < 0.01$).

Variable	Year	Statistics					Mean Square Error	
		R2	DIC	AIC	BIC	RMSE	Training	Validation
CGC	2011	0.81	11945	753.51	776.6	10.55	115.5	114.4
	2010	0.81	11945	753.51	776.59	11.15	118.7	117.4
	2009	0.72	10090	630.01	649.07	11.23	92.0	91.3
Grass Height	2011	0.82	0.38	−265.1	−244.1	0.060	50.2	48.7
	2010	0.79	0.49	−215.8	−192.7	0.071	69.5	68.7
	2009	0.73	6027.2	588.8	607.8	0.067	67.3	65.9

DIC = Deviance Information Criterion, AIC = Akaike Information Criterion, BIC = Bayesian Information Criterion, RMSE = Root Mean Square Error, CGC = Coverage of Grass Canopy.

For CGC, the best selected models for 2010 and 2011 used the bands in combination with the B element of the TC. For 2009, the model only used bands (Table 3). Similarly, the grass height models of 2009 and 2011 used only bands. Conversely, the models of 2010 used both the bands and the W element of TC (Table 4).

Table 3. Estimators of the regression models for coverage grass canopy (CGC).

Year	Intercept	B1	B2	B3	B4	B5	B7	TCb
2011	−8.5	2.2	−6.9 *	1.9	0.9	0.3	−2.3 *	0.7
2010	29	1.6	−1.8	3.0	2.0	2.3	−2.8 *	−2.4
2009	−5.4	2.5	−4.9	3.8	−1.9	1.7 *	−3.4 *	N/A

(B = band; 1,2,4,5,7 = band number; TCb = b element of the Tassaled Cap band; * $p < 0.01$).

Table 4. Estimators of the regression models for grass height.

Year	Intercept	B1	B2	B3	B4	B5	B7	TCw
2011	0.51 *	0.02	−0.07 *	0.04 *	−0.01	0.01 *	−0.02 *	
2010	0.87 *	0.03 *	−0.03	0.06 *	0.02	−0.06	−0.06 *	−0.09
2009	88.4 *	−2.04	4.1	0.12	−0.61	0.87 *	−2.6 *	

(B = band; 1,2,4,5,7 = band number; TCw = w element of the Tassaled Cap band; * $p < 0.01$).

3.3. *Distribution Maps of CGC, Grass Height and Kappa Index*

The selected models allowed the obtainment of CGC maps for three years (Figures 4a, 5a and 6a). According to the Kappa Index, among all the CGC maps a moderate concordance was presented, which was slightly higher between 2009 and 2010 and lower between 2010 and 2011 (Table 5). This was perhaps due to the fact that in 2009 and 2010 much of the surface of the study area presented a higher CGC (50–75%), while in 2011 the predominant interval was mainly located in low or intermediate levels (25–50%), as can be seen in Figure 4a. In 2009, the areas with higher CGC were in the zones of grassland and in the desert scrub. For 2010, the highest CGC was found in the northwest, a zone in which grassland areas and oak-pine forest are predominant. A similar pattern of distribution of the classes was seen in 2010 and 2011, although in a lower and more dispersed way for the latter.

As in the case of CGC, three maps of grass height were obtained from the selected models (Figures 4b, 5b and 6b). Based on the Kappa index only, a moderate concordance was presented between the maps of 2010 and 2011, while 2009 had a weak concordance for the two years (Table 5). According to the graph of the surface of the grass height ranges (Figure 7), the basic difference among the years with moderate concordance lay mainly in the fact that during 2009 the interval with the highest frequency was 0–25 cm, and for 2010 and 2011 it was 50–75 and 25–50 cm, respectively. Regarding 2009, it was characterized by grasses of low grass height along almost all of CUZA, only 8% of the area had a height of grass in the range of 25–50 cm, located primarily in the oak-pine areas. In 2010, the taller-height range (50–75 cm) was distributed generally almost along the entire area, except for the southeastern region, where grasslands with a lower grass height prevailed. Grasslands of greater grass height (>75) were conserved in both forest-oak and grassland areas. In 2011, the height ranges had a behavior similar to 2010, except that in the former, the range with the greatest presence was 25–50 cm in 57.2% of the area. In terms of distribution, the lowest grasses were in the oak-pine forests, whereas the middle and taller heights were distributed more or less uniformly through the area.

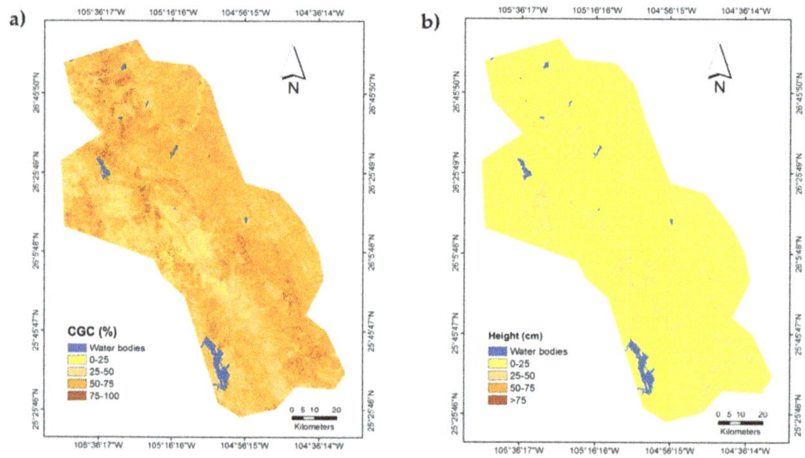

Figure 4. (**a**) Coverage of grass canopy (CGC) and (**b**) grass height distributions for 2009.

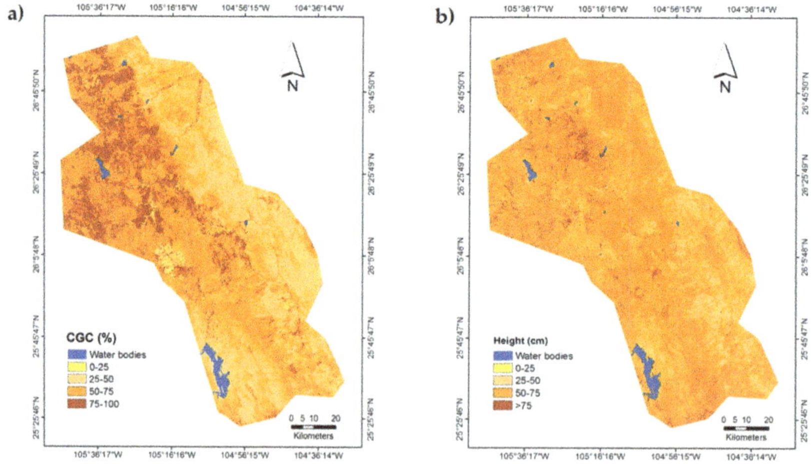

Figure 5. (**a**) Coverage of grass canopy (CGC) and (**b**) grass height distributions for 2010.

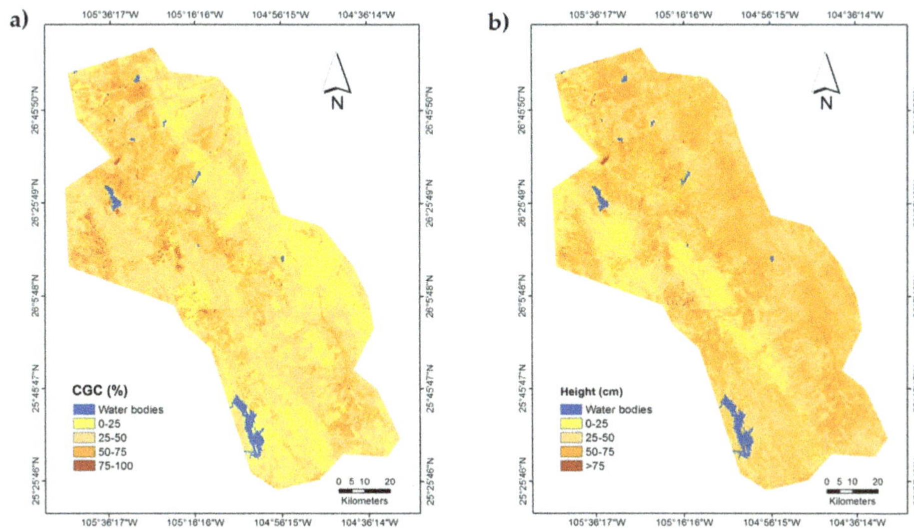

Figure 6. (**a**) Coverage of grass canopy (CGC) and (**b**) grass height distributions for 2011.

According to the Kappa index, the CGC images exhibit a moderate agreement among them. In the case of the variable grass height, the highest agreement occurred between 2010 and 2011 (Table 5). In the case of the variable CGC, the greatest agreement was observed the years 2009 and 2010. The lowest agreement between the images was obtained for the variable grass height when comparing 2009 and 2010, as well as 2009 and 2011, which can be explained by the large surface area with a low grass height showed in the map of 2009. The field data showed lower mean and statistical differences for this year, as mentioned above. However, the best agreement between maps of 2010 and 2011 does not necessarily mean a spatial correspondence between the variables. Thus, an area with high CGC does not necessarily correspond to an area with a high grass height of the grassland. Moreover, although the descriptions of the variable distribution were generally completed, various degrees of structural heterogeneity can be appreciated with a greater detail in the maps.

Table 5. Kappa index between the maps for Coverage of Grass Canopy (CGC) and grass height during the three years studied.

	CGC		
2009	**2010**	**2011**	
1	0.59 **	0.56 **	2009
	1	0.49 **	2010
	Grass Height		
2009	**2010**	**2011**	
1	0.36 *	0.40 *	2009
	1	0.59 **	2010

* weak and ** moderate agreements.

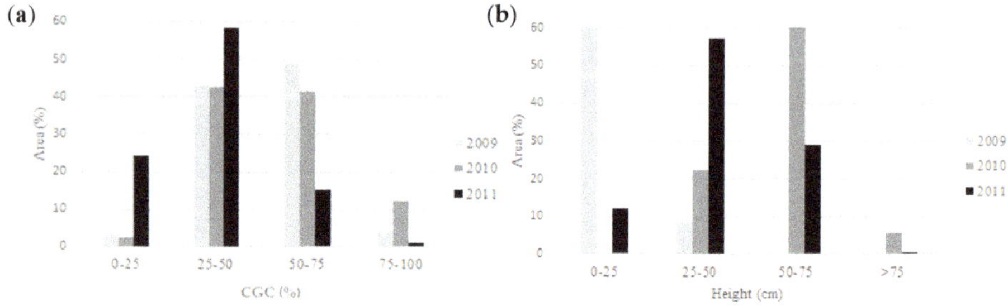

Figure 7. Area in percentages occupied by the ranges of (**a**) Coverage of Grass Canopy (CGC) and (**b**) grass height in the region Cuchillas de la Zarca.

4. Discussion

In the literature, there is a diversity of papers similar to this study in which the coefficients of determination are lower or similar, especially in cases where the same sensor is used [23,25]. The precision of the results may be partially due to the method of obtaining information in the field, the characteristics of the sensor (i.e., temporal and spatial resolution) and its age, the type of ecosystem, as well as the statistical analyses carried out. The latter appears to substantially affect the fit of the models between the data from the field and the sensor data [40–43]. Although some studies have obtained a good fit with the use of a limited number of bands [25,42], others studies [23] have required the interaction of a greater number of bands, indices or synthetic bands to achieve a better fit, as it was the case in this study. It is worth mentioning that the models selected for one year may not be appropriate to be used for a different year, because the images from different years have different digital values [44]. Possibly, a good fit on the models was not achieved with the data from the images due to an unsatisfactory correction of the data, which did not allow the use of a single model for each variable for the three years [45].

To explain the slight quantitative differences between the CGC maps, the study of Khumalo and Holechek [46] was considered. They found that the precipitation of the previous year is one of the major factors that are associated with biomass production in grasslands of the Chihuahuan Desert in southern New Mexico. Their results showed that individually, precipitation in August is the variable most highly correlated to the grassland condition. When analyzing the drought data for the three years, the presence of this phenomenon was not observed in the area during this month [38]. This may explain why there were basically no differences in the Kappa index of CGC for the three years. Therefore, for these variations among the years, other aspects such as spatial variation in grazing, forest fires and the phenology of the grass species should be considered. The CGC levels recorded during the three winter seasons studied could have enabled different species of wildlife, such as migratory grassland birds, to find this important habitat requirement [47].

The high reflectance values for the middle-infrared and near-infrared bands correspond to the general pattern of spectral behavior in vegetated surfaces [42]. The highest values for bands 2–4, 7 in 2011 may indicate high moisture contents in the soil as well as vegetation vigor. The latter factor may be attributed to the photosynthetic activity of perennial grasses, which grow and develop during the rains of February. However, according to the comparison of the maps, this difference does not seem to have greater repercussions with respect to the marked differences between the levels of CGC and grass height between 2010 and 2011. The decline in reflectance values of the bands 1–3 in some of the quadrants in the central and the southeastern zones, seems to be more associated with the particular characteristics of the composition and structure of the vegetation in some transects, since the resulting maps did not reflect the same pattern.

The CGC and grass height showed quantitative inter-annual fluctuations and spatio-temporal variations without an apparent specific pattern. Jin et al. [48] found similar results for biomass, which is directly related to CGC and grass height. The greatest quantitative variation was registered for the variable height, which was possibly due to the fact that this variable is more susceptible to changes than CGC. This may be related to the grasses' characteristics, such as senescence [49]. In addition, when grazing is practiced in an area, the height of the grassland is the first variable affected. However, and similar to the variable CGC, other abiotic factors should be discarded, such as climate variables, which may affect the characteristics of the grassland [50,51].

The spatial heterogeneity observed in all the maps on a larger scale, agreed with the results found in other studies [48,52]. Even though in our study we found evidence of spatial gradients, these were not as marked as the ones found at the larger scale. Among other things, this element allows the presence of various species of grassland birds that have different CGC and grass height requirements during the winter season [14]. Hence, it has been found that the CUZA region is an area of high diversity of not only bird species, but also of some others, such as grasses and forbs [53,54]. According to McGranahan et al. [55], additional factors that could influence this heterogeneity are fires and grazing.

The maps generated for CGC and grass height showed moderately high accuracies that can be representative and used as a reference to help assess current and future land use development, grasslands capacities, animal production potentials and the status of grassland avian habitats [41,56–58].

5. Conclusions

Through the use of images of the Landsat TM5 sensor, it was possible to predict the spatial distribution of CGC and grass height of a grassland, using linear models in combination with field data. Although a field validation was not performed, the maps contributed to the understanding of the characteristics and dynamics of the Chihuahuan Desert grasslands. The information generated may serve to establish better strategies for monitoring grasslands and determine those areas that are most relevant for conservation. In addition, the methodology implemented in this study can be applied to perform a program for monitoring the grassland condition.

A benefit of the methodology applied in this study is the ability to identify and focus fragmented and degraded areas. Other benefits include the capacity to extrapolate from limited ground monitoring locations with increased confidence to the possibility of monitoring inaccessible areas, as well as to analyze changes across the landscape over time.

The maps obtained with this approach can be used as layers to model the potential distribution of migratory grassland birds given that the structural characteristics of vegetation are among the main factors that determine their distribution in the area.

Acknowledgments: We greatly thank the CONACYT (Consejo Nacional de Ciencia y Tecnología) for the support given as a scholarship to pursue the PhD program of the main author. Likewise, we appreciate the comments from reviewers of the manuscript, which really increased the quality of the final document.

Author Contributions: José Hugo Martínez-Guerrero, Alfredo Pinedo-Alvarez and Alberto Rodríguez-Maturino conceived and designed the methodology. Alberto Rodríguez-Maturino performed the field sampling and produced the first draft. Emilio Pereda-Solis and Isaías Chairez-Hernández analyzed the data. Federico Villarreal-Guerrero and Marusia Rentería-Villalobos substantially contributed to the discussions, revisions and editions of the manuscript.

Conflicts of Interest: The authors declare no conflict of interest.

References

1. Adams, J.M.; Hugues, F.; Denard, F.L.; McGlade, J.; Woodward, I.F. Increases in Terrestrial Carbon Storage from the Last Glacial Maximum to the Present. *Nature* **1990**, *348*, 711–714. [CrossRef]
2. Peng, J.; Zhenhuan, L.; Yinghui, L.; Jiansheng, W.; Yinan, H. Trend Analysis of Vegetation Dynamics in Qinghai. Tibet Plateau using Hurst Exponent. *Ecol. Indic.* **2012**, *14*, 28–39. [CrossRef]
3. Herrera, A.Y.; Cortes, O.A. Listado Florístico y Aspectos Ecológicos de la Familia Poaceae Para Chihuahua, Durango y Zacatecas, México. *J. Bot. Res. Inst. Tex.* **2010**, *4*, 711–738.

4. Pool, D.B.; Panjabi, A.O.; Macias, D.; Solhjem, M. Rapid Expansion of Croplands in Chihuahua, Mexico Threatens Declining North American Grassland Bird Species. *Biol. Conserv.* **2014**, *170*, 274–281. [CrossRef]
5. Manjarrez, D.C.; Pinedo, A.A.; Pinedo, A.C.; Villarreal, G.F.; Cortes, P.L. Vegetation Landscape Analysis due to Land Use Changes on Arid Lands. *Pol. J. Ecol.* **2015**, *63*, 272–279. [CrossRef]
6. Hoekstra, M.J.; Boucher, T.M.; Taylor, H.R.; Carter, R. Confronting a Biome Crisis: Global Disparities of Habitat Loss and Protection. *Ecol. Lett.* **2005**, *8*, 23–29. [CrossRef]
7. USGS (United States Geological Survey). The North American Breeding Bird Survey, Results and Analysis 1966–2012. Available online: http://www.mbr-pwrc.usgs.gov/bbs/ (accessed on 5 May 2014).
8. Macias, D.A.; Montoya, A.B.; Gonzalez, M.C.E.; Rodriguez, S.J.R.; Hunt, W.G.; Krannitz, P.G. Factors Influencing Habitat Use by Migratory Grassland Birds in the State of Chihuahua, Mexico. *Auk* **2009**, *126*, 896–905. [CrossRef]
9. Guo, Q.; Zhongmin, H.; Shenggong, Li.; Xuanran, L.; Xiaomin, S.; Guirui, Y. Spatial Variations in Aboveground Net Primary Productivity Along a Climate Gradient in Eurasian Temperate Grassland: Effects of Mean Annual Precipitation and Its Seasonal Distribution. *Glob. Chang. Biol.* **2012**, *18*, 3624–3638. [CrossRef]
10. Reichmann, G.L.; Sala, E.O.; Debra, P.C. Precipitation Legacies in Desert Grassland Primary Production Occur through Previous-Year Tiller Density. *Ecology* **2013**, *94*, 435–443. [CrossRef] [PubMed]
11. Granados, S.D.; Sanchez, G.A.; Granados, R.L.V.; De la Rosa, B.A. Ecología de la Vegetación del Desierto Chihuahuense. *Revista Chapingo. Serie Ciencias Forestales y del Ambiente* **2011**, *17*, 111–130. [CrossRef]
12. Arellano, E.C.; Vidal, L.J.C.; García, H.L.; Laundré, W.J.; Reza, C.F.; Mejia, M.F.M.; Vargas, R.M.; Galaviz, D.L.F.; Romero, G.A.; Spilsbury, M.A. Registro de Presencia y Actividades de Algunos Mamíferos en el Desierto Chihuahuense, México. *THERYA* **2014**, *5*, 793–816. [CrossRef]
13. Manzano, P.; List, R. Grasslands of Mexico: A Perspective on Their Conservation. In Proceedings of the Grasslands Ecosystems, Endangered Species, and Sustainable Ranching in the Mexico-U.S. Borderlands, Conference Proceedings, Fort Collins, CO, USA, 2006; Basurto, X., Hadley, D., Eds.; United States Department of Agriculture: Erie, KS, USA, 2006; pp. 43–47.
14. Martínez, G.J.H.; Pereda, S.M.E.; Wehenkel, C. Association of Ammodramus Bairdii A. 1844, and Other Species of grassland Granivorous Birds in Winter Time in Northwestern Mexico. *Open J. Ecol.* **2014**, *4*, 281–288. [CrossRef]
15. Czerwinski, C.J.; Douglas, J.K.; Scott, W.M. Mapping Forest Growth and Decline in a Temperate Mixed Forest Using Temporal Trend Analysis of Landsat Imagery, 1987–2010. *Remote Sens. Environ.* **2014**, *141*, 188–200. [CrossRef]
16. Vittek, M.; Brink, A.; Donnay, F.; Simonetti, D.; Desclée, B. Land Cover Change Monitoring Using Landsat MSS/TM Satellite Image Data over West Africa between 1975 and 1990. *Remote Sens.* **2014**, *6*, 658–676. [CrossRef]
17. Goward, S. N.; Masek, G.J.; Williams, L.D.; Irons, R.J.; Thompson, R.J. The Landsat 7 Mission: Terrestrial Research and Applications for the 21st Century. *Remote Sens. Environ.* **2001**, *78*, 3–12. [CrossRef]
18. Wulder, A.M.; Masek, G.J.; Cohen, B.W.; Loveland, R.T.; Woodcock, E.C. Opening the Archive: How Free Data Has Enabled the Science and Monitoring Promise of Landsat. *Remote Sens. Environ.* **2012**, *122*, 2–10. [CrossRef]
19. Hansen, M.C.; Loveland, T.R. A Review of Large are Monitoring of Land Cover Change Using Landsat data. *Remote Sens. Environ.* **2012**, *122*, 66–74. [CrossRef]
20. Coppin, P.; Jonckheere, I.; Nackaerts, K.; Muys, B.; Lambin, E. Digital Change Detection Methods in Ecosystem Monitoring: A Review. *Int. J. Remote Sens.* **2004**, *25*, 1565–1596. [CrossRef]
21. Krofcheck, D.J.; Eitel, U.H.J.; Vierling, A.L.; Schulthess, U.; Hilton, M.T.; Dettweiler, R.E.; Pendleton, R.; Litvak, E.M. Detecting Mortality Induced Structural and Functional Changes in a Piñon-Juniper Woodland Using Landsat and RapidEye Time Series. *Remote Sens. Environ.* **2014**, *151*, 102–113. [CrossRef]
22. Schmitt, H.M.; Sweeney, P.S.; Evans, P.T. Classification of Coffee-Forest Landscapes Using Landsat TM Imagery and Spectral Mixture Analysis. *Photogramm. Eng. Remote Sens.* **2013**, *79*, 457–468. [CrossRef]
23. Sivanpillai, R.; Ewers, E.B. Relationship between Sagebrush Species and Structural Characteristics and Landsat Thematic Mapper data. *Appl. Veg. Sci.* **2013**, *16*, 122–130. [CrossRef]
24. Pucheta, E.E.; Heil, F.L.; Schneider, C. Modelos de Regresión Para la Estimación de la Biomasa Aérea en un Pastizal de Montaña de Pampa de Achala. (Córdoba, Argentina). *AGRISCIENTIA* **2004**, *21*, 23–30.

25. Manrique, S.; Nuñez, V.; Franco, J.; Seghezzo, L. Prediccion de Biomasa Natural a Partir de Sensores Remotos en el Valle de Lerma. *Avances en Energías Renovables y Medio Ambiente* **2010**, *14*, 63–70.
26. Calvert, M.A.; Walde, J.S.; Taylor, D.P. Non-breeding Drivers of Population Dynamics in Seasonal Migrants: Conservation Parallels Across Taxa. *Avian Conserv. Ecol.* **2009**, *4*, 5. [CrossRef]
27. Chaneton, J.E. Factores que Determinan la Heterogeneidad de la Comunidad Vegetal en Diferentes Escalas Espaciales. In *La Heterogeneidad de la Vegetación de los Agroecosistemas*; Oesterheld, M., Aguiar, M.R., Ghersa, C.M., Paruelo, J.M., Eds.; Editorial Facultad de Agronomía-UBA: Buenos Aires, Argentina, 2005; pp. 19–42.
28. Hoyt, A.C. The Chihuahuan Desert: Diversity at Risk. *Endanger. Species Bull.* **2002**, *27*, 16–17.
29. Rzedowski, J. *Vegetación de México*; Limusa: Mexico City, México, 1981; p. 504.
30. De León, M.D.; Pinedo, A.A.; Martínez, G.J.H. Aplicación de Sensores Remotos en el Análisis de la Fragmentación del Paisaje en Cuchillas de la Zarca, México. *Investigaciones Geográficas, Boletín del Instituto de Geografía* **2014**, *84*, 42–53. [CrossRef]
31. CONABIO. (Comisión Nacional para el Conocimiento y Uso de la Biodiversidad). Catálogo de Metadatos Geográficos. Red de Carreteras Escala 1:10,000,000. Available online: http://www.conabio.gob.mx/informacion/metadata/gis/carre1mgw.xml?_xsl=/db/metadata/xsl/fgdc_html.xsl&_indent=no (accessed on 16 October 2017).
32. Crist, P.E.; Richard, C.C. A Physically Based Transformation of Thematic Mapper Data-The TM Tasseled Cap. *IEEE Trans. Geosci. Remote Sens.* **1984**, *22*, 256–263. [CrossRef]
33. Chander, G.; Markham, L.B.; Helder, L.D. Summary of Current Radiometric Calibration Coefficients for Landsat MSS, TM, ETM+, and EO-1 ALI Sensors. *Remote Sens. Environ.* **2009**, *113*, 893–903. [CrossRef]
34. Chuvieco, S.E. *Teledetección Ambiental: La Observación de la Tierra Desde el Espacio*; Ariel Ciencia: Barcelona, Spain, 2002.
35. Dobson, J.A.; Barnett, A. *An Introduction to Generalized Linear Models*; Taylor & Francis Group: New York, NY, USA, 2008.
36. Rosenfield, H.G.; Fitzpatric, K.L. A Coefficient of Agreement as a Measure of Thematic Classification Accuracy. *Photogramm. Eng. Remote Sens.* **1986**, *52*, 223–227.
37. Eastman, R.J. *Idrisi Selva*; Version 17; Clark Labs, Clark University: Worcester MA, USA, 2012.
38. SMN. Monitor de Sequía de México. Available online: http://smn.cna.gob.mx (accessed on 15 January 2014).
39. Wiles, L.J.; Dunn, G.; Printz, J.; Patton, B.; Nyren, A. Spring Precipitation as a Predictor for Peak Standing Crop of Mixed-Grass Prairie. *Rangel. Ecol. Manag.* **2011**, *64*, 215–222. [CrossRef]
40. Cohen, B.W.; Maiersperger, K.T.; Gowerc, T.S.; Turner, P.D. An Improved Strategy for Regression of Biophysical Variables and Landsat ETM+ Data. *Remote Sens. Environ.* **2003**, *84*, 561–571. [CrossRef]
41. Xie, Y.; Sha, Z.; Yu, M.; Bai, Y.; Zhang, L. A Comparison of Two Models with Landsat Data for Estimating above Ground Grassland Biomass in Inner Mongolia, China. *Ecol. Model.* **2009**, *220*, 1810–1818. [CrossRef]
42. Gasparri, I.; Parmuchi, M.G.; Bono, J.; Karszenbaum, H.; Montenegro, C.L. Assessing Multi-Temporal Landsat ETM+ Images for Estimating Above-Ground Biomass in Subtropical Dry Forests of Argentina. *J. Arid Environ.* **2010**, *74*, 1262–1270. [CrossRef]
43. Karl, W.J. Spatial Predictions of cover Attributes of Rangeland Ecosystems Using Regression Kriging and Remote Sensing. *Rangel. Ecol. Manag.* **2010**, *63*, 335–349. [CrossRef]
44. Song, C.; Woodcock, E.; Seto, C.K.; Lenney, P.M.; Macomber, A.S. Classification and Change Detection Using Landsat TM data: When and How to Correct Atmospheric Effects? *Remote Sens. Environ.* **2001**, *75*, 230–244. [CrossRef]
45. Lencinas, D.J.; Bell, M.D. Estimación de Clases de Edad de las Plantaciones de la Provincia de Corrientes, Argentina, con Base en Datos Satelitales Landsat. *Bosque* **2007**, *28*, 106–118. [CrossRef]
46. Khumalo, G.; Holechek, J. Relationships between Chihuahuan Desert Perennial Grass Production and Precipitation. *Rangel. Ecol. Manag.* **2005**, *58*, 239–246. [CrossRef]
47. Yahner, H.R. *Wildlife Behavior and Conservation*; Springer Science: New York, NY, USA, 2012.
48. Jin, Y.; Yang, X.; Qiu, J.; Li, J.; Gao, T.; Wu, Q.; Zhao, F.; Ma, H.; Yu, H.; Xu, B. Remote Sensing-Based Biomass Estimation and Its Spatio-Temporal Variations in Temperate Grassland, Northern China. *Remote Sens.* **2014**, *6*, 1496–1513. [CrossRef]
49. Duru, M.; Ducrocq, H. Growth and Senescence of the Successive Grass Leaves on a Tiller. Ontogenic Development and Effect of Temperature. *Ann. Bot.* **2000**, *85*, 635–643. [CrossRef]

50. Westoby, M.; Walker, B.; Noy-Meir, I. Opportunistic Management for Rangelands Not at Equilibrium. *J. Range Manag.* **1989**, *42*, 266–274. [CrossRef]
51. Si, Y.; Schlerf, M.; Zurita, M.R.; Skidmore, A.; Wang, T. Mapping Spatio-Temporal Variation of Grassland Quantity and Quality Using MERIS Data and the PROSAIL Model. *Remote Sens. Environ.* **2012**, *121*, 415–425. [CrossRef]
52. Gao, T.; Yang, X.; Jin, Y.; Ma, H.; Li, J.; Yu, H.; Yu, Q.; Zheng, X.; Xu, B. Spatio-Temporal Variation in Vegetation Biomass and Its Relationships with Climate Factors in the Xilingol Grasslands, Northern China. *PLoS ONE* **2013**, *8*, e83824. [CrossRef] [PubMed]
53. Martínez, G.J.H.; Wehenkel, C.; Pereda, S.E.; Panjabi, A.; Levandoski, G.; Corral, R.J.; Díaz, M.R. Relación Entre la Cobertura del Suelo y Atributos de la Vegetación Invernal con *Ammodramus Bairdii*, Audubon 1844, en el Noroeste de México. *Agrociencia* **2011**, *45*, 443–451.
54. Hovick, J.T.; Elmore, D.R.; Fuhlendorf, D.S. Structural Heterogeneity Increases Diversity of Non-Breeding Grassland Birds. *Ecosphere* **2014**, *5*, 1–13. [CrossRef]
55. McGranahan, A.D.; Raicovich, M.G.; Wilson, N.W.; Smith, K.C. Preliminary Evidence that Patch Burn-Grazing Creates Spatially Heterogeneous Habitat Structure in Old-Field Grassland. *Southeast. Nat.* **2013**, *12*, 655–660. [CrossRef]
56. Moreau, S.; Bosseno, R.; Gu, X.; Baret, F. Assessing the Biomass Dynamics of Andean Bofedal and Totora high-Protein Wetland Grasses from NOAA/AVHRR. *Remote Sens. Environ.* **2003**, *85*, 516–529. [CrossRef]
57. Martin, L.M.; Moloney, K.A.; Wilsey, B.J. An Assessment of Grassland Restoration Success Using Species Diversity Components. *J. Appl. Ecol.* **2005**, *42*, 327–336. [CrossRef]
58. Kawamura, K.; Akiyama, T.; Yokot, H.; Tsutsumi, M.; Yasuda, T.; Watanabe, O.; Wang, S. Quantifying Grazing Intensities Using Geographic Information Systems and Satellite Remote Sensing in the Xilingol Steppe Region, Inner Mongolia, China. *Agric. Ecosyst. Environ.* **2005**, *107*, 83–93. [CrossRef]

land

MDPI

Article

Satellite Monitoring of Vegetation Response to Precipitation and Dust Storm Outbreaks in Gobi Desert Regions

Yuki Sofue [1], Buho Hoshino [2,*], Yuta Demura [3], Kenji Kai [4], Kenji Baba [2], Eunice Nduati [1], Akihiko Kondoh [1] and Troy Sternberg [5]

1 The Graduate School of Sciences, Chiba University, Chiba 263-8522, Japan; yuki.candy.s126@gmail.com (Y.S.); mnduati.eunice@gmail.com (E.N.); kondoh@faculty.chiba-u.jp (A.K.)
2 College of Agriculture, Food and Environment Sciences, Rakuno Gakuen University, Ebetsu 069-8501, Japan; kbaba@rakuno.ac.jp
3 Field Researchers Corporation Co., Ltd., Sapporo, Hokkaido 060-0007, Japan; demura.yuta@gmail.com
4 Graduate School of Environmental Studies, Nagoya University, Nagoya 464-8601, Japan; kai@info.human.nagoya-u.ac.jp
5 School of Geography University of Oxford, South Parks Road, Oxford OX1 3QY, UK; troy.sternberg@geog.ox.ac.uk
* Correspondence: aosier@rakuno.ac.jp; Tel.: +81-11-388-4913

Received: 12 December 2017; Accepted: 27 January 2018; Published: 1 February 2018

Abstract: Recently, droughts have become widespread in the Northern Hemisphere, including in Mongolia. The ground surface condition, particularly vegetation coverage, affects the occurrence of dust storms. The main sources of dust storms in the Asian region are the Taklimakan and Mongolian Gobi desert regions. In these regions, precipitation is one of the most important factors for growth of plants especially in arid and semi-arid land. The purpose of this study is to clarify the relationship between precipitation and vegetation cover dynamics over 29 years in the Gobi region. We compared the patterns between precipitation and Normalized Difference Vegetation Index (NDVI) for a period of 29 years. The precipitation and vegetation datasets were examined to investigate the trends during 1985–2013. Cross correlation analysis between the precipitation and the NDVI anomalies was performed. Data analysis showed that the variations of NDVI anomalies in the east region correspond well with the precipitation anomalies during this period. However, in the southwest region of the Gobi region, the NDVI had decreased regardless of the precipitation amount, especially since 2010. This result showed that vegetation in this region was more degraded than in the other areas.

Keywords: vegetation response to precipitation; dust storm outbreak; cross correlation analysis; the Hovmoller diagram; environmental regime shift; Gobi desert of Mongolia

1. Introduction

Located in Central Asia, the Gobi includes a great desert and semi-arid region that stretches across huge portions of both Mongolia and China. The characteristic vegetation constitutes mixtures of grasslands, shrubs, saltwort and thorny trees. The Mongolian Gobi is a source for the formation of dust storms that sweep across East Asia [1]. Dust storms frequently occur in arid and semi-arid regions and may have contributed to the desertification observed in recent decades as well as the accelerated occurrence of more arid conditions over the drylands of Asia [2]. In the arid regions of continental East Asia, dust storms frequently occur in spring [3,4]. Vegetation coverage is one of the most important factors for the reduction of dust storm occurrence [5–7]. In these regions, the summer season is the highest season of vegetation activity. Previous studies showed that not only spring vegetation plays an important role in reducing dust storm frequencies but the previous summer vegetation in these arid and semi-arid regions also plays a role. For example, Kurosaki et al. [8] proposed a hypothesis in which dead leaves of grasses in spring, which are residue from the preceding summer's vegetation, were the main factor suppressing dust events. In a more recent study, Nandiatsetseg and Shinoda [9] also pointed out that standing dead grasses had the most significant impact on spring dust outbreaks. In addition, it is known that spring dust storm frequency in China appears to be more correlated with NDVI from the prior summer than that in March to May of the same year [10]. The land surface memory in the Mongolian grasslands was investigated [11]. They showed that soil moisture and vegetation anomalies determined by summer precipitation were maintained during winter, affecting dust occurrence in the following spring. Furthermore, perennial plants such as shrub, dominant in arid and semi-arid Gobi region, may remain alive as NDVI memory and retain as a memory of belowground biomass anomalies under the snow cover and in the frozen soil [12]. They also pointed out that the belowground biomass anomalies appeared to affect the aboveground anomalies at the initial vegetation growth during the spring as detected by NDVI [12]. This shows that perennial plants can easily affect vegetation coverage.

Water is the main limiting factor for vegetation growth over southern Mongolia [13]. However, both observation and modelling studies have indicated that an aridity trend is occurring and will occur most significantly in the semi-arid regions with droughts becoming more widespread in the Northern Hemisphere, including Asia, and particularly in Mongolia e.g., [14–17]. Furthermore, Huang et al., 2016 [18] point out that the warming trends over drylands, particularly in arid regions, are twice as great as those over humid regions. In this study, we used time series satellite vegetation measurements from the National Oceanic and Atmospheric Administration (NOAA) Advanced Very High Resolution Radiometer (AVHRR) sensor to examine the variability and trends of land surface conditions in the Gobi region as represented by vegetation index data from 1985 to 2013.

2. Materials and Methods

2.1. Data of Analysis

2.1.1. NDVI Data

In this study, we use the Normalized Difference Vegetation Index (NDVI) to estimate vegetation variation. NDVI is given by

$$\mathrm{NDVI} = \frac{\mathrm{NIR} - \mathrm{RED}}{\mathrm{NIR} + \mathrm{RED}} \tag{1}$$

where, RED and NIR are the surface reflectance bands in the 550–700 nm (visible) and 730–1000 nm (infrared) regions of the electromagnetic spectrum, respectively. The NDVI3g data set used in this study is derived from measurements made by the AVHRR sensor aboard NOAA polar orbiting satellite series (NOAA-7, 9, 11, 14, 16). The NDVI3g data set is provided by the GIMMS group at NASA's Goddard Space Flight Center, as described by Tucker et al. [19] and cover the period from 1981 to 2013, with a spatial resolution of 8 km by 8 km. The NDVI data were generated from processed 15-day NDVI composites using the maximum value compositing procedure to minimize effects of cloud contamination, varying solar zenith angles and surface topography [20]. For this study, we subset the Gobi region covering the geographical domain 90° E–117.5° E and 40° N–47.5° N, from the continental data set for the period from January 1985 to December 2013 (Figure 1). Figure 1 shows the map of the NDVI average of all data for the study period from 1985 to 2013.

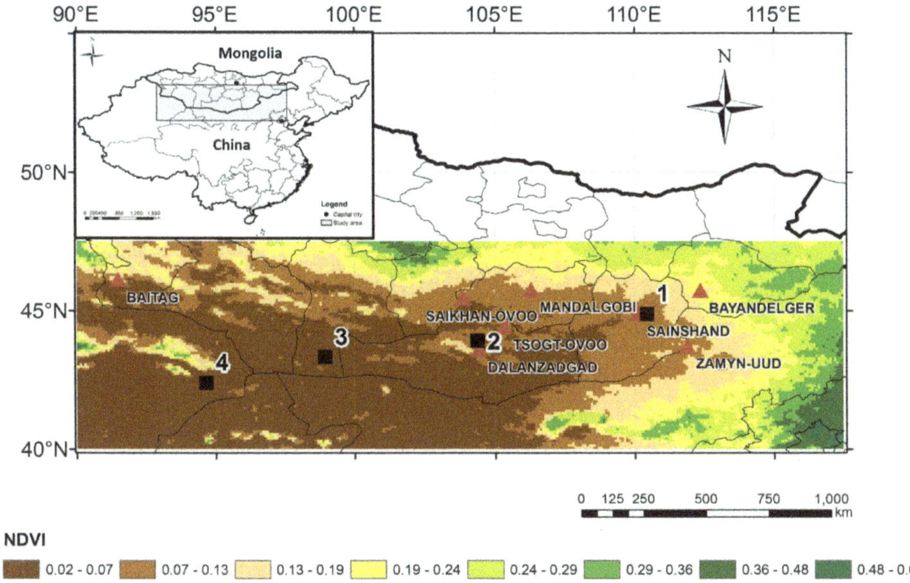

Figure 1. Long-term mean NDVI for the Gobi region (1985–2013) showing the transition from the eastern region with NDVI values of 0.6 to the west with values 0.02. Where, Long-term mean NDVI was averaged monthly maximum NDVI from 1985 to 2013. The numbered locations indicate sites where NDVI data were extracted to examine the temporal variations and trends in NDVI from 1985 to 2013. Furthermore, the red triangles show that the distribution of WMO (the World Meteorological Organization) synoptic stations in Mongolia used for this study. All stations are located in the desert steppe zone of the study area.

2.1.2. Precipitation Data

The Global Precipitation Climatology Project (GPCP) data was derived from a joint analysis of satellite data, and gauge data [21]. This precipitation data has daily and monthly data. Daily data has ($1^\circ \times 1^\circ$) spatial resolution acquired between October 1996 and May 2015. Monthly data has ($2.5^\circ \times 2.5^\circ$) spatial resolution acquired between January 1979 and May 2015. Since previous applications of NDVI in the Gobi region were focused mainly on the rainy season, NDVI patterns during the Growing Season (GS) were analyzed. In addition, we used daily precipitation data obtained from WMO synoptic data from 2001 to 2010. This data was used to evaluate the relationship with the number of dust storm days and NDVI using single regression analysis.

2.1.3. Dust Storm Monitoring Data

The WMO SYNOP surface weather data was used as the number of dust storm days. This data is daily data and recorded as 1 when dust storms are observed visually in a day. The data used in this study was obtained from 8 stations (Figure 1). The available period of this data is between 2001 and 2010.

2.2. Methods

We examined the spatiotemporal and seasonal variations, as well as the anomaly patterns for the monthly time series from 1985 to 2013. The growing season was defined by examining the long-term mean patterns of both precipitation and NDVI as shown in Figure 2a,b. Respectively, and with reference to long-term patterns of annual average precipitation distribution [22]. The months of May through

September were selected to represent the average start and end of the Growing Season, referred to here as GS using Figure 2. These figures were created from the results of calculating the monthly average of precipitation and NDVI during the study period. This shows the long-term mean for this region. Interannual variability in the NDVI pattern was examined by calculating yearly GS anomalies as follows;

$$\mathrm{NDVI}\sigma = [((\mathrm{NDVI}\alpha)/(\mathrm{NDVI}\mu) - 1) \times 100] \quad (2)$$

where NDVIσ are the respective GS percent anomalies, NDVIα are individual seasonal GS means and NDVIμ is the long-term GS mean. We also examined the precipitation anomalies during GS using the same method as that used for NDVI anomalies. Then we used the cumulative values of precipitation during GS. In addition, we performed cross correlation analysis between the cumulative precipitation and averaged NDVI for 15 days by averaging pixels in (1° × 1°) and verified the results of comparison with both trends. Cross-correlation is a generalization of the correlation measure as it takes into account the lag of NDVI relative to precipitation. Here, 1 time lag shows that there is 15 days lag between NDVI and precipitation. Cross-correlation is particularly important to assess the causal relationship between two signals in time. The analysis period is determined by the period of daily precipitation data from 1996 to 2013. Furthermore, we performed a single regression analysis to confirm the relationship between vegetation in GS with the number of days in which dust storms occurred during the following spring and also precipitation in GS at each of the 8 stations.

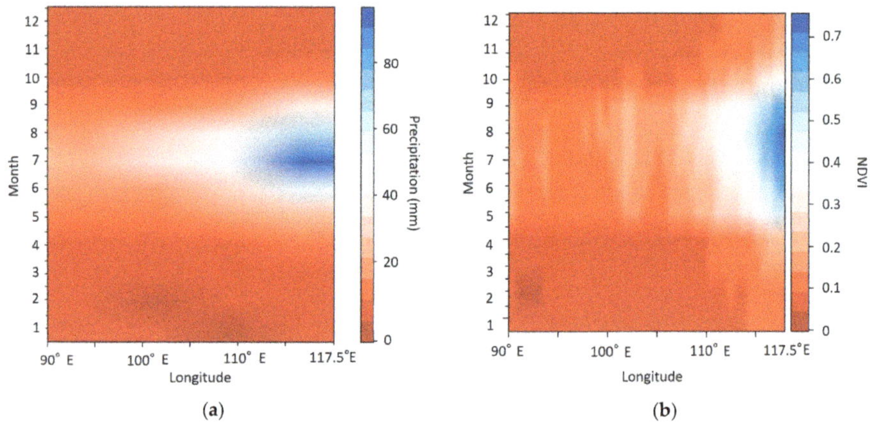

Figure 2. Hovmoller diagrams: (**a**) monthly precipitation and (**b**) monthly NDVI for the Gobi region averaged between 90° E–117.5° E and 40° N–47.5° N.

3. Results

3.1. Spatial Patterns and Trends for Vegetation Response to Precipitation

The time series anomaly for the region are depicted by the Hovmoller diagram for the period from January 1985 to December 2013 (Figure 3a,b). It was considered that variation in vegetation arises due to a difference in distribution of climatological conventional precipitation in the monsoon season. The amount of precipitation, which is supplied by monsoons from the Pacific and Indian Oceans [23], differs greatly between east and west. Following Figures 2 and 3, vegetation in the eastern region (from 110° E to 117.5° E), which had higher conventional precipitation, had a higher response to the precipitation than the other regions. In the eastern region, relatively low amounts of precipitation had been reported between 1999 and 2011. However, vegetation anomaly responds to a little increased precipitation such as 2003 and 2008. In contrast, in the central part of the study area (100° E from 110° E), a high response of vegetation to higher precipitation was observed such as in 1994, 1995 and

2003, but there was not high response to precipitation after 2005. The western region beginning from 100° E to 90° E showed a low response of vegetation to precipitation as compared to the east and central parts starting from 1985 due to the lower amount of conventional precipitation. However, there was no response of vegetation to precipitation especially since 2005. Fluctuations in precipitation anomaly with the whole region have increased especially since 2000. During the period 1985–1992, precipitation anomalies show almost normal precipitation conditions ranging between −20% and 10%. And during the period 1993–1999, precipitation anomalies show above normal precipitation conditions with positive anomalies ranging between 30% and 70% in the central region and part of western region. However, the period 2000–2013 increases a pattern of below normal precipitation conditions. In this period, the extreme negative value was less than 40% below normal. According to Figure 3, from 1985 to 2000, the NDVI patterns almost agreed with precipitation patterns. NDVI in the eastern region had kept this pattern after 2000 as well. Contrary to this, a part of the center region and whole western region had a negative trend especially from 2010 and did not recover in the following years with greater values of precipitation. It is assumed that this area had become desertified. This region includes Xinjiang, which is very sensitive to climate change [24].

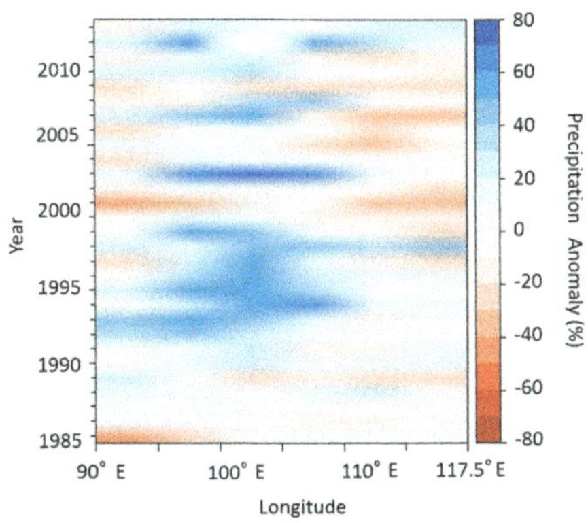

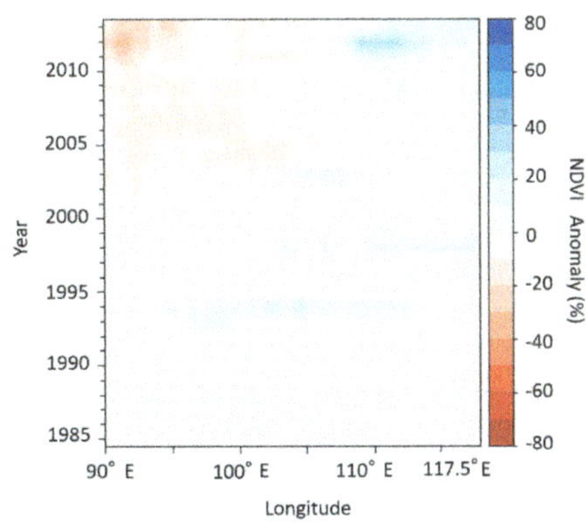

Figure 3. Hovmoller diagrams: (**a**) precipitation anomaly and (**b**) NDVI anomaly for the Gobi region averaged between 90° E–117.5° E and 40° N–47.5° N.

Time series of NDVI and precipitation for selected locations across the Gobi region for the period from 1985 to 2013 are shown in Figure 4. The data presented here are averaged NDVI values and cumulative precipitation for GS at each point. Sites 1 and 2 showed no change in trends of NDVI through the time series. On the other hand, site 3 showed a positive trend from 2003, and site 4 showed a negative trend from around 2009. There was a big difference in variation in NDVI values for sites 1 to 4. Sites 1 and 2 had a relatively large variation of NDVI year to year following precipitation variation. Conversely, sites 3 and 4 had a small variation. Site 4 especially shows a declining trend since 2010.

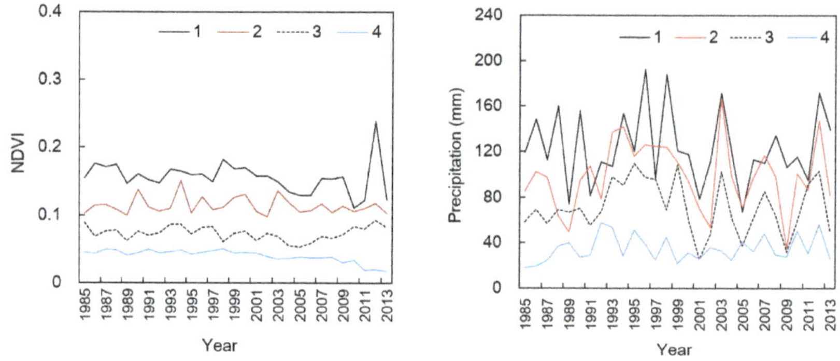

(a) (b)

Figure 4. Time series of average GS (the Growing season for plants; from May to September) NDVI for selected sites across the Gobi region and cumulative GS precipitation during the same period (**a**,**b**). All sites, excluding site 4, exhibit a similar trend of precipitation variation over time (**a**). Site 4 shows minimal variation. On the one hand, the distribution patterns of NDVI have indicated a decreased amount of precipitation and a shift from the Northeast to Southwest region (**b**).

3.2. Cross Correlation Analysis

The results of the cross-correlation analysis across the Gobi region for the period 1996 to 2013 are shown in Figure 5a. All number are correlation coefficient (r) and the values of probability p showed these are significant ($p < 0.05$). The distribution of correlation coefficient is shown in Figure 5b. In the eastern region, there was a relatively high trend of correlation coefficient. By contrast, the time lag was larger and the correlation coefficient was very low in the western region, especially in the southwest area. The time lag was almost 0. It showed that the vegetation responded to precipitation within 15 days after precipitation events. The positive relationship between NDVI and precipitation during GS in the Saikhan-ovoo site is shown as an example (Figure 6). The highest correlation coefficient value was 0.72 ($R^2 = 0.52$, $p < 0.05$) at time lag 0 locations. The vegetation had decreasing trends, but we postulate that it would recover in most locations during seasons with sufficient precipitation.

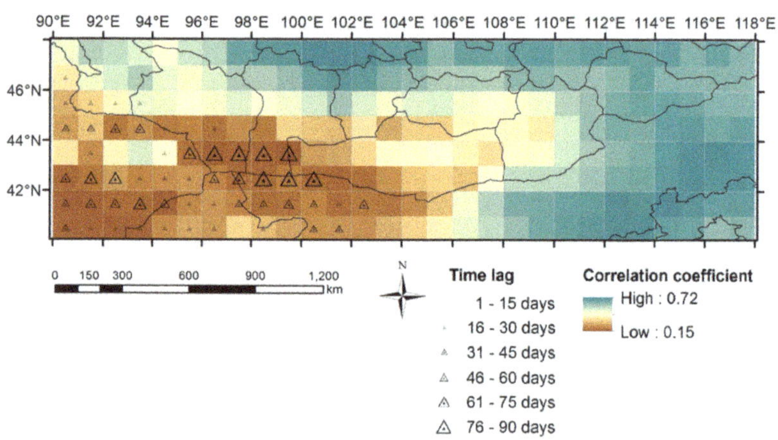

(a)

Figure 5. *Cont.*

	90°E	91°E	92°E	93°E	94°E	95°E	96°E	97°E	98°E	99°E	100°E	101°E	102°E	103°E
48°N	0.47	0.56	0.55	0.51	0.43	0.53	0.56	0.65	0.67	0.68	0.71	0.70	0.72	0.66
47°N	**0.38**	0.48	0.51	0.56	0.49	0.46	0.50	0.56	0.56	0.59	0.65	0.67	0.66	0.61
46°N	**0.29**	**0.34**	**0.40**	**0.48**	0.48	0.43	0.47	0.50	0.44	0.48	0.52	0.58	0.50	0.46
45°N	**0.26**	**0.32**	**0.21**	**0.24**	**0.21**	0.22	**0.20**	0.24	0.22	0.34	0.32	0.30	0.37	0.28
44°N	0.37	**0.24**	0.40	0.51	**0.43**	**0.19**	**0.17**	**0.18**	**0.17**	**0.16**	0.27	0.28	0.34	0.42
43°N	**0.18**	**0.20**	**0.26**	**0.23**	**0.24**	**0.25**	**0.23**	**0.19**	**0.19**	**0.22**	**0.19**	0.18	0.19	0.25
42°N	**0.17**	**0.17**	**0.18**	**0.19**	**0.15**	**0.23**	**0.25**	**0.23**	**0.19**	**0.19**	**0.23**	**0.19**	**0.21**	0.20
41°N	**0.17**	**0.17**	**0.17**	**0.17**	**0.24**	**0.28**	**0.26**	0.36	**0.30**	0.27	**0.25**	**0.25**	0.24	0.29

	104°E	105°E	106°E	107°E	108°E	109°E	110°E	111°E	112°E	113°E	114°E	115°E	116°E	117°E
48°N	0.66	0.61	0.65	0.65	0.66	0.66	0.63	0.65	0.65	0.64	0.66	0.61	0.64	0.61
47°N	0.60	0.57	0.57	0.56	0.60	0.61	0.59	0.58	0.60	0.64	0.61	0.60	0.60	0.60
46°N	0.50	0.51	0.47	0.45	0.46	0.48	0.55	0.58	0.62	0.63	0.64	0.63	0.60	0.61
45°N	0.33	0.33	0.40	0.41	0.41	0.38	0.51	0.58	0.62	0.61	0.65	0.66	0.62	0.64
44°N	0.43	0.44	0.40	0.36	0.42	0.46	0.52	0.58	0.60	0.61	0.66	0.68	0.68	0.68
43°N	0.32	0.34	0.31	0.43	0.54	0.55	0.55	0.58	0.62	0.62	0.66	0.68	0.67	0.67
42°N	0.24	0.31	0.48	0.56	0.62	0.63	0.65	0.65	0.69	0.67	0.69	0.67	0.65	0.66
41°N	0.34	0.40	0.49	0.54	0.59	0.63	0.65	0.67	0.69	0.69	0.65	0.64	0.59	0.60

(**b**)

Figure 5. (**a**) Map of the correlation between precipitation and NDVI, and distribution of time lag in response to precipitation; (**b**) Correlation coefficient matrix. Each bold number shows that the correlation coefficient has some time lag.

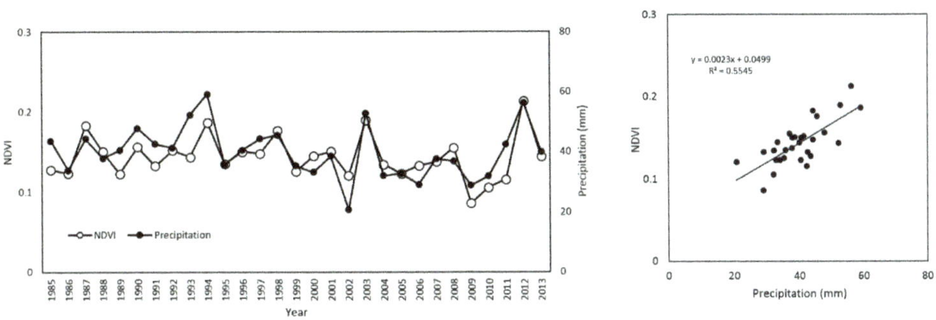

(**a**) (**b**)

Figure 6. (**a**,**b**) Relationship between NDVI and precipitation during GS at Saikhan-ovoo site.

3.3. Single Regression Analysis

As shown in Figure 7, the highest frequency of dust storm occurs in the spring season. Figures 7 and 8 show that the year of low vegetation agreed with relatively high frequency of dust storm occurrence, such as in 2009. Figure 8 shows the relationship between summer vegetation and the number of days in which dust storms occurred during the following spring at each meteorological station. These stations are located in the desert steppe zone of our study area. We analyzed this relationship using single regression analysis. From these results, the correlation coefficients were negative and the values were relatively high at Mandalgobi, Bayandelger, and Sainshand. These 3 sites have more plant species than places such as Tsogtovoo, Dalanzadgad and so on. Annual plant species are especially unstable, and the amount of biomass can have an effect on the frequency of dust storm occurrence. Therefore, the conditions of vegetation coverage during GS might influence dust storm frequency. It has been suggested that maintaining vegetation coverage during this period could reduce outbreaks of dust storms during the following spring. It also indicates that the vegetation condition in the southwest region of the Gobi should be monitored more carefully in the future. Figure 9 shows the relationship between NDVI and precipitation in GS at each station. The correlation coefficients were positive and the values were relatively high at Mandalgobi, Saikhan-ovoo, and Sainshand. These results show that precipitation can affect NDVI and also dust storm occurrences in some regions.

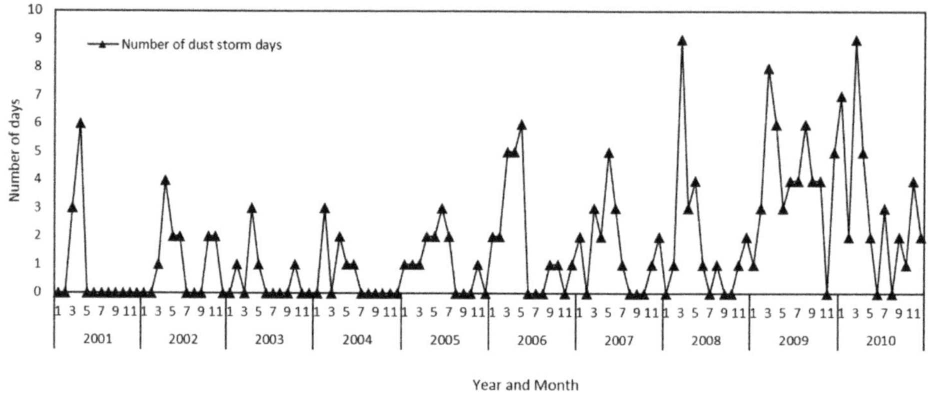

Figure 7. The annual and monthly distribution of dust storm days at Saikhan-ovoo site.

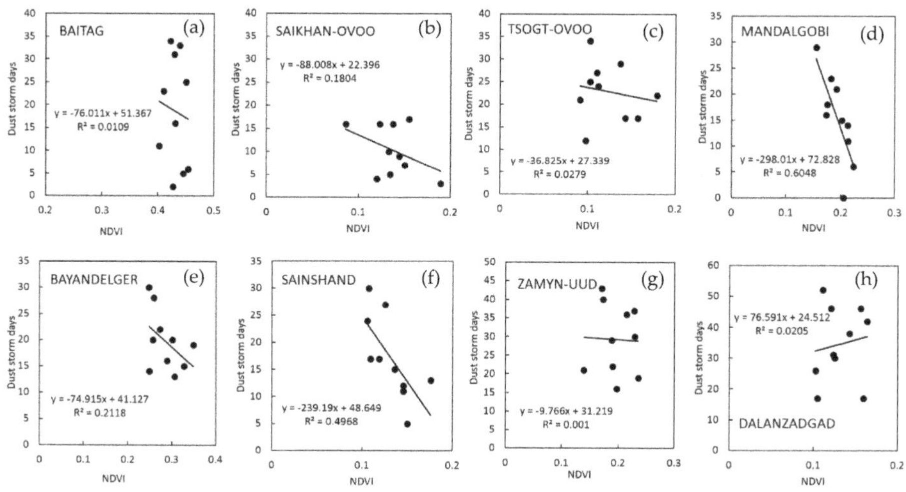

Figure 8. (**a**–**h**) Relationship between dust storm days and NDVI at each station.

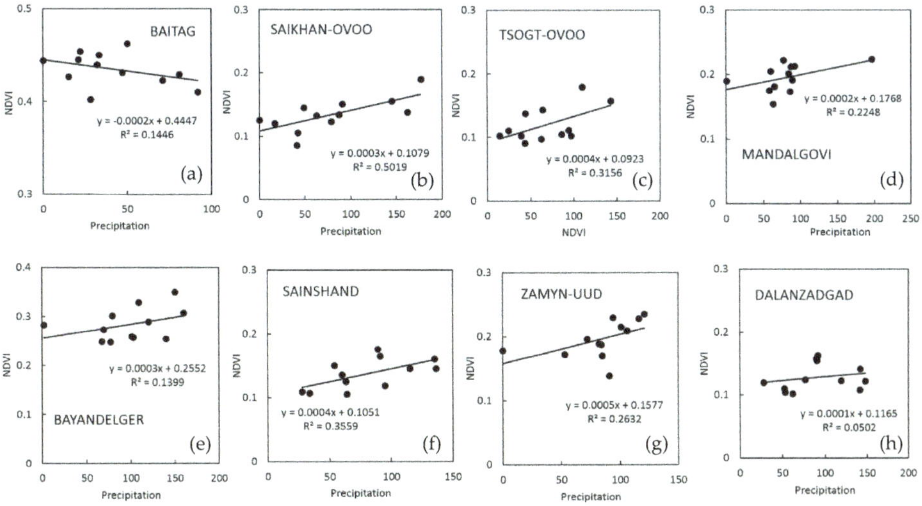

Figure 9. (**a**–**h**) Relationship between NDVI and precipitation at each station.

4. Discussion

In the arid and semi-arid Gobi region, vegetation cover is mainly constituted of annual and perennial plants. For example, *Suaeda aralocaspica* is a monoecious annual species commonly found in the Gobi desert and many perennial plants are found in this region, especially shrubs typified as *Haloxylon ammodendron*. Annual plants exist as dead grass in the spring, but are not reflected in NDVI. However, rainfall encourages the growth of annually herbaceous plants and is recorded as a memory of biomass (Dry Matter Productivity) in summer, and in the following year they suppress dust emission as dry grass. The differences in dead grass coverage rates may increase or decrease the outbreak of dust storms. On the other hand, perennial plants have very deep roots and this type of vegetation are effective in extracting water from their bare surroundings and therefore survive [25], so the effect of precipitation would be minimal. Furthermore, they can survive winter into the following spring and affect the frequency of dust storm outbreaks. However, once perennial plants, e.g., shrubs are in a dormant state, they need a substantial amount of time to recover. This is one of the contributory factors to the occurrence of desertification. In our study area, the typical plants are mainly perennial plants because of low precipitation and we found that the vegetation coverage had decreased in our study period.

Real world systems occasionally undergo substantial changes triggered by minor disturbances. A major theoretical finding when it comes to regime shifts is that ecosystems recover slowly from small perturbations in the vicinity of tipping points (Figure 10b). However, indicators of critical slowing down are not manifested in all cases where regime shifts occur, because not all regime shifts are associated with tipping points. In recent years, this challenge has gained importance as it is unclear how grassland ecosystems will respond to current trends in climate patterns and anthropogenic pressures [26]. The provided water such as rainfall is one of the most important factors for vegetation especially in arid regions (Figure 10a). As shown in Figures 10 and 11, the areas highlighted with the circle and square respectively, are situated in China, such as Gansu, Xinjiang uygur and Bayannaoer. A previous study showed that there was a declining trend in Bayannaoer in Inner Mongolia during 1999–2012 [27]. In 2001 and 2002, very low precipitation with anomaly value of −40% or less was found and vegetation tendency changed around the same time, after that, the negative trend continued (Figure 10c,d). From these results, it is considered that one cause of vegetation degeneration is likely to be change in precipitation. In semi-arid ecosystems, water availability is the dominant factor regulating soil respiration [28]. Water availability moderates the effects of other factors such as temperature and substrate supply, on soil respiration. And its interannual variations is directly linked to both the intensity and the frequency of precipitation. Shifts in precipitation regime will alter not only the size of individual rainfall events, but also the length of the dry-spell duration and thus the antecedent soil water condition [29]. When the pasture land undergoes desertification, the seed bank in the soil disappears and the soil layer is destroyed. We found that vegetation does not respond to precipitation at all after a certain time in some regions (Figure 10). We found that since 2003, vegetation no longer responded to precipitation in F2 region.

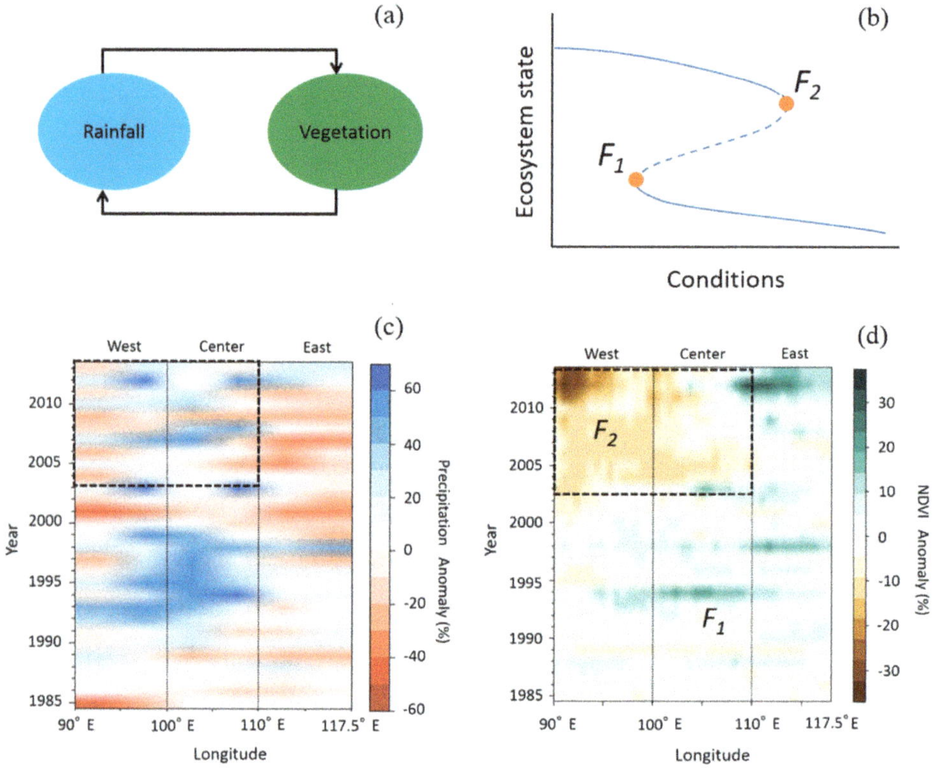

Figure 10. Environmental regime shifting in North China during 2003–2013. (**a**) shows that the concept of relationship between precipitation and vegetation. And (**b**) shows that the concept of regime shift. (**c**,**d**) are enhanced the difference Figure 3 (Hovmoller diagrams: where, (**c**) precipitation anomaly and (**d**) NDVI anomaly for the study area).

Desertification can increase the occurrence of dust storms as has been observed in the Tibetan Plateau and Hexi Corridor in recent years [30]. This includes areas located in Northwest China, within the Tarim Basin. Perennial plants are dominant in this area due to low precipitation and desertification is therefore more likely to occur when there are drought conditions. This study focuses on the dynamic interaction between precipitation, vegetation (NDVI) and dust emissions, however, only in the growing season (GS) are the annual grasses reflected in NDVI. Desertification is a kind of environment regime shift. While environment regime shift was not found in our Mongolian sites, it was observed in our Chinese sites [31] (Figure 11). As shown in Figure 11, the area highlighted with the circle where vegetation cover was degraded, is situated in China and includes Gansu, Xinjiang

Uygur and Bayannaoer. According to previous studies, due to its unique geographical position and vulnerable ecological environment, Xinjiang is very sensitive to climate change [24].

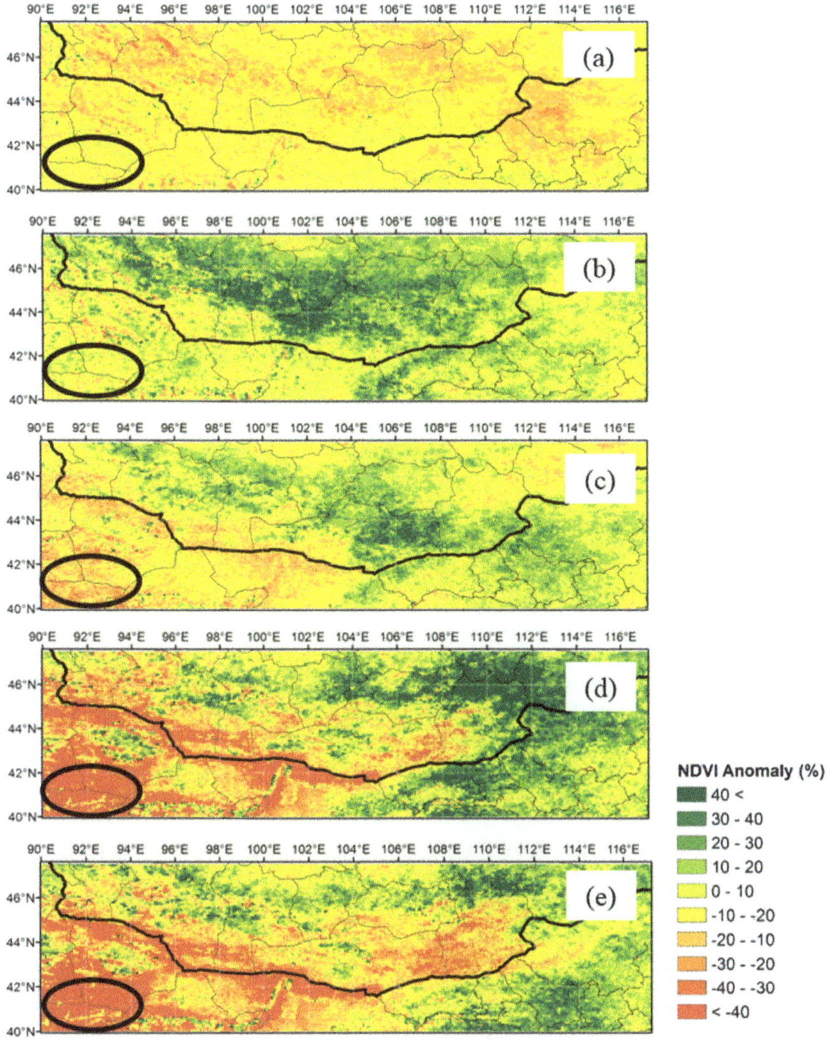

Figure 11. Anomaly NDVI in GS in (**a**) 1989; (**b**) 1994; (**c**) 2003; (**d**) 2012 and (**e**) 2013 [31].

5. Conclusions

Satellite measurements of vegetation dynamics in the Gobi region for a period of 29 years showed interannual variation and trends. In the Gobi region, precipitation is confined to the period from May to September. The variations of NDVI anomalies in the eastern region correspond well with the documented precipitation anomalies during this period. However, some parts, especially those in the southwest region of the Gobi region showed that the NDVI had decreased regardless of the precipitation amount.

As a result of this study, it was unlikely that desertification would occur in our Mongolian sites. However, as a result of this study, precipitation variation became significantly higher in the period after the year 2000. Furthermore, we found that vegetation in low precipitation areas was more degraded than that in high precipitation areas. Therefore, it is necessary to carefully continue monitoring the vegetation condition, especially in this region, in future studies.

Acknowledgments: This work was supported by JSPS KAKENHI Grant Numbers JP24340111, JP25550079, JP26281003 and the MEXT Supported Program for the Strategic Research Foundation (S1391001) at Rakuno Gakuen University. We are grateful to P. Tsedendamba and D. Munkhjargal for their support on field survey in Mongolia. In addition, we would like to thank B. Alsaaideh for giving some kind advice about this paper.

Author Contributions: Y.S., B.H., E.N. and A.K. analyzed the data and wrote the draft of manuscript and designed the study. Y.D., K.K., Y.S., K.K., K.K. and T.S. participated in the field measurements and participated in the satellite images analysis. Y.S., Y.D. and B.H. also established the GIS special database.

Conflicts of Interest: The authors declare no conflict of interest.

Data Availability: The NDVI 3g dataset used in this paper can be accessed and freely downloaded from the ECOCAST homepage (the GIMMS group at NASA's Goddard Space Flight Center): https://ecocast.arc.nasa.gov/data/pub/gimms/. Global Precipitation Climatology Project (GPCP) data can be downloaded from The NOAA/ESRL Physical Sciences Division (PSD) home page. https://www.esrl.noaa.gov/psd/data/gridded/data.gpcp.html. Number of livestock data can be downloaded from Mongolian Statistical Information Service home page. http://1212.mn/en/contents/stats/.

References

1. Natsagdorj, L.; Jugder, D.; Chung, Y.S. Analysis of dust storms observed in Mongolia during 1937–1999. *Atmos. Environ.* **2003**, *37*, 1401–1411. [CrossRef]
2. Huang, J.; Wang, T.; Wang, W.; Li, Z.; Yan, H. Climate effects of dust aerosols over East Asian arid and semiarid regions. *J. Geophys. Res. Atmos.* **2014**, *119*, 11398–11416. [CrossRef]
3. Littmann, T. Dust storm frequency in Asia: Climatic control and variability. *Int. J. Climatol.* **1991**, *11*, 393–412. [CrossRef]
4. Parungo, F.; Li, Z.; Li, X.; Yang, D.; Harris, J. Gobi dust storms and the great green wall. *Geophys. Res. Lett.* **1994**, *21*, 999–1001. [CrossRef]
5. Ishizuka, M.; Mikami, M.; Yamada, Y. An observational study of soil moisture effects on wind erosion at a gobi site in the Taklimakan Desert. *J. Geophys. Res.* **2005**, *110*, D18S03. [CrossRef]
6. Lee, E.H.; Shon, B.J. Examining the impact of wind and surface vegetation on the Asian dust occurrence over three classified source regions. *J. Geophys. Res.* **2009**, *114*, D06205. [CrossRef]
7. Lee, E.H.; Shon, B.J. Recent increasing trend in dust frequency over Mongolia and Inner Mongolia regions and its association with climate and surface condition change. *Atmos. Environ.* **2011**, *45*, 4611–4616. [CrossRef]
8. Kurosaki, Y.; Shinoda, M.; Mikami, M. What caused a recent increase in dust outbreaks over East Asia? *Geophys. Res. Lett.* **2011**. [CrossRef]
9. Nandintsetseg, B.; Shinoda, M. Land surface memory effects on dust emission in a Mongolian temperate grassland. *J. Geophys. Res. Biogeosci.* **2015**, *120*, 414–427. [CrossRef]
10. Zou, X.K.; Zhai, P.M. Relationship between vegetation coverage and spring dust storms over northern China. *J. Geophys. Res.* **2004**, *109*. [CrossRef]
11. Shinoda, M.; Nandintsetseg, B. Soil moisture and vegetation memories in a cold, arid climate. *Glob. Planet. Chang.* **2011**, *79*, 110–117. [CrossRef]
12. Shinoda, M.; Nachinshonhor, G.U.; Nemoto, M. Impact of drought on vegetation dynamics of the Mongolian steppe: A field experiment. *J. Arid Environ.* **2010**, *74*, 63–69. [CrossRef]
13. Liu, H.; Tian, F.; Hu, H.C.; Hu, H.P.; Sivapalan, M. Soil moisture controls on patterns of grass green-up in Inner Mongolia: An index based approach. *Hydrol. Earth Syst. Sci.* **2013**, *17*, 805–815. [CrossRef]
14. Fu, C.; Diaz, H.F.; Dong, D.; Fletcher, J.O. Changes in atmospheric circulation over Northern Hemisphere oceans associated with the rapid warming of the 1920s. *Int. J. Climatol.* **1999**, *19*, 581–606. [CrossRef]
15. Barlow, M.; Cullen, H.; Lyon, B. Drought in central and southwest Asia: La Nina, the warm pool, and Indian Ocean precipitation. *J. Clim.* **2002**, *15*, 697–700. [CrossRef]
16. Dai, A.; Trenberth, K.E.; Karl, T.R. Global variations in droughts and wet spells: 1900–1995. *Geophys. Res. Lett.* **1998**, *25*, 3367–3370. [CrossRef]
17. Lotsch, A.; Friedl, M.A.; Anderson, B.T.; Tucker, C.J. Response of terrestrial ecosystems to recent Northern Hemispheric drought. *Geophys. Res. Lett.* **2005**, *32*, L06705. [CrossRef]

18. Huang, J.; Wang, T.; Wang, W.; Li, Z.; Yan, H. Accelerated dryland expansion under climate change. *Nat. Clim. Chang.* **2016**, *6*, 166–171. [CrossRef]
19. Tucker, C.J.; Pinzon, J.E.; Brown, M.E.; Slayback, D.A.; Pak, E.W.; Mahoney, R.; Vermote, E.F.; Saleous, N.E. An extended AVHRR 8-km NDVI dataset compatible with MODIS and SPOT vegetation NDVI data. *Int. J. Remote Sens.* **2005**, *26*, 4485–4498. [CrossRef]
20. Holben, B.N. Characteristics of maximum-value composite images from temporal AVHRR data. *Int. J. Remote Sens.* **1986**, *7*, 1417–1434. [CrossRef]
21. Huffman, G.J.; Adler, R.F.; Bolvin, D.T.; Gu, G. Improving the global precipitation record: GPCP Version 2.1. *Geophys. Res. Lett.* **2009**, *36*, L17808. [CrossRef]
22. Anyamba, A.; Tucker, C.J. Analysis of Sahelian vegetation dynamics using NOAA-AVHRR NDVI data from 1981–2003. *J. Arid Environ.* **2005**, *63*, 596–614. [CrossRef]
23. Wang, B.; Fan, Z. Choice of South Asian summer monsoon indices. *Bull. Am. Meteorol. Soc.* **1999**, *80*, 629–638. [CrossRef]
24. Hu, R.-J.; Fan, Z.-L.; Wang, Y.-J. Assessment about the impact of climate change on environment in Xinjiang since recent 50 years. *Arid Land Geogr.* **2001**, *24*, 97–103. (In Chinese)
25. Hardenberg, J.; Meron, E.; Shachak, M.; Zarmi, Y. Diversity of vegetation patterns and desertification. *Phys. Rev. Lett.* **2001**, *87*, 198101. [CrossRef] [PubMed]
26. Dakos, V.; Carpenter, S.R.; van Nes, E.H.; Scheffer, M. Resilience indicators: Prospects and limitations for early warnings of regime shifts. *Philos. Trans. R. Soc. B Biol. Sci.* **2015**, *370*, 20130263. [CrossRef]
27. Buhe, B. Factor Analysis and Vegetation Change in China Inner Mongolia through Satellite Remote Sensing. Ph.D. Thesis, Chiba University, Chiba, Japan, 2015.
28. Yan, L.; Chen, S.; Huang, J.; Lin, G. Water regulated effects of photosynthetic substrate supply on soil respiration in a semiarid steppe. *Glob. Chang. Biol.* **2011**, *17*, 990–2001. [CrossRef]
29. Yan, L.; Chen, S.; Xia, J.; Luo, Y. Precipitation regime shift enhanced the rain pulse effect on soil respiration in a semi-arid steppe. *PLoS ONE* **2014**, *9*, e104217. [CrossRef] [PubMed]
30. Wang, X.; Chen, F.; Dong, Z. The relative role of climatic and human factors in desertification in semiarid China. *Glob. Environ. Chang.* **2006**, *16*, 48–57. [CrossRef]
31. Sofue, Y.; Hoshino, B.; Nduati, E.; Kondoh, A.; Kai, K.; Purevsuren, T.; Baba, K. Remote sensing methodology for detection of environmental regime shifts in semi-arid region. In Proceedings of the 37th Geoscience and Remote Sensing Symposium, Fort Worth, TX, USA, 23–28 July 2017; pp. 5113–5116.

land

MDPI

Article

Determining the Frequency of Dry Lake Bed Formation in Semi-Arid Mongolia From Satellite Data

Yuta Demura [1,2], Buho Hoshino [1,*], Kenji Baba [1], Christopher McCarthy [3], Yuki Sofue [4], Kenji Kai [5], Tsedendamba Purevsuren [1], Katsuro Hagiwara [6] and Jun Noda [6]

1 Graduate School of Dairy Sciences, Rakuno Gakuen University, Ebetsu, Hokkaido 069-8501, Japan; demura.yuta@gmail.com (Y.D.); kbaba@rakuno.ac.jp (K.B.); pujee_ts@yahoo.com (T.P.)
2 Field Researchers Corporation Co. Ltd, Sapporo, Hokkaido 060-0007, Japan
3 Graduate School of Global Environmental Studies, Kyoto University, Kyoto 606-8501, Japan; mccarthy.ch@gmail.com
4 Graduate School of Sciences, Chiba University, Chiba 263-8522, Japan; yuki.candy.s126@gmail.com
5 Graduate School of Environmental Studies, Nagoya University, Nagoya 464-8601, Japan; kai@info.human.nagoya-u.ac.jp
6 School of Veterinary Medicine, Rakuno Gakuen University, Hokkaido 069-8501, Japan; k-hagi@rakuno.ac.jp (K.H.); jnoda@rakuno.ac.jp (J.N.)
* Correspondence: aosier@rakuno.ac.jp; Tel.: +81-011-388-4913

Received: 25 October 2017; Accepted: 6 December 2017; Published: 8 December 2017

Abstract: In the Mongolian Plateau, the desert steppe, mountains, and dry lake bed surfaces may affect the process of dust storm emissions. Among these three surface types, dry lake beds are considered to contribute a substantial amount of global dust emissions and to be responsible for "hot spots" of dust outbreaks. The land cover types in the study area were broadly divided into three types, namely desert steppe, mountains, and dry lake beds, by a classification based on Normalized Difference Water Index (NDWI) calculated from MODIS Terra satellite images, and Digital Elevation Model (DEM). This dry lake beds extracting method using remote sensing offers a new technique for identifying dust hot spots and potential untapped groundwater in the dry lands of the Gobi region. In the study area, frequencies of dry lake bed formation were calculated during the period of 2001 to 2014. The potential dry lake area corresponded well with the length of the river network based on hydrogeological characterization ($R^2 = 0.59$, $p < 0.001$). We suggest that the threshold between dry lake bed areas and the formation of ephemeral lakes in semi-arid regions is eight days of total precipitation.

Keywords: dry lake beds; dust storm emission; remote sensing; Gobi Desert region

1. Introduction

Dust storms are a common phenomenon that may negatively affect human and animal health. In severe cases of dust storms, deaths of livestock have been reported in East Asia [1]. Because a large quantity of soil particles are produced by erosion and sedimentation transportation and deposit in dry lake beds [2], such dry lake beds represent areas where dust emissions occur [3]. Ginoux et al. (2012) found dust sources to have natural and anthropogenic origins, with natural dust sources accounting for 75% of global emissions and anthropogenic sources accounting for 25% [4]. Satellite images of dust storms in the Gobi region often show that large dust clouds are comprised of many well-defined plumes that emerge from "point" sources [5]. Farebrother et al. (2017) found a strong power relationship between the lake area and the mass of deflated lake bed sediments in Southeastern Australia [6]. In Mongolia's Gobi Desert, dry lake beds have been identified as a source of high-frequency dust storms. In fact, the Gobi Desert has been recognized as an important dust storm source region [7].

Dry lake beds are considered to contribute a substantial amount of global dust emissions and to be responsible for "hot spots" of dust outbreaks. Existing studies investigate potential water resources in the Gobi for use in many sectors, including regional planning, agriculture, and nomadic grazing. Water resources are identified through topographic and optical images obtained from remote sensing satellites and integrated into GIS to investigate and identify potential shallow water correlated to paleolakes and dry lakes beds [8]. These fields require an understanding of when and how water collects and evaporates across a given area.

This study applies the use of remote sensing to identify potential dry lake beds, taking into account the context of dry lake bed formation.

2. Materials and Methods

2.1. Study Area

The study area is located in Dornogobi, Dundgobi and Umnugobi provinces in semi-arid Mongolia. It is part of the drylands of the eastern Gobi, and has been identified as a source of dust storms (Figure 1) [1]. The semi-arid desert climate was determined from the Aridity Index of Millennium Ecosystem Assessment [9]. The annual average precipitation in Sainshand (Station number: 443540) is 111 mm. Precipitation from June to November accounts for approximately 82% of the annual precipitation. In contrast, the precipitation from December to May is 18% of the annual precipitation.

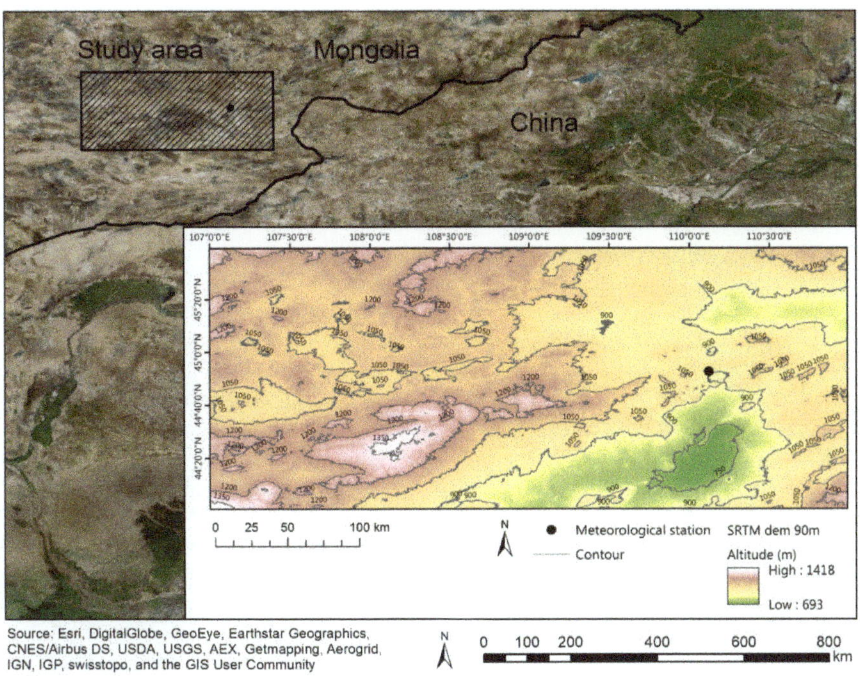

Figure 1. Map of the study area. The altitude in the area is 1030 ± 130 m. Mountains and valleys are common geographical features in the area.

2.2. Classification of the Land Cover

Within the study area there are few variations of land use and land cover. Therefore, the land cover classification was divided into mountainous regions, dry lake beds, and desert steppe using satellite remote sensing. We define a mountain as an area where the slope is greater than 1.8°. The dry lake beds are lakes that form seasonally depending on precipitation levels. Desert steppe is a typical area of the Gobi where there are a few plants and dry grasses.

The classification method for the land cover (mountains, dry lake beds, desert steppe) is shown as follows. The mountains are defined by an inclination threshold value of more than 1.8° and a higher altitude than the surrounding areas as calculated from the Digital Elevation Model (DEM). The threshold value for the angle of inclination was determined from a field survey conducted in Sainshand. The DEM (Spatial resolution: 90 m) was obtained from the Shuttle Radar Topography Mission (SRTM) and analyzed in ESRI ArcGIS10.3 and Spatial Analyst tool (Extension tool of ArcGIS10.3).

Using satellite remote sensing technology, the identification of dry lake beds has not been firmly established until now. In the case of an optical satellite, identification by spectral characteristics is difficult because the soil is bare in semi-arid regions such as Mongolia. Therefore, previous identification methods of dry lake beds have focused on geographical characteristics (such as low lying areas and plains) and the watershed topography of dry lake beds [10]. This method identifies watershed areas during a target period, with the identified watershed area defined as a potential dry lake bed. Figure 2 shows a flow chart for the identification of potential dry lake beds. Here, the Normalized Difference Water Index (NDWI) is a numerical indicator that uses the visible red and near-infrared bands of the electromagnetic spectrum to analyze remote sensing measures of water areas and wet surfaces. The NDWI can be written as Equation (1) [11].

$$\mathrm{NDWI} = (\varrho_{\mathrm{Red}} - \varrho_{\mathrm{SWIR}})/(\varrho_{\mathrm{Red}} + \varrho_{\mathrm{SWIR}}) \quad (1)$$

where, ϱ_{Red} and ϱ_{SWIR} refer to the spectral reflectance measurements acquired in the visible red and short-wavelength infrared regions, respectively. The NDWI ranges from 1 to -1. NDWI > 0 is classified as water areas and NDWI ≤ 0 defines non-water areas [12]. In our study, MODIS-Terra eight-day 500 m atmospheric corrected reflectance products (MOD09A1) were used to evaluate the water area conditions from 18 February 2000 to 26 June 2015. The total number of images is 704, which are used in the eight-day composite image.

To remove the influence of snow and cloud cover, a mask processing using the QA (Quality Assessment) band of MOD09A1 was conducted. Firstly, the area of potential dry lake beds extracted maximum NDWI from all pixels in NDWI images from September 2000 through July 2014 (Figure 3). Then, to determine elevation and ground level, a mask processing of the maximum NDWI pixels with the angle of inclination less than 1.8° calculated by DEM was performed. In addition, the algorithm for re-sampling used the Nearest Neighbor method, which maintained an original pixel level. Finally, all NDWI pixels greater than 0 were extracted as an area of potential dry lake beds. The precision of potential dry lake beds location was compared with the result of aerial photointerpretation where Spot image (space resolving power: 2.5 m) was used to inspect the classification accuracy. Spot image is a data source acquired from the French National Space Research Center used for Google Earth of Google Inc. Positional information of the aerial photointerpretation included 454 data points within the study area. Furthermore, extracted potential dry lake beds were clarified by comparing the geographical characteristics of river networks and watersheds. There are no perennial streams found within the river networks of semi-arid regions such as lake beds and wadis, and following a rainfall occurrence, intermitted surface flows appear within the lower portions. The areas form as sheet flow and conveyance (channel) areas [13]. The area of length of a simple linear regression is one way to model a relationship between river networks and potential dry lake beds because it is supposed that potential dry lake beds are connected with river networks for hydrology. The river network 15 s HydroSHEDS (Hydrological data and maps based on Shuttle Elevation Derivatives at multiple scales) was provided by WWF (World Wildlife Fund).

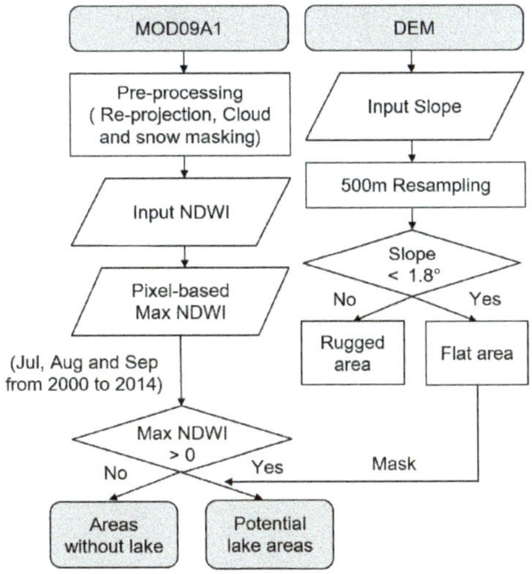

Figure 2. The identification method of potential dry lake beds based on Terra/MODIS and Digital Elevation Model (DEM).

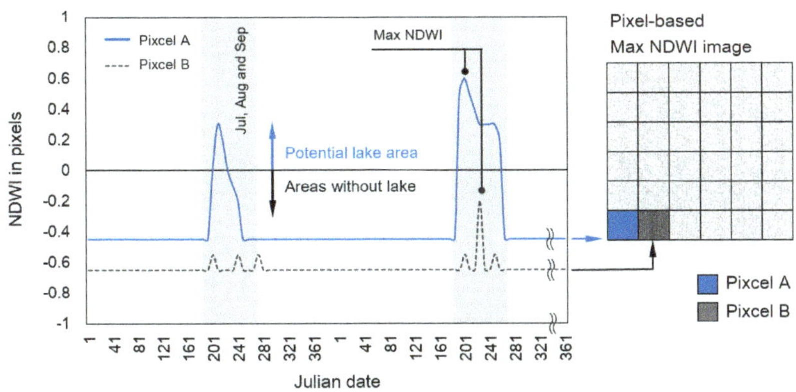

Figure 3. The pattern diagram of potential lake areas without lakes based on pixel-based maximum Normalized Difference Water Index (NDWI) images. In the schematic diagram of NDWI shown in Figure 2, pixel A is extracted as a potential lake area, and pixel B is extracted as an area without a lake from September 2000 through July 2014.

2.3. Formation of Dry Lake Beds

Precipitation in drylands is characterized by large fluctuations throughout the year [14]. In desert steppe regions such as the Gobi, the drop of permeability of the topsoil caused by rain has been confirmed [15]. From such topsoil and characteristics of rainfall patterns, precipitation may contribute to the formation of dry lake beds. Therefore, the changing tendencies of dry lake beds were identified as time series data of precipitation from 2000 through 2014. Precipitation data was extracted when ephemeral lakes were formed for the first time in each year. Then, because the scale of time and space is large, precipitation data from the GPCP (Global Precipitation Climatology Project) 1° Daily dataset was used. The total precipitation across eight days was aligned with the temporal resolution of dry lake beds data. The precipitation across eight days added up to the GPCP daily precipitation as the time scale of MOD09A1 product is eight days. The precipitation data was created from daily observations taken from 1 January 2000 to 31 July 2014.

3. Results

3.1. Classification of Land Cover

Using an aerial photointerpretation classification technique, 363 of 454 dry lake beds were extracted. This means that our classification method succeeded in identifying dry lake beds with an accuracy of approximately 80%. Mountainous areas were found from the southwestern reaches of the study area (44°10′~44°40′ N, 107°00′~109°00′ E) (Figure 4). Dry lake beds tended to be distributed in the northern mountainous areas (44°40′ N). On the other hand, desert steppe constituted the greatest percentage of land cover. In addition, mountainous areas, dry lake beds, and desert steppe varied greatly within our study area. Desert steppe and mountain regions represented 69% and 30%, respectively, with dry lake beds accounting for 1% of our study area. The area of desert steppe was approximately 56 times the area of dry lake beds.

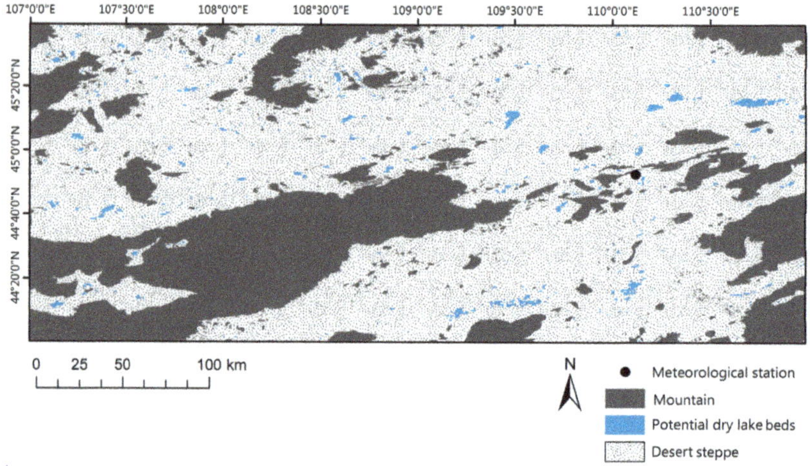

Figure 4. The distribution of three land cover types in the study area.

We then collated the identified dry lake beds with the river network found within our study site. The surface flows provide runoff within the lower dry lake beds through the river network (Figure 5). Dry lake beds in lower elevations can be understood as being part of a well-developed river network. In addition, the area of dry lake beds showed a tendency to increase (Figure 6) ($R^2 = 0.59$, $p < 0.001$, N = 109). Because of rainfall occurrence, intermitted surface flows appear within the lower portions [13], with the area of potential dry lake beds increasing as a result of an extended river network. Thus, the geographical characteristics of dry lake beds became clear when found in low elevations, which developed the river network.

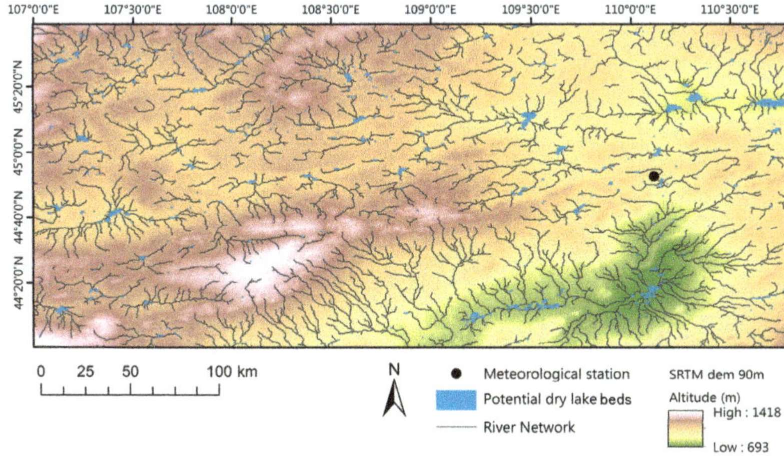

Figure 5. The distribution of potential dry lake beds and the river network.

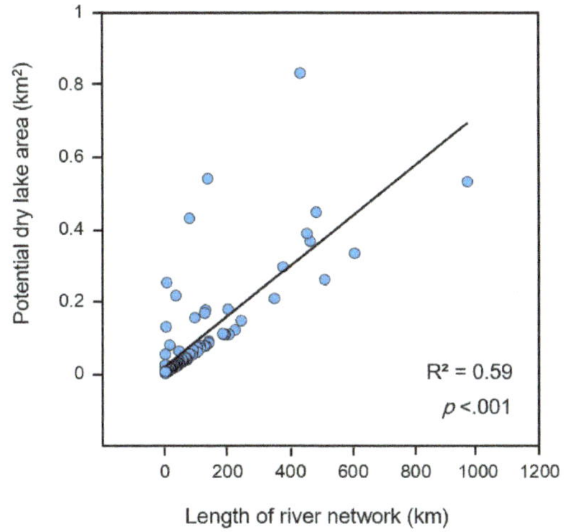

Figure 6. Correlation between potential dry lake bed areas and the length of the river network.

3.2. Formation of the Dry Lake Beds

Rashki et al. (2013) found that the dryness of lakes in the Hamoun basin was strongly associated with changes in land–atmosphere fluxes, soil moisture, and the frequency and intensity of dust storms [16]. In this study, the frequency of dry lake beds was compared with the interannual variability of precipitation to identify the time when dry lake beds in the study area appeared. Figure 7 shows the frequency of dry lake beds of the study area from 2000 through 2014 in addition to precipitation levels. The time series data shows total precipitation of the GPCP and the number of dry lake beds in the study area. The monthly mean frequency of the dry lake beds increased from 95.2% to 98.5% in the period of June to September. The monthly mean frequency of the dry lake beds increased from 97.2% to 99.9% in the period of October to May. The frequency of dry lake beds decreased from 361 to 266 in 2000 and from 361 to 301 in 2012. During this time, precipitation increased from 369 mm in the period between August 5 and August 20, 2000 to 639 mm in the period between July 4 and July 27, 2012 (Table 1). The mean precipitation of the same period was 60 mm and 88 mm, respectively.

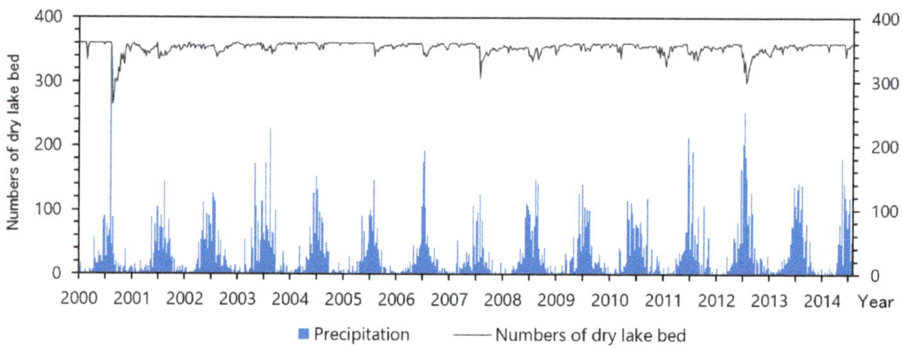

Figure 7. The correlation between the precipitation and the frequency of dry lake bed formation at each of the eight levels of pixels in study area.

Table 1. Difference in precipitation when dry lake beds decreased.

Period	Total Precipitation	Mean Precipitation
5 August 2000 to 20 August 2000	369 mm	60 mm
4 July 2012 to 27 July 2012	639 mm	88 mm

Potential dry lake beds were clarified from the differences in the relationship between precipitation and the formation of lakes (Figure 8). Precipitation was extracted when ephemeral lakes were formed for the first time in each year. The total precipitation over eight days was 369 mm in 2000 and at least 73 mm in 2010. In addition, ephemeral lakes were the largest in area in 2000 (Figure 7). It has also been suggested that the threshold for ephemeral lake formation in dry lake bed areas in semi-arid regions is eight days of precipitation.

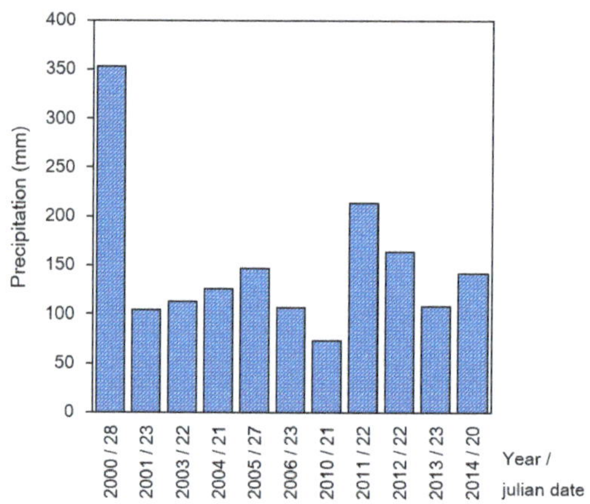

Figure 8. The total precipitation over eight days transformed dry lake beds into ephemeral lakes in potential dry lake bed areas.

4. Discussion

This study has shown that the frequency of dry lake bed formation has a strong correlation with precipitation. Dry lake beds are one important source as well as a hotspot of Asian dust emission. Dust affects Earth's climate, ecology, and economies across a broad range of scales, both temporally and spatially, and is an integral part of the Earth's climate system [6]. Dust sources, of natural and anthropogenic origins, were calculated with their respective contributions to emissions, and extensively compared against existing literature. Natural dust sources account for 75% of global emissions; anthropogenic sources account for 25% [4]. In this study we suggest a new method to identify dry lake beds using remote sensing. The classification accuracy was 80%. We believe that our accuracy was influenced by low resolution images and the limitations of aerial photointerpretation. However, it is a decent accuracy to be used in large areas such as semi-arid Mongolia. It is assumed that dry lake beds are affected by the full length of the river network and other geographic characteristics (such as elevation) as well as heterogeneous precipitation patterns. Lakes formed between June 2000 and September 2014 show drastic changes year by year, but it is suggested that lake beds dry up between October and May. In other words, the state of the Earth's surface in the area of potential dry lake beds dries during a period with no to little precipitation, and it is thought that water areas form during times of precipitation. To better understand the relationship between precipitation and the formation of dry lake beds, future studies should employ the use of high-resolution satellite images. In addition, it is necessary for the relationship of precipitation and the formation of dry lake beds to consider the geographic characteristics of the region and the influence of river networks.

5. Conclusions

This study suggests a new method to identify dry lake beds using hydrological analysis and remote sensing. The method may be used effectively in large areas such as drylands. Dry lake beds are affected by the full length of the river network and geographic characteristics (such as elevation) and precipitation patterns. Dry lake beds, which formed between June and September from 2000 through 2014, show year to year changes, but it is suggested that dry lake beds in semi-arid Mongolia dry up from October to May, which corresponds with periods of low precipitation. In our study area, the distribution of potential dry lake beds was 1%, with mountainous areas and desert steppe accounting for 30% and 69%, respectively.

Acknowledgments: This work was supported by JSPS KAKENHI Grant Numbers JP24340111, JP25550079, JP26281003 and the MEXT Supported Program for the Strategic Research Foundation (S1391001) at Rakuno Gakuen University. We are grateful to Zoljargal Enkhtuvshin and Dorji Munkhjargal for their support on the field survey in Mongolia. And also we would like to thank Dulam Jugder for giving some kind advice about this paper.

Author Contributions: Y.D. and B.H. analyzed the data and wrote the draft of manuscript and designed the study. K.B., C.M., Y.S., K.K., K.H., and J.N. participated in the field measurements and participated in the experiment of scattering soil particles. Y.D., Y.S., and T.P. established the GIS special database.

Conflicts of Interest: The authors declare that they have no competing interests.

References

1. Natsagdorj, L.; Jugder, D.; Chung, Y.S. Analysis of dust storms observed in Mongolia during 1937–1999. *Atmos. Environ.* **2003**, *37*, 1401–1411. [CrossRef]
2. Shao, Y. *Physics and Modelling of Wind Erosion*, 2nd ed.; Springer: Berlin/Heidelberg, Germany, 2008.
3. Zhang, B.; Tsunekawa, A.; Tsubo, M. Identification of Dust Hot Spots from Multi-Resolution Remotely Sensed Data in Eastern China and Mongolia. *Water Air Soil Pollut.* **2015**, *226*, 1–18. [CrossRef]
4. Ginoux, P.; Prospero, J.; Gill, T.E.; Hsu, N.C.; Zhao, M. Global-scale attribution of anthropogenic and natural dust sources and their emission rates based on MODIS Deep Blue aerosol products. *Rev. Geophys.* **2012**, *50*, 2012RG000388. [CrossRef]
5. Prospero, J.M.; Ginoux, P.; Torres, O.; Nicholson, S. Environmental characterization of global sources of atmospheric soil dust derived from the NIMBUS & TOMS absorbing aerosol product. *Rev. Geophys.* **2002**, *40*, 1002. [CrossRef]
6. Farebrother, W.; Hesse, P.P.; Chang, H.C.; Jones, C. Dry lake beds as sources of dust in Australia: A volumetric approach based on Lake Basin and deflated dune volumes. *Quat. Sci. Rev.* **2017**, *161*, 81–98. [CrossRef]
7. Kurosaki, Y.; Mikami, M. Threshold wind speed for dust emission in East Asia and its seasonal variations. *J. Geophys. Res. Atmos.* **2007**, *112*, D17202. [CrossRef]

8. Sternberg, T.; Paillou, P. Mapping potential shallow groundwater in the Gobi Desert using remote sensing: Lake Ulaan Nuur. *J. Arid Environ.* **2015**, *118*, 1–27. [CrossRef]
9. Millennium Ecosystem Assessment. Chapter 22. Dryland systems. In *Ecosystems and Human Well-Being: Current State and Trends*; Islands Press: Washington, DC, USA, 2005; pp. 623–662.
10. Tooth, S. Process, form and change in dryland rivers: A review of recent research. *Earth-Sci. Rev.* **2000**, *51*, 67–107. [CrossRef]
11. Gao, B. NDWI—A normalized difference water index for remote sensing of vegetation liquid water from space. *Remote Sens. Environ.* **1996**, *58*, 257–266. [CrossRef]
12. Ji, L.; Zhang, L.; Wylie, B. Analysis of dynamic thresholds for the normalized difference water index. *Photogramm. Eng. Remote Sens.* **2009**, *75*, 1307–1317. [CrossRef]
13. Sen, Z. *Wadi Hydrology*; CRC Press, Taylor & Francis Group: Boca Raton, FL, USA, 2008.
14. Davi, N.K.; Jacoby, G.C.; D'Arrigo, R.D.; Baatarbileg, N.; Li, J.; Curtis, A.E. A tree-ring-based drought index reconstruction for far-western Mongolia: 1565–2004. *Int. J. Climatol.* **2009**, *29*, 1508–1514. [CrossRef]
15. Kitamura, Y. *Desert of the Encyclopedia*; Maruzen Publishing: Tokyo, Japan, 2009; p. 192. (In Japanese)
16. Rashki, A.; Kaskaoutis, D.G.; Goudie, A.S.; Kahn, R.A. Dryness of ephemeral lakes and consequences for dust activity: The case of the Hamoun drainage basin, southeastern Iran. *Sci. Total Environ.* **2013**, *463–464*, 552–564. [CrossRef] [PubMed]

MDPI
St. Alban-Anlage 66
4052 Basel
Switzerland
Tel. +41 61 683 77 34
Fax +41 61 302 89 18
www.mdpi.com

Land Editorial Office
E-mail: land@mdpi.com
www.mdpi.com/journal/land

Lightning Source UK Ltd.
Milton Keynes UK
UKHW020848081222
413492UK00005B/49

9 783039 213474